# Innovative Development in Micromanufacturing Processes

*Innovative Development in Micromanufacturing Processes* details cutting edge technologies in micromanufacturing processes, an industry which has undergone a technological transformation in the past decade. Enabling engineers to create high performance, low cost, and long-lasting products, this book is an essential companion to all those working in micro and nano engineering.

As products continue to get smaller and smaller, the field of micromanufacturing has gained an international audience. This book looks at both approaches of micromanufacturing: top-down and bottom-up. The top-down approach includes subtractive micromanufacturing processes such as microturning, micromilling, microdrilling, laser beam micromachining, and magnetic abrasive finishing. The bottom-up approach involves additive manufacturing processes such as microforming, micro deep drawing, microforging, microextrusion, and microwelding. Additionally, microjoining and microhybrid manufacturing processes are discussed in detail. The book also aids engineers and students in solving common manufacturing issues such as choice of materials and testing.

The book will be of interest to those working in micro and nano engineering and machining, as well as students in manufacturing engineering, materials science, and more.

## Manufacturing Design and Technology

*Series Editor J. Paulo Davim*

*Industrial Tribology: Sustainable Machinery and Industry 4.0*
Edited by Jitendra Kumar Katiyar, Alessandro Ruggiero, T V V L N RAO and J. Paulo Davim

For more information about this series, please visit: https://www.routledge.com/Manufacturing-Design-and-Technology/book-series/CRCMANDESTEC

# Innovative Development in Micromanufacturing Processes

Edited by

Pawan Kumar Rakesh
J. Paulo Davim

CRC Press
Taylor & Francis Group
Boca Raton London New York

CRC Press is an imprint of the
Taylor & Francis Group, an **informa** business

Designed cover image: Dr. Pawan Kumar Rakesh

First edition published 2024
by CRC Press
2385 NW Executive Center Drive, Suite 320, Boca Raton FL 33431

and by CRC Press
4 Park Square, Milton Park, Abingdon, Oxon, OX14 4RN

*CRC Press is an imprint of Taylor & Francis Group, LLC*



ISBN: 978-1-032-42918-2 (hbk)
ISBN: 978-1-032-42926-7 (pbk)
ISBN: 978-1-003-36494-8 (ebk)

DOI: 10.1201/9781003364948

Typeset in Times
by MPS Limited, Dehradun

# Contents

# Preface

Micromanufacturing processes are a highly challenging field that produces miniaturized products such as mobile phones, laptop, watches, and jet engines. As per consumer demands, the product needs to be manufactured to a precise shape and size. The selection of materials for the fabrication of miniaturized products is still exhilarating. This book provides an outline of the different micromanufacturing processes including microgrinding, micro-EDM, micro-electrochemical, fiber laser, surface coating, direct ink writing, ultrasonic molding, spark plasma sintering, magnetorheological polishing, microfinishing, nanotechnology, and their applications.

This book has been edited to provide the course contents of micromanufacturing processes for undergraduate/postgraduate students and is aimed to be the best reference book for this course. We are highly thankful to all the contributors for their wisdom in writing the individual chapters. We would like to thank, namely, Sharpe Nicola, Shatakshi Singh and his entire team at CRC Press, and Taylor and Francis group for their support and guidance throughout the editing process. We wish to express our truthful thanks to God almighty "Maharaj Dhiraj Shri Ladesar Bhagwan" who has given us courage and vision to edit this book.

**Dr. Pawan Kumar Rakesh**
*National Institute of Technology Uttarakhand, India*

**Prof. J P Davim**
*University of Aveiro, Portugal*

# Editor Biography

**Dr. Pawan Kumar Rakesh** is serving as an Assistant Professor, Department of Mechanical Engineering, National Institute of Technology Uttarakhand (NIT Uttarakhand), Srinagar Garhwal, India. He has also served as the Head of Department, Mechanical Engineering (2013–2016), and an Associate Dean (Faculty Welfare) from 2016 to 2020. Working in a growing organization, he has substantial exposure to several lab developments, academic structure, and student welfare. He is also the Coordinator of the Design Innovation Center, NIT Uttarakhand. He has authored 52 articles published in reputed international/national journals and conferences. He has also organized two international conferences and fifteen workshops at the Institute. He has edited four books, namely, Processing of Green Composites, Advances in Engineering Design, Joining Processes for Dissimilar and Advanced Materials, and Cellulose Composites. He has contributed 12 book chapters in different books. He has also supervised three PhD and six M.Tech students.

**Prof. J P Davim** is a Full Professor at the University of Aveiro, Portugal. He is also distinguished as honorary professor in several universities/colleges/institutes in China, India and Spain. He received his Ph.D. degree in Mechanical Engineering in 1997, M.Sc. degree in Mechanical Engineering (materials and manufacturing processes) in 1991, Mechanical Engineering degree (5 years) in 1986, from the University of Porto (FEUP), the Aggregate title (Full Habilitation) from the University of Coimbra in 2005 and the D.Sc. (Higher Doctorate) from London Metropolitan University in 2013. He is Senior Chartered Engineer by the Portuguese Institution of Engineers with an MBA and Specialist titles in Engineering and Industrial Management as well as in Metrology. He is also Eur Ing by Engineers Europe FEANI-Brussels and Fellow (FIET) of IET-London. He has more than 35 years of teaching and research experience in Manufacturing, Materials, Mechanical and Industrial Engineering, with special emphasis in Machining & Tribology. He has also interest in Management, Engineering Education and Higher Education for Sustainability. He has guided large numbers of postdoc, Ph.D. and master's students as well as has coordinated and participated in several financed research projects. He has received several scientific awards and honors. He has worked as evaluator of projects for ERC-European Research Council and other international research agencies as well as examiner of Ph.D. thesis for many universities in different countries. He is the Editor in Chief of several international journals, Guest Editor of journals, books Editor, book Series Editor and Scientific Advisory for many international journals and conferences.

# Contributors

**Muhammad Abas**
University of Engineering and Technology
Peshawar, Pakistan

**A. Alaboodi**
Qassim University
Qassim, Saudi Arabia

**K. Algadah**
Technical and Vocational Training Corporation
Riyadh, Saudi Arabia

**Luis D. Cedeño-Viveros**
Institute of Advanced Materials for Sustainable Manufacturing
Tecnológico de Monterrey
Monterrey, México

**Ying Chen**
The Affiliated Stomatological Hospital of Nanjing Medical University
Jiangsu Province Key Laboratory of Oral Diseases
Jiangsu Province Engineering Research Center of Stomatological Translational Medicine
Nanjing, China

**Manas Das**
Indian Institute of Technology
Assam, India

**Dina V. Dudina**
Lavrentyev Institute of Hydrodynamics
Siberian Branch of the Russian Academy of Sciences
Novosibirsk, Russia
and
Novosibirsk State Technical University
Novosibirsk, Russia
and
Institute of Solid State Chemistry and Mechanochemistry
Siberian Branch of the Russian Academy of Sciences
Novosibirsk, Russia

**Erika García-López**
Tecnológico de Monterrey, Escuela de Ingeniería y Ciencias
Monterrey, México

**Rajiv Kumar Garg**
Dr. B R Ambedkar National Institute of Technology
Punjab, India

**Bikash Ghoshal**
Guru Gobind Singh Educational Society's Technical Campus
Jharkhand, India

**Ranjitsinha R. Gidde**
SVERI's College of Engineering Pandharpur
Maharashtra, India

**Sandeep Gour**
Indian Institute of Technology
Madhya Pradesh, India

**Ce Guo**
Taiyuan University of Technology
Shanxi, China

**Shivani Gupta**
Mechanical and Industrial Engineering Department
Indian Institute of Technology
Uttrakhand, India
and
Dr. Vishwanath D. Karad MIT World Peace University
Pune, India

**Fei Han**
The Affiliated Stomatological Hospital of Nanjing Medical University
Jiangsu Province Key Laboratory of Oral Diseases
Jiangsu Province Engineering Research Center of Stomatological Translational Medicine
Nanjing, China

**Gaoying Hong**
Stomatology Hospital
School of Stomatology
Zhejiang University School of Medicine
Clinical Research Center for Oral Diseases of Zhejiang Province
Key Laboratory of Oral Biomedical Research of Zhejiang Province
Cancer Center of Zhejiang University
Hangzhou, China

**Xiao-long Hu**
Hunan University
Changsha, China

**Ahmed Mohamed Mahmoud Ibrahim**
Minia University
El-Minya, Egypt

**Neelesh Kumar Jain**
Indian Institute of Technology
Madhya Pradesh, India

**Ziaullah Jan**
GIK Institute
Topi, Pakistan

**Marcel Janer**
Eurecat, Technological Center of Catalonia
Barcelona, Spain

**Sadasivam Kannan**
Government Arts College for Men
Tamil Nadu, India

**Amarjit P. Kene**
SVERI's College of Engineering
Maharashtra, India

**Vyacheslav I. Kvashnin**
Lavrentyev Institute of Hydrodynamics
Siberian Branch of the Russian Academy of Sciences
Novosibirsk, Russia
and
Novosibirsk State Technical University
Novosibirsk, Russia

**Wei Li**
Hunan University
Changsha, China

**Wenhui Li**
Taiyuan University of Technology
Shanxi, China

**Xiuhong Li**
Taiyuan University of Technology
Shanxi, China

**Mallikarjun Mahajanshetti**
RV College of Engineering
Karnataka, India

**Ranajit Mahanti**
Indian Institute of Technology
Assam, India

**I.V. Manoj**
Nitte Meenakshi Institute of Technology
Karnataka, India

**Cong Mao**
Changsha University of Science and Technology
Changsha, China

**Anbazhagan Murugan**
Karpagam College of Engineering
Tamil Nadu, India

**S. Narendranath**
National Institute of Technology
Karnataka, India

**Luis H. Olivas-Alanis**
Tecnológico de Monterrey, Escuela de Ingeniería y Ciencias
Monterrey, México

**Debasish Panigrahi**
National Institute of Technology
Odisha, India

**Rajni Patel**
SVERI's College of Engineering
Maharashtra, India

**S.K. Patel**
National Institute of Technology
Odisha, India

**Xavier Plantà**
Eurecat, Technological Center of Catalonia
Barcelona, Spain

**V. Radhakrishnan**
Formerly associated with
Indian Institute of Space Science and Technology
Kerala, India
and
Indian Institute of Technology
Tamil Nadu, India

**Khalid Rahman**
GIK Institute
Topi, Pakistan

**Rithwik Shankar Raj**
RV College of Engineering
Karnataka, India

**Arumugam Raja**
CNR-SPIN, University of Salerno
Salerno, Italy

**Pawan Kumar Rakesh**
National Institute of Technology
Uttarakhand, India

**Jinka Ranganayakulu**
RV College of Engineering
Karnataka, India

**K. Venkata Rao**
Vignan's Foundation for Science
Karnataka, India

**Ying-hui Ren**
Hunan University
Changsha, China

**Maria Dolors Riera**
Polytechnical University of Catalonia
Barcelona, Spain

**Ciro A. Rodríguez**
Tecnológico de Monterrey, Escuela de Ingeniería y Ciencias
Monterrey, México

**Sweta Rout**
National Institute of Technology
Odisha, India

**Abdul Samad Shameem**
Karpagam Academy of Higher Education
Tamil Nadu, India

**Apurbba Kumar Sharma**
Mechanical and Industrial Engineering Department
Indian Institute of Technology
Uttrakhand, India

**Vadivel Siva**
Karpagam Academy of Higher Education
Tamil Nadu, India

**V. S. Sooraj**
Department of Aerospace Engineering
Indian Institute of Space Science and Technology
Kerala, India

**P.V. Srihari**
RV College of Engineering
Karnataka, India

**Shijia Tang**
The Affiliated Stomatological Hospital of Nanjing Medical University
Jiangsu Province Key Laboratory of Oral Diseases
Jiangsu Province Engineering Research Center of Stomatological Translational Medicine
Nanjing, China

**Arumugam Thangamani**
Karpagam Academy of Higher Education
Tamil Nadu, India

**Subramanian Thangarasu**
Kalasalingam Academy of Research and Education
Tamil Nadu, India

**Elisa Vázquez-Lepe**
Tecnológico de Monterrey, Escuela de Ingeniería y Ciencias
Monterrey, México

**Yumin Wu**
The Affiliated Stomatological Hospital of Nanjing Medical University
Jiangsu Province Key Laboratory of Oral Diseases
Jiangsu Province Engineering Research Center of Stomatological Translational Medicine
Nanjing, China

**Haifeng Xie**
The Affiliated Stomatological Hospital of Nanjing Medical University
Jiangsu Province Key Laboratory of Oral Diseases
Jiangsu Province Engineering Research Center of Stomatological Translational Medicine
Nanjing, China

**Shengqiang Yang**
Taiyuan University of Technology
Shanxi, China

**Ran You**
The Affiliated Stomatological Hospital of Nanjing Medical University
Jiangsu Province Key Laboratory of Oral Diseases
Jiangsu Province Engineering Research Center of Stomatological Translational Medicine
Nanjing, China

**Qiyue Zhou**
The Affiliated Stomatological Hospital of Nanjing Medical University
Jiangsu Province Key Laboratory of Oral Diseases
Jiangsu Province Engineering Research Center of Stomatological Translational Medicine
Nanjing, China

# 1 Micromanufacturing of Miniature Non-Circular Gears by WSEM Process

*Sandeep Gour and Neelesh Kumar Jain*

## 1.1 INTRODUCTION TO μ-MANUFACTURING

Technological development has converted the large-size components into micro and miniaturized components for their use in various smaller-size machines and devices. Rapid miniaturization of the products which are extensively used in biomedical, military, aerospace, robotics, agriculture, precision engineering, and other fields of engineering has increased the demand for manufacturing of products in very small sizes. Products in the size range from 0.5 mm to 10 mm are known as meso-products whereas μ-products fall in the size range from 1 to 500 μm [1]. Miniaturization leads to various economic advantages through a reduction in cost, weight, and volume of products, as well as space and power requirements to operate the devices. Manufacturing of micro-size products is called μ-manufacturing. It is very different than the manufacturing of larger-size products using conventional manufacturing processes. Special types of manufacturing processes, machines, equipment, tools, and technically skilled manpower are required for μ-manufacturing to achieve high dimensional and geometrical accuracy with good quality surface finish. Selection of suitable μ-manufacturing process, correct setting of the process parameters, work material, proper cooling system, and μ-product handling tools are some important aspects of μ-manufacturing.

## 1.2 DIFFERENT μ-MANUFACTURING PROCESSES

Very small size, complicated shape, nano-scale level dimensional and geometrical tolerances, nano-finishing requirement, and poor machinability of materials of miniaturized products are the major challenges in μ-manufacturing. Growing emphasis on the product miniaturization has led to evolution of many μ-manufacturing processes. Non-conventional processes are very effective in μ-manufacturing of different types of micro-products. Conventional processes are also popular and used according to the requirement. Different types of processes such as mechanical, chemical, and hybrid are available for μ-manufacturing. These processes can

DOI: 10.1201/9781003364948-1

be classified into μ-additive, μ-machining, μ-deforming, μ-joining, and hybrid μ-manufacturing processes. Their details are described as follows.

### 1.2.1 μ-Additive Manufacturing Processes

Processes in which material is added layer-by-layer or some other specified manner for μ-manufacturing of parts are known as μ-additive manufacturing processes. The material can be supplied in different forms such as wire, pallets, powder, sheet, and their combinations. Some examples of μ-additive manufacturing processes are stereolithography, chemical vapor deposition (CVD), physical vapor deposition (PVD), polymer deposition, electrochemical micromanufacturing, photo-electro-forming, μ-casting, and μ-injection molding. Electrochemical micromanufacturing process is used for precision manufacturing of large-scale 2D and quasi-3D micro-sized geometries from metallic materials [2]. μ-laser and electron beams based μ-additive manufacturing processes are used to manufacture μ-parts and μ-devices with high resolution as compared to other additive manufacturing processes such as fused deposition modeling [3].

### 1.2.2 μ-Machining Processes

μ-machining processes remove a very small volume of material from the workpiece to produce μ-products. Processes such as μ-milling, μ-turning, μ-grinding, and μ-polishing are the conventional processes belonging to this category. Whereas μ-electrochemical machining (μ-ECM), μ-spark erosion machining (μ-SEM), electron beam machining (EBM), laser beam machining (LBM), plasma arc machining (PAM), photo-chemical machining (PCM), and jet electrolyte micromachining (JEMM) are the advanced processes belonging to the family of μ-machining processes. The JEMM process uses spray of pressurized electrolyte supplied through a nozzle for electrochemical reaction at the interface between the electrolyte and the metallic workpiece causing its very controlled electrolytic dissolution [4].

### 1.2.3 μ-Deforming Processes

μ-deforming processes include μ-stamping, μ-incremental forming, μ-superplastic forming, μ-hydro forming, μ-hot embossing, micro and nanoimprinting, μ-extrusion, μ-forging, μ-bending, and μ-deep drawing. In these processes, a very small amount of deforming force is applied to the workpiece for μ-manufacturing of different miniaturized products.

### 1.2.4 μ-Joining Processes

Different μ-parts can be joined together by μ-joining processes using very small amount of controlled heat input or even glue. μ-joining processes employ various non-invasive heating sources such as laser beam, electron beam, microwave, infrared, and ultrasound. Examples of μ-joining processes are μ-mechanical-assembly, laser welding, infrared welding, microwave welding, ultrasonic welding, vacuum soldering, bonding, and gluing.

### 1.2.5 Hybrid μ-Manufacturing Processes

When two or more μ-manufacturing processes are used simultaneously for μ-manufacturing of parts, it is called as hybrid μ-manufacturing processes. These processes impart advantages to all the constituent processes. Micro-laser-ECM, LIGA (x-ray lithography, electroplating, and molding) and LIGA combined with laser-machining, μ-SEM and laser-machining, laser-assisted μ-forming, μ-assembly injection molding, shape deposition and laser-machining, and combined μ-machining and casting are some examples of hybrid μ-manufacturing processes. Aerosol jet printing is a non-traditional hybrid μ-manufacturing process which uses the directed aerosol stream to provide consistent deposition using a 1–5 mm offset between nozzle and substrate. It enables it patterning of more complex surfaces such as spiral pattern on the surface of a golf ball [5].

## 1.3 APPLICATIONS OF μ-MANUFACTURING PROCESSES

Applications of μ-manufacturing processes include as follows: Manufacturing of μ-reactors, μ-sensors, μ-actuators, μ-valves, μ-pumps, μ-motors, μ-fans, μ-gears, μ-tools, μ-molds, μ-fuel cells, μ-biomedical devices, μ-fluidic devices, μ-products used in various micro-electro-mechanical systems (MEMS), nano-electro-mechanical systems (NEMS), and micro-optical-electronic-mechanical systems (MOES), μ-mechanical and other μ-devices used in automotive, aviation, space applications, telecommunication, and IT industries. Specific μ-fluidic devices are used in biomedical industry for drug production and delivery, and in the rapid diagnostic kits for diseases [6]. Different μ-components such as μ-switches, μ-sensors and probes, memory chips, μ-actuators, μ-motors, μ-fuel cells, μ-mechanical devices, electrical connectors, optical devices, biomedical implants, magnetic hard drive heads, computer processors, and MEMS devices used in electronics, automotive, biotechnology, energy, communications, optics, and other applications are made by different μ-manufacturing processes [7]. Many industrial applications use miniature and precise products with arrays of μ-holes [8]. Demand of μ-components is continuously increasing in almost all fields of engineering as manufacturers are descaling the product size for their economic and efficient manufacturing.

## 1.4 μ-GEARS

Gears are the rotating machine elements in which teeth are made on external or internal periphery of gear blanks. A gear meshes with another gear for transmission of power and motion. Gears having outer diameter in the size range from 0.1 to 1.0 mm are referred to as μ-gears [9]. They have ultra-light weight, smaller size, compactness, higher dimensional accuracy, better functional and operating characteristics, longer service life, and zero backlash as compared to macro-gears with ability to sustain their performance under risky ecological environments. They are widely used as the key elements of different types of μ-systems [9]. Very tight tolerances, precision, and need of very high geometrical accuracy are some important factors required in the manufacturing of μ-gears. Accurate designing and effective manufacturing of μ-gears allow them to use in many critical areas of applications such as biomedical, military, aviation, robotics, unmanned aerial vehicles (UAV), sophisticated cameras, high-precision instruments, and automobiles.

### 1.4.1 Classification of μ-Gears

Classification of μ-gears is similar to the classification of macro-gears. In μ-gears, the measurement scale is reduced to a few μm. They can categorize as μ-gear (having outside diameter 0.1–1.0 mm) and meso-gears (having outside diameter 1.0–10 mm). A major classification based on the shape of μ-gear is as follows:

1. Micro circular gears (μ-CG).
2. Micro non-circular gears (μ-NCG).

Generally, the term gear represents circular gear but if the shape of a gear is different than circular then it is known as non-circular gear. A μ-NCG can be of elliptical, triangular, square, pentagonal, hexagonal, and other designed irregular shapes. Based on gear teeth designs, μ-gears can be classified as μ-spur gear, μ-helical gear, and μ-bevel gear. The teeth are straight and parallel to the axis of μ-spur gear whereas they are inclined at some angle with the axis of μ-helical gear. μ-bevel gear is a cone shape gear and teeth are made on the tapered surface of the gear. They are used to transmit power and motion in case of intersecting shafts.

### 1.4.2 Applications of μ-Gears

Different types of μ-gears are widely used in MEMS, NEMS, and MOES as important actuating components. Various applications of μ-gears are in μ-motors, μ-pumps, mini-copter, μ-robots, μ-spy robot, etc. Transmission systems in which μ-gears are used have many industrial applications such as in biomedical engineering for minimal-invasive surgery equipment and in the field of industrial automation for hexapod actuators [10]. To achieve a high torque density, μ-geared ultrasonic motors are used in which μ-planetary gears are used for amplification of the motor torque to several times [11].

### 1.4.3 Manufacturing Processes for μ-Gears

Manufacturing of μ-gear is a task of high technical skills and accuracy. It requires special manufacturing processes, machines, tools, and highly skilled technical manpower. Following some processes can be used for manufacturing of μ-gears:

- UV-LIGA.
- Laser punching.
- Laser shock punching.
- μ-powder injection molding.
- μ-Wire Spark Erosion Machining (WSEM).
- μ-3D printing.
- μ-extrusion.
- μ-molding.
- Bio-etching.
- Combination of μ-reciprocated WSEM and precision forging.

## 1.5 MINIATURE NON-CIRCULAR GEARS

Miniature gears having shapes other than circular and outer diameter in the range of 0.1 to 10 mm are known as miniature non-circular gears (MNCGs). They are manufactured to perform special type of tasks. In MNCGs, teeth are made on a pre-designed non-circular gear blank. Rotation of MNCGs is different than rotation of circular gears. They work to produce variable output speed with constant input speed. The functioning of MNCGs is not only to transfer power from one shaft to another but to generate the pre-specified motion also. They are balanced, compact, easy to assemble, and provide exact kinematic solutions for different mechanisms. Many heavy and large kinematic mechanisms used in different machineries can be replaced by suitable MNCGs. Manufacturing of MNCGs is very difficult due to their irregular shapes and complex teeth designs. Their miniaturization imparts more complexities in their manufacturing. MNCGs can be designed as per the motion requirements and manufactured using suitable manufacturing process. MNCGs are assembled in such a manner that they will produce desired motion output while working.

MNCGs are capable of producing stop-and-dwell motion, non-uniform rotation, quick return motion, conversion of constant input speed into variable output speed, increase and decrease the value of torque as per the requirement, and can produce complex mechanical functions which are useful in different machineries such as robots and automatic machines. MNCGs are used to reduce speed fluctuations in rotating shaft of automobiles and automatic machineries. Cams and linkages, intermittent mechanisms, quick return mechanism, and other similar mechanism can be replaced by MNCGs to get more efficient operation. Practical applications of MNCGs include as follows: Textile machines, printing presses, rice planting machines, fertilizer distributor, trans-planters, window shade panel drives, winding machines, mechanical presses, hydraulic presses, high torque hydraulic engines, automatic feed machines, conveyers, potentiometers, continuously variable transmissions (CVTs), toys, and home appliances.

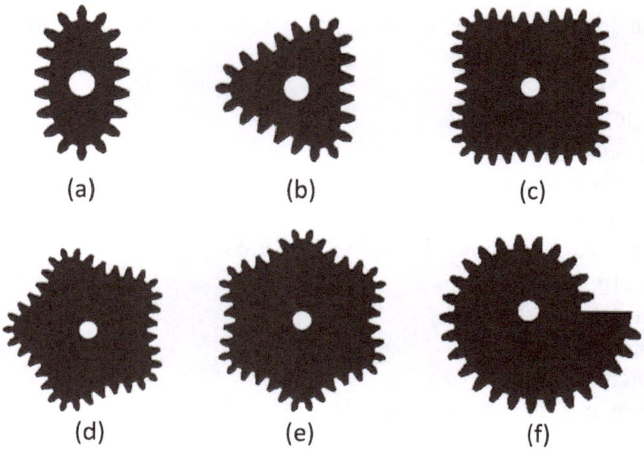


**FIGURE 1.1** Different types of MNCGs: (a) Miniature elliptical gear, (b) Miniature triangular gear, (c) Miniature square gear, (d) Miniature pentagonal gear (e) Miniature hexagonal gear, and (f) Miniature logarithmic gear.

### 1.5.1 Classification of MNCGs

According to the outer shape of MNCGs, they can be classified in various types such as miniature elliptical gear, miniature triangular gear, miniature square gear, miniature pentagonal gear, miniature hexagonal gear, and so on. MNCGs can be classified based on the number of lobes such as uni-lobe and bi-lobe MNCGs. Other types of MNCGs are miniature eccentric gear and miniature oval gear. Figure 1.1 shows different types of MNCGs.

As per the requirements, MNCGs are manufactured in their designed irregular shapes also. MNCGs can be produced with straight teeth, helical teeth, and bevel teeth, and can be termed as straight MNCGs, helical MNCGs, and bevel MNCGs. A combination of bevel and helical teeth MNCGs known as bevel-helical MNCGs, can also be made for special requirements. Miniature logarithmic gears can also be considered as a type of MNCGs.

### 1.5.2 Challenges in μ-Manufacturing of MNCGs

Accurate and precise manufacturing of MNCGs has many difficulties due to their non-standard shapes and smaller size. The following are the challenges involved in the manufacturing of MNCGs:

- Selection of suitable gear material.
- Selection of suitable manufacturing process.
- Requirement of special miniature tools.
- Achieving high dimensional and geometrical accuracy.
- Achieving nano-finishing.

- Attaining very tight tolerances.
- Reduction of manufacturing time.
- Flexibility for manufacturing of different types of MNCGs.
- Suitability for mass production.
- High level of manufacturing skills.

### 1.5.3 Past Work on Manufacturing of MNCGs

Xiao et al. [12] proposed fundamental design of miniature pure-rolling non-circular line gear (PNLGs) mechanism and prototypes of miniature PNLGs pair were manufactured by 3D printing process. To the best knowledge of authors, no other process has been used yet for the manufacturing of MNCGs.

## 1.6 WSEM PROCESS

Figure 1.2 shows a schematic diagram of WSEM process. It uses small diameter electrically conductive wire as the tool electrode (cathode), workpiece as the anode, and pressurized deionized (DI) water as the dielectric fluid. Spark energy generated due to ionization of DI water is used for machining of electrically conductive material of workpiece through its melting and vaporization which takes place due to high temperature (10,000 to 12,000°C) generated for a fraction of a second in the machining zone when the direct current (DC) is supplied in it. The DI water also flushes away the removed small debris from the workpiece material. The wire travels in the machining zone through a pre-defined path which creates the desired shape on the workpiece. According to the supply method of the dielectric fluid to the machining zone, WSEM process can be

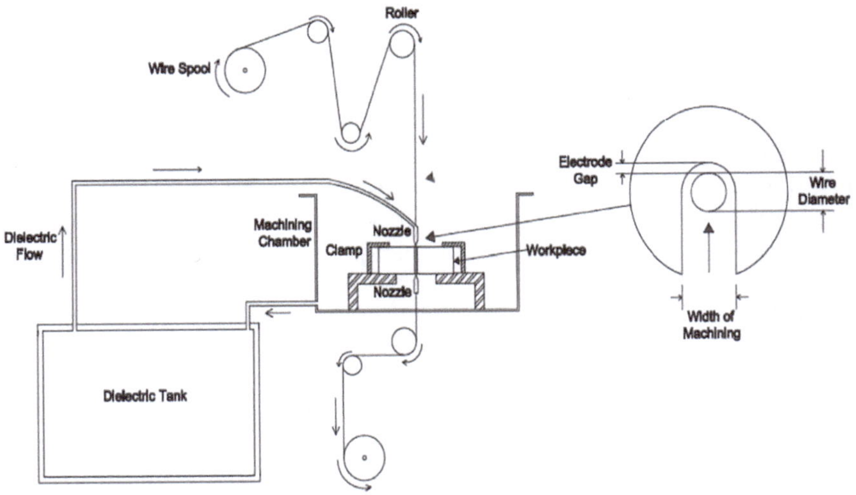


**FIGURE 1.2** Schematic of working principle of WSEM process.

classified as submerged type, non-submerged (coaxial flushing), and dry and near-dry type [13]. In submerged WSEM process, both the wire electrode and workpiece are submerged in the dielectric liquid during machining. In the non-submerged type WSEM process, dielectric fluid flushes at the machining zone using suitable nozzles. In dry and near-dry type WSEM process, gas is used as a dielectric making this process less corrosive and more economical and environment friendly [14].

### 1.6.1 μ-WSEM Process

The working principle of μ-WSEM process is same as WSEM process on the reduced scale of machining. In μ-WSEM process, an electrically conductive wire of 20 to 250 μm diameter range is used as an electrode [15]. Values of the process parameters are also reduced considerably as compared to WSEM process. Machining of hard alloys and tempered steel is possible by μ-WSEM process to produce precision micro-tooling such as μ-injection and μ-punching for μ-mechanical, electronic, and optical devices [16]. μ-WSEM process is widely used in manufacturing of μ-gears, μ-air turbines, μ-grabbers, μ-electrodes, μ-rector system, μ-injection needles, μ-milling tools, μ-molds, and μ-cutting tools [17]. Highly accurate finishing cuts can be made using dry μ-WSEM process due to narrow gap between the wire and the workpiece, and negligible reaction force during the machining [18].

## 1.7 NET-SHAPE μ-MANUFACTURING OF MNCGS BY WSEM PROCESS

WSEM was used in the presented research as an effective process for net-shape μ-manufacturing of MNCGs. Net-shape manufacturing refers to elimination of any further post-processing of the manufactured part. Net-shape manufactured products can directly be used in their applications. The WSEM process was found capable of producing net-shape products with high dimensional accuracy and high degree of surface finish. Four-axis computer numerical controlled (CNC) WSEM machine (Ecocut model from Electronica India Ltd, Pune) was used for net-shape μ-manufacturing of MNCGs. Plain brass wire of 250 μm diameter was used as a tool electrode and deionized water was used as a dielectric fluid. Part-programs for the MNCGs were prepared using the *AutoCAD* software and its data were imported to the *Elcam* software which generates wire path for the WSEM machine. After confirmation of the correctness of the part-programs through simulation, MNCGs were μ-manufactured.

### 1.7.1 Experimentation

Size of MNCGs is very small so their material should have enough strength to withstand the load applied while operation. In many applications MNCGs are

**TABLE 1.1**
**Values of WSEM Process Parameter Used in Net-Shape μ-Manufacturing of MNCGs from SS 304**

| Name of the Parameter | Value |
|---|---|
| Pulse-on Time | 1.3 μs |
| Pulse-off Time | 50.5 μs |
| Peak current | 2 A |
| Servo voltage | 13 V |
| Wire feed rate | 3 m/min |
| Wire tension | 660 g |
| Water pressure | 15 kg/cm$^2$ |
| Cutting speed overwrite | 100% |

used in corrosive environment also therefore their material should be corrosion resistant. Consequently, stainless steel of 304 grade (SS 304) was selected as a material for MNGCs due to its high strength, high resistance to corrosion and wear, and easy availability. Selection of suitable values of WSEM process parameters is a very important aspect in manufacturing of MNCGs. Table 1.1 presents the WSEM process parameters used in net-shape μ-manufacturing of MNCGs from SS 304. These parameters were decided after extensive experiments using wire breakage frequency, deflection of wire, workpiece material, machining width, and cutting time as the criteria.

Figure 1.3 schematically depicts different activities involved in net-shape μ-manufacturing of MNCGs by WSEM process. It involved the following procedure:

- Preparation of the workpiece for μ-manufacturing of MNGCs by making its top and bottom surfaces perfectly flat through removal of surface peaks and unevenness by grinding and buffing processes.
- Drilling of holes at the calculated distances in the workpiece plate by the CNC milling machine for inserting the wire to execute the part-program from the origin defined in it.
- Preparation of CNC part-programs for different types of MNCGs using *AutoCAD* and *Elcam* software. Transferring the prepared part-program to the WSEM machine. Confirming correctness of wire travel path through dry-run simulation of the prepared part-program.
- Loading of a spool of 250 μm diameter fresh wire on the WSEM machine and filling of deionized water up to the required level in the dielectric tank.

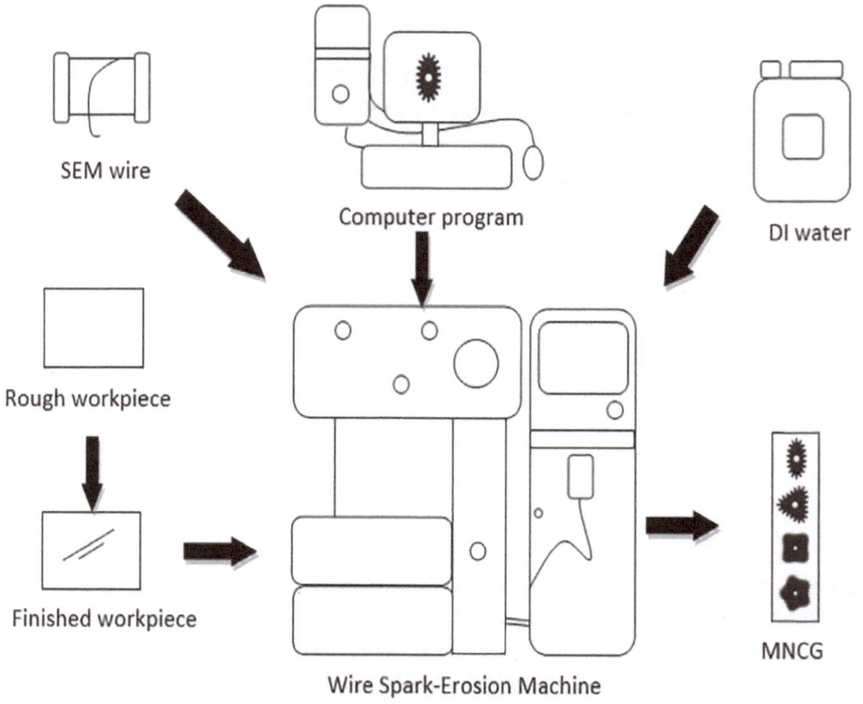


**FIGURE 1.3** Schematic of different activities involved in net-shape μ-manufacturing of MNCGs by the WSEM process.

- Clamping of the workpiece properly on the worktable of WSEM machine using clamps and bolts. Positioning of the workpiece should be perfect with respect to the machine axes and traveling direction of the wire which is defined in the part-program.
- Setting the decided values of WSEM parameters to start μ-manufacturing of desired MNCGs. Positioning of the wire at the initial coordinates from where the part-program starts executing. Dry-run of the part-program to check axis limits and uninterrupted movement of the wire in the machining zone.
- Actual execution of the part-program and continuously monitoring the WSEM machine status for timely control of process parameters, if required, and to avoid wire breakage.
- Removal of the workpiece plate and the manufactured MNCGs very carefully due to their very small size and ensuring that they are not flushed by the high-pressure dielectric fluid. Cleaning and drying of the manufactured MNCGs. Figure 1.4 shows the workpiece plate after μ-manufacturing of different MNCGs from it.

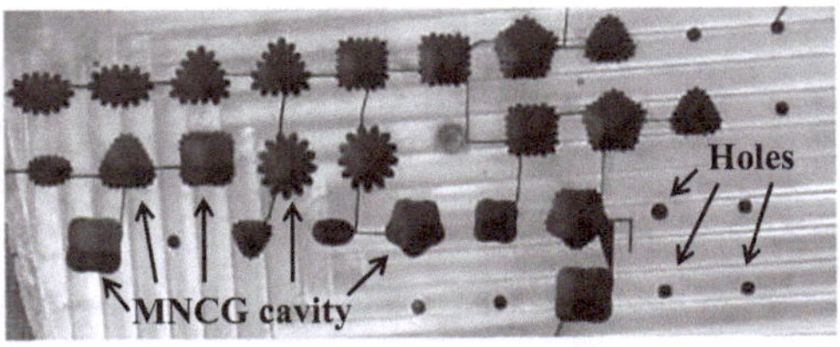


**FIGURE 1.4** Workpiece plate after μ-manufacturing of MNCGs from it.

### 1.7.2 Quality of MNCGs Manufactured by WSEM Process

Figure 1.5 shows photographs of different MNCGs manufactured by WSEM process showing their size with respect to a steel ruler. Figure 1.6 shows optical microscopic images of two teeth of the manufactured miniature triangular gear depicting their shape and size. It can be observed in Figure 1.6 that the teeth of the WSEM manufactured miniature triangular gear are well in their designed shapes and size, and edges of the gear teeth are straight and corners are sharp. Tool marks generated during the flattening of the workpiece can also be clearly seen on the top surfaces of the teeth.

#### 1.7.2.1 Topography of Flank Surfaces

A tooth was cut from the miniature triangular gear for further study by stereo zoom microscope. Figure 1.7 shows different microscopic views of the teeth. Erosion of the workpiece material can be seen in these images. Tooth shape and other surfaces are good. The machined surfaces are clean, no metallic chips are present, neither any machining residue is adhered to the surfaces, no oil or other liquid is present, and no surface distortion can be seen.

Scanning electron microscope (SEM) images are shown in Figure 1.8. It can be seen from these images that the machined surfaces are free from micro-cracks and other micro-damages. The teeth cut are well in their shape. No micro-voids can be seen.

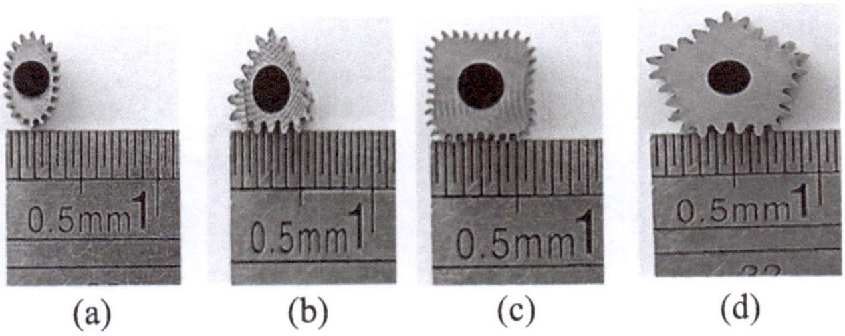


**FIGURE 1.5** Different MNCG net-shapes manufactured by WSEM process: (a) Elliptical, (b) Triangular, (c) Square, and (d) Pentagonal.

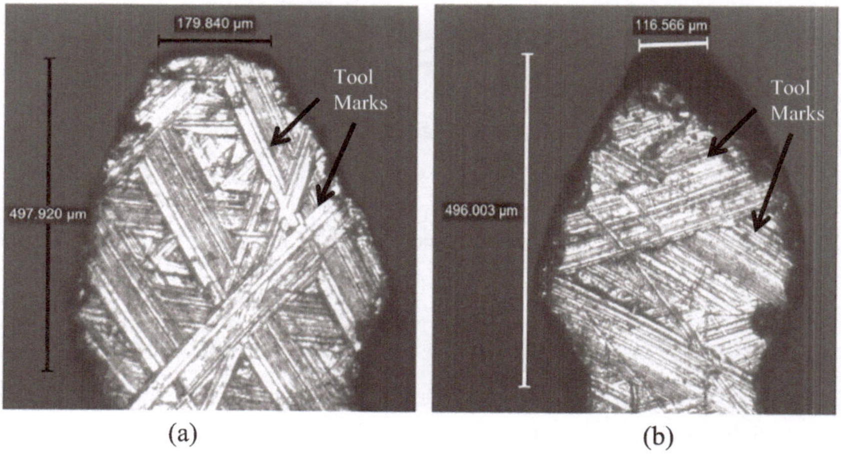


**FIGURE 1.6** Optical microscopic images of two teeth of WSEM manufactured miniature triangular gear. (a) First tooth, (b) Second tooth.

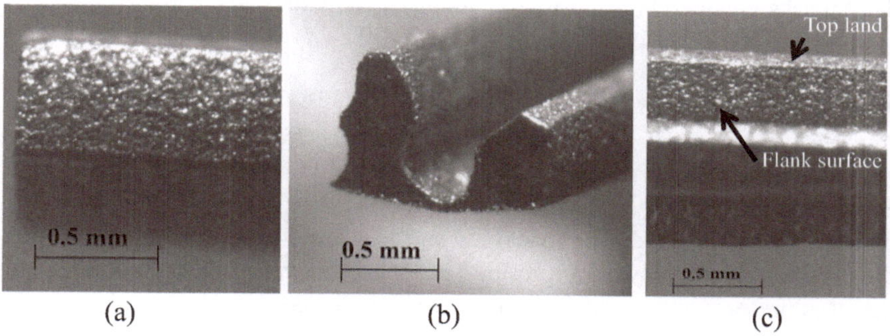


**FIGURE 1.7** Stereo zoom microscopic images of the teeth of the MNCGs: (a) Tooth flank surface, (b) Tooth profile, and (c) Top land and flank surface.

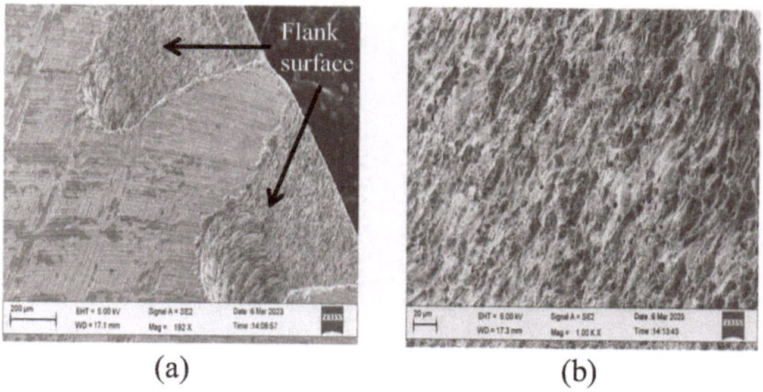


**FIGURE 1.8** SEM images of MNCGs: (a) Tooth profile, and (b) Flank surface.

#### 1.7.2.2 Surface Roughness of Flank Surfaces

Roughness of tooth flank surface is a very important factor which affects performance, efficiency, and service life of gears significantly. It affects the ability of gear to resist corrosion, wear, and fatigue. Surface roughness of the WSEM manufactured MNCGs was measured using 3D surface roughness-cum contour tracing equipment Marsurf LD 130 using the evaluation length of 0.80 mm. Ranges of measured maximum and average surface roughness are found to be 10.1 to 11.3 μm, and 1.42 to 1.56 μm respectively which are satisfactory from their functioning point of view.

It can be concluded from the quality observations of the WSEM manufactured MNCGs that they are (i) accurate in their designed shapes, (ii) their shape and flank surfaces uniform thought, (iii) free from any surface damage, (iv) no burrs, adhesive or any other unwanted material are present on their machined surfaces, (v) high degree of geometrical and dimensional accuracy is achieved, and (vi) net-shape and can directly be used to their applications without any post-processing or post-finishing operations.

## 1.8 FUNCTIONING OF DIFFERENT TYPES OF MNCGS

Functioning of MNCGs can be understood by meshing tests of different combinations of MNCGs of same module irrespective of their shapes. Tooth of each type of manufactured MNCGs were marked by ink to identify its rotation. MNCGs of same shape such as elliptical with elliptical, triangular with triangular, and square with square were meshed together. MNCGs of same module and different shapes such as elliptical with triangular, elliptical with square, triangular with square, triangular with pentagonal, and square with pentagonal, were rolled against each other. Figure 1.9 shows the meshing of MNCGs of different shapes. Meshing of all types of the MNCGs occurred smoothly without any jerks and slipping indicating satisfactory functioning of all types of manufactured MNCGs. Table 1.2 summarizes angles of rotations for driving and driven miniature gears for different meshings shown in Figure 1.9.

For all types of meshing tests, leftward gear is driver gear and rightward gear is driven gear. At the initial stage, the marked teeth are at 0° angular position for every test. Meshing of the miniature elliptical gears is shown in Figures 1.9a1 to 1.9a5. The driver gear starts rotating in clockwise direction which rotates the driven gear in anticlockwise direction. During the first 90° rotation of driver gear, the driven gear rotates 45° covering small distance. After 180° of driver gear rotation, driven gear reaches to 180° position covering more distance. Rotation of the driver gear by 270° quickly moves the driven gear to 315°. Complete 360° rotation of the driver gear rotates the driven gear by 360° and both the gear reaches to their initial condition. Meshing of miniature triangular gears is depicted in Figures 1.9b1 to 1.9b5. Clockwise rotation of the driver gear rotates the driven gear anticlockwise in equal angles so in this case both the meshing gears travel same angular distance, and their rotation is also same. Both the gears reach to their initial position after completing 360° rotation of the

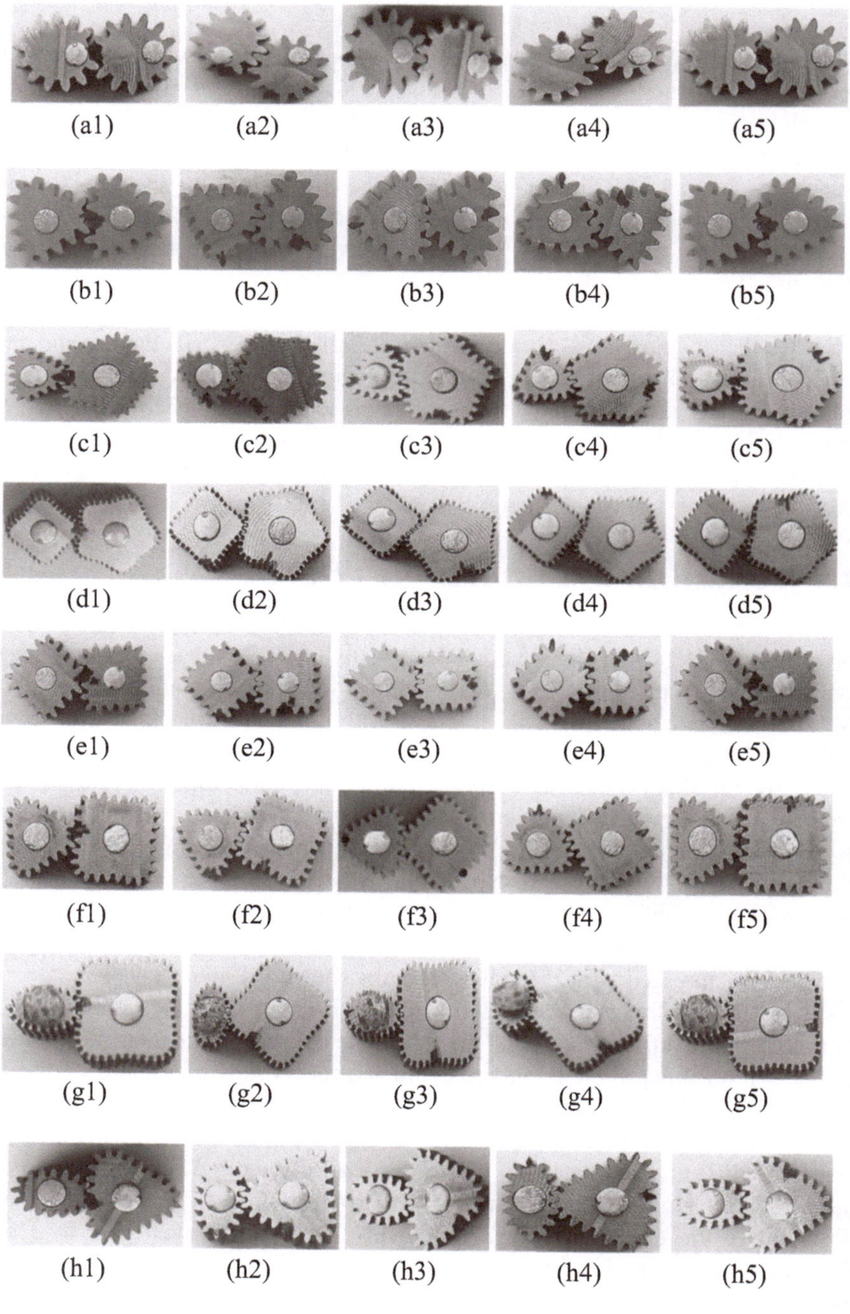


**FIGURE 1.9** Meshing of different types of WSEM manufactured MNCGs: (a1 to a5) elliptical with elliptical, (b1 to b5) triangular with triangular, (c1 to c5) triangular with pentagonal, (d1 to d5) square with pentagonal, (e1 to e5) square with square, (f1 to f5) triangular with square, (g1 to g5) elliptical with square, and (h1 to h5) elliptical with triangular.

**TABLE 1.2**
**Summary of Rotation of Driven Miniature Gear with Respect to Driver Miniature Gear**

| Meshing Type | Rotation of the Driver (Degrees) | | | | Rotation of the Driven (Degrees) | | | |
|---|---|---|---|---|---|---|---|---|
| Elliptical with elliptical (Figures 1.9a1 to 1.9a5) | 90 | 180 | 270 | 360 | 45 | 180 | 315 | 360 |
| Triangular with triangular (Figures 1.9b1 to 1.9b5) | 90 | 180 | 270 | 360 | 90 | 180 | 270 | 360 |
| Triangular with pentagonal (Figures 1.9c1 to 1.9c5) | 90 | 180 | 270 | 360 | 45 | 90 | 135 | 225 |
| Square with pentagonal (Figures 1.9d1 to 1.9d5) | 90 | 180 | 270 | 360 | 45 | 135 | 225 | 270 |
| Square with square (Figures 1.9e1 to 1.9e5) | 90 | 180 | 270 | 360 | 90 | 180 | 270 | 360 |
| Triangular with square (Figures 1.9f1 to 1.9f5) | 90 | 180 | 270 | 360 | 45 | 135 | 225 | 270 |
| Elliptical with square (Figures 1.9g1 to 1.9g5) | 90 | 180 | 270 | 360 | 45 | 90 | 135 | 180 |
| Elliptical with triangular (Figures 1.9h1 to 1.9h5) | 90 | 180 | 270 | 360 | 45 | 135 | 180 | 225 |

driver gear. Figures 1.9c1 to 1.9c5 show meshing of miniature triangular gear with miniature pentagonal gear. The triangular gear is driver gear and the pentagonal gear is driven gear in this case. When 90° rotation is given to the driver gear, it rotates the driven gear by 45°. Rotation of the driver gear by 180° moves the driven gear by 90°. After 270° rotation of the driver gear the driven gear rotates by 135°. A complete 360° rotation of the driver gear moves the driven gear by 225°. These differences in the angular positions of the driven gear with respect to the driver gear are due to meshing of different shapes MNCGs. Meshing of the miniature square and miniature pentagonal gears is shown in Figure 1.9d1 to 1.9d5. In this, the square gear is driver gear and the pentagonal gear is driven gear. First 90° rotation of the square gear makes 45° rotation to the pentagonal gear. Rotation of the driver gear by 180°, 270°, and 360° rotates the driven gear by 135°, 225°, and 270°, respectively. Meshing of one miniature square gear with another is shown in Figures 1.9e1 to 1.9e5. Every 45° rotation of the driver gear produces 45° rotation of the driven gear. In this case, MNCGs of same shapes are meshing so their rotation is same. After 360° rotation both the gears reach to their initial position. As shown in Figures 1.9f1 to 1.9f5 the miniature triangular gear is driver gear and the miniature square gear is driven gear. When the driver gear rotates by 90° the driven gear rotates by 45°. The driver rotation of 180° rotates the driven by 135°. When the driver reaches to 270°, the driven reaches to 225°. Complete 360° rotation of the driver gear moves the driven gear by 270°. Meshing of miniature elliptical gear with miniature square gear is shown in Figures 1.9g1 to 1.9g5. In this pair the miniature elliptical gear is working as driver gear and the miniature square is working as driven gear. 90° rotation of the driver gear rotates the driven gear by 45°. Further rotation of the driver gear by 180° and 270° rotates the driven gear by 90° and 135°, respectively. Complete 360° rotation of the driver gear rotates the driven gear by 180°. Figures 1.9h1 to 1.9h5 is showing meshing of miniature elliptical gear with miniature triangular gear. The miniature elliptical is driver gear and the miniature triangular is driven gear. When the driver gear rotates by 90° the driven rotates by 45°. Rotation of the driver gear by 180° moves the driver gear by 135°. The driven gear rotates 180° when the driver gear rotates 270°. A complete 360° revolution of the driver gear makes 225° rotation of the driven gear.

## 1.9 CONCLUDING REMARKS

This chapter presented a brief introduction of different μ-manufacturing processes, their applications, details of μ-manufacturing of extremely complicated MNCGs by WSEM process, and quality and functioning assessment of the manufactured MNCGs. The following are some important conclusions from the presented details in this chapter:

- μ-manufacturing is an essential tool to manufacture miniaturized machines components and devices. Close tolerances, high accuracy, precision, and work on nano and μ-scale are some major challenges associated with μ-manufacturing. There are various μ-machining processes are available such as μ-milling, μ-WEDM, LIGA, etc.
- WSEM and μ-WSEM processes are capable of machining smaller-size components with high degree of accuracy. WSEM process was found suitable for net-shape μ-manufacturing of MNCGs with very good quality and surface finish.
- Extremely complicated shapes of MNCGs are successfully produced by WSEM process. The MNCGs have good dimensional and geometrical accuracy, better surface topography, and are net-shaped, i.e., any kind of post-processing is not required before their final use.
- WSEM process is flexible therefore it is suitable for manufacturing of different types of MNCGs. The manufacturing time consumed in the process is less and no machine setting time is required. Mass production of different types of MNCGs is possible only by changing the part-programming of the gears. No need to change the part-program for manufacturing of same gear number of times.
- Flank surfaces of the manufactured MNCGs are smooth and free from burr or any other unwanted material adhered to them. Gear flank surface distortion was also not seen. No μ-cracks and μ-voids were present on the surfaces.

- Meshing tests of same shape MNCGs such as miniature triangular gear with miniature triangular gear, and meshing of different shapes MNCGs such as miniature triangular gear with miniature pentagonal gear were performed to understand their functioning. All types of MNCGs were working fine. Their meshing was observed satisfactory. During the meshing no jerk and jamming were seen.
- Fluctuations in gear speed such as sudden increase, sudden decrease, and constant speed have been seen in one complete rotation of the driver MNCG. If a driver gear is rotating by some angle and its driven gear is rotating by smaller angle it means that the speed of the driver gear is more than the driven gear. Shape of MNCGs is an important factor which determines their position when they mesh. Instantaneous speed of MNCGs is also governed by the shape of the gear.
- WSEM has found an effective manufacturing process for MNCGs. μ-WSEM process can be used for net-shape μ-manufacturing of μ-NCG.

## REFERENCES

[1] SPL Kumar, J Jerald, S Kumanan, R Prabakaran, 2014, A review on current research aspects in tool-based micromachining processes, *Materials and Manufacturing Processes*, 29(11-12), 1291–1337, 10.1080/10426914.2014.952037

[2] X Li, P Ming, S Ao, W Wang, 2022, Review of additive electrochemical micro-manufacturing technology, *International Journal of Machine Tools and Manufacture*, 173, 103848, 10.1016/j.ijmachtools.2021.103848

[3] G Konstantinou, E Kakkava, L Hagelüken, PVW Sasikumar, J Wang, MG Makowska, G Blugan, N Nianias, F Marone, HV Swygenhoven, J Brugger, D Psaltis, C Moser, 2020, Additive micro-manufacturing of crack-free PDCs by two-photon, polymerization of a single, low-shrinkage preceramic resin, *Additive Manufacturing*, 35, 101343, 10.1016/j.addma.2020.101343

[4] T Li, X Yan, X Fang, P Jin, J Li, KF Rabbi, N Miljkovic, 2021, In-situ jet electrolyte micromachining and additive manufacturing, *Applied Physics Letters*, 119, 171602, 10.1063/5.0067988

[5] NJ Wilkinson, MAA Smith, RW Kay, RA Harris, 2019, A review of aerosol jet printing-a non-traditional hybrid process for micro-manufacturing, *The International Journal of Advanced Manufacturing Technology*, 105, 4599–4619, 10.1007/s00170-019-03438-2

[6] L Orazi, V Siciliani, R Pelaccia, K Oubellaouch, B Reggiani, 2022, Ultrafast laser micromanufacturing of microfluidic devices, *Procedia CIRP*, 110, 122–127, 10.1016/j.procir.2022.06.023

[7] B Nagarajan, Z Hu, X Song, W Zhai, J Wei, 2019, Development of micro selective laser melting: the state of the art and future perspectives, *Engineering*, 5, 702–720, 10.1016/j.eng.2019.07.002

[8] L Ouyang, D Zhou, Y Ma, Y Tu, 2018, Ensemble modeling based on 0–1 programming in micro-manufacturing process, *Computers & Industrial Engineering*, 123, 242–253, 10.1016/j.cie.2018.06.020

[9] SK Chaubey, NK Jain, 2018, State-of-art review of past research on manufacturing of meso and micro cylindrical gears, *Precision Engineering*, 51, 702–728, 10.1016/j.precisioneng.2017.07.014

[10] B Haaefner, M Biehler, R Wagner, G Lanza, 2018, Meta-model based on artificial neural networks for tooth root stress analysis of micro-gears, *Procedia CIRP*, 75, 155–160, 10.1016/j.procir.2018.04.031
[11] T Mashimo, T Urakubo, Y Shimizu, 2018, Micro geared ultrasonic motor, *IEEE/ASME Transactions on Mechatronics*, 23(2), 781–787, 10.1109/TMECH.2018.2792462
[12] X Xiao, Y Chen, X Xie, Y Shao, 2021, A miniature non-circular line gear with pure-rolling contact, *Bulletin of the JSME: Journal of Advanced Mechanical Design, Systems, and Manufacturing*, 15(6), 1–14, 10.1299/jamdsm.2021jamdsm0071
[13] L Slatineanu, O Dodun, M Coteata, G Nagit, IB Bancescu, A Hrituc, 2020, Wire electrical discharge machining- a review, *Machines*, 8, 69, 10.3390/machines8040069
[14] A Banu, MY Ali, MA Rahman, M Konneh, 2019, Investigation of process parameters for stable micro dry wire electrical discharge machining, *The International Journal of Advanced Manufacturing Technology*, 103, 723–741, 10.1007/s00170-019-03603-7
[15] RK Porwal, K Vinod, PS Chauhan, RK Shukla, 2021, An overview to micro-wire-electrical discharge machining, *IOP Conference Series: Materials Science and Engineering*, 1136, 012021, 10.1088/1757-899X/1136/1/012021
[16] P Sivaprakasam, P Hariharan, S Gowri, 2019, Experimental investigations on nano powder mixed micro-wire EDM process of Inconel-718 alloy, *Measurement*, 147, 106844, 10.1016/j.measurement.2019.07.072
[17] P Sivaprakasam, JU Prakash, P Hariharan, S Gowri, 2021, Micro-electric discharge machining (micro-EDM) of aluminum alloy and aluminum matrix composites - a review, *Advances in Materials and Processing Technologies*, 8(2), 1699–1714, 10.1080/2374068X.2020.1865127
[18] MY Ali, A Banu, M Shaffiq, MA Rahman, M Konneh, M Salehan, 2019, Investigation of taper angle in dry micro wire EDM, *International Journal of Mechanical Engineering and Robotics Research*, 8(5), 725–728, 10.18178/ijmerr.8.5.725-728

# 2 Research Status on Microgrinding Technology

*Wei Li, Xiao-long Hu, Cong Mao, Ying-hui Ren, and Ahmed Mohamed Mahmoud Ibrahim*

## 2.1 BACKGROUND OF MICROGRINDING TECHNOLOGY

The modern application of microcomponents in industries such as aviation, aerospace, communication, medicine, and bioengineering has been expanding. Therefore, silicon-based materials represented by silicon carbide and fused silica require higher precision and efficiency microtechnologies, such as microchannel plates, micromechanical vibrating gyroscopes, micromechanical optical switches, microaccelerometers, etc. [1,2]. The current processing of silicon-based micro-components is dominated by MEMS manufacturing technology, which includes a series of processes such as etching, lithography, and bonding. However, this technology is limited to the simple two-dimensional geometry of the part and the equipment used to manufacture microcomponents is expensive [3,4]. At present, the military and civil fields have put forward the requirements of structural specialization and dimensional precision for silicon-based microcomponents. The overall size of microcomponents is reduced to sub-millimeter and nanometer structural features, furthermore, it is often dominated by grooves, thin walls, shaped surfaces, or their combined structures. These microcomponents use different materials, complex structures, and high precision requirements, The corresponding manufacturing technology has put forward high requirements [5]. On the other hand, the contradiction between the machining accuracy and efficiency of microcomponents and the actual demand is becoming more and more prominent. Compared with laser, ion beam, UV embossing, and other technologies, micro-cutting and microgrinding processing technologies (Figure 2.1) can realize the three-dimensional processing of tiny parts of various materials. Micromachining equipment is small in size, has low energy consumption, and has low cost, which is one of the development directions of green manufacturing [6,7].

As shown in Figure 2.2, the complex three-dimensional surface structure of the micro jet engine turbine, the non-spherical structure of the microlens mold, the complex surface structure of the microchannel plate diameter is less than 8 mm [8,9]. These components have small machined feature sizes and a large

DOI: 10.1201/9781003364948-2

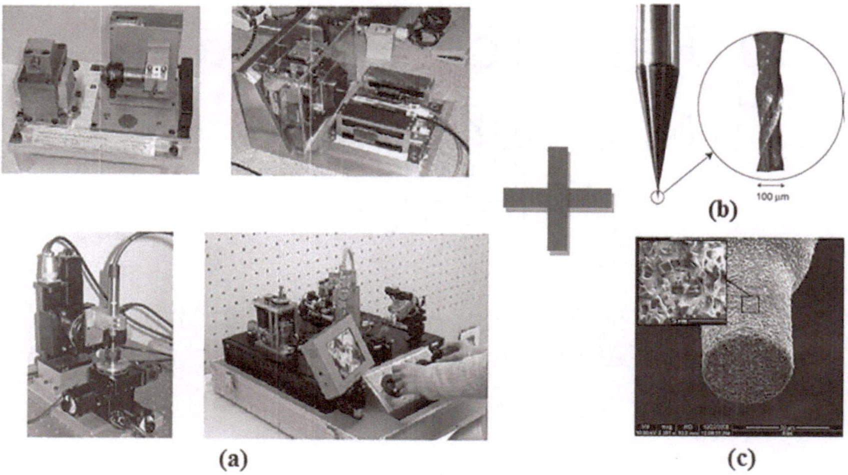


**FIGURE 2.1** Microfabrication equipment: (a) micromachine tool; (b) micromill; (c) micromill tool.

**Sources: References [6].**

overall dimensional span and complex structure. The surface quality and dimensional accuracy of the components are required to be high, which is typical of meso/microscale machinery manufacturing technology [10]. Therefore, there is an urgent need to develop micromachining technologies that are economically viable and capable of machining silicon-based parts with micro and complex structures in a one-time operation. Currently, it is already possible to machine a variety of microcomponents or microstructures using micromachining (including micromachining milling) (Figure 2.2) [11]. The dimensional accuracy of many components is able to meet the requirements despite a lack of surface quality. It is worth to be mentioned that for the microcomponents of hard and brittle materials, the processing damage such as microcracks and micro pits on the surface is obvious. Not only that, in the processing of such components as microgroove, microhole, microrib, and other three-dimensional microstructure, the edges are very easy to produce cracks, chipped edges, chipped corners, and other defects. It seriously affects the use of such parts' performance and promotion of applications. Therefore, in order to achieve high dimensional accuracy and high surface quality requirements for microcomponents of hard and brittle materials, the micromachining technology of micro/nano-scale ultra-precision grinding using microabrasives (microabrasive rods) with ultra-fine abrasives has been proposed in recent years.

Although the use of microfabrication has fulfilled the requirements of achieving nano-level processing surface roughness of microcomponents of hard and brittle materials [12]. It still has not solved the problem of processing defects such as cracks, chipping, and missing corners of the angular

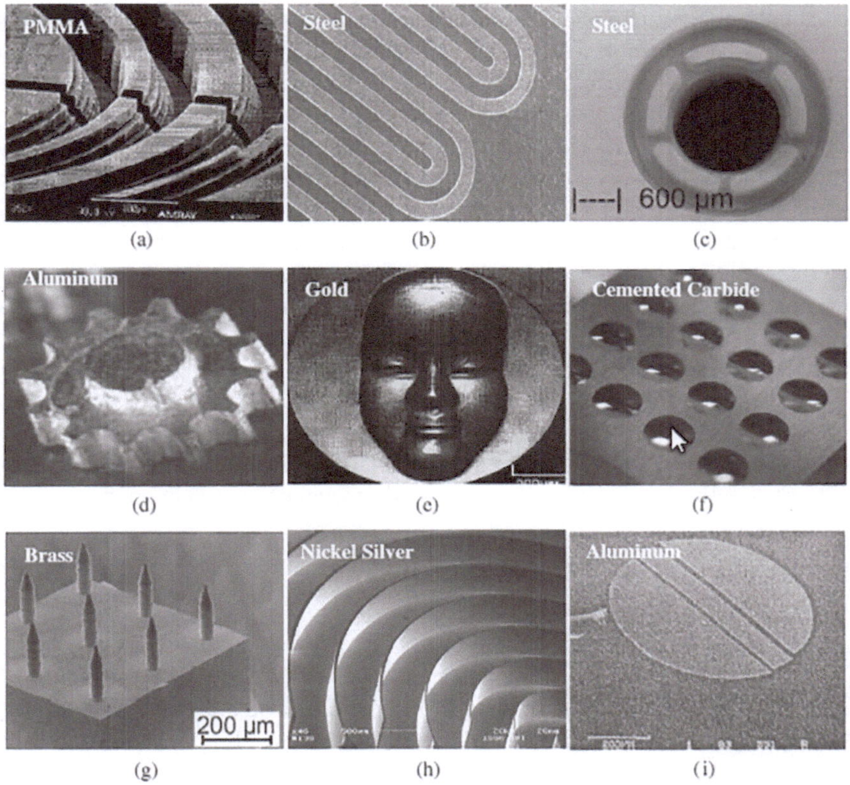


**FIGURE 2.2** Microcomponents and microstructures machined by microfabrication technique: (a) microcable groove; (b) microreactor; (c) micro mold; (d) micro gear; (e) micro face; (f) microprojector; (g) micro pin array; (h) microthin wall; (i) fusion target platinum.

**Sources: References [6].**

edges of microstructures. The reason that stands behind is that the current microspindle speed of micromachines is generally less than 200,000 r/min, and the effective grinding part of the microgrinding rod is between a few microns and several hundred microns in diameter, which leads to low processing efficiency of microfabrication and thus increases the wear and replacement speed of the microgrinding rod. For instance, Aurich et al. [13] used a 4 μm diameter microgrinding tool to microgrind single-crystal silicon material at 160,000 r/min. The microgrinding speed is only 0.03 m/s, which severely limited the material removal rate. The machined microgrooves showed obvious cracks and chipping. The main reason is the inability to achieve ductile material removal due to the low grinding speed and the coupling size effect between the abrasive grains and the workpiece.

At present, for micro parts of hard and brittle materials, the use of microgrinding technology has been able to achieve a nano-level surface roughness processing. But it is still difficult to break through the microstructure of the angular edge of the crack, chipping, and other processing damage defects, as well as processing accuracy, processing efficiency, and the increasingly prominent contradiction of the actual demand. The key components of the micromachine tool and the development of the machine are still an issue of the matter. Furthermore, system design and performance evaluation need further deep investigations. The microabrasives used are mainly electroplated diamond microabrasives, with a small holding force and few abrasive grains involved in grinding at the same time. The microabrasives wear and vastly deteriorate which affects negatively the tool life. Therefore, it is very necessary to carry out research on micromachine tool equipment and new processing methods for high precision and high-efficiency processing of tiny parts of hard and brittle materials, which is of great scientific significance and application value to promote the development and application of microfabrication technology and enhance the processing and manufacturing level of microcomponents.

## 2.2 ANALYSIS OF THE RESEARCH STATUS OF MICROGRINDING TECHNOLOGY

### 2.2.1 Microgrinding Process

During microfabrication of hard and brittle materials, the material is mainly removed by brittle fracture under the extrusion or scratching action of abrasive grains, and the corner edges and surfaces are highly susceptible to machining damage such as macrocracks, microcracks, and micro pits (Figure 2.3a). In this regard, Gong [14,15] pointed out that ductile removal of silicon can be achieved by controlling the maximum undeformed chip thickness less than 20 nm in single-crystal silicon microgrinding (Figure 2.3b).

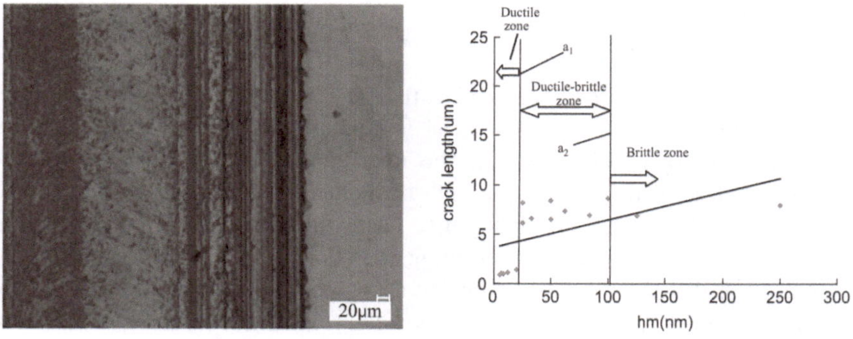


(a) Edge surface (b) Analysis result

**FIGURE 2.3** Shift in the removal mode of single-crystal silicon microfabrication.

**Sources: References [15].**

Although “brittle/ductile removal mode conversion” and “perfect plastic deformation in ductile domain machining mode only” for brittle materials are still controversial. In contrast, it has been widely established that the machined surface damage produced in the ductile removal mode is very minimized [16]. Gassilloud [17] derived the scratch depth of single-crystal silicon elastic-plastic deformation within 100 nm by nano-scratch test (Figure 2.4).

Zhang et al. [18] concluded that the scratch depth during plastic deformation of 6H-SiC is about 110 nm. With the existing micromachines and micro-abrasives, it is practically difficult to control the microgrinding thickness to an undeformed chip thickness of 110 nm or even 20 nm. For example, Aurich et al. [13] developed an ultra-precision microgrinder that integrates micro-grinding tool shaping and microfabrication and used a Ø4 μm microgrinding tool to process microstructures, and still showed significant cracking and chipping (Figure 2.5). This is mainly due to the failure of the material to achieve ductile removal due to the low grinding speed and the coupling size effect between the tool, abrasive, and workpiece. Zhou et al. [19] studied the microgrinding quality of single-crystal silicon by using Ø0.9 mm microgrinding head at different spindle speeds, feed rate, and grinding thickness parameters, and measured the surface roughness Ra between 0.5 and 1.5 μm. It can be seen that the micro-grinding process to achieve complete ductile removal or ductile removal dominated by controlling the thickness of undeformed chips and then to achieve high edge accuracy and high surface quality of silicon-based microcomponents, requires too demanding conditions and is very difficult to apply in practice.

In fact, it is difficult to achieve damage-free machining with a single mechanical process. For this reason, researchers have proposed the following composite micromachining processes: One is the electrochemical-assisted micromachining process. Cao [20] used this technology to process micro glass components, which not only take less than 30% of the traditional grinding process but also achieve a surface roughness of 0.05 μm Ra. Meanwhile, the laser-assisted microfabrication process, which is based on the principle of microfabrication after softening the material through the laser-induced, forms a

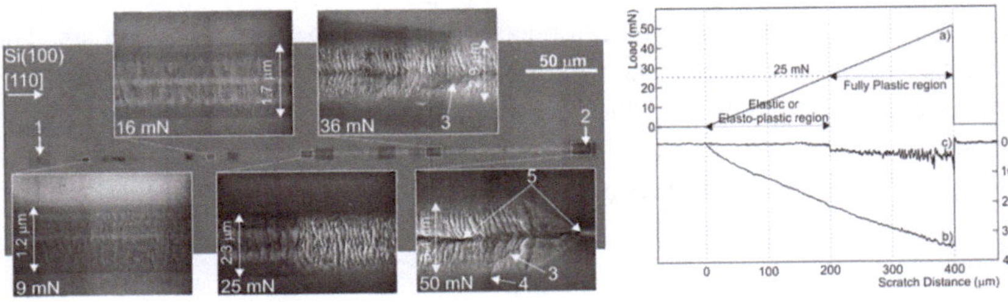


(a) Scratch (b) Measurement result

**FIGURE 2.4** Single-crystal silicon nano-scratch test.

**Sources: References [17].**

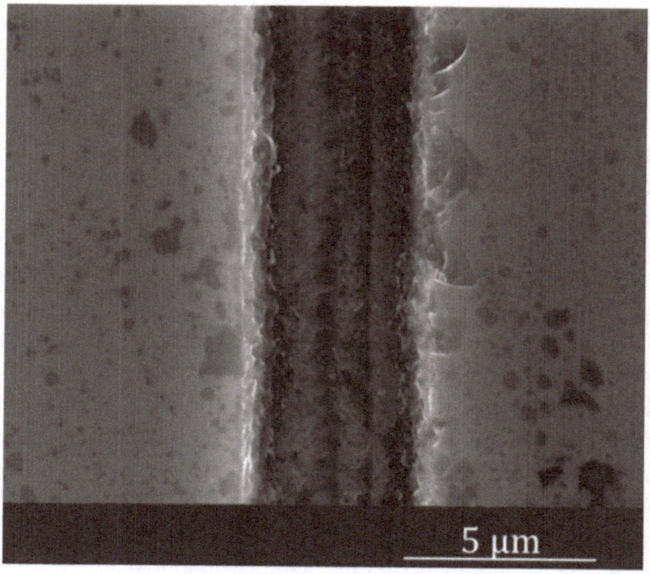


**FIGURE 2.5** Single-crystal silicon microgrinding processing microchannel.

**Sources: References [13].**

group of a thermal cracks layer. For instance, Kumar [21] and Chang [22] used this technique for machining microcomponents made of ceramic materials, and the results show that the laser-assisted microgrinding process not only reduces grinding forces and tool wear and improves surface quality, but also increases material removal rates. More techniques have been used to assist the microgrinding. The ultrasonic vibration-assisted microgrinding process is one of these techniques. Onikura [23], Egashira [24], and Zhang [25] applied the technique to the processing of microcomponents of stainless steel, glass, and ceramics, and the results show that the scale of abrasive chips and crack damage scale were reduced, and the microgrinding tool load and abrasive grain shedding were also reduced. However, the electrochemical and laser-assisted micromachining processes proposed so far are more complex and have higher processing costs. The ultrasonic vibration-assisted microfabrication requires the microgrinding tool diameter to be too small, and the processing damage control effect is limited. Therefore, there are difficulties in the practical application of the above three composite microfabrication processes.

### 2.2.2 Mechanism of Microgrinding

Due to the dramatic reduction of the diameter of the microgrinding head, the existence of neglected factors in the conventional grinding process, such as plowing force, grinding wheel deformation, and size effect. These factors largely affect the quality of microgrinding. When the grinding wheel diameter decreases to within millimeters, the contact arc length gradually decreases to a small range of variation, the grinding temperature shows a trend of first decreasing and then increasing, and the plowing force increases significantly [26]. In addition, the grinding depth of the abrasive grains in microfabrication is usually smaller than the grain size of the workpiece being machined, indicating that the abrasive grains finish cutting inside the crystal or at the crystal boundary. The size effect of the material and the recrystallization of the microstructure will significantly change the material removal mechanism. Therefore, the microfabrication mechanism is significantly different from the conventional grinding mechanism. Thus, many governing factors significantly affect the performance of the microgrinding. These factors can be discussed in detail as follows:

a. **Ductile domain removal mechanism for hard and brittle materials.** For plastic materials, subsurface damage caused by dislocation slip and plastic deformation cannot be ignored. In conventional grinding processes, the grinding grains interact with the workpiece at a large negative rake angle and high strain rate. However, in microfabrication, slip deformation of the material, lattice mutation, intergranular fracture, and tiny indentations and dents are more prominent and cannot be ignored. Chen et al. [27] established a ductile domain removal model for abrasive processing by considering the effect of abrasive action on ductile domain processing. When the abrasive size is reduced from 18 μm to 1.6 μm. The surface dominated by brittle fracture gradually transforms into a fully plastic surface without brittle fracture and cracks, and this action realizes a "grinding instead of polishing" processing technique. The plastic deformation involved in the scratching process is caused by the formation of polycrystalline nanocrystals, high-angle lattice dislocations, and near-atomic-scale defects. Xie et al. [28] investigated the material nano-removal mechanism considering the effect of micro-abrasive edge radius. The material removal mechanism is dominated by dislocation motion, and the cutting appears as a dislocation slip zone, chip formation zone, and elastic deformation zone, and both dislocation slip zone and chip formation zone are suppressed with the increase of cutting edge radius. As the radius of the cutting edge increases, the dislocation slip zone and chip formation zone are suppressed, while the elastic deformation zone tends to grow continuously. When the cutting edge radius increases to a value greater than a threshold, the elastic deformation zone shifts further. As the grinding depth increases, the effect of elastic deformation gradually decreases and chips are formed. At the same time, the surface roughness value of the workpiece decreases and the surface quality starts to become better. Not only plastic materials but also hard and brittle materials are affected by the minimum chip thickness effect and the phenomenon of elastic reversion exists. Interestingly, Manea [29] established a theoretical model of elastic-plastic transition by analyzing the geometric model of K9 glass grinding and measured the elastic recovery during the grinding of K9 glass in the experiment.

For hard and brittle materials, microcomponents or micro geometries with high precision and surface quality can be obtained by controlling the relevant critical conditions in microgrinding processes. Guo [30] observed the occurrence and extension of cracks in quartz glass at different grinding depths (0.1–1 μm), and the critical depth of the brittle-ductile transition is 0.36 μm. Due to the complexity of the micromechanical structure and the limitations of the microgrinding rod manufacturing limits, ductile domain removal grinding is not possible by controlling the grinding depth alone, so other relevant process parameters must be determined. To this end, Yin [31] investigated the ductile domain removal mechanism of YAG polycrystalline ceramic materials. At shallower scratch depths, the scratched grooves are completely moldable. The grooved surface is uniform and smooth, with no cracking occurring, indicating a ductile removal mechanism. As the scratch depth increases, the material is removed in a manner that not only flakes off at the edges of the scratched grooves but also produces tears on the smooth groove surface. A series of brittleness phenomena including notch fracture, radial crack extension, transverse crack interweaving, twisting and tearing of the scratch groove, large-scale flaking, and brittle fracture gradually appear due to the "orange peel effect" unique to polycrystalline materials. When the grinding depth reaches a certain value and continues to increase, the fracture occurring between the material grains in microfabrication is significantly reduced. More grinding chips are generated between grain boundaries, so the fracture burrs are greatly reduced. Although studies such as the literature [32] have given the ductile domain of microfabrication of hard and brittle materials, the critical condition model of the ductile-brittle combination, and the grinding process parameters. However, it is still a complex integrated process with both plastic and brittle removal due to the influence of micromechanical vibration, microgrinding bar deflection, material micro defects, and material inhomogeneity. It is still difficult to realize a microfabrication process where plastic removal is dominant.

b. **Size effect and recrystallization effect.** When ultra-fine abrasive grains are cut inside the grain or at the grain boundary, the thickness of undeformed chips is very small and there is a size effect in material removal, which can lead to a significant increase in specific grinding energy or unit grinding force. De Oliveira [33] reported that factors contributing to dimensional effects include tool edge radius, subsurface plastic deformation, material processing enhancement, and material separation. Sahoo [34] proposed an improved undeformed chip thickness algorithm by considering tool runout, minimum chip thickness, and the trajectory of all the teeth passed in one full turn of the tool to elucidate the size effect due to material strain gradients at the material nano-cutting scale. Park [35] developed a predictive model for the microgrinding process by integrating mechanical and thermal effects in a single grit interaction model at the microscopic level of material removal, together with the size effect of micromachining. The results show that the plowing force or the indentation force generated is also an important influence on the size effect. In addition, the effect of work-piece material recrystallization effect cannot be neglected when microfabrication grinding is performed inside the grain or at the boundary. Moriwakim [36] and Turkovich [37] pointed out that when processing polycrystalline materials, the processing mechanism is highly dependent on the grain size and its orientation. Suzuki [38] found that the cutting mechanisms of single-crystal, polycrystalline, and amorphous materials differ greatly, and their cutting forces vary continuously with grain boundaries. The subsurface damage of the materials consists of amorphous layers, dislocation layers with grain boundaries, and microcracks. In addition, the stresses generated in the workpiece due to heat in microfabrication can seriously affect the material properties. However, the duration of thermal shock caused by a single abrasive grain is much smaller than that in the contact area of the workpiece, and if the thermal conductivity of the workpiece material such as Al6061-T6 is high, then the thermal damage will be less. But for hard and brittle materials, the effect of thermal effect is still relatively large.

### 2.2.3 Microgrinding Tool

Microgrinding tools are another key tool factor affecting the quality of microfabrication. Two main types of microgrinding tools have been reported: One is coated microgrinding tool, which is prepared by electroplating [39], chemical plating, and chemical vapor deposition [40,41] (CVD) processes (Figure 2.6). The coated microgrinding tool has only a single layer of abrasive, which is difficult to recondition and reuse after wear. The other is the integral sintered PCD microgrinding tool, which can be reused after reconditioning [42,43], but the cost is high. In microfabrication, the wear of the microgrinding tool is more sensitive to the quality of the process than in conventional grinding [44]. Wen [45] investigated the wear of microabrasive heads used for microfabrication of soda lime glass and showed that the wear of microabrasive heads increased significantly with increasing grinding speed and feed rate. After grinding 15 microchannels, almost all of the abrasive grains were vanished (Figure 2.7). Due to the drastic reduction in the outer diameter size of the microgrinding head and the contact area with the workpiece, fewer abrasive grains are involved in the microfabrication and the impact load of individual grains on the material is enhanced. Simultaneously, the smaller abrasive grains have reduced holding power, resulting in severe abrasive grain wear.

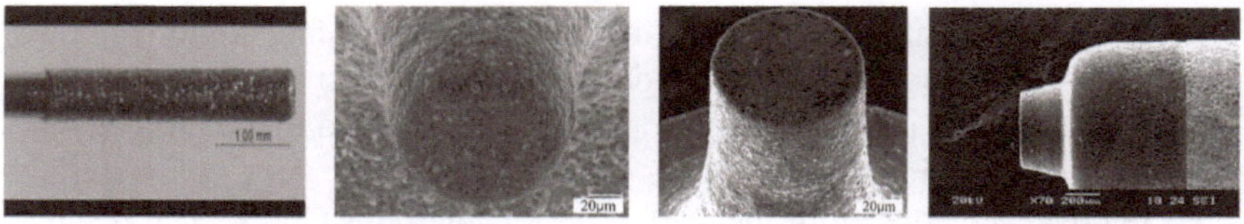


(a) Electroplating (b) Chemical plating (c) CVD (d) PCD

**FIGURE 2.6** Various types of microgrinding heads.

**Sources: References [39,41] and [43].**

**FIGURE 2.7** Wear of microabrasive tool.

**Sources: References [45].**

Cemented carbide is commonly used as the substrate for microgrinding tools because of its high toughness, high strength, wear resistance, and heat resistance. The abrasive grains selected for microgrinding tools are mainly diamond and CBN, and diamond is more widely used in microgrinding tools because of its low wear rate, high shape stability, and fast heat transfer. However, due to the irregularity of the grain shape, the randomness of the direction, and the inconsistency of the projection height. The load-bearing capacity of the microabrasives changes dramatically, thus easily causing deformation of the microabrasives. For this reason, Butler-Smith [46] proposed a new type of microgrinding wheel with an ordered arrangement of regularly shaped abrasive grains (Figure 2.8). However, it is difficult to apply to microabrasive rods with a diameter of less than 1 mm at present due to the limitation of processing conditions.

**FIGURE 2.8** Microgrinding wheel developed by Butler-Smith (3.2 mm).

**Sources: References [46].**

With the development of microfabrication technology, the preparation of ultra-micro tools with a diameter of less than 1 μm has become a reality [47]. However, as the diameter decreases sharply, the influence of grinding rod stiffness, abrasive type, particle size, and hardness on the grinding performance of microgrinding rods becomes more prominent. Therefore, the manufacturing of microabrasives requires higher technical requirements. There are two main types of existing microabrasive heads: One is coated microabrasive rods, which are mainly prepared by electroplating, chemical vapor deposition, and cold spraying processes. Compared to the chemical vapor deposition coating process, diamond microabrasives prepared by the electroplating process have poorer machinability. This is attributed to that the chemical vapor deposition process is capable of coating smaller and sharper super microgrinding particles. The microabrasive rods prepared by the cold spraying process have less abrasive bonding force and faster wear, which are still difficult to apply in practical processing. Another kind of microabrasive rod is the integral sintered one. Masaki [48] used sintered PCD microgrinding rods for microfabrication tests on carbide materials with the surface roughness of 28 nm Rz and 5 nm Ra. The microabrasive rod is coated with only the outermost layer of abrasive grains, so the microabrasives wear quickly and need to be replaced. In contrast, the sintered metal-based microabrasive rod has a large number of embedded abrasive grains, which can be repaired, and their service life is extended by regular dressing using a suitable dressing process. Due to the drastic reduction of the outer diameter and the contact area with the workpiece, the microabrasive rod has very few abrasive grains involved in microfine grinding, and the abrasive grains are easily blunted or broken. Meanwhile, the abrasive grains are selected to be finer, the contact area of the individual abrasive jointing agent is reduced, the abrasive grains are held with less force, and the abrasive grains fall off intensifies. However, the dressing of the microabrasive rod is essential. However, due to the size limitation, traditional contact dressing methods such as turning, diamond roller, rolling, and grinding are difficult to carry out. Therefore, non-contact dressing methods such as EDM and electrolysis can only be used. Wei et al. [49] proposed electrochemical discharge dressing (ECDD) for metal-based microgrinding rods. However, the high cost and long dressing time of these non-traditional dressing methods have made it difficult to promote their application. As a result, most microgrinding rods are discarded after being used and worn, especially coated microabrasive rods, which are rarely reused after reconditioning.

### 2.2.4 Micromachine Tools

As a carrier of microfabrication technology, the overall performance of micromachines is closely related to the machining quality and machining efficiency of components. Researchers have carried out relevant research and developed prototype machines in micromachines and their key components. Okazaki [50] developed micromachines, the maximum speed of its microspindle is up to 200,000 r/min, a maximum feed rate of 50 mm/s. Jahanmir [51] developed a test system for micromilling machining with a microspindle with a maximum speed of nearly 500,000 r/min and a maximum table feed rate of 12.5 mm/s. Representative ultra-precision micromachines include the ultra-precision micromilling machine developed by Balasubramanian et al. [52]. This milling machine has an X-axis positioning accuracy of 87 nm and a stroke of 16 μm, and a Y-axis positioning accuracy of 92 nm and a stroke of 8 μm. The machining accuracy can reach 170 nm, but the size range of this micromilling machine is small. The micromilling machine developed by the Harbin Institute of Technology [53] has a maximum microspindle speed of 140,000 r/min and table travel of 30 mm × 30 mm × 25 mm. The microgrinding machine constructed by Shenyang University of Technology [54] has a maximum microspindle speed of 160,000 r/min, a table stroke of 200 mm, and a resolution range of 0.008–2.000 μm. Although micromachines for micromilling and microgrinding have been developed worldwide. However, micromachines have not yet fully entered the practical stage for the following reasons. First, the processing efficiency is low. For example, the microgrinding rod of φ4 μm has been used for microgrinding processing [55], but its grinding speed is only 0.03 m/s at the maximum spindle speed of 160,000 r/min, which seriously limits the material removal rate and the processing efficiency of the microgrinding machine. According to the required microgrinding speed of commonly used engineering materials, the microspindle speed should reach or even exceed 500,000 r/min. In addition, the micro ultra-precision tables developed so far have been able to achieve submicron level positioning accuracy, and some of them have even reached the nanometer level. However, the small stroke and low feed rate of micromachining machines have led to low machining efficiency of micromachining machines. Secondly, the machining accuracy of micromachines is low, and the grinding thickness is generally at the micron level. However, in the existing spindle of micromachine tools, the runout error of the tool can reach 3–20 times the grinding chip thickness, which seriously affects the machining quality of the parts. In order to machine microcomponents with submicron precision, the runout error of the tool should be within submicron. In addition, due to the miniaturization of system components, many of the miniature machine tools developed to date cannot meet the requirements of high rigidity, vibration resistance, and high precision. Therefore, it is very difficult to achieve ductile removal processing of silicon-based microcomponents by strictly controlling the undeformed chip thickness within 110 nm due to microgrinding head machining errors.

## 2.3 RELATED WORK DONE BY THIS RESEARCH GROUP

### 2.3.1 Microspindle

#### 2.3.1.1 Ultra-High-Speed Precision Microspindle Development

The existing microspindles and their tools suffer from two problems: One is the high rotary speed with low rotary accuracy, and the other is the high rotary accuracy with low rotary speed. It is difficult to realize the problem of free tool replacement under high rotary accuracy. We proposed a variety of new structural design ideas and methods to address the mentioned problems. The first idea is a new one-piece micro flexible coupling structure. The new connection structure contains a micro flexible coupling and a shape memory alloy (SMA)-based microchuck, which utilizes the dual-range shape memory effect of the SMA ring to achieve clamping at room temperature (above 15°C) and disassembly at low temperature (below 0°C). The structure is simple and symmetrical, easy to operate, and can effectively compensate for manufacturing and mounting errors of the microspindle as well as runout errors at ultra-high speeds, etc. The second design is a new microspindle design method with a flexible connection between the power shaft and tool and shank-rotor integration. The micro pneumatic turbine drive is used to achieve ultra-high-speed rotation. Consistently, the micro flexible connection structure is used to eliminate the turbine shaft error. The tool shank is used as the rotor directly supported by the gas hydrostatic bearing to realize the ultra-high speed and precise rotation of the tool. Using the dual-range shape memory effect of the SMA ring to achieve free tool replacement. A microspindle principle prototype (Figure 2.9a) was trialed with a maximum speed of 120,000 r/min and a tool radial runout error of 3.31 μm, which was well applied. The test results showed that the novel design could effectively solve the problem that the tool, preserving has high rotary accuracy with a high degree of freedom to replace it under the ultra-high speed of the microspindle. Third, a double-row nozzle impact pneumatic microturbine and equal diameter turbine rotor structure were proposed, and an improved microspindle prototype was trialed (Figure 2.9b). This prototype reached a maximum speed of 245,000 r/min with a tool radial runout error of 2.79 μm, much smaller than the 10.18 μm of the turbine shaft.

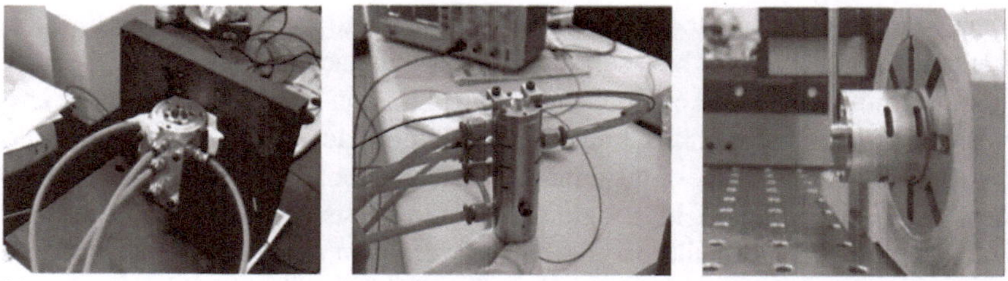

(a) First-generation micro-machines, (b) Second-generation micro-machines, (c) Third-generation micro-machines

**FIGURE 2.9** The developed ultra-high-speed precision microspindle.

Considering the poor performance of the small-bore throttle-type gas hydrostatic bearing in terms of load-carrying capacity and stiffness. The third-generation ultra-high-speed pneumatic microspindle driven by a turbine and supported by a porous gas hydrostatic bearing (Figure 2.9c) was piloted with an external structure size of Ø28 mm × 65 mm and a maximum speed of 400,000 r/min. The developed series of microspindles are designed for practical applications and marketability with a simple and compact structure. A large technological-level breakthrough can be achieved under the existing general machining conditions, so the manufacturing difficulty and cost are also low.

### 2.3.2 Micromachine Tool

First, the use of the above self-developed ultra-high-speed precision microspindle to solve the problem of low cutting speed of microtool ($\Phi < 1$ mm). Second, the use of a macro-micro combination of ultra-precision table drives to solve the problem of high precision and rapid feed. Furthermore, the high-speed, large-stroke, low-resolution motion of the system is accomplished by the motor-driven macro positioning stage. The piezoelectric ceramic-driven micro-positioning stage has numerous benefits for small stroke and high resolution, which is suitable for compensating the position error of macro positioning and suppressing residual vibration. Therefore, the whole ultra-precision positioning stage achieves the performance of large-stroke, high speed, high accuracy, and high-frequency response. Third, the combination of high-stiffness bed structure design and integration optimization of the whole system ensures the machining accuracy of the micromachine tool. The multi-layer ribbed composite reinforcement structure is proposed to enhance the stiffness of the micromachine tool, while the load performance of the guide rail is optimized to ensure the machining accuracy and stability of the micromachine tool. Fourth, the combination of error prediction and in-situ compensation to further improve the machining accuracy of micromachine tools. On the hand, the microspindle system stiffness chain model and cutting load versus tool tip displacement model under

**FIGURE 2.10** Development of high-performance micromachines.

static conditions are combined to predict the error of the microspindle tool. On the other hand, a comprehensive analysis of the factors influencing the dynamic error of the tool and the table positioning error of the ultra-high-speed precision microspindle at ultra-high speed is presented. Thus, the governing law of the errors is revealed, and the error in-situ compensation mechanism is established to significantly improve the micromachining accuracy. A prototype of a high-performance micromachine with external structural dimensions of 365 mm × 765 mm × 600 mm was developed (Figure 2.10) and tested for micromilling and microgrinding machining. The machining quality of the microcomponents was comparable to that of existing machining machines, but with high machining efficiency and low utility consumption, and floor space.

### 2.3.3 Chemical-Mechanical Microgrinding

A newly designed chemical-mechanical microgrinding process for ultra-precise meso/microcomponents for single-crystal silicon fabrication is proposed. The proposed technique utilizes chemical modification solutions to modify and soften the surface layer, followed by microgrinding to fabricate the meso/microstructures. The first step is to formulate a chemical modification solution (Table 2.1) to modify the surface of the workpiece with a special chemical solution before microgrinding. Silicon is firstly oxidized to form silicon dioxide ($SiO_2$). The formed silica reacts with hydroxide ions ($OH^-$) to form silicates to soften the surface layer. The workpiece is then microground with a microgrinding tool (<1 mm diameter) to obtain the desired microscopic features. It is obvious that the quality of the microchannels obtained by the chemical-mechanical grinding technique on the modified workpiece is much better than on the original workpiece. On the modified workpiece, the crack lengths of edge chipping damage in the microchannels were significantly reduced and even in some areas, there was no chipping damage (Figure 2.11). The quality of the microchannels on the pristine workpiece was very poor. In contrast, surface chemical modification can effectively improve the quality of the microchannels [55].

**TABLE 2.1**
**Reagents for the Chemical Modification Solution**

| Name | Molecular/Chemical Formula | Function | Proportion (wt %) |
|---|---|---|---|
| Deionized water | $H_2O$ | Base solution | 60–80% |
| Sodium hydroxide | $Na_2CO_3$ | Reactant | 7–12% |
| Triethanolamine | $C_6H_{15}NO_3$ | pH modifier | 3–8% |
| Hydrogen peroxide | $H_2O_2$ | Oxidizing agent | 2–4% |
| EDTA ferric sodium salt | $C_{10}H_{12}FeN_2NaO_8$(EDTA-Fe-Na) | Catalyst | 2–4% |
| Polyethylene glycol | $HO(CH_2CH_2O)_nH$ | Additive | 1.5–3% |
| Benzotriazole | $C_6H_5N_3$ | – | 1.5–3% |
| Hexamethylenetetramine | $C_6H_{12}N_4$ | – | 1.5–3% |
| Sodium molybdate | $Na_2MoO_4$ | – | 1.5–3% |

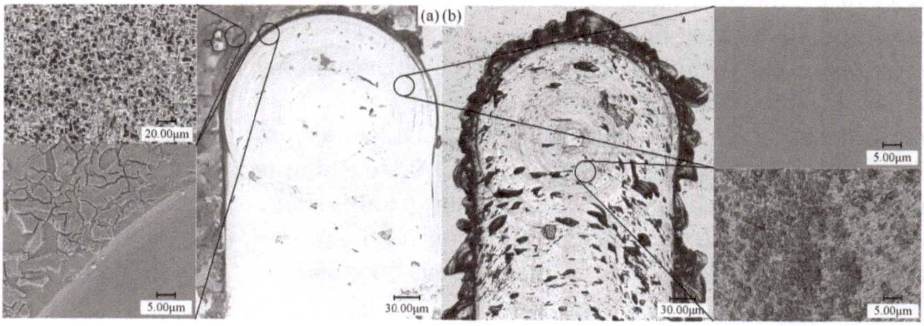


**FIGURE 2.11** Microchannels machined on (a) modified and (b) original workpieces.

## 2.4 CHALLENGES AND DEVELOPMENT TRENDS

Microgrinding has attracted widespread attention for its high efficiency, and certain research results have been achieved. In recent years, microgrinding technology and its multidisciplinary collaborative manufacturing technology are still being explored. There are many fundamental problems to be solved in grinding processes and mechanisms, microgrinding grinding tools, and micro-machine tool development. The theoretical models established in the micro-fabrication mechanism and process differ greatly from the actual ones, indicating that sufficient basic theories have not been accumulated. Nevertheless, dis-continuous processing is more common in microfabrication due to the influence of abrasive grain size. Microscopic defects in the material have a great impact on microfabrication, and the thermodynamic properties of the microfabrication process need to be further explored. The application of composite processes in microfabrication is not yet widespread, and there is a lack of theoretical systems and experimental investigations toward the application of multiple materials and composite processing. For this reason, the field of microgrinding processing will continue to expand, and the development of composite materials and new materials will become an inevitable trend.

Microgrinding tools are equivalent to the "teeth" of the micromachine tools industry, and the service life of microabrasives is an important core factor that restricts the development of microgrinding technology to a deeper level. In addition, the non-standardization of microgrinding rods needs further in-depth and systematic research. As the carrier of microgrinding rods, micromachine tools lack design theory and performance evaluation. For example, it is difficult to achieve high speed for the key spindle with the use of micromachine tools, and the runout error will be greater. Moreover, the micromotion mechanism of micromachine tools also limits the application of micromachining, and its positioning accuracy and feed accuracy are lower. For this reason, it has become an inevitable trend for material surface processing to develop toward high-efficiency nano-scale removal. Diversification of microcomponents makes a constant drive to the development of high precision and complex structure. In order to meet the actual needs, both processing efficiency and accuracy need to be improved, which requires vigorous development of ultra-precision machining components and intelligent processing systems. Highly composite, integrated, and flexible processing process direction continues to move forward. The composite processing process using multi-energy field, laser, electrochemical, ion beam, plasma, ultrasonic, and other technologies will effectively improve the processing efficiency and processing quality. Composite machining can also effectively solve the microfine grinding processing of precision, complex, and small batch-shaped structure parts.

## 2.5 CONCLUSION

The following conclusions are drawn from the present study:

1. Microgrinding tools continue to evolve toward manufacturing limits in order to meet the complex and precise, heterogeneous, and diverse requirements of microcomponents. Microgrinding technology is also gradually being combined with another micro/nanomanufacturing technologies.
2. In recent years, the comprehensive performance of miniature machine tools and their microcomponents of the processing quality has improved greatly. This is due to the continuous development of machine tool components, such as some porous mass hydrostatic gas bearings, whose structure is simpler than the small hole throttling type. And it has a higher load-bearing capacity and greater stability, with improved machinability.
3. The proposed chemical-mechanical microgrinding process achieves a larger-scale ductile domain processing technique compared to the conventional microgrinding process.

## ACKNOWLEDGMENTS

The presented work was funded by the Natural Science Foundation of China (52275421, 51875192, 52005174), the Excellent Youth Project of Educational Committee of Hunan Province (2022JJ10010), the key research and development program of Hunan Province (2022WK2003), the Natural Science Foundation of Hunan Province (Nos.2021JJ40064, 2020JJ4193), and the Natural Science Foundation of Changsha (No.kq2014048). The authors acknowledge the financial support.

## REFERENCES

[1] Chae J, Park SS, Freiheit T Investigation of micro-cutting operations. *International Journal of Machine Tools & Manufacture*, 2006, 46(3-4): 313–332.
[2] Uhlmann E, Mullany B, Biermann D, et al. Process chains for high-precision components with micro-scale features. *CIRP Annals-Manufacturing Technology*, 2016, 65(2): 549–572.
[3] Brousseau E, Dimov S, Pham D Some recent advances in multi-material micro- and nano-manufacturing. *The International Journal of Advanced Manufacturing Technology*, 2010, 47 (1): 161–180.
[4] Zhou ZX, Li W, Song TJ, et al. Performance requirements of microspindles for microfabrication and their research status. *Journal of Mechanical Engineering*, 2011, 47(19): 149–157.
[5] Perveen A, Jahan MP, Rahman M, Wong YS A study on microgrinding of brittle and difficult-to-cut glasses using on-machine fabricated poly crystalline diamond (PCD) tool. *Journal of Materials Processing Technology*, 2012(212): 580–593.
[6] Huo DH, Cheng K, Wardle F Design of a five-axis ultra-precision micro-milling machine-ultra mill. Part 1: holistic design approach, design considerations and specifications. *The International Journal of Advanced Manufacturing Technology*, 2010, 47 (9): 867–877.
[7] Qin Y, Brockett A, Ma Y, Razail A, Zhao J, Harrison C, Pan W, Dai X, Loziak D Micro-manufacturing: research, technology outcomes and development issues. *The International Journal of Advanced Manufacturing Technology*, 2010, 47 (9): 821–837
[8] War ZB Fundamental research on the design and manufacturing of PCD micro milling cutters. *Nanjing University of Aeronautics and Astronautics*, 2015.
[9] Hoffmeister HW, Wenda A Novel grinding tools for machining precision micro parts of hard and brittle materials. *Proceedings of the 15th Annual Meeting of the ASPE*, 2000: 152–155.
[10] Sun YZ, Liang Yi C, Cheng K Micrometer and mesoscale mechanical engineering. *Journal of Mechanical Engineering*, 2004, 40(5): 1–6.
[11] Bang Y, Lee K, Oh S 5-axis micro milling machine for machining micro parts. *The International Journal of Advanced Manufacturing Technology*, 2005, 25 (9-10): 888–894.
[12] Masaki T, Kuriyagawa T, Yan J, Yoshihara N Study on shaping spherical poly crystalline diamond tool by micro-electro-discharge machining and micro-grinding with the tool. *International Journal of Surface Science and Engineering*, 2007, 1(4): 344–359.
[13] Aurich JC, Carrella M, Walk M Micro grinding with ultra-small micro pencil grinding tools using an integrated machine tool. *CIRP Annals - Manufacturing Technology*, 2015, 64(1): 325–328.

[14] Cheng J, Wu J, Gong YD, et al. Experimental study on the single grit interaction behaviour and brittle–ductile transition of grinding with a diamond micro-grinding tool. *International Journal of Advanced Manufacturing Technology*, 2017, 91(1-4): 1209–1226.

[15] Cheng J, Gong YD Experimental study of surface generation and force modeling in micro-grinding of single crystal silicon considering crystallographic effects. *International Journal of Machine Tools and Manufacture*, 2014, 77: 1–15.

[16] He W, Liu C, Xu G, et al. Effect of temperature on ductile-to-brittle transition in diamond cutting of silicon. *The International Journal of Advanced Manufacturing Technology*, 2021, 116: 3447–3462.

[17] Gassilloud R, Ballif C, Gasser P, et al. Deformation mechanisms of silicon during nanoscratching. *Physica Status Solidi*, 2005, 202(15): 2858–2869.

[18] Zhang FH, Meng BB, Geng YH, et al. Study on the machined depth when nanoscratching on 6H-SiC using Berkovich indenter: modelling and experimental study. *Applied Surface Science*, 2016, 368: 449–455.

[19] Zhou YG, Gong YD, Gao Q, et al. Experimental study on the factors influencing the surface quality of single crystal silicon micro-scale side grinding. *Journal of Northeastern University*, 2017, 38(7): 983–988.

[20] Cao XD, Kim BH, Chu CN Hybrid micromachining of glass using ECDM and micro grinding. *International Journal of Precision Engineering and Manufacturing*, 2013, 14(1): 5–10.

[21] Kumar M, Melkote S, Lahoti G Laser-assisted microgrinding of ceramics. *CIRP Annals-Manufacturing Technology*, 2011, 60(1): 367–370.

[22] Chang WL, Luo XC, Zhao QL, et al. Laser assisted micro grinding of high strength materials. *Advanced Materials Research*, 2012, 496: 44–49.

[23] Onikura H, Inoue R, Okuno K, et al. Fabrication of electroplated micro grinding wheels and manufacturing of microstructures with ultrasonic vibration. *Advanced Materials Research*, 2003, 238: 9–14.

[24] Egashira K, Okina R, Yamaguchi K, et al. Drilling of microholes using diamond grinding tools. *Advanced Materials Research*, 2016, 1136: 435–439.

[25] Zhang JH, Li H, Zhang ML, et al. Study on force modeling considering size effect in ultrasonic-assisted micro-end grinding of silica glass and $Al_2O_3$ ceramic. *The International Journal of Advanced Manufacturing Technology*, 2017, 89(1-4): 1173–1192.

[26] Pratap A, Patra K, Dyakonov AA Manufacturing miniature products by micro-grinding: a review. *Procedia Engineering*, 2016, 150: 969–974.

[27] Li C, Piao Y, Meng B, et al. Phase transition and plastic deformation mechanisms induced by self-rotating grinding of GaN single crystals. *International Journal of Machine Tools and Manufacture*, 2022, 172: 103827.

[28] Xie W, Fang F Effect of tool edge radius on material removal mechanism in atomic and close-to-atomic scale cutting. *Applied Surface Science*, 2020, 504: 144451.

[29] Manea H, Cheng X, Ling S, et al. Model for predicting the micro-grinding force of k9 glass based on material removal mechanisms. *Micromachines*, 2020, 11(11): 969.

[30] Guo X, Li M, Luo X, et al. Smoothed-particle hydrodynamics investigation on brittle–ductile transition of quartz glass in single-grain grinding process. *Nanomanufacturing and Metrology*, 2020, 3: 299–306.

[31] Yin H, Wang S, Guo B, et al. Effects of scratch depth on material-removal mechanism of yttrium aluminium garnet ceramic. *Ceramics International*, 2022, 48(19): 27479–27485.

[32] Cheng J, Gong YD Experimental study on ductile-regime micro-grinding character of soda-lime glass with diamond tool. *International Journal of Advanced Manufacturing Technology*, 2013, 69(1-4): 147–160.

[33] De Oliveira FB, Rodrigues AR, Coelho RT, et al. Size effect and minimum chip thickness in micromilling. *International Journal of Machine Tools and Manufacture*, 2015, 89: 39–54.

[34] Sahoo P, Patra K Mechanistic modeling of cutting forces in micro-end-milling considering tool run out, minimum chip thickness and tooth overlapping effects. *Machining Science and Technology*, 2018.

[35] Park HW, Liang SY Force modeling of micro-grinding incorporating crystallographic effects. *International Journal of Machine Tools and Manufacture*, 2008, 48(15): 1658–1667.

[36] Moriwaki T Experimental analysis of ultra-precision machining. *International Journal of the Japan Society for Precision Engineering*, 1995, 29(4): 287–290.

[37] Von Turkovich BF, Black JT Micro-machining of copper and aluminum crystals. *Journal of Manufacturing Science and Engineering*, 1970, 92(1): 130–134.

[38] Suzuki T, Nishino Y, Yan J Mechanisms of material removal and subsurface damage in fixed-abrasive diamond wire slicing of single-crystalline silicon. *Precision Engineering*, 2017, 50: 32–43.

[39] Chen ST, Tsai MY, Lai YC, et al. Development of a micro diamond grinding tool by compound process. *Journal of Materials Processing Technology*, 2009, 209(10): 4698–4703.

[40] Kirsch B, Bohley M, Arrabiyeh PA, et al. Application of ultra-small micro grinding and micro milling tools: possibilities and limitations. *Micromachines*, 2017, 8(9): 261.

[41] Denkena B, Hoffmeister HW, Hahmann D, et al. Machining of micro-systems. *Microsystem Technologies*, 2008, 14(12): 1909–1916.

[42] Wei C, Hu D, Xu K, et al. Electrochemical discharge dressing of metal bond micro-grinding tools. *International Journal of Machine Tools and Manufacture*, 2011, 51(2): 165–168.

[43] Perveen A, San WY, Rahman M Fabrication of different geometry cutting tools and their effect on the vertical micro-grinding of BK7 glass. *The International Journal of Advanced Manufacturing Technology*, 2012, 61(1): 101–115.

[44] Feng J, Kim BS, Shih A, et al. Tool wear monitoring for micro-end grinding of ceramic materials. *Journal of Materials Processing Technology*, 2009, 209(11): 5110–5116.

[45] Wen XL, Gong YD Modeling and prediction research on wear of electroplated diamond micro-grinding tool in soda lime glass grinding. *International Journal of Advanced Manufacturing Technology*, 2017, 91(9-12):3467–3479.

[46] Butler-Smith PW, Axinte DA, Daine M Solid diamond micro-grinding tools: from innovative design and fabrication to preliminary performance evaluation in Ti–6Al–4V. *International Journal of Machine Tools and Manufacture*, 2012, 59: 55–64.

[47] Ohmori H, Katahira K, Naruse T, et al. Microscopic grinding effects on fabrication of ultra-fine micro tools. *CIRP Annals-Manufacturing Technology*, 2007, 56(1): 569–572.

[48] Masaki T, Kuriyagawa T, Yan J, et al. Study on shaping spherical poly crystalline diamond tool by micro-electro-discharge machining and micro-grinding with the tool. *International Journal of Surface Science and Engineering*, 2007, 1(4): 344–359.

[49] Wei CJ, Hu DJ, Xu KZ, et al. Electrochemical discharge dressing of metal bond micro-grinding tools. *International Journal of Machine Tools and Manufacture*, 2011, 51(2): 165–168.
[50] Okazaki Y, Mori T, Morita N Desk-top NC milling machine with 200 krpm spindle, ASPE Annual Meeting, Altamonte Springs, Washington D.C., USA, Raleigh： ASPE, 2001: *192*–195.
[51] Jahanmir S, Ren Z, Heshmat H, et al. Design and evaluation of an ultrahigh speed micro-machining spindle[J]. *Machining Science and Technology*, 2010, 14(2): 224–243.
[52] Balasubramanian A, Jun MB, Devor RE, et al. A submicron multiaxis positioning stage for micro- and nanoscale manufacturing processes. *Journal of Manufacturing Science and Engineering*, 2008, 130(3): 1–8.
[53] Yazhou S, Yingchun L, Shen D Development of a miniaturized machine. *Journal of Harbin Institute of Technology*, 2005, 37(5): 591–593.
[54] Zhao J, Tang J Development of a miniaturized numerical controlled grinding machine tool. *Development & Innovation of Machinery & Electrical Products*, 2014, 27(3): 168–170.
[55] Li W, Chen Q, Ren Y, et al. Hybrid micro-grinding process for manufacturing meso/micro-structures on monocrystalline silicon. *Materials and Manufacturing Processes*, 2020, 36(1): 17–26.

# 3 Numerical Investigation and Simulation of Micro-EDM Process

*Ranajit Mahanti and Manas Das*

## 3.1 INTRODUCTION

*Micro-electrical discharge machining* (μ-EDM) process is one of the most well-known non-traditional machining techniques which finds use in a wide range of industrial applications including the fabrication of dies and molds, biomedical components, injectors with high-aspect-ratio micro-holes, etc. [1]. By using *micro-electro discharge drilling* (μ-EDD), Kurafuji and Masuzawa created first-time micro-holes on a carbide material plate of 50 μm thickness [2]. The process capabilities of μ-EDM, along with-it variants and applications, have since been expanded by a variety of studies conducted by a large number of researchers. The mechanism of material erosion is the same for all μ-EDM variants and the erosion rate depends upon the process parameters (electrical and non-electrical) of the variants. The electrothermal models are the most common and widely used to recognize material erosion mechanisms in conventional EDM and μ-EDM [3–9]. The material removal process in μ-EDM is intricate and probabilistic. This process involves the fields of electrodynamics, hydrodynamics, thermodynamics, and electromagnetism [10]. It is nearly impossible to offer a simple, organized explanation that can adequately explain the mechanism's fundamentals. Therefore, various μ-EDM models (analytical and numerical simulation) based on some assumptions have been developed to explain the discharge phenomena in the *inter-electrode gap* (IEG) and material erosion from the electrodes.

The development of a reliable analytical model to predict material erosion during machining operation is crucial for the μ-EDM process. This kind of model is considered a reference to select the important process parameters and also predict the responses including *material removal rate* (MRR) from the work-piece and the tool electrode (also known as "tool wear rate"), and surface topography. The analytical models of the μ-EDM process are classified as plasma channel models that explain the formation of the plasma in the IEG during discharge using fluid dynamics and kinetics approach, and material erosion models [11]. However, the accuracy of analytical models depends upon the assumptions based on the model developed. The oversimplified models are less


DOI: 10.1201/9781003364948-3

reliable and models with fewer assumptions are more computationally challenging and rich [11]. Therefore, various numerical simulation-based models have been established to explain the electrodynamics in the IEG (formation of plasma), hydrodynamics in the melt pool, and temperature distribution (crater profile) in μ-EDM process, which is part of the investigation throughout the chapter.

This chapter presents an electrostatic and electrothermal-based simulation to explain the variation of the electric field in the IEG with the μ-EDM parameters and temperature distribution or crater profile of the workpiece material, respectively. This chapter comprises two different sections. In the first section, the fundamentals of the μ-EDM process are stated to understand the material removal mechanism, process variables, variants with process capabilities, and applications which helps to gain ideas for developing μ-EDM models. The second section presents a comprehensive numerical analysis (electrostatic and electrothermal-based simulation) of the μ-EDM process. The numerical background along with model formulation and their results is discussed in this section.

## 3.2 FUNDAMENTALS OF MICRO-EDM PROCESS

This section includes the material erosion mechanism, process variables, and μ-EDM variants with applications for getting ideas about the fundamentals of the process to develop numerous numerical models.

### 3.2.1 Material Erosion Mechanism

The material removal/erosion mechanism of μ-EDM is the same as conventional EDM. The materials are eroded from the electrode through a series of sparks/discharges. With spark generation phenomena, two electrically conducting electrodes are brought under a potential difference to generate sparks. Then, the thermal energy of the sparks is used to remove the material from both electrode surfaces. When a potential difference between the two electrodes is applied, then the electron tends to move from cathode to anode, as shown in Figure 3.1. Electrons move from cathode to anode after it overcomes the work potential. However, the electrons encounter dielectric fluid in the period of travel toward the positively charged electrode (anode/workpiece). This causes dielectric molecules to split into electrons and ions. Thus, the effective resistance reduces, and conductivity increases in the dielectric, which causes a continuous flow of electrons. The vaporization of dielectric fluid results in the formation of expansion bubbles that further reduces the effective resistance [12,13]. This facilitates the plasma channel formation, and the electron flow channelizes to the minimum spark gap due to the minimum resistance at this point [14]. This flow of electrons for a short time period is known as spark/discharge.

The electron bombardment with the workpiece results in the conversion of kinetic energy into thermal energy. This induces sudden melting and evaporation of the workpiece material in the local region and results in material erosion from

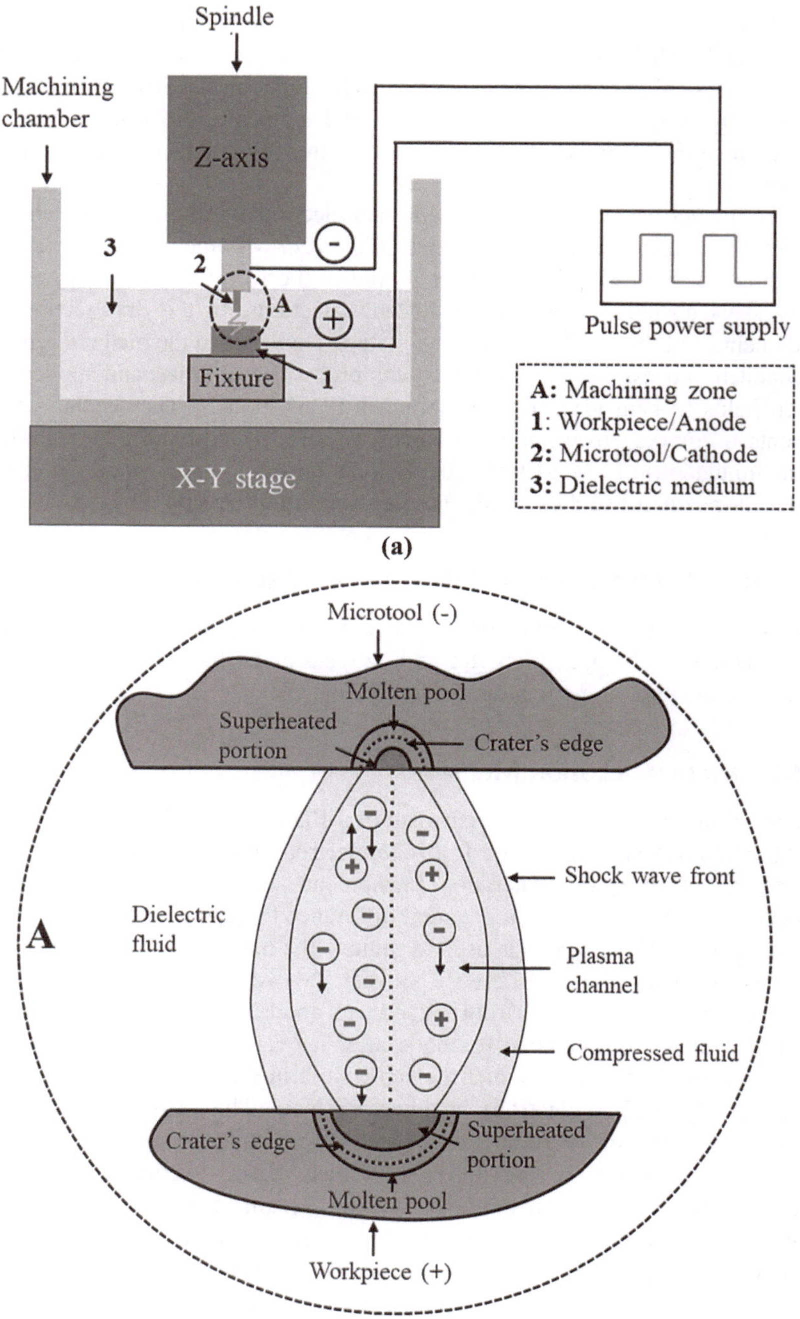


**FIGURE 3.1** Schematic representation of (a) μ-EDM process and (b) microscopic view of machining zone/crater formation.

the electrodes as a crater (shown in Figure 3.1(b)). As the material is removed from the workpiece and the tool, the "minimum gap point" changes to the next minimum gap region. Likewise, the process occurs, and the material removes forms the surfaces of the workpiece and the tool.

The ions travel toward the cathode and erode the material from the tool surface as a result of a crater which is considered "tool wear" (shown in Figure 3.1(b)). Mallik and Ghosh in 1985, stated positive polarity electrodes experience more material removal compared to negative polarity [15]. The pulsed power supply ensures uniform distribution of discharges over the surface, enabling uniform material removal from the workpiece surface than the arc discharge.

### 3.2.2 Process Variables

The variables of the μ-EDM process are the input parameters and responses. Figure 3.2 illustrates a schematic representation that details the process variables of the μ-EDM process.

The input variables are classified as electrical parameters and non-electrical parameters. Electrical parameters include source voltage, peak current, and capacitance for RC-based circuits; pulse on-time/pulse duration, pulse off-time/pulse interval, and gap voltage for transistor-based circuits. The non-electrical parameters include spindle speed (rpm), tool feed rate, the flow rate of dielectric, flushing types, dielectric types, and tool and workpiece material properties. The performances of the μ-EDM process or responses include MRR, surface integrity (surface quality/topology, microhardness, residual stress, etc.), *tool wear rate* (TWR), *recast layer thickness* (RCT), and geometrical distortions (taper, overcut, circularity error, etc.) due to tool wear. As each machining parameter has a unique

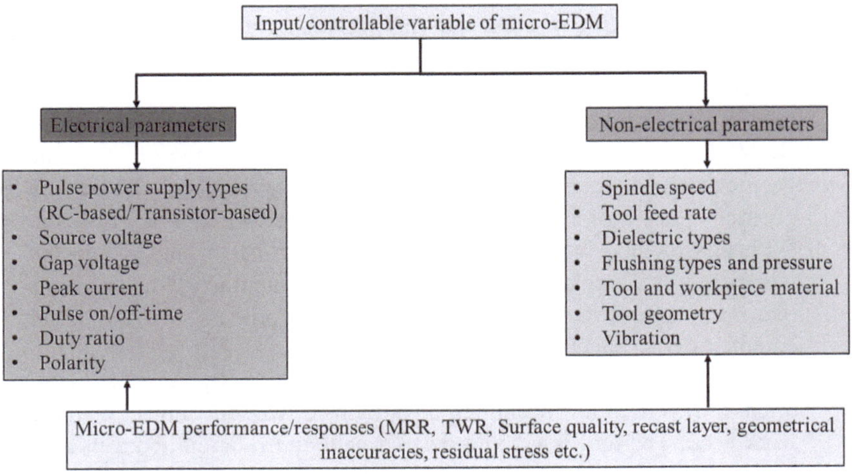


**FIGURE 3.2** Schematic representation of input parameters and performance of the μ-EDM process.

**TABLE 3.1**
**Difference between Macro and Micro Version of EDM [16,17]**

| Parameters | Macro-EDM | Micro-EDM |
|---|---|---|
| Tool electrode | > 999 μm | < 999 μm |
| IEG | 10–500 μm | < 10 μm |
| Peak current | > 3 A | < 3 A |
| OCV | 4–400 V | 10–120 V |
| Pulse duration | 0.5–8 ms | 0.05–100 μs |
| Energy per mass | High | Low |
| MRR | 0.3–30 $mm^3$/min | 0.000125–0.02 $mm^3$/min |
| RCT | 1–8 μm | Less than 1 μm |
| Surface quality | 0.8–3.1 μm | 0.015–0.8 μm |

relationship with the performance parameters and meticulous selection of optimum parameters resulting in higher productivity with minimal defects. Furthermore, the process variables of the μ-EDM and their ranges are different from the conventional EDM, as illustrated in Table 3.1.

### 3.2.3 Micro-EDM Variants with Applications

Micro-EDM is classified into various variants or configurations depending upon the microtool geometry, relative movement between tool and workpiece, and the intended use or applications. These variants are:

a. ***Micro-EDM drilling (μ-EDD):*** In this variant, a rotating microelectrode with simple geometry is plunged into the workpiece to produce shallow or deep micro-holes. Components like micro-holes for inkjet nozzles, reverse EDM tool electrodes, etc., can be manufactured by this process.
b. ***Die-sinking micro-EDM:***In this version, a mirror image of the same geometry tool is required for the machining of workpiece with desired geometry. The tool is plunged into the workpiece without any rotation. With this method, complex-shaped profiles can be machined into the workpiece in a single step. It can be used to manufacture components like micro gears, molds for the micromolding process, and any other complex (3D) shaped microfeatures.
c. ***Micro-EDM milling:*** In μ-EDM milling, simple tool geometry (cylindrical rod) is used for machining layer by layer without contact with the workpiece. The tool is given a desired path and rotation about its axis by a program to trace and scan the geometry over the workpiece. Microfluidic channels, micropillars, and other microfeatures can be manufactured by this machining process.

d. ***Microwire EDM:*** Microwire as a tool is used to slice out the micro-features. A continuous movement is given to the tool by a *computer numeric control* (CNC) program for the machining process. Forming tools, stamping tools, spinning nozzles, etc., can be manufactured from this machining process.
e. ***Micro-electro discharge grinding (EDG):*** In μ-EDG, the polarity of the tool is reversed to positive during machining. This process is mainly used for the on-machine fabrication of microtool electrodes. Different μ-EDG configurations are available for various operations, such as stationary block EDG, revolving EDG, and wire-EDG.

### 3.2.4 Summary

This section discusses the general principles or material erosion mechanism of μ-EDM, process variables, and all the types of μ-EDMs variants along with their applications. A general understanding/principle of micro-EDM is required to develop the various numerical models and for experimental investigation. The process variables of μ-EDMs along with a comparison of macro and micro-EDM present to understand the μ-EDM parameters and their ranges which helps to select the process parameters during development of models. Different electric and non-electric parameters are also highlighted which are enforced to control the μ-EDM performance.

## 3.3 NUMERICAL ANALYSIS OF MICRO-EDM PROCESS

This section includes the background of numerical models to find research potential, various model formulation/modeling approaches, and results of developed models.

### 3.3.1 Background of Numerical Models

In order to discover an adequate and precise solution to the physics-dependent governing equations of the μ-EDM process, most analytical investigations now employ numerical techniques like the finite element (FE), boundary element (BE), finite volume (FV), and finite difference (FD) approaches. Mostly FE based analyses have been performed using different commercial software to present the single-discharge and multi-discharge phenomenon of the μ-EDM process. Extensive research has been done on μ-EDM to study the underlying principles of material removal. Various theoretical and computational models have been created and are being compared to the experimental results. Some of the prior research based on this is included in this section.

Allen and Chen [18] utilized thermal and structural models to analyze material erosion or crater generation through the concept of temperature distribution over the workpiece surface and the development of residual stress, respectively. Gaussian heat flux, a symmetrical heat source, has been used to give heat input over the workpiece surface [19]. Furthermore, the single-discharge phenomenon is simulated using MATLAB® software to illustrate the material erosion from the anode (workpiece) and cathode (tool electrode) and the effect of the process parameters on material removal. It has been concluded that the crater radius is increased with the pulse duration and tensile residual stress is generated near the crater's edges. This tensile stress is increased with the increase of time. Murali and Yeo [20] used a similar concept to develop a thermo-mechanical model of titanium alloy. Transient thermal analysis has been carried out to determine the crater profile, temperature gradient, and residual stress using a Gaussian plasma heat source, time-dependent plasma radius, and temperature-dependent thermophysical and mechanical properties of titanium alloy. The experimental crater diameter is equal to 3.99 times the crater depth which has been found using an atomic force microscope (AFM) and this is in close agreement with simulation results. The induced tensile residual stress is well less than the ultimate strength of the material. Wang et al. [21] investigated the single spark μ-EDM process using a developed numerical model where a Gaussian heat source, thermal boundary condition, and distinct energy fraction transferred to electrodes (0.5 to anode and 0.2 to cathode) for deposition. A micro-cylinder of brass material is then deposited on the steel surface once the transient thermal study has determined the best machining conditions. Mujumdar et al. [22] developed the plasma discharge model in deionized water considering the global model approach along with the dynamics of the plasma bubbles and their growths. The plasma is considered spatially uniform and it conserved the mass and energy during growths. The parameters including plasma composition, electron and ions temperature, bubble's radius, and pressure are evaluated upon the application of voltage and current from a power supply. The applied spark gap and electric field are varied from 0.5 to 20 μm and 10 to 2000 MV/m, respectively, to model the plasma discharge. The transient electron density and temperature (taken as average) are in the range of $3.88 \times 10^{24}$–$30.33 \times 10^{24}$ /m$^3$ and 11013–29864 K, respectively. Finally, the condition of experiment is evaluated based on the developed simulation model, and outcomes from this are considered the heat flux available to the electrodes for material erosion. Again, Mujumdar et al. [23] established the material removal model using the heat flux, found from the previously developed model to observe the molten pool and crater generation in the μ-EDM process. This model solved the heat transfer, dielectric fluid flow, and mass conservation equation in COMSOL® Multiphysics software to observe the effect of hydrodynamic force on the melt pool formation. This electrothermal model estimates the crater dimension of 78–96 μm diameter and 8–9 μm depth for a pulse on-time of 2 μs. Shao and Rajurkar [24] explored the material erosion phenomena from the anode in the μ-EDM process using the developed model through electrothermal analysis. This model considered the heat input as Gaussian heat source, temperature-dependent thermophysical properties of 304 stainless steel, and time-varying plasma source. The energy partition at electrodes has been calculated to be 7.37% for the anode (workpiece) and 6.78% for the cathode (tool electrode) using the analytical expression for temperature profile at electrodes, the expanding plasma model, and empirically measured crater dimensions. The simulation results are the close agreement with the experimental after considering the recast layer (re-solidified layer). The measured crater radius is overestimated experimentally and a deviation of 11% has been found from their model. A step forward from earlier studies, the researcher [25] also simulated the μ-EDM process while considering the combined effects of heat transfer and fluid flow (molten material considered as fluid). The developed model utilized the momentum, mass, and energy conservation equation to estimate the crater profile (shape and size) for a single discharge. The flow of melt material is radially outward and the Marangoni force (surface tension gradient force) is the key contributor to melt pool development and crater profile. The model also predicted the formed recast layer thickness based on the reduction of molten cavity depth after cooling cycle. Similarly, the changes in crater geometry with the Marangoni effect have been analyzed using CFD simulation by Kliuev et al. [26]. The simulation showed that the Marangoni stress has an effect on the crater's shape but has no impact on material removal.

Most of the existing literature on the electrothermal model is based on different assumptions and researchers have attempted to develop a realistic material erosion model throughout the years. Still, the development of material removal model and electric field variation with IEG for different assumptions, workpiece material, and process parameters combination is a protentional research field. Some developed models are assumed constant thermophysical properties, constant heat flux radius, and different heat source models like point, disk, and Gaussian. From the literature survey, it is found that few works have been done on modeling and simulation of crater shapes for different discharge energies at a fixed pulse on-time. In order to address the aforementioned research gap, accurate modeling of the process using COMSOL® Multiphysics software has been attempted in the current study.

### 3.3.2 Research Methodology

For reducing computational efforts, the geometry of the workpiece is considered as a 2D axisymmetric geometry which is shown in Figure 3.3. The proposed electrothermal model has solved the energy balance equation with appropriate initiation and thermal boundary conditions according to the process' thermal behavior. Apart from the electrothermal model, an electrostatics simulation has been performed to observe the distribution of the electric field in the IEG using electric potential. The following assumptions have been made during the electrothermal simulation.

- The workpiece is made of homogeneous isotropic material and properties are taken as temperature dependent.
- A single spark is considered in the analysis.
- The workpiece has no internal stresses before μ-EDM operation.

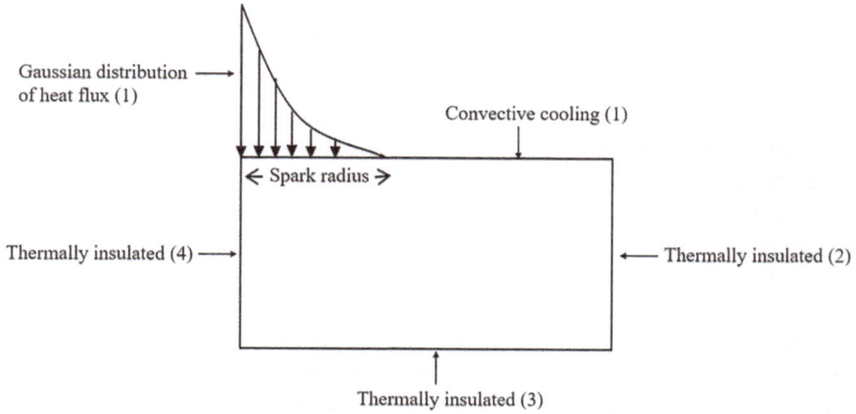


**FIGURE 3.3** Smart meshing approach utilized for the (a) electrostatics simulation and (b) electrothermal model.

- The plasma heat source is considered as Gaussian heat flux which is a function of time-varying plasma radius.
- Conduction within the workpiece and convection between the workpiece and dielectric fluid are the two methods of heat transfer.
- During a whole μ-EDM cycle, the capacitor is fully charged and subsequently discharged.

#### 3.3.2.1 Governing Equation

The Fourier heat conduction equation (Eq. (3.1)) in the cylindrical coordinate system is solved for simulating the μ-EDM process for temperature distribution.

$$\frac{1}{r}\frac{\partial}{\partial r}\left(\frac{\partial T}{\partial r}\right)+\frac{\partial}{\partial z}\left(\frac{\partial T}{\partial z}\right)+Q_{in}=\frac{1}{\alpha}\frac{\partial T}{\partial t} \tag{3.1}$$

where $r$ is the radial distance ($m$), $k$ is thermal conductivity ($W/m\text{-}K$), $t$ is time ($s$), $T$ is the temperature ($K$), $\alpha$ is thermal diffusivity ($m^2/s$), and $Q_{in}$ is the net heat flux ($W/m^2$) associated with this process which represents the difference between heat influx and heat losses.

The electrostatics simulation solved Gauss's law (Eq. (3.2)) and Faraday's law of electrostatics (Eq. (3.3)) in the stationary domain using the electric potential and dielectric shielding conditions.

$$\nabla . D=\rho_s \tag{3.2}$$

$$E=-\nabla V \tag{3.3}$$

where $\rho_s$ is the space charge density, $D$ is the electric displacement field, $V$ is the electric potential, and $E$ is the electric field.

#### 3.3.2.2 Calculation of Energy

Micro-EDM uses the RC-based relaxation circuit where the discharge energy, $E_d$ (J) is given by Eq. (3.4).

$$E_d = \frac{1}{2} C_d V_d^2 \tag{3.4}$$

where $C_d$ is the capacitance value ($F$), $V_d$ is the discharge energy ($J$), and $V_d$ is the discharge voltage ($V$). The total heat flow rate ($q$) is calculated from Eq. (3.5) and $t_{on}$ is the pulse on-time in µs. A fraction of this heat input goes to the dielectric fluid, cathode, and anode. The fraction of heat transferred to the anode is shown by Eq. (3.6) and the fraction value $F$ has been taken as 0.08, as suggested by Patel et al. [5].

$$q = E_d / t_{on} \tag{3.5}$$

$$Q_a = Fq \tag{3.6}$$

The uniform heat flux which is transferred to the anode (workpiece) is given by Q ($W/m^2$) as shown in Eq. (3.7) where $r_p$ is the plasma radius ($m$).

$$Q = \frac{Q_a}{\pi r_p^2} \tag{3.7}$$

#### 3.3.2.3 Heat Flux

The majority of researchers have assumed that heat flux is distributed uniformly. This presumption is not realistic. The real crater that's been created during the experiment clearly demonstrates this aspect. In the current work, it is assumed that the heat flux is Gaussian distribution, as given by Eq. (3.8).

$$Q_r = Q_{max} \exp\left(-3\left(\frac{r}{r_p}\right)^2\right) \tag{3.8}$$

where $Q_{max}$ is the maximum heat flux density ($= 3.1572Q = 3.1572(\eta q/\pi r_s^2)$) at $r = 0$ [20], $Q$ is the uniform heat flux stated in Eq. (3.7), $r_s$ is plasma radius ($m$), and $Q_r$ is rate of heat absorption ($W/m^2$) with $r$ following Gaussian distribution. The time-varying plasma/spark radius is expressed in Eq. (3.9) [20]:

$$r_p = \kappa t_{on}^{0.75} \tag{3.9}$$

where $\kappa$ is the proportionality constant and the value of $\kappa$ is taken as 0.788 [24,25]. However, Murli and Yeo [20] considered the constant value of spark radius (2.1 μm).

### 3.3.2.4 Boundary Condition

For the electrothermal model, the workpiece temperature before the operation is comparable to the ambient temperature ($T_a$). of the dielectric fluid where the workpiece is immersed. Consequently, the initiation for this analysis is as follows:

$$T(r, z, t = 0) = T_a \tag{3.10}$$

Spark discharge may happen over the work surface during operation in places where the IEG is smallest. The heat produced during discharge is transported to the workpiece through conduction and loses some of its energy due to convection that reflects in the boundary conditions. Figure 3.3 shows the two-dimensional axi-symmetric process with all the boundaries. The boundary condition is as follows:

#### *3.3.2.4.1 Boundary 1*

This condition arises from the Gaussian heat input and convectional heat loss at the domain's upper surface. The boundary condition is as follows for the pulse on-time:

$$k\frac{\partial T}{\partial z} = \begin{cases} Q_r & r \le r_p \\ h_c(T - T_a) & r \ge r_p \end{cases} \tag{311}$$

#### *3.3.2.4.2 Boundary 2, 3, 4*

$$\frac{\partial T}{\partial z} = 0 \tag{3.12}$$

where $h_c$ is the convection heat transfer coefficient (*W/m²-K*). No heat transfer takes place from boundaries 2, 3, and 4, hence insulated boundary condition is applied at these boundaries.

For the electrostatics simulation, an electric potential of 200 V is applied as a boundary condition through the cathode and anode. Apart from this, a dielectric shielding condition is applied by defining the relative permittivity and surface thickness of the dielectric medium.

### 3.3.2.5 Meshing and Numerical Parameters

An axisymmetric 2D domain has been selected for analysis of material erosion and a 2D geometry of varying IEG with the combination of the cathode (tool), anode (workpiece), and dielectric medium has been formed to analyze electric field variation in the spark gap.

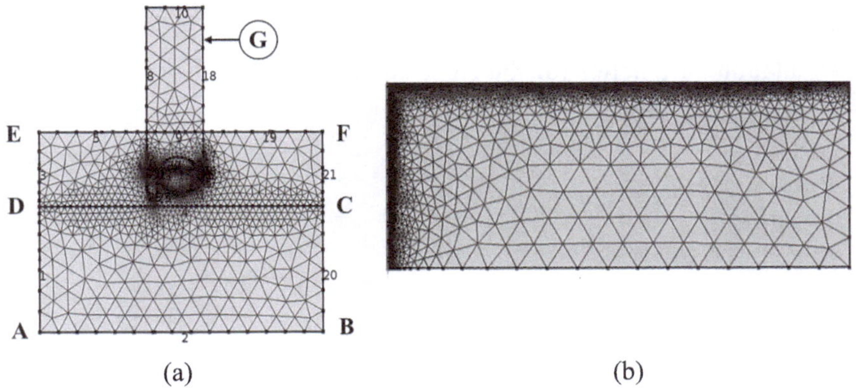


**FIGURE 3.4** Meshing for (a) electrostatics and (b) electrothermal simulation.

In electrostatics simulation, a relatively dense mesh in between the cathode and anode are selected to observe the electric field variation in the IEG more accurately. The corresponding mesh and its parameters are presented in Figure 3.4(a) and Table 3.2, respectively. The domain "ABCDA," "CDEFC," and "G" are considered as the workpiece, dielectric medium, and tool electrode, respectively, for this analysis.

In the electrothermal analysis, the normal predefined mesh is selected for the whole computational domain (size of 50 × 20 $\mu m^2$) with the fine boundary element (500 triangular elements) as the net heat flux is applied on the boundary. The simulation parameters for this model are presented in Table 3.3. The initial temperature has been taken the same as the temperature of the dielectric which is the ambient temperature (293.15 K). The piecewise temperature-dependent thermophysical properties of titanium alloy have been used for more realistic results [27], which are shown in Figure 3.5.

**TABLE 3.2**
**Mesh Parameters for Electrostatics Simulation**

| Parameters | Value |
|---|---|
| Maximum size of the element | 0.348 |
| Minimum size of the element | 0.00156 |
| Curvature's resolution | 0.3 |
| Maximum growth rate | 1.3 |
| Average and minimum mesh quality | 0.9441and 0.571 |
| Triangular, edge, and vertex elements | 14607, 956, and 30 |

**TABLE 3.3**
**Simulation Parameters for the Electrothermal Model**

| Parameters | Value |
|---|---|
| Analysis type | 2D axisymmetric thermal transient |
| Workpiece material | Ti-6Al-4V |
| Domain size | 50 × 20 μm$^2$ |
| Initial/ ambient temperature | 293.15 K |
| Capacitance | 3300 pF |
| Spark radius | 2.1 μm |
| Melting point of workpiece | 1923 K |
| Voltage | 110 V |
| Pulse duration | 100 ns |
| Convective heat transfer coefficient | 1000 W/m$^2$-K |

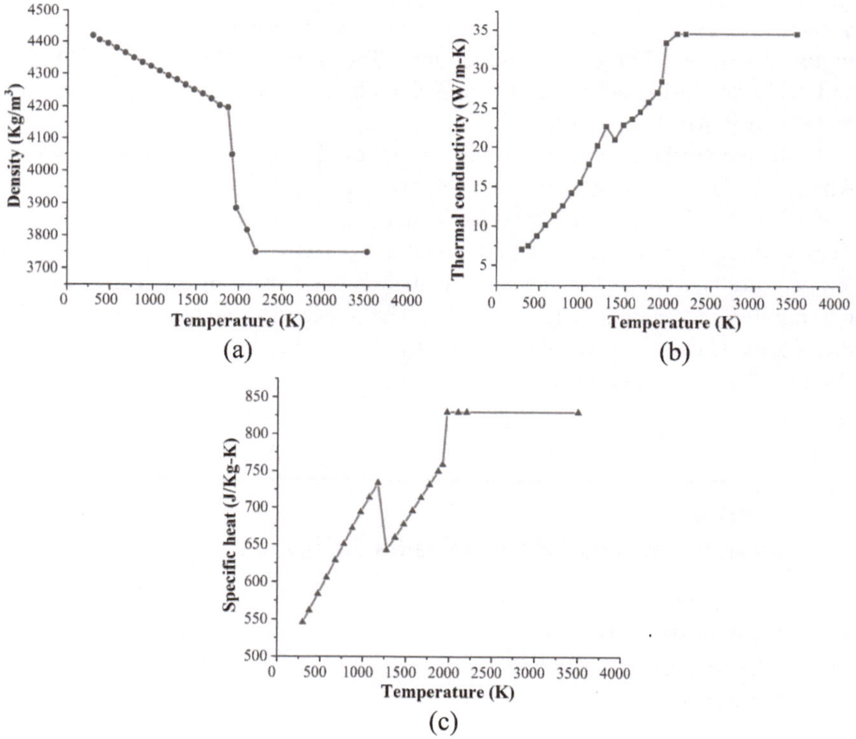


**FIGURE 3.5** Thermophysical properties of Ti-6Al-4V (a) density ($\rho$), (b) thermal conductivity ($K$), (c) specific heat ($C_p$) [27].

## 3.4 RESULTS AND DISCUSSION

The developed electrostatics and electrothermal models are solved in stationary and transient time domains using COMSOL Multiphysics® software to estimate the electric field variation, and crater profile, respectively. The results of these two models are presented in this section.

### 3.4.1 Electrostatics Simulation

To understand the process of discharge plasma formation it is necessary to understand the distribution of electric potential and field intensity between tool and workpiece because it plays a key role in determining the size of the plasma channel. In μ-EDM, the microtool and workpiece are separated by a very small gap. A voltage is applied to them and when the applied potential becomes greater than the breakdown potential of the dielectric fluid, a spark is generated between them. But the discharge takes place at the minimum gap between the tool and workpiece and this minimum gap can be anywhere between them. In the actual process, tool and workpiece surfaces are not perfectly smooth and consist of many undulations of very small size owing to the variation of minimum gap point throughout the IEG. Therefore, a simulation of the μ-EDM process is done in order to understand the distribution of electric field density between the electrodes.

In this simulation, the workpiece is attached to the negative terminal and the microtool to the positive end (reverse polarity). Tool material, workpiece, and dielectric fluid are tungsten, iron, and ethanol with temperature-independent properties, respectively. There can be any number of undulations present on the bottom surface of the tool but in the present simulation, only three undulations are considered on the tool's bottom surface which are not perfectly hemispherical. A voltage of +200 V and −200 V is applied to the tool and workpiece, respectively and simulation is done only to prove the theory that spark is generated in minimum gap distance because of maximum electric field intensity present there. Under static condition, relationship between the electric field and the applied voltage is presented in Eq. (3.3).

Circular undulations are present at the bottom of the tool and it can be clearly seen that the machining gap is minimum between the first undulation and the workpiece where the intensity is maximum (shown in Figure 3.6). The magnified view of the machining gap which shows the direction of the electric field in the gap is shown in Figure 3.7. The length of the arrow is directly proportional to the electric field intensity. On the bottom surface of the tool, different-sized undulations are considered in the simulation because, if more than one undulation is present, the electric field will be at its maximum on the section of the larger undulation, where the distance between the microtool and the workpiece will be smallest.

Figure 3.8 illustrates the electric field's variation in the Y-component along the bottom surface of the tool electrode where electric field intensity is highest for the section of largest undulation and least for the section of smallest

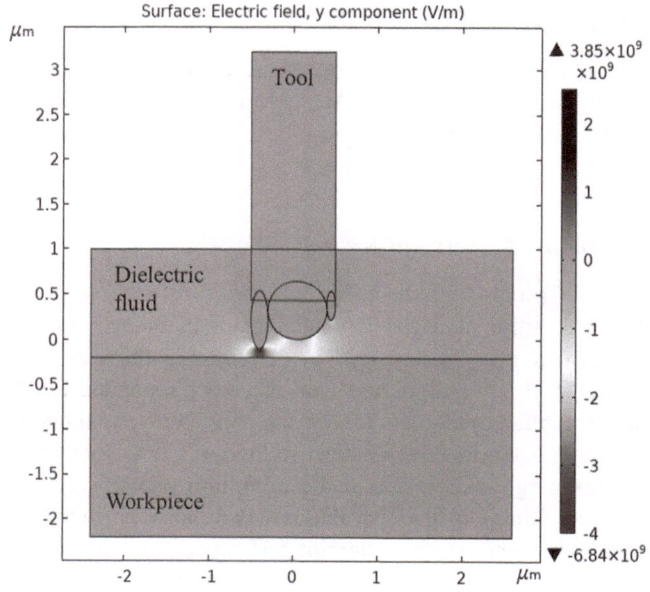


**FIGURE 3.6** Distribution of the electric field intensity in the machining gap.

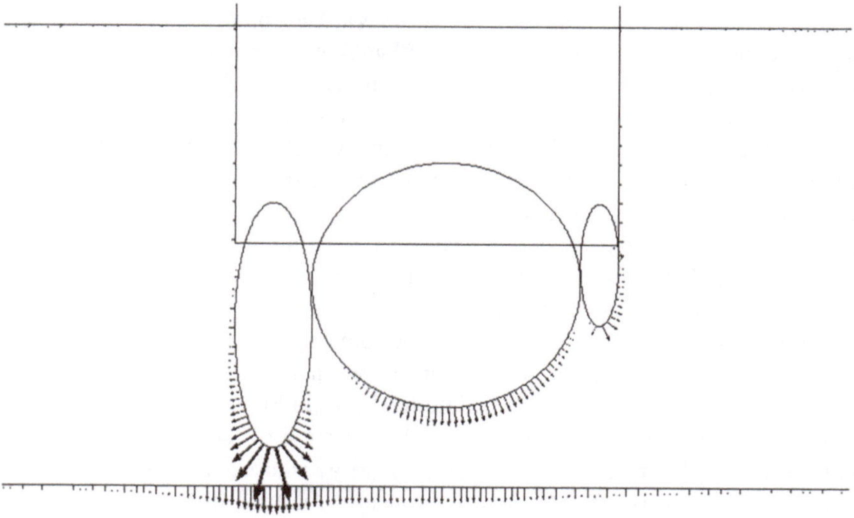

**FIGURE 3.7** Vector plot of the electric field intensity variation throughout the IEG (Scale factor: 4.5E-11).

undulation. It shows only variation of the vertical component of the electric field (i.e. Y-component) while the arrow diagram shows vector distribution of electric field intensity throughout the IEG/ machining gap. Therefore, the formation of plasma takes place at the section of larger undulation only.

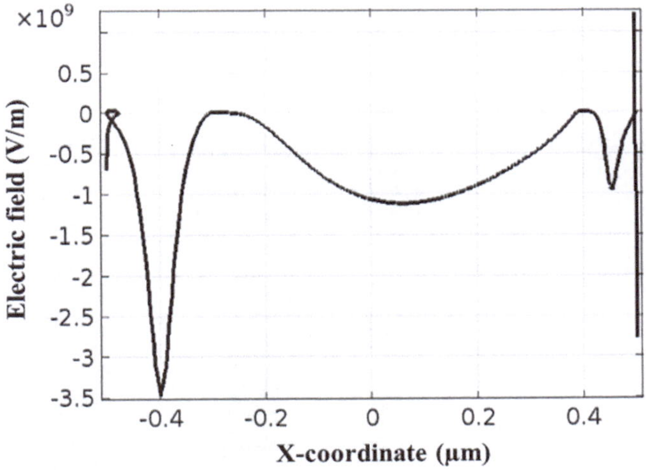


**FIGURE 3.8** Electric field distribution for different undulation along the bottom surface of the tool.

### 3.4.2 Electrothermal Simulation

The electrothermal/material erosion model of titanium alloy has been developed based on the mentioned assumptions, temperature-dependent thermophysical properties (shown in Figure 3.5), and thermal analysis only. The workpiece surface develops a temperature profile as a result of the electrothermal energy conversion, and the crater profile (diameter and depth) is assessed based on the temperature distribution. The temperature distribution achieved at the spark end with its magnified view is shown in Figure 3.9. Based on the temperature distribution and melting isotherm, the crater geometry has been found.

From Figure 3.10(a), it has been found that temperature first reduces rapidly from the center and after 4.5 μm it reaches ambient temperature. As approach from the top corner of the workpiece towards the bottom corner, the temperature first drops quickly, and after 3.3 μm it reaches ambient temperature, as illustrated in Figure 3.10(b). This temperature distribution helps to evaluate the expected melt pool dimensions. The crater profile is evaluated based on the assumption that all the material in the workpiece above the melting point has been eradicated. The crater radius and depth are found as 2.55 μm and 1.82 μm, respectively, based on this assumption as shown in Figure 3.11.

#### 3.4.2.1 Model Validation

To compare the accuracy of this model, results found from the model have been validated using published results by Murali and Yeo [20]. Figure 3.12(a) and (b) compare the temperature variation over the top surface and depth of the crater for the developed model as well as for the model proposed by Murali and Yeo. The crater radius and depth have been found as 3.15 μm and 1.83 μm, respectively,

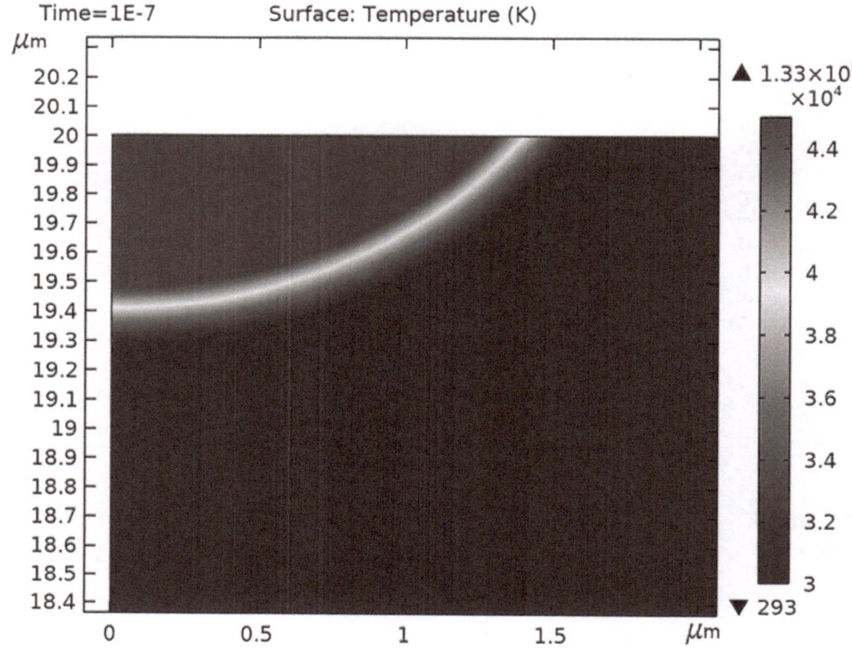


**FIGURE 3.9** Magnified view of temperature distribution obtained at the spark end.

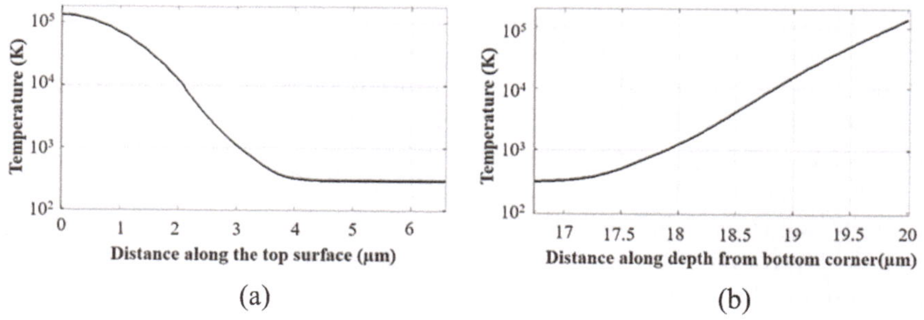


**FIGURE 3.10** Temperature variation along (a) top surface and (b) depth of the workpiece.

from the model developed by Murali and Yeo [20], which is the close agreement with the present model.

#### 3.4.2.2 Parametric Study

Since the results generated by the present model were found in good agreement with published results [20]. Therefore, the present model has been further used to investigate the material erosion (crater geometry/ volume) in the μ-EDM process for different workpiece material (copper) and discharge energy. The spark radius is taken as time-varying as given in Eq. (3.9) and the discharge energy ($E_d$) as

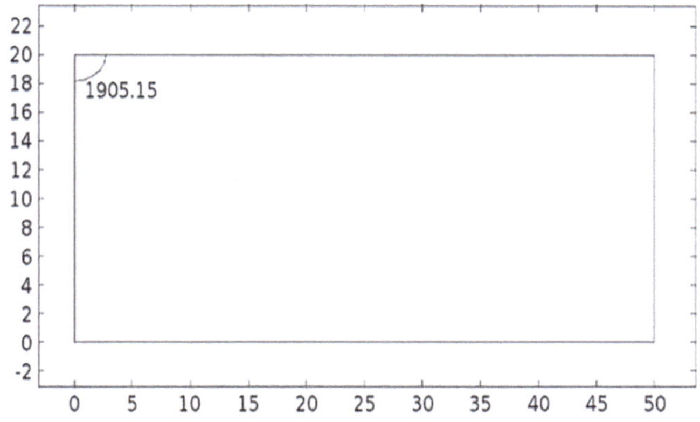


**FIGURE 3.11** Crater profile formed on the workpiece after a single discharge.

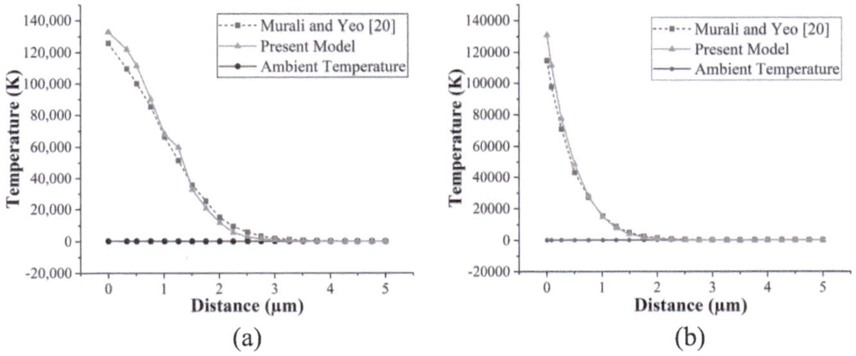


**FIGURE 3.12** Validation of temperature distribution along (a) top surface and (b) depth of the workpiece.

shown in Eq. (3.4), is varied for the parametric investigation. The temperatures along the top surface from the center and depth from the top surface center have been increased with the increase of discharge energies as shown in Figure 3.13. The heat input i.e. Gaussian heat flux on the workpiece surface is enhanced with the increased discharge energy resulting in high temperature. Crater diameter and depth are also constantly increased with the increase in energy as shown in Figure 3.14.

### *3.4.2.2.1 Crater Volume Calculation*

For the calculation of volume from the crater geometry obtained from simulation a rectangle of length dy is assumed at a depth y from the center and the width of the rectangle is taken as x. Now when the elementary rectangle is rotated by an angle of 360°, it will generate a cylinder of height dy and radius x, as shown in Figure 3.15. The elementary volume dV is given by Eq. (3.13).

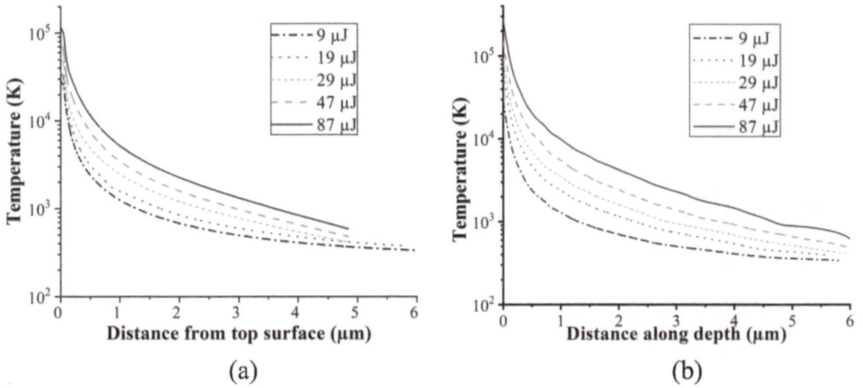


**FIGURE 3.13** Temperature variation along (a) top surface and (a) depth of the workpiece for different discharge energies.

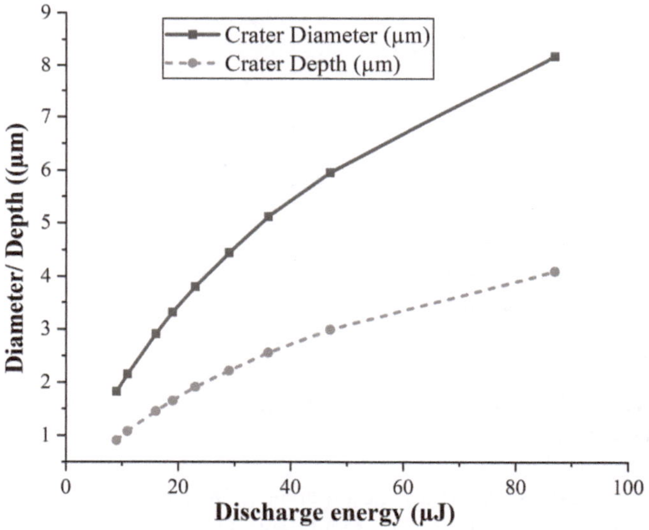


**FIGURE 3.14** Crater diameter and depth with different discharge energies.

$$dV = \pi x^2 dy \tag{3.13}$$

The crater shape is assumed to be perfectly hemispherical with a flat face. Therefore, to calculate the total crater volume, the elementary volume is integrated from 0 to the obtained depth H. The volume of the crater formed (V) is expressed as:

$$V = \int_0^H \pi x^2 dy \tag{3.14}$$

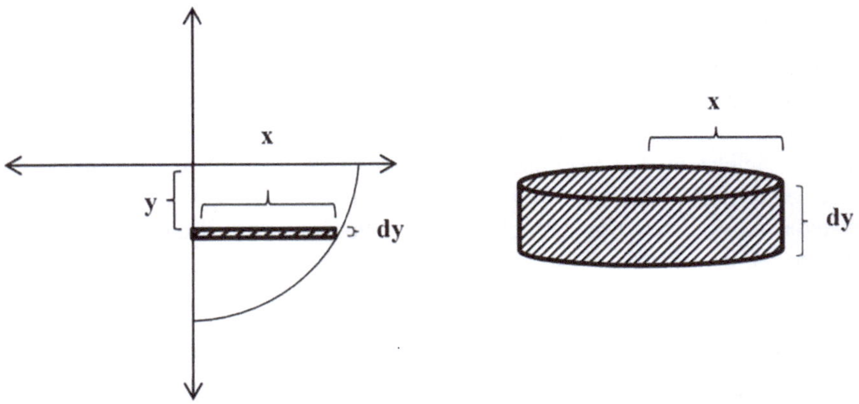


**FIGURE 3.15** Schematic presentation of elementary volume calculation of the formed crater.

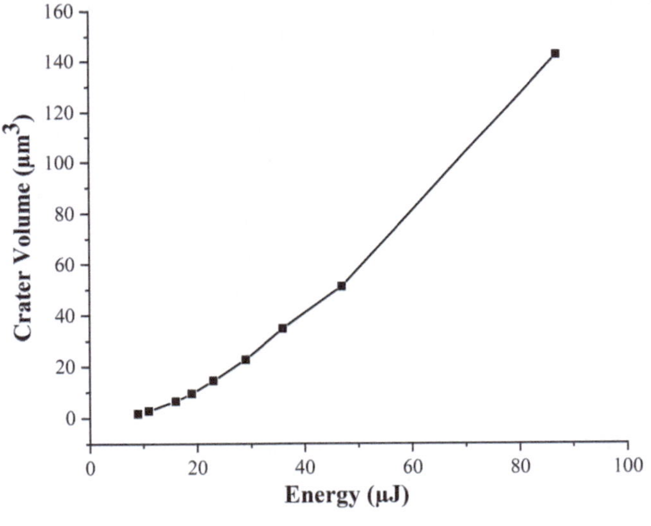


**FIGURE 3.16** Calculated crater volume based on predicted crater profile with different discharge energies.

The calculated crater volume formed with different discharge energies is presented in Figure 3.16. As crater diameter and depth are increased with the energy (as shown in Figure 3.14), resulting in enhanced material erosion (crater volume) per discharge. The experimental crater volume per discharge may be underestimated the value estimated by simulation as re-solidification of molten material is not considered in this model [24]. However, the estimated material erosion per discharge in this model is the close agreement with the literature for moderate to the high level of discharge energy [24].

## 3.5 CONCLUSIONS

This chapter presents the fundamentals of the μ-EDM process including the material erosion mechanism, process variables, various configurations of μ-EDM, and numerical investigation of the μ-EDM process such as electrostatic and electrothermal simulation. The electrostatics simulation reveals the theory of high electric field generation in the narrowest IEG. The electrothermal simulation reveals the material erosion/ removal through the generation of craters from the workpiece surface. Further, the developed electrothermal model is validated by the literature. However, from the discussion, the following key conclusions can be made.

- Micro-EDM is a thermoelectric type non-contact machining process where the electrical energy from the pulsed power supply is converted to thermal (heat) energy through a series of spark/ discharges leading to material erosion via melting and vaporization of electrodes. The μ-EDM variants including drilling, die-sinking, milling, and EDG are used to produce micro-holes, micro molds, micro-slots or channels, and other 3-D microfeatures. Numerical studies or models are required to understand the role of the electric field in sparks generation, the material erosion mechanism from the electrodes, and the suitable selection of machining parameters for cost-effective production.
- Simulation of electric field intensity proved the theory that a spark will be generated at a minimum gap between the tool and workpiece because of the maximum electric field intensity present there. The electric field goes outwards from the anode (tool) and inwards into the cathode (workpiece), as reversed polarity is used in the electrostatics simulation.
- The electrothermal model analyses the conversion of electric energy from an RC-based circuit to thermal energy and leads to material removal through a formation of a crater on the titanium alloy surface. Temperature distributions along the top surface and depth fall rapidly after the spark radius boundary. The crater radius and depth are 2.55 μm and 1.82 μm, respectively, which is the close agreement with the published literature [20]. Crater profile (diameter and depth) or crater volume increases with the increase in discharge energy at a given pulse duration. This study provides an easy understanding of complex theoretical concepts as well as predicts the trial parameters for the experiment in a quick and cost-effective way.

## REFERENCES


[1] I. Fassi and F. Modica, *Micro-electro discharge machining: Principles, recent advancements and applications*, vol. 12, no. 5. MDPI AG, 2021.

[2] H. Kurafuji and T. Masuzawa, "Micro-EDM of camented carbide alloys," *J. Japan Soc. Electr.-Mach. Eng.*, vol. 2, no. 3, pp. 1–16, 1968.

[3] F. V. Dijck and R. Snoyes, "Plasma channel diameter growth affects stock removal in EDM," *CIRP Ann.*, vol. 21, no. 1, pp. 39–40, 1972.

[4] S. H. Yeo, W. Kurnia, and P. C. Tan, "Critical assessment and numerical comparison of electro-thermal models in EDM," *J. Mater. Process. Technol.*, vol. 203, no. 1–3, pp. 241–251, Jul. 2008, doi: 10.1016/j.jmatprotec.2007.10.026.
[5] M. R. Patel, M. A. Barrufet, P. T. Eubank, and D. D. DiBitonto, "Theoretical models of the electrical discharge machining process. II. The anode erosion model," *J. Appl. Phys.*, vol. 66, no. 9, pp. 4104–4111, Nov. 1989, doi: 10.1063/1.343995.
[6] S. N. Joshi and S. S. Pande, "Thermo-physical modeling of die-sinking EDM process," *J. Manuf. Process.*, vol. 12, no. 1, pp. 45–56, Jan. 2010, doi: 10.1016/j.jmapro.2010.02.001.
[7] S. H. Yeo, P. C. Tan, and W. Kurnia, "Effects of powder additives suspended in dielectric on crater characteristics for micro electrical discharge machining," *J. Micromechanics Microengineering*, vol. 17, no. 11, 2007, doi: 10.1088/0960-1317/17/11/N01.
[8] S. H. Yeo, W. Kurnia, and P. C. Tan, "Electro-thermal modelling of anode and cathode in micro-EDM," *J. Phys. D. Appl. Phys.*, vol. 40, no. 8, pp. 2513–2521, Apr. 2007, doi: 10.1088/0022-3727/40/8/015.
[9] R. Dijck and F. V. Snoeys, "Investigations of EDM operations by means of thermomathematical models," *CIRP Ann.*, vol. 20, no. 1, pp. 35–36, 1971.
[10] T. Ni, Q. Liu, Y. Wang, Z. Chen, and D. Jiang, "Research on material removal mechanism of micro-edm in deionized water," *Coatings*, vol. 11, no. 3, pp. 1–9, 2021, doi: 10.3390/coatings11030322.
[11] R. Mahanti and M. Das, "Micro-EDM: Modeling and optimization," in *Advanced Machining Science*, V. K. Jain, Ed. Boca Raton: CRC Press, 2022, pp. 79–116.
[12] A. Banu and M. Y. Ali, "Electrical discharge machining (EDM): A review," *Int. J. Eng. Mater. Manuf.*, vol. 1, no. 1, pp. 3–10, 2016, doi: 10.26776/ijemm.01.01.2016.02.
[13] E. C. Jameson, *Electrical Discharge Machining*. Dearborn: Society of Manufacturing Engineers, 2001.
[14] M. Kunieda, B. Lauwers, K. P. Rajurkar, and B. M. Schumacher, "Advancing EDM through fundamental insight into the process," *CIRP Ann. - Manuf. Technol.*, vol. 54, no. 2, pp. 64–87, Jan. 2005, doi: 10.1016/s0007-8506(07)60020-1.
[15] A. Mallik and A. K. Ghosh, *Manufacturing Science*. Harlow: Ellis Horwood Ltd, Publisher, 1986.
[16] B. Bhattacharyya and B. Doloi, "Micromachining processes," in *Modern Machining Technology*, 2020, pp. 593–673.
[17] L. Raju and S. S. Hiremath, "A state-of-the-art review on micro electro-discharge machining," *Procedia Technol.*, vol. 25, pp. 1281–1288, 2016, doi: 10.1016/j.protcy.2016.08.222.
[18] P. Allen and X. Chen, "Process simulation of micro electro-discharge machining on molybdenum," *J. Mater. Process. Technol.*, vol. 186, no. 1–3, pp. 346–355, May 2007, doi: 10.1016/j.jmatprotec.2007.01.009.
[19] V. Yadav, V. K. Jain, and P. M. Dixit, "Thermal stresses due to electrical discharge machining," *Int. J. Mach. Tools Manuf.*, vol. 42, no. 8, pp. 877–888, Jan. 2002, doi: 10.1016/S0890-6955(02)00029-9.
[20] M. S. Murali and S. H. Yeo, "Process simulation and residual stress estimation of micro-electrodischarge machining using finite element method," *Japanese J. Appl. Physics, Part 1 Regul. Pap. Short Notes Rev. Pap.*, vol. 44, no. 7 A, pp. 5254–5263, Jul. 2005, doi: 10.1143/JJAP.44.5254.
[21] Y. K. Wang, B. C. Xie, Z. L. Wang, and Z. L. Peng, "Micro EDM deposition in air by single discharge thermo simulation," *Trans. Nonferrous Met. Soc. China (English Ed.*, vol. 21, no. SUPPL. 2, Aug. 2011, doi: 10.1016/S1003-6326(11)61623-3.

[22] S. S. Mujumdar, D. Curreli, S. G. Kapoor, and D. Ruzic, "A model of micro electro-discharge machining plasma discharge in deionized water," *J. Manuf. Sci. Eng. Trans. ASME*, vol. 136, no. 3, pp. 1–12, 2014, doi: 10.1115/1.4026298.
[23] S. S. Mujumdar, D. Curreli, S. G. Kapoor, and D. Ruzic, "Modeling of melt-pool formation and material removal in micro-electrodischarge machining," *J. Manuf. Sci. Eng. Trans. ASME*, vol. 137, no. 3, pp. 1–9, 2015, doi: 10.1115/1.4029446.
[24] B. Shao and K. P. Rajurkar, "Modelling of the crater formation in micro-EDM," *Procedia CIRP*, vol. 33, pp. 376–381, 2015, doi: 10.1016/j.procir.2015.06.085.
[25] B. Shao and K. P. Rajurkar, "Modelling of the crater formation in micro-EDM," 2015, doi: 10.1016/j.procir.2015.06.085.
[26] M. Kliuev, K. Florio, M. Akbari, and K. Wegener, "Influence of energy fraction in EDM drilling of Inconel 718 by statistical analysis and finite element crater-modelling," *J. Manuf. Process.*, vol. 40, pp. 84–93, Apr. 2019, doi: 10.1016/j.jmapro.2019.03.002.
[27] M. Jamshidinia, F. Kong, and R. Kovacevic, "Numerical modeling of heat distribution in the electron beam melting® of Ti-6Al-4V," *J. Manuf. Sci. Eng.*, vol. 135, no. 6, pp. 1–14, 2013, doi: 10.1115/1.4025746.

## NOMENCLATURE

| | |
|---|---|
| *AFM* | Atomic Force Microscope |
| $C_d$ | Capacitance value of RC pulse generator (μF) |
| $C_p$ | Specific heat capacity (J/kg K) |
| *CFD* | Computational Fluid Dynamics |
| $D$ | Electric displacement field |
| $E$ | Electric field (V/m) |
| $E_d$ | Discharge energy |
| *EDD* | Electro Discharge Drilling |
| *EDG* | Electro Discharge Grinding |
| $F$ | Fraction of energy at anode |
| $h_f$ | Convective heat transfer coefficient ($W/m^2$-K) |
| *IEG* | Inter-electrode Gap (μm) |
| $K$ | Thermal conductivity (W/m-K) |
| *MRR* | Material Removal Rate |
| *OCV* | Open circuit voltage (V) |
| $q$ | Total heat flow rate (W) |
| $Q$ | Uniform heat flux, transferred to the anode ($W/m^2$) |
| $Q_a$ | Fraction of heat transferred to the anode (W) |
| $r_p$ | Plasma/ spark radius (μm) |
| $T_a$ | Ambient temperature (K) |
| *TWR* | Tool Wear Rate |
| $V$ | Electric potential (V) |
| $V_d$ | Discharge voltage (V) |
| *alpha* | Thermal diffusivity ($m^2/s$) |
| k | Proportionality constant for considering spark radius |
| *rho* | Density of the workpiece ($kg/m^3$) |
| $\mathrm{rho}_s$ | Space charge density ($C/m^3$) |

# 4 Micro-Electrochemical Machining

*Bikash Ghoshal*

## 4.1 INTRODUCTION

Micromachining means machining of small dimensions ranging from submicron to micron. Such a smaller amount of machining is not easily achievable by a conventional technique where machining is done by the contact of tool and workpiece. Achieving good surface finish is the major problem in conventional as well as some of the non-conventional machining methods such as laser beam machining (LBM), electron beam machining (EBM), and electro discharge machining (EDM) where thermal distortion of the machined surface is common. Abrasive jet machining (AJM) and Ultrasonic machining (USM) etc. are also not suitable for achieving high dimensional accuracy and good surface finish. Chemical machining (CM) and electrochemical machining (ECM) processes are thermal free and material removal takes place due to atomic level dissociation resulting in stress free and excellent surface finish. The basic principle of ECM is based on Michael Faraday's principle of electrolysis. ECM is an anodic dissolution process where workpiece and tool are respectively anode and cathode which are separated by a small electrolytic gap, called machining gap or inter-electrode gap (IEG). When the electric current is passed through the electrolyte the anode workpiece dissolves locally so that the generated shape on the workpiece is approximately negative mirror image of the tool. When the ECM process is used in microscopic domain for manufacturing of ultra-precision shapes, it is known as Electrochemical Micromachining (EMM).

### 4.1.1 ECM Process Overview

Gusseff patented ECM in 1929. This was subsequently followed by its establishment as major technology in 1960 for finishing operations in aircraft and aerospace industries. Electroplating and electropolishing were rampantly used in manufacturing industries at that time. Electrochemical machining is dissolution of anode metal based on Faraday's laws of electrolysis. Reactions take place at the anodic workpiece, electrolyte, and cathode tool. Electrochemical reactions for ECM of low carbon steel or iron in the electrolyte of sodium chloride (NaCl) aqueous solution are shown in Figure 4.1. When voltage is applied, ionization of iron, sodium chloride, and water takes place. Iron atoms from anode dissolve in electrolyte as iron ion leaving behind electrons as shown below:

DOI: 10.1201/9781003364948-4

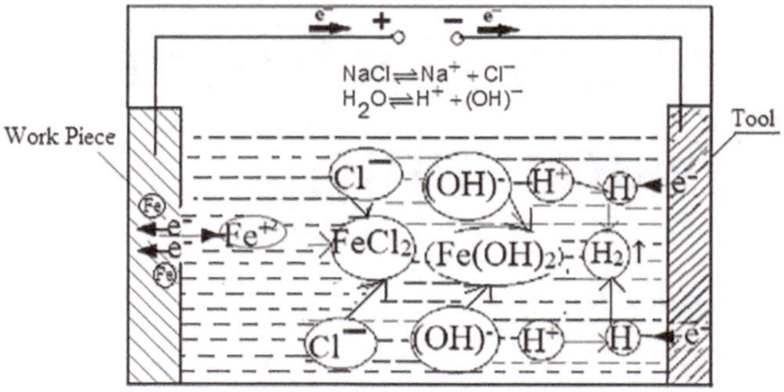


**FIGURE 4.1** Schematic illustration of electrochemical reactions in ECM.

$$Fe \rightarrow Fe^{++} + 2e$$

NaCl is ionized to $Na^+$ and $Cl^-$ and water is ionized to $H^+$ and $(OH)^-$. Electrons released by anodic iron move to the cathode to neutralize hydrogen ions ($H^+$) to produce hydrogen gas at cathode. Iron ions react with $Cl^-$ and $(OH)^-$ to form $FeCl_2$ and $Fe(OH)_2$ respectively. Finally, these are transformed to sludge. To maintain the dissolution process, sludges are removed by circulating electrolyte at high pressure between the electrodes. With the passing of machining time, inter-electrode gap (IEG) increases and thereby electrical resistance to current flow increases. To maintain steady dissolution process, IEG is kept constant either by moving forward the cathode towards the anode or vice versa. Anodic workpiece gradually takes a shape of negative mirror image of the cathode tool as explained in Figure 4.2. The figure explains that dissolution rate is more where the gap is less due to higher current density at lesser inter-electrode gap (IEG).

Important process parameters of ECM are voltage, current, current density distribution, nature of electrolyte, flow rate, concentration and conductivity of electrolyte, atomic weight of the workpiece, inter-electrode gap (IEG), shape of tool and machining time, etc. These process parameters influence machining output like metal removal rate (MRR), surface roughness, accuracy, profile shape, etc. [1,2].

### 4.1.2 Classification of Advanced Electrochemical Machining

Some of the advanced ECM processes are presented in Figure 4.3 and briefly described as follows:

**Electrochemical turning:** Workpiece is rotated as in conventional turning and connected to positive terminal. Tool is mounted on feed motion system and connected to negative terminal. Both workpiece and tool dipped in electrolyte need potential difference for machining to start.

**Electrochemical drilling:** Anodic workpiece and cathode tool are immersed in electrolyte. Generally, tool moves in vertical (z axis) direction with the help of precise tool feeding system.

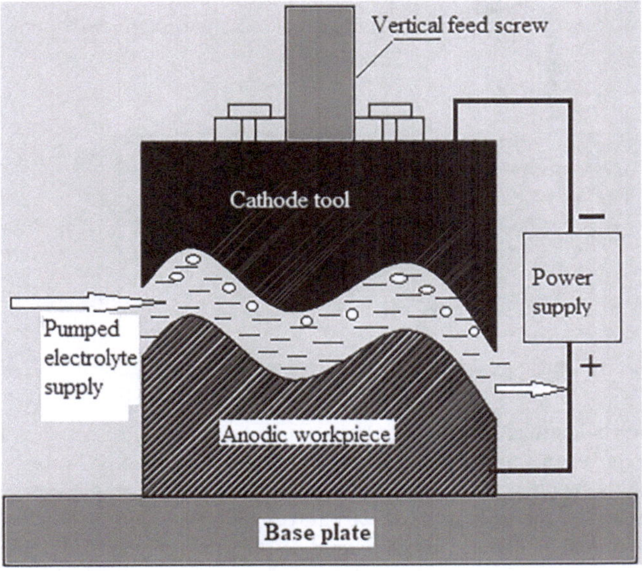


**FIGURE 4.2** Schematic of ECM system.

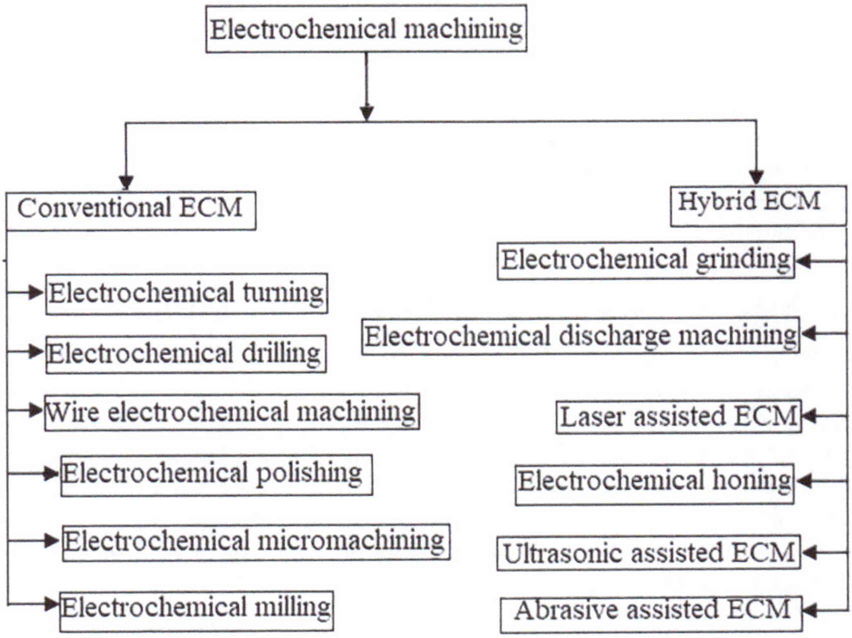


**FIGURE 4.3** Different advanced electrochemical machining.

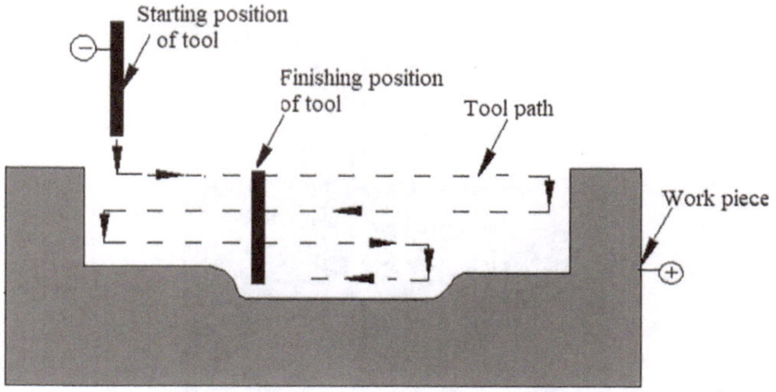


**FIGURE 4.4** 3D shape generation by electrochemical milling.

**Wire electrochemical machining:** Very small diameter wire is used as tool and generally moves in vertical direction while workpiece is fixed on horizontal table. Electrolyte is supplied at machining zone through nozzle. Workpiece is cut as per the movement of the table creating wire travel path.

**Electrochemical polishing:** Electrochemical polishing is preferably suitable for 3D curved surfaces which mechanical polishing can hardly polish, Kim and Park (2019) [3].

**Electrochemical milling:** Machining process is same as conventional end milling. Simple geometric tool is used and large area is milled by controlling the movement of tool and workpiece resulting in 3D shape generation as shown in Figure 4.4.

**Hybrid ECM:** ECM is utilized to augment the machining process by other methods so that machined surface quality is improved.

**Electrochemical Grinding:** A metallic wheel and workpiece are separated by small inter-electrode gap (IEG). Negative potential is connected to wheel and positive potential is applied to workpiece. A jet of electrolyte is supplied between rotating wheel and workpiece. X-Y stage is utilized for horizontal motion of workpiece under rotating wheel. This process is utilized for generation of smooth flat surface.

## 4.2 ELECTROCHEMICAL MICROMACHINING (EMM)

Nowadays, manufacturing of microproducts has drawn attention for serving to the needs of huge population. Electrochemical micromachining (EMM) is solving existing difficulties for the fabrication and production of micro-parts. The term micromachining means material removal in the range of 1 to 999 μm [2,4,5]. But in general, it means material removal in the micron or submicron range. EMM is an anodic dissolution of metal by electrochemical reaction where, anodic workpiece and cathode tool are separated by an appropriate electrolyte and pulsed power is applied for better localization of current. Current passes through the narrow gap generally below 15 μm between micro tool and workpiece. The material removal rate (MRR) is very small compared to conventional ECM process. MRR is mainly controlled by pulse duration and pulse frequency in addition to Ohmic resistance and other phenomena in the double layer. EMM is carried out at very low voltage and current with the precise movement of micro tool and workpiece. Accuracy and precision of machining are higher than that in conventional ECM. It requires knowledge of electrical, chemical, thermal, hydrodynamic phenomena and process control involving electronics instruments. The process parameters of EMM include frequency of pulse power supply, duty ratio, pulse width, inter-electrode gap, average voltage, current density distribution, dissolution current, amplitude of micro tool vibration, frequency of micro tool vibration, and transient phenomena of electrical double layer which exists on boundaries across workpiece—electrolyte and tool—electrolyte interface.

## 4.3 NECESSITY OF ELECTROCHEMICAL MICROMACHINING

Realizing of stress free and excellent surface finish is the biggest advantage in EMM. Chemical machining cannot be controlled properly in this micromachining domain. ECM machining process is applied to the micromachining range of applications for manufacturing of ultra-precision shapes and it is called Electrochemical Micromachining (EMM). EMM is a very promising micromachining technology due to its advantages such as no tool wear, absence of stress/burr, high MRR, ability to machine complex shapes regardless of hardness, better precision and control, rapid machining, reliable, flexible, environmentally acceptable (electrolyte is less pollutant). It also permits machining of chemically resistant materials like copper alloys, stainless steel, titanium, and superalloys which are widely used in aerospace, biomedical, electric, and MEMS applications [6–8]. EMM is used widely nowadays in manufacturing semiconductor devices and thin metallic films.

## 4.4 FUNDAMENTALS OF MICRO-ELECTROCHEMICAL MACHINING

In EMM, material removal takes place under ultra-short pulses where pulse width, duty ratio, and frequency of pulse play important roles in accuracy of micromachining. Various types of potential drops (resistances) in addition to electrolyte resistance are there and hence, current efficiency comes into play to account for double-layer capacitance, Warburg impedance, and charge transfer resistance.

The double-layer capacitance refers to two parallel layers of charges surrounding the metal dipped in electrolyte as shown in Figure 4.5. The first layer, the surface charge (either positive or negative), comprises ions adsorbed directly into the metal due to a host of chemical interactions. The second layer is composed of ions attracted to the surface charge via the Coulomb force, electrically screening the first layer. This second layer is loosely associated with the metal

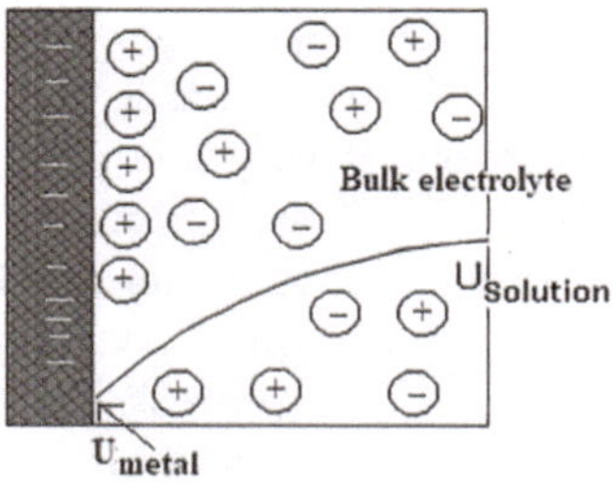


**FIGURE 4.5** Gouy-Chapman double-layer model.

surface, because it is made of free ions which move in the fluid under the influence of electric attraction and thermal motion rather than being firmly anchored. It is thus called the diffuse layer. The potential difference U is developed due to transfer of metal ions into the solution.

At the electrode-electrolyte interface, the metal goes to the electrolyte as ions, but cannot cross energy barrier till applied voltage is greater than activation potential. Charges remain polarized as double-layer capacitor. There is no electrolyte flow in EMM operation and hence, there is not sufficient transfer of mass from one electrode to the other electrode. This gives rise to diffusion component which creates impedance known as the Warburg impedance. Diffusion is the movement of a species under the influence of a gradient of chemical potential (i.e. a concentration gradient). The concentration of dissolved metal ions at the anode surface will be higher than that at the bulk solution. Like resistance, impedance is a measure of the ability of a circuit to resist the flow of electrical current. Unlike resistance, impedance is not limited by the simple concept of resistance. Electrochemical impedance is usually measured by applying an AC potential to an electrochemical cell and measuring the current through the cell. This impedance depends on the frequency of the potential perturbation. In the immediate vicinity of the anode, the concentration of metal ions pulsates with the frequency of pulse power. It increases during the pulse on time and relaxes during the rest of cycle time. Thus, diffusion layer pulsates and in the case of ultra-short pulses, it does not build up. At high frequencies, the Warburg impedance is small since diffusing reactants do not have to move very far. At low frequencies the reactants have to diffuse farther, thereby increasing the Warburg impedance.

Whenever a metal is dipped in electrolyte, the metal molecules can dissolve into the electrolyte. In the forward reaction, electrons enter the metal and metal ions diffuse into the electrolyte. This is called charge transfer. Again in backward reaction metal ions are discharged to metal. Thus equilibrium condition is reached and an equilibrium rate of ion movement called the exchange current density continues. For the dissolution of anodic metal, the reaction must be forward only i.e. irreversible. As potential is applied between anode and cathode for anodic dissolution, the resistance overcome in irreversible charge transfer is called charge transfer resistance.

Since electrochemical micromachining (EMM) is performed with ultra-short pulses, the machining rate is not only controlled by electrolyte resistance, double-layer capacitance, Warburg impedance, and charge transfer resistance but also influenced by pulse duration and pulse frequency. Under pulse power supply, charging and discharging of double layer is must in a cycle. EMM is carried out with small voltage, very small current, highly precise motion of micro tool or workpiece, and small inter-electrode gap. The epoch making progress in the accuracy of micro machined products by electrochemical micromachining (EMM) started with the introduction of high-frequency ultra-short voltage pulses by Schuster et al. [9]. Dissolution of metal is confined to the resolution of machining limited by the charging time constant, $\tau = \rho_s C_d g$ of the double layer. Where $\rho_s$ is specific resistance or resistivity of electrolyte, $C_d$ is double-layer capacitance, and g is inter-electrode gap. Electrolyte resistance along with electrical double-layer capacitance constitutes an equivalent circuit as shown in Figure 4.6. The active electrolyte resistance $R_S$ is in series with the parallel combination of the double-layer capacitance $C_d$ and polarization resistance $R_P$. If the duration of pulse (on time) is longer than the charging time constant for a machining zone of IEG (g), the double layer is charged enough for dissolution of metal. The regions of higher IEG for which charging time constant, τ is obviously greater than pulse duration, do not get opportunity for dissolution due to inadequate charging of double layer. Pulse frequency is also another parameter controlling electrode kinetics. The charging time constant per cycle remains same and with increase in frequency nonFaradaic time (non-dissolution time)

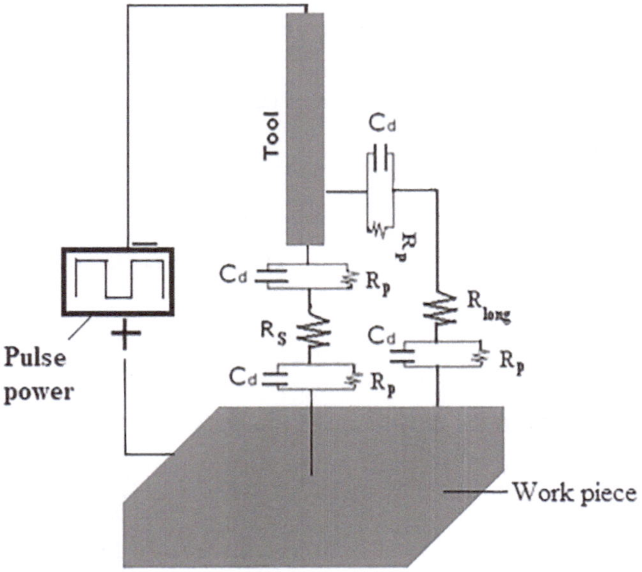


**FIGURE 4.6** Equivalent circuit model in EMM cell.

increases. Thus, effective time of dissolution per unit time decreases resulting in decreased machining rate. Depending on the operating conditions and metal electrolyte combinations, different anodic reactions take place when sufficient pulse power is applied. Rate of these reactions are dependent on the ability of the system to remove the reaction products as soon as they are formed and supply of fresh electrolyte to the inter-electrode gap. All of these factors influence the machining performance namely dissolution rate, shape control, and surface finish of the workpiece.

### 4.4.1 Electrochemistry of Micro-Electrochemical Machining

When sufficient voltage is applied to overcome the overpotentials involved in EMM, electrochemical reactions occur at the cathode, the anode, and in the electrolyte. The kind of reaction depends on the following factors:

i. Nature of metal being machined.
ii. Type of electrolyte.
iii. Temperature of electrolyte etc.

The cathode and the anodic reactions are described below.

a. ***Cathode reactions***

Possible reactions causing the evolution of hydrogen gas at the cathode are:

$$2H^+ + 2e^- \rightarrow H_2 \uparrow \quad \text{(when the electrolyte is acidic in nature)} \tag{4.1}$$

$$2H_2O + 2e^- \rightarrow H_2 \uparrow + 2(OH)^- \text{(when the electrolyte is alkaline/neutral in nature)} \tag{4.2}$$

b. ***Anode reactions***

At the anode i.e. workpiece two possible reactions may occur:

i. Dissolution of metal ions.
ii. Evolution of oxygen or halogen gas.

The reaction leading to the dissolution of the metal

$$M \rightarrow M^{n+} + ne^- \tag{4.3}$$

At the appropriate voltage, metal ions get the energy required to cross the double-layer barrier to go to the bulk electrolyte and the following reaction may occur by Jain (2002) [10].

$$M^{n+}n(OH)^- \rightarrow M(OH)_n \quad (4.4)$$

The reactions leading to the evolution of oxygen or halogen gas are as follows:

$$2H_2O \rightarrow O_2 \uparrow + 4H^+ + 4e^-(\text{in acidic electrolyte}) \quad (4.5)$$

$$4(OH)^- \rightarrow 2H_2O + O_2 \uparrow + 4e^-(\text{in alkaline electrolyte}) \quad (4.6)$$

Reactions occurring in general during EMM for machining of stainless steel, SS-304 workpiece with $H_2SO_4$ as electrolyte can be represented as following:

$$\text{Anode:} \quad Fe \rightarrow Fe^{3+} + 3e^- \quad (4.7)$$

$$Cr \rightarrow Cr^{3+} + 3e^-$$

$$\text{Cathode:} \quad 2H_2O + 2e^- \rightarrow 2(OH)^- + H_2\uparrow \quad (4.8)$$

$$\text{Electrolyte:} \quad H_2SO_4 \rightarrow 2H^+ + (SO_4)^{2-} \quad (4.9)$$

$$2Fe^{3+} + 3(SO_4)^{2-} = Fe_2(SO_4)_3(\text{red colored}) \quad (4.10)$$

$$2Cr^{3+} + 3(SO_4)^{2-} = Cr_2(SO_4)_3(\text{green colored}) \quad (4.11)$$

Possibility of presence of $Fe_2(SO_4)_3$ (red colored) and $Cr_2(SO_4)_3$ (green colored) have been validated during EMM. The electrolyte at the machining zone often becomes yellowish after a substantial time of micromachining. Red colored $Fe_2(SO_4)_3$ and green colored $Cr_2(SO_4)_3$ create yellow color according to RGB color code.

Anodic reactions occurring in general during EMM for machining of iron (Fe) workpiece with $NaNO_3$ as electrolyte can be represented as following:

$$\text{Anode:} \quad Fe \rightarrow Fe^{++} + 2e^- \quad (4.12)$$

$$\text{Electrolyte:} \quad NaNO_3 \rightarrow Na^+ + (NO_3)^- \quad (4.13)$$

$$Fe^{++} + 2(OH)^- \rightarrow Fe(OH)_2 \quad (4.14)$$

$$4Fe(OH)_2 + 2H_2O + O_2 \rightarrow 4Fe(OH)_3 \downarrow (\text{Sludge}) \quad (4.15)$$

From the above reactions, it is clear that only anodic metal is dissolved at the expense of electrical energy. Cathode remains unaltered while electrolyte carries current through the movement of ions between tool and workpiece. Removal of heat and sludge from the machining zone is also performed by electrolyte. It is worth mentioning that outside the electrolyte, current flows due to movement of free electrons in metal.

## 4.5 MECHANISM OF METAL REMOVAL

Faraday's two laws of electrolysis are as follows:

i. The mass of any material dissolved or deposited is directly proportional to the amount of electricity which is passed through the electrolyte.
ii. The amounts of different substances dissolved or deposited by the same quantity of electricity are proportional to their chemical equivalent weights.

The two laws are combined to give the equation

$$\mathrm{m} = \frac{Mit}{FZ} \tag{4.16}$$

Where m is the mass (in grams) dissolved or deposited, i is the DC current (in amperes) passed for time t (in sec), M is the atomic weight (grams) of metallic ion, F is Faraday's constant, and Z is the valency of cation.

Volumetric dissolution rate is given by

$$\mathrm{Q} = \frac{Mi}{F\rho Z} \ \mathrm{cm^3/sec} \tag{4.17}$$

Where ρ is the density of metal undergoing dissolution (g/cm$^3$).

When the anode is superalloy where atomic weights, valencies, and weight % of the composition are $M_i$, $Z_i$, and $X_i$ respectively then mass removal rate (MRR) of alloy per unit charge is as follows:

$$\frac{Q\rho X_1}{100} + \frac{Q\rho X_2}{100} + \frac{Q\rho X_3}{100} + \ldots\ldots = \frac{M_1}{Z_1 F} + \frac{M_2}{Z_2 F} + \frac{M_3}{Z_3 F} + \ldots\ldots\ldots$$

$$or \ \sum_{i=1}^{n} \frac{Q\rho X_i}{100} = \sum_{i=1}^{n} \frac{M_i}{FZ_i}$$

Thus, volume removal rate of alloy per unit charge is given by

$$Q = \frac{100}{\rho F}\left(\frac{1}{\sum_{i=1}^{n} \frac{X_i Z_i}{M_i}}\right) \mathrm{cm^3/sec} \tag{4.18}$$

Here, ρ is the overall density of the alloy (g/cm$^3$) and n is the number of elements in the alloy.

### 4.5.1 MRR Equation for EMM for Single Metal

Mathematical modeling is helpful in predicting MRR and shape profile such that corrective approach can be initiated before actual experimentation. Experimental planning for variation of the parameters involved in the modeling can also be done beforehand and subsequently modified to achieve the desired results.

In the case of EMM, pulsed power is supplied. There is pulse on time, $T_{on}$ and pulse off time, $T_{off}$ during each cycle as shown in Figure 4.7. The total current supplied during a pulse period is the sum of faradaic current and nonFaradaic current [11]. At the very beginning of the pulse, instantaneous voltage is $V_{max.}$ Total current is used for the time $t_c$ to charge the double-layer capacitance without dissolution of metal from anodic workpiece and this current is called nonFaradaic current. While the double layer is being charged, the potential drops due to overpotential, and after the proper charging of double layer, voltage becomes flat ($V_{faradaic}$) and faradaic current flows throughout the rest of pulse on time resulting in dissolution of anodic metal.

Since the material removal occurs by the faradaic currents only during each cycle, the MRR per cycle can be calculated utilizing Eq. (4.16):

$$MRR_{cycle} = \int_{t_c}^{T_{on}} \frac{i}{Z}\frac{M}{F}dt \tag{4.19}$$

Resistance, R in the inter-electrode gap can be expressed by Ohm's law as given below:

$$R = \frac{\rho_s\, g}{A}$$

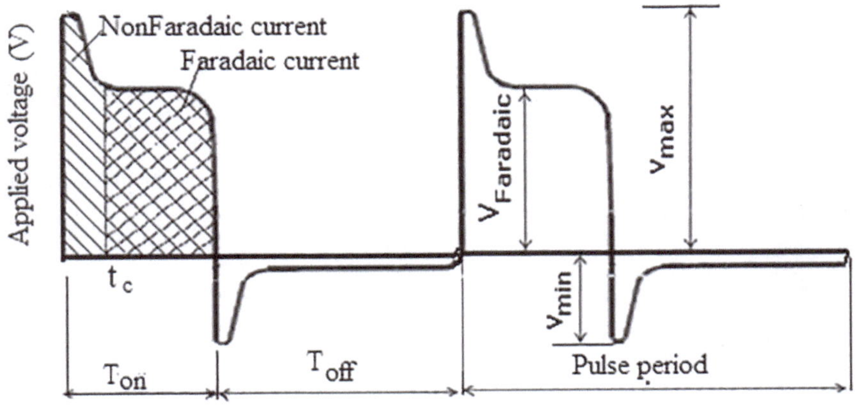


**FIGURE 4.7** Charging and discharging pulse wave during EMM.

Where g is inter-electrode gap, $\rho_s$ is specific resistance of electrolyte, and A is active electrode area. Therefore, Eq. 4.19 can be expressed as given below:

$$\begin{aligned} MRR_{cycle} &= \int_{t_c}^{T_{on}} \frac{V_{faradaic}\ A\ M}{\rho_s\ g\ Z\ F} dt \\ or,\ \ MRR_{cycle} &= \frac{V_{faradaic}\ A\ M}{\rho_s\ g\ ZF}(T_{on} - t_c) \\ \mathrm{MRR} &= \frac{\mathrm{V_{faradaic}\ A\ M}}{\rho_s\ \mathrm{g\ Z\ F}}(\mathrm{T_{on}} - \mathrm{t_c})\mathrm{f} \end{aligned} \tag{4.20}$$

Where f is pulse frequency in Hertz. $V_{faradaic}$, $t_c$, and $T_{on}$ can be obtained from the acquired data by oscilloscope during EMM operation.

### 4.5.2 MRR Equation for EMM of Alloy

For EMM of single metal, volume removal per sec is given by

$$Q = \frac{V_{faradaic}\ A\ M}{\rho\ \rho_s\ g\ Z\ F}(fT_{on} - ft_c)$$

Where ρ is density of the metal (g/cm$^3$).

$$fT_{on} = \frac{1}{T} \times T_{on} = Duty\ \ ratio\,(D)$$

Where D is duty ratio and T is cycle time.

$$Therefore,\ \ Q = \frac{V_{faradaic}\ A\ M}{\rho\ \rho_s\ g\ Z\ F}(D - ft_c) \tag{4.21}$$

Eq. (4.21) can be combined with Eq. (4.18) for calculation of volume removal rate of alloy per unit charge as given below:

$$Q = \frac{100\ V_{faradaic}\ A}{\rho\rho_s g\ \ F\Sigma_i \frac{X_i Z_i}{M_i}}(D - ft_c)\ \ \mathrm{cm^3/sec} \tag{4.22}$$

Where ρ is overall density of the alloy (g/cm$^3$). Eq. (4.22) can be used for the estimation of machining time of alloy.

### 4.5.3 Current Density Distribution and Shape Evolution

Let the small elemental area dA on anode be dissolved. Eq. (4.17) can be expressed as

$$\frac{dAdx}{dt} = \frac{Mi}{F\rho Z} \tag{4.23}$$

Where dx is elemental dissolution of anodic metal normal to dA.

Rate of recession in two-dimension is given by

$$\frac{dx}{dt} = \frac{E\ M}{F\rho z} \tag{4.24}$$

Where E is the net current density and the most significant parameter in shaping of anode. Current density is governed by convection, migration, and diffusion of ions as well as electrode reaction kinetics. Mathematically net current density can be given by

$$E = E_{mt} + E_{et} \tag{4.25}$$

Ghoshal, Bhattacharyya (2014) [12].

$E_{mt}$ is current density due to mass transfer and $E_{et}$ is current density due to electron transfer current for charging double-layer capacitance.

Butler-Volmer equation provides $E_{et}$ as given below:

$$E_{et} = E_0\left[exp\left\{\frac{(1-\beta)ZF\ \Delta V}{RT}\right\} - exp\left(\frac{-\beta ZF\Delta V}{RT}\right)\right] \tag{4.26}$$

Where $E_0$ is exchange current density, $\Delta V$ is activation overpotential and equal to ($V_{max} - V_{Faradaic}$), $\beta$ is charge transfer coefficient, R is universal gas constant, and T is absolute temperature. $E_{mt}$ is given by

$$E_{mt} = Z^2dc\left(\frac{F^2}{RT}\right) \times \frac{\partial V}{\partial x} + FZd \times \frac{\partial c}{\partial x} \tag{4.27}$$

Bard and Faulkner (2001) [13].

Where d is diffusivity, c is concentration of anodic ions, R is universal gas constant, and T is temperature in K. $\frac{\partial V}{\partial x}$ is rate of change of potential in the direction of dissolution of metal. $\frac{\partial c}{\partial x}$ is rate of change of concentration of ions in the direction of dissolution of metal. The concentration of metal ions is supposed to rise for the duration of the pulse on time and to reduce during pulse off time. Generally, high pulse frequency (MHz) is applied during EMM for better accuracy and thereby built up of concentration gradient is not there. Hence, Eq. (4.27) is modified to

$$E_{mt} = Z^2dc\left(\frac{F^2}{RT}\right) \times \frac{\partial V}{\partial x} \tag{4.28}$$

Electric field distribution in the IEG is controlled by Laplace's equation

$$\nabla^2 V = 0 \tag{4.29}$$

Solving Laplace's equation is necessary for calculating current density due to mass transfer utilizing Eq. (4.28). Current density distribution gives recession utilizing Eq. (4.24) at different discretized points on the anodic metal for a given time step. Simulation of a few time steps predicts shape of anode [12,14]. Purcar et al. [15] solved the potential model numerically and applied boundary element method to simulate model of 3D anodic shape change during electrochemical machining. Finite element simulation method can also be used for shape prediction in EMM [16]. Simulation process of inter-electrode gap between cylindrical tool and workpiece is shown in Figure 4.8 with the following assumptions:

i. The process of EMM is in stable condition.
ii. The conductivity of the electrolyte does not change with time.
iii. The concentration gradient in the bulk electrolyte is negligible.

The electrical potential V in the inter-electrode gap is described by Laplace's equation

$$\nabla^2 V = 0$$

Boundary conditions are as follows:

Anode surface$|1,2,3,4,5 = V$ (applied voltage) (i)
Cathode tool surface$|\ 6,7,8 = 0$ (ii)
$\frac{\partial V}{\partial N}\ |9,10,11,12 = 0$ (iii)

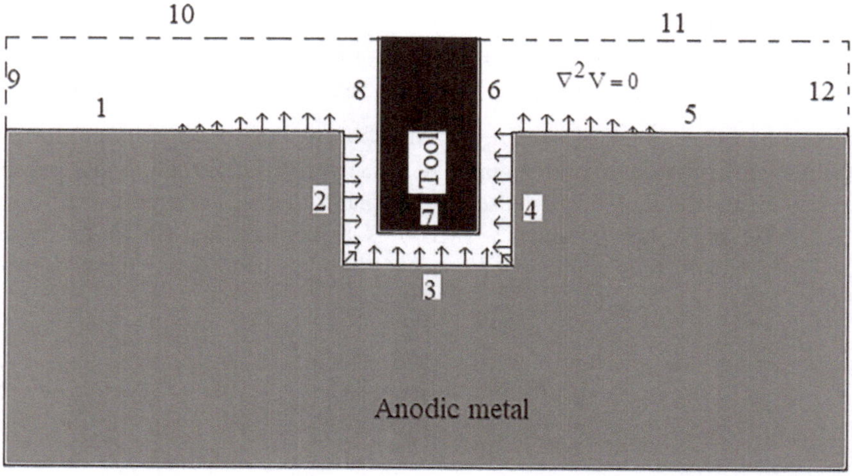


**FIGURE 4.8** Current density distribution in the IEG.

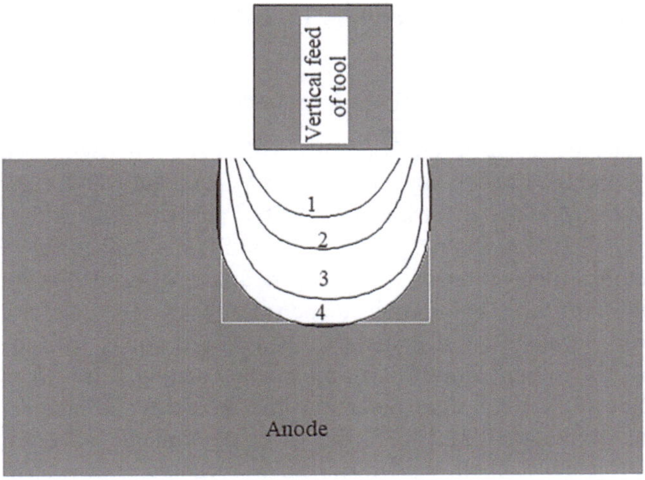


**FIGURE 4.9** Generation of profile during EMM.

Where N is the surface normal on the surface indicated by 9,10,11,12. Laplace's equation is solved subject to the given boundary conditions. Equipotential lines are parallel to the surface normal. Various discretized points on the anodic surface are marked and current density is calculated based on electric field distribution. Size of arrows in the figure indicates approximate current density distribution. The recession of discretized points along the direction of the current is calculated utilizing Eq. (4.24). After a definite time step, $\Delta t$ profile shape can be predicted to be approximately profile 1. Next simulation gives profile 2 and likewise profile 3 and profile 4 as shown in Figure 4.9.

The shape can be produced by three techniques. Designed tool of complicated shape is moved in vertical axis for realizing the job. Sinking and milling method [17] can be applied to produce the same shape as produced by the designed tool. A simple cylindrical micro tool is drilled to the required depth of workpiece and then designed path is followed by milling. This technique is appropriate for generation of micro features where no of products are small since design of tool involves cost. Desired shape can also be generated by third technique of through-mask EMM. Metal dissolution at the workpiece surface is restricted within the cavity created by the photoresist mask. Anode is photoresist patterned and uncovered metal surface is removed but undercut and shape evolution are controlled by aspect ratio, gaps to opening ratio, and mask depth ratio.

Taper angle of cross-sectional profile of blind microchannel generated by conical micro tool is lesser than that by cylindrical micro tool. Taper angle of blind microchannel generated by the sinking and milling method is always lesser than that of microchannel generated by the scanning machining layer-by-layer method [18].

## 4.6 MACHINING CONDITIONS IN MICRO-ELECTROCHEMICAL MACHINING

### 4.6.1 Power Supply in Micro-Electrochemical Machining

Pulsed DC or pulsed AC power is useful in EMM for improving accuracy and enhanced resolution. Low voltage of 4 V to 10 V and thereby low current density is applied in EMM to avoid detrimental effect on surface finish of the workpiece [19]. Very small IEG in the range of 5 to 15 μm is maintained for better accuracy. Sometimes, deposition of machining debris occurs on the micro tool if sufficient pulse off time is not provided and thereby, causes loss of accuracy due to non-uniform dissolution of material. When pulsed power is used instead of DC voltage, IEG is kept almost clean by suitably setting pulse-off time during pulse duration [20]. Very short pulse on time increases the precision, surface quality, and accuracy [9,21,22]. Shape of pulse voltage waveform may be rectangular, triangular, or sinusoidal. Generally, rectangular waveform is used in EMM but recent development suggests that triangular waveform is best in terms of kerf width (overcut) for the same frequency and same pulse width. Kerf width is lowest when triangular waveform is used and lesser in sinusoidal waveform than that in rectangular voltage waveform [23]. Energy input is least in triangular pulse followed by sinusoidal pulse and then rectangular pulse for same frequency, duty ratio, and peak voltage. This is evident from Figure 4.10 which shows that area of pulse width in rectangular pulse is highest followed by sinusoidal pulse and then triangular pulse. Figure 4.10(a) shows the rectangular waveform before application to EMM. Figure 4.10(b) exhibits triangular waveform and Figure 4.10(c) exhibits sinusoidal waveform with same pulse width and off time. Pulse width means pulse on time and off time means rest time for the cycle. The shape of the rectangular pulse wave changes as shown in Figure 4.7 during EMM. Improvement in surface finish is also reported in triangular pulse waveform compared to rectangular waveform.

Figure 4.11 shows the variation of overcut with applied pulse voltage during EMM. Machining conditions are 0.1 M $H_2SO_4$ electrolyte, 35% duty ratio, 5 MHz pulse frequency, and 78.125 μm/sec feed rate during scanning method of

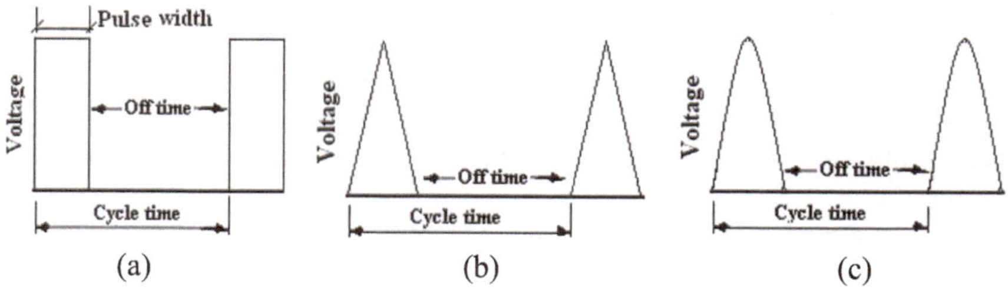


**FIGURE 4.10** (a) Rectangular waveform (b) Triangular waveform (c) Sinusoidal waveform.

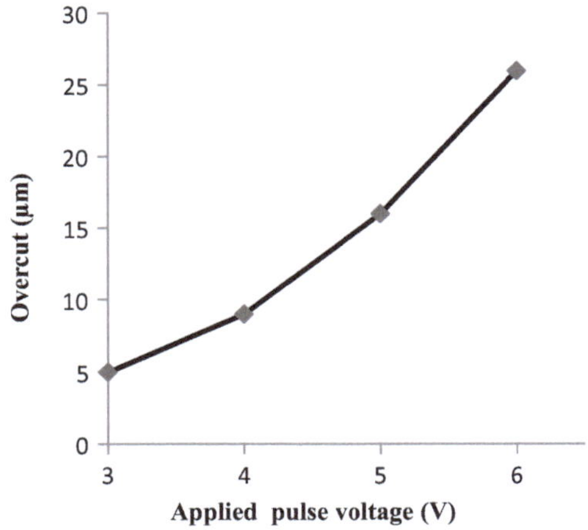


**FIGURE 4.11** Variation of overcut with pulse voltage.

microchannel generation on stainless steel sheet. Eq. (4.20) explains that volume removal rate is proportional to the voltage. Moreover, hydrogen bubbles at higher voltage make a non-conducting atmosphere resulting in stray current effect which in turn creates more sidewall dissolution. Thus, overcut increases almost linearly with pulse voltage. Figure 4.12 shows that the overcut of both micro hole and microchannel decreases with the increase in pulse frequency. The overcut of microchannel generated by sinking and milling method is slightly lower than the overcut of drilled micro hole over the whole ranges of pulse frequency. Micro holes and microchannels are generated on a 35 µm thick SS-304 stainless steel sheet. Machining conditions are 0.2 M $H_2SO_4$ electrolyte, 35% duty ratio, and optimum feed rate.

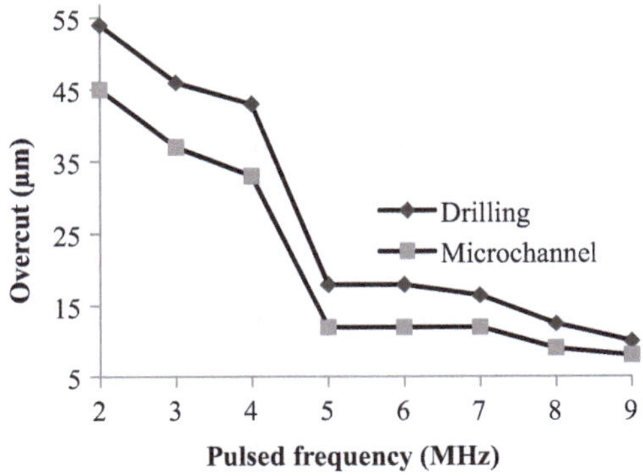


**FIGURE 4.12** Overcut versus pulsed frequency.

### 4.6.2 Electrolytes in EMM

An electrolyte in EMM cell carries current between tool and workpiece along with removing heat and sludge from the machining zone. Function of suitable electrolyte is to dissolve anodic metal actively as metal ion but not to discharge at cathode. Also, it must be chemically stable for continuous machining. Additionally, electrolyte should be cheap, easily available, and non-corrosive as far as possible. In general, the following three types of electrolytes are used for ECM or EMM:

a. Neutral aqueous salt.
b. Aqueous acids of low concentration.
c. Aqueous alkalis.

Neutral salts are cheap, environment friendly, and easily available but one disadvantage of using neutral aqueous salt is the formation of sludge in the machining zone creating frequent short circuit during machining. Thereby, narrow IEG for better accuracy cannot be maintained. Aqueous acids of low concentration can maintain very small IEG as reaction products remain dissolved and hence, suitable for better accuracy of machining. Aqueous alkalis form passive layer on workpieces and metal hydroxides cause congestion in the IEG [9]. Thus, its use as electrolyte is restricted. Strong NaOH or KOH solutions are used for machining of tungsten or tungsten carbide. Sometimes, passive film of oxides is formed on the workpiece surface by reaction with oxygen. A passive layer does not allow machining to continue until higher potential is applied [7,24]. NaCl solution can break the passive layer and reported as non-passivating electrolyte [25,26]. Electrolyte is chosen keeping in mind MRR, accuracy, and surface finish. A list of electrolytes used by various researchers is shown in Table 4.1. Sometimes, mixed electrolytes are more useful in machining of alloy [27–29]. Different combinations of mixed electrolytes produce different desired machining output for EMM of Titanium alloy. In the case of aqueous solution of NaCl and $NaNO_3$, maximum MRR occurs while NaCl and Glycerol electrolyte produces optimal overcut and best circularity is gained from the mixture of NaCl and citric acid [30].

Concentration of electrolyte highly influences overcut and surface finish. Figure 4.13 shows the variation of entry overcut with concentration during scanning method of microchannel generation on SS-304 stainless steel with $H_2SO_4$ as electrolyte [31]. Overcut increases with increase in concentration of electrolyte. Increase in electrolyte concentration enhances concentration of ions taking part in machining and thereby resistance to current flow is decreased. Localization of machining deteriorates with the increase in concentration of electrolyte due to decrease in charging time constant,$\tau$. EMM parameters are 35% duty ratio, 5 MHz pulse frequency, 3V average voltage, 78.125 µm/sec feed rate, and 44 µm depth of machining. Surface finish improves with increase in concentration as shown in Figure 4.14. However, in the case of aqueous salts as electrolyte, surface finish

**TABLE 4.1**
**Metal Electrolyte Combination for EMM**

| Single Metal/Alloy | Electrolyte |
|---|---|
| Iron | Aqueous solution of NaCl |
| Copper | Aqueous solution of $NaNO_3$ |
| Nickel | Low concentration of $H_2SO_4$ |
| Titanium | Sodium bromide (NaBr) solution |
| Stainless steel | Low concentration of $H_2SO_4$ |
| Tungsten / Tungsten carbide | Strong NaOH or KOH solution |
| Tungsten carbide with cobalt binder | Aqueous solution of $H_2SO_4$ mixed with $HNO_3$ |
| Alloy steel $20MnCr_5$ | Aqueous NaCl and potassium dichromate ($K_2Cr_2O_7$) |
| Nickel based superalloy | Aqueous solution of NaCl and HCl |
| | Aqueous solution of NaCl and $NaNO_3$ for Max. MRR |
| Titanium alloy | NaCl and Glycerol |

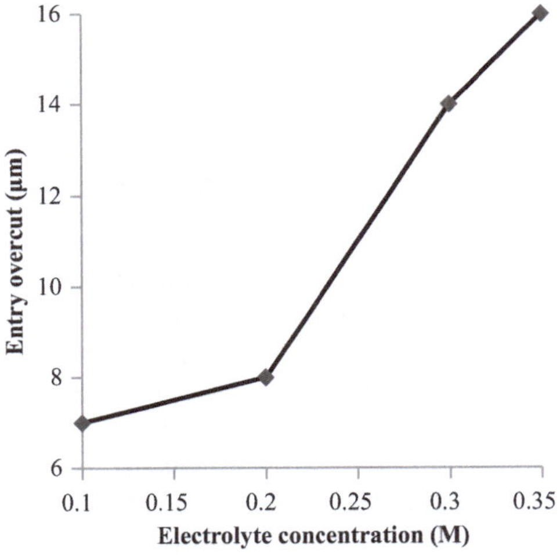


**FIGURE 4.13** Overcut versus concentration.

may deteriorate with higher concentration due to excess crystallization of sludge in IEG. MRR increases with increase in concentration of electrolyte due to the fact that more current is carried by more ions present within IEG, R. J. Leese and Ivanov (2016), Kozak and Rajurkar (2004) [32,33].

Heat generated by Ohmic resistance in the IEG is to be removed by electrolyte to escape micro sparks triggered by bubbles. Otherwise, surface of anode is damaged resulting in poor surface finish [9,27,34,35].

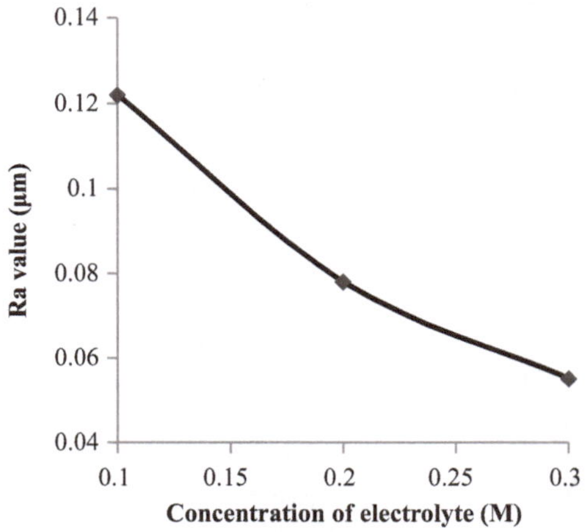


**FIGURE 4.14** Effect of concentration on surface finish.

The electrolyte is supplied to the machining chamber by pump fitted with filter. Delivery system may be in the form of jet or array of nozzles. When the diameter of micro tool is very small (below 100 μm), stagnant electrolyte is better due to the fact that tool may not withstand flow pressure.

### 4.6.3 Role of Inter-Electrode Gap in Machining Accuracy

Resolution of machining is controlled by charging time constant, $\tau = \rho_s\ C_d\ g$ of the double layer as discussed in section 4. High accuracy and good surface finish in EMM are dependent on charging time constant which in turn depends on IEG (g). IEG is generally maintained below 15 to 20 μm for better localization of resolution, Silva et al. (2000) [36]. Maintaining of extremely small IEG is a challenge as direct measurement of IEG is still not available. Oscilloscope is used to observe the abnormality of pulse signal and accordingly gap size is rectified for stable machining. Figure 4.7 shows the pulse shape during normal steady dissolution. Voltage drops down too much during micro sparking and takes the shape as shown in Figure 4.15. Immediate increase in gap is necessary for stable machining. Figure 4.16 shows the pulse shape during short circuit. Micro tool touches the workpiece and no machining occurs. Tool feed rate is approximately decided based on derived equations of volume removal rate (Eq. 4.21 & Eq. 4.22). In each cycle, circuit must be charged and then discharged for dissolution to take place. Steady Faradaic current shown in Figure 4.7 may be used for the measurement of IEG. Boundary limit of gap may be detected with current sensor at the IEG and appropriately set at CNC programming controlling the machining. For stable machining, the tool feed rate must be equal to the dissolution rate. If tool feed rate is more compared to machining rate, the gap reduces and current increases.

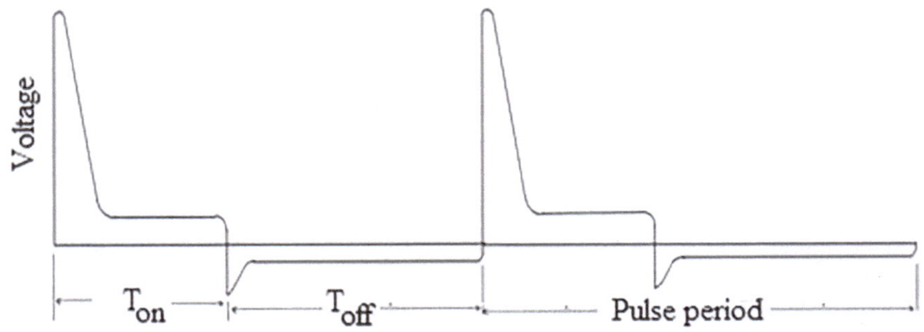


FIGURE 4.15 Pulse shape during sparking.

FIGURE 4.16 Pulse shape during short circuit.

The tool moves back when maximum current limit is reached and the tool moves forward when lower limit of current value is reached due to increased IEG. Thus, equilibrium gap is maintained.

## 4.7 TYPES OF EMM

Classification of various types of EMM is shown in Figure 4.17. When metal dissolution is allowed only through mask or photoresist pattern, it is known as through-mask EMM. In Maskless EMM, dissolution of metal from the anodic workpiece is not constrained by mask but localized by high-frequency pulse power or by sidewall coating of micro tool.

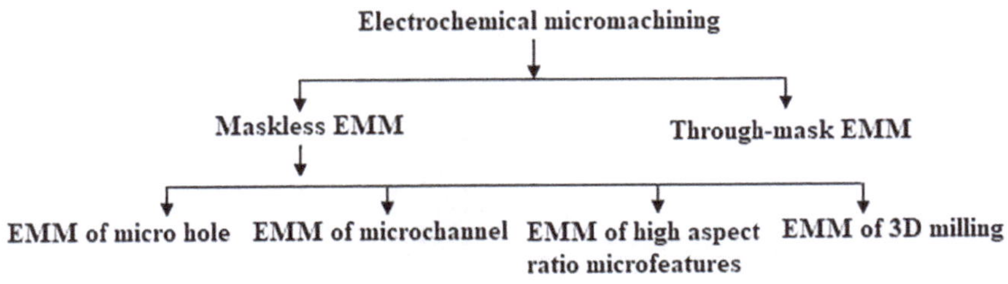


FIGURE 4.17 Different types of EMM.

### 4.7.1 Through-Mask EMM

Inert patterned photoresist mask is generally attached or bonded to the anodic workpiece. Electrolyte flows through the inter-electrode gap at high speed. When voltage is applied between anode and cathode, dissolution takes place only from the exposed metal surface at the workpiece surface. Patterns in the mask are replicated to the anode. Sometimes, a patterned insulation mask plate is coated with cathode metal to be used repeatedly for mass production [37]. Metal removal may be operative from a single side or both sides of anodic plate with necessary arrangements. Figure 4.18 shows a schematic diagram of through-mask EMM. "D" represents exposed anode surface diameter/width and D′ represents entry diameter/width of the machined feature. Undercut is the phenomena associated with through-mask EMM and defined as D′-D. If current density vector diagram at the exposed anode surface is analyzed it would be observed that due to the current aggregation effect, current density is much higher at the mask edge [16]. As a result, rate of dissolution of metal is more at the mask edge. Shape evolution of machined profile is influenced by mask aspect ratio and spacing to opening ratio. Aspect ratio = h/D, where h is the thickness of mask. In the case of small aspect ratio, undercut is more and when aspect ratio is greater than 1, current density distribution becomes more uniform resulting in lesser undercut. But, a higher aspect ratio creates difficulty in removal of reaction products.

### 4.7.2 Maskless EMM

Photoresist mask is not used in this type of EMM. Localized dissolution of metal is possible by controlling machining parameters and design of suitable micro tool. Stray current effect is minimized by narrow inter-electrode gap and using passivating electrolyte. Maskless EMM may be classified into following categories based on machining features:

i. EMM of micro hole.
ii. EMM of microchannel.
iii. EMM of high aspect ratio micro feature.
iv. 3D EMM.

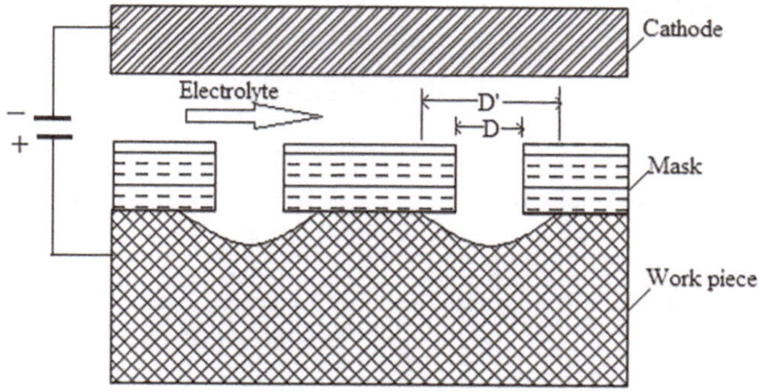


**FIGURE 4.18** Schematic of through-mask EMM.

#### 4.7.2.1 EMM of Micro Hole

Micro hole is the basic feature of micro-components and most basic operation of micromanufacturing. Tool feed is required in vertical direction only for micro hole generation. Shapes of borehole may be stepped holes, tapered hole (nozzle), and taper free holes. Reduction of overcut is a challenge and optimization is done by controlling process parameters. Most important parameter is high frequency of pulsed power which plays major role for reduction of overcut. Highest possible feed rate without short circuit also reduces overcut due to lesser machining time and lesser time of stay of tool at inlet. Sometimes, sidewall of micro tool is coated with insulation layer to reduce overcut. The taper in a through hole can be reduced by few minutes of machining after perforation [21]. When machining zone is simulated with conical tool it is found that current density is lower at entry side of conical tool than that at the exit side and as a result, taper angle decreases with extra time machining after perforation (2014) [12]. Reversed taper and conical tool can be used for generation of taper free boreholes.

#### 4.7.2.2 EMM of Microchannel

Micro profiles or microchannels are necessary parts of microfluidic systems, sensors, chemical micro-reactors, heat exchangers of computer chips etc. Microchannels can be generated by micro mechanical milling, micro electro discharge machining, micro laser machining, and micro-electrochemical machining. EMM has specific advantages over the other micromachining methods such as stress free polished machined surface, no heat affected zone, machining capability of metals irrespective of hardness, and no tool wear. Shaping strategy of micro borehole in EMM can be extended to three-dimensional micromachining for microchannels and profiles. Parameters controlling the shape and accuracy of micro hole in EMM definitely influence the shape and accuracy of microchannel. Micro profile or microchannel can be generated by (a) scanning method or (b) sinking and milling method.

a. Scanning method: The micro tool is set on a to and fro motion while machining goes on between the front face of the micro tool and anodic workpiece. Before each scan movement, downward feed of micro tool is given. The micro tool travels rapidly and repeatedly. Metal is removed layer by layer basis as shown in Figure 4.19(a). This method has disadvantages like higher machining time, higher overcut, and deviation on both ends of microchannel.
b. Sinking and milling method: Sinking or micro drilling to final depth or required depth of machined feature is done at the starting point followed by milling along the path of the micro profile. Front lateral surface of the micro tool takes part in micromachining as shown in Figure 4.19(b). The micro tool moves slowly over the path of micro profile during EMM. Advantages of this method are lower machining time, smaller overcut, and deviation on pilot hole only i.e. greater width at the sink point [17].

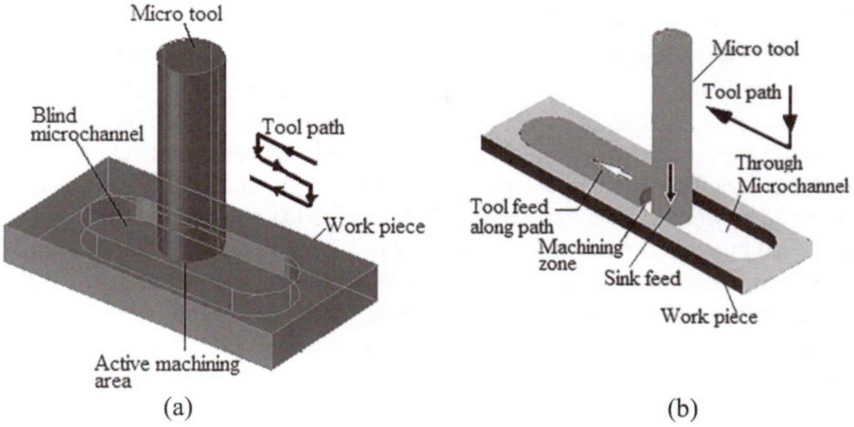


**FIGURE 4.19** (a) Scanning method. (b) Sinking and milling method.

### 4.7.2.3 EMM of High Aspect Ratio Micro Feature

Microchannels and micro holes of high aspect ratio can be fabricated by micro EDM, electron beam machining (EBM), and laser beam machining (LBM) and EMM. Thermal distortion of the machined surface with surface cracks and residual stresses in the workpiece are the disadvantages of EDM, EBM, and LBM. Problems encountered during generation of high aspect ratio micro features by EMM are non-availability of fresh electrolyte at high depth narrow machining zone. Another most crucial problem is increase in width (overcut) of micro features due to long time of stay of micro tool. Conical side coated micro tool is very much suitable to restrict overcut and also to assist flow of electrolyte at high depth machining zone. Flow of electrolyte and thereby disposal of sludge from high depth machining zone becomes easy by application of vibration of micro tool, Ghoshal and Bhattacharyya (2015) [38]. The micro tool of 2.86° sharp taper angle is used to generate high aspect ratio microchannel on SS-304 stainless steel plate by sinking 720 µm vertically downward followed by horizontal milling along the path of 500 µm length as shown in Figure 4.20. Micro tool is vibrated with amplitude 0.5 µm at 82 Hz frequency for renewal of electrolyte at high depth of machining. Other machining parameters are 3 MHz pulse frequency, 0.2 M $H_2SO_4$ electrolyte, 0.139 µm/s tool feed rate for the horizontal milling, and 0.288 µm /s sink feed rate. As the time of stay of micro tool is higher at sink point of microchannel, the width is larger with entry width 156 µm and exit width 153 µm. Figure 4.20(a) exhibits entry side view with average entry width 145.2 µm and Figure 4.20(b) shows the exit side with average exit width 144.7 µm. Approximate aspect ratio is 5:1. High aspect ratio micro features have wide application in the area of surgery, biotechnology, inkjet printers, higher volume flow in microfluidic devices, and microelectronic cooling system.

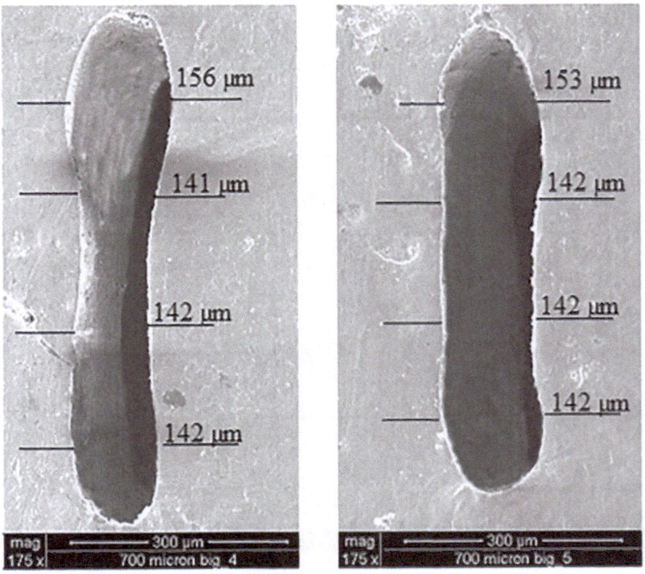


(a) (b)

**FIGURE 4.20** (a) Perspective entry view (b) Perspective exit view.

#### 4.7.2.4 3D EMM

3D features are machined on electrically conductive material by utilizing EMM techniques. Stray current effect is controlled by micro tool coating, high pulse frequency, narrow inter-electrode gap, and using disk type micro tool. Simple cylindrical micro tool can also be utilized for generation of 3D features by controlled movement of micro tool along designed tool path. Computer aided design helps in generation of correct tool paths for required 3D shape generation. Vertical feed of micro tool in Z direction is given step by step after scanning in XY plane. Figure 4.4 explains generation of 3D shape by electrochemical milling.

## 4.8 SETUP FOR MICRO-ELECTROCHEMICAL MACHINING

The EMM setup consisting of various sub-components such as electrical power and controlling system, stepper motor along with microprocessor for motion control of tool and workpiece, mechanical unit for holding the tool, workpiece and vibratory unit for vibrating the tool longitudinally. Figure 4.21 exhibits the schematic diagram of developed EMM setup. The mechanical machining unit is most essential part of the EMM system. Precision tool movement is required in micrometer range for EMM. The mechanical machining unit comprises of XYZ movement stage controlled through personal computer, machining chamber along with work holding devices, mechanical structure of the machine for holding arrangement of piezoactuator mounted with micro tool. The movement

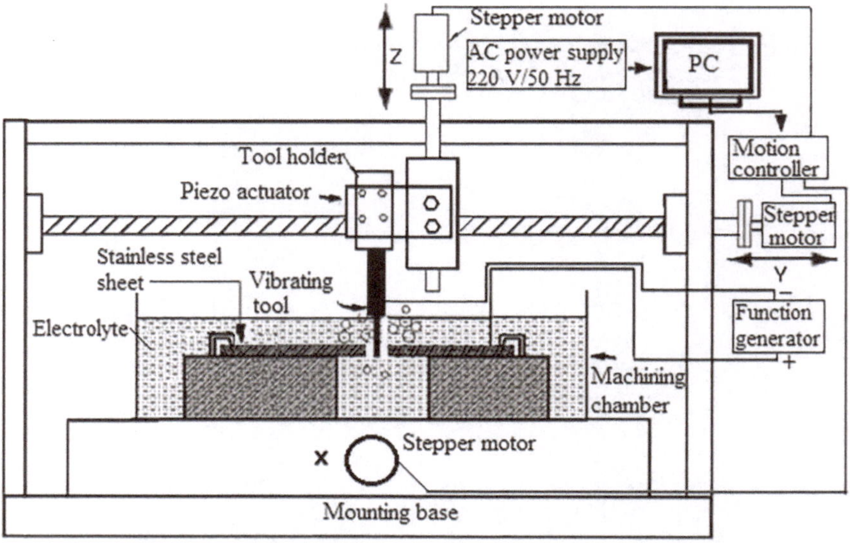


**FIGURE 4.21** Schematic view of EMM system.

of the three axes is controlled by programming software as well as by manually. Suitable programs are to be developed for machining of microchannels, micro holes, or 3D micro features and stored in computer.

The machining chamber rests on the base plate of XYZ stage. It is made of Perspex sheet of different thickness. Perspex is a non-conductive and non-corrosive and transparent material. That's why it is selected for developing the machining chamber of EMM setup. Inside the machining chamber, a workpiece holding arrangement is made. An outlet for the used electrolyte is made at the front side of the machining chamber at the base level.

Micro tool is brazed with the copper tool holder, which is fixed to the Z-axis slide by a fixture or mounted on piezoactuator for vibration of micro tool. The fixture is made of Teflon, which is a good insulating material so that the mechanical structure is protected from the power. The piezoactuator (PU65HR) is actually a vibrator cum Nano positioning equipment. It has high mechanical stability and backlash free vertical translation of 66 μm. A modulation input from a function generator is fed to controlling unit of piezoactuator for the control of amplitude of vibration which varies with the variation of voltage. Highest amplitude of vibration of 66 μm can be applied with the model: PU65HR. Frequency of vibration of the piezoactuator is controlled by the frequency of input voltage and maximum frequency below resonant frequency of 1320 Hz can be applied.

The main supply of 230V/50 Hz single phase is connected to an Agilent make function generator, model: 33250A used for the supply of rectangular pulsed AC (slightly negative bias during off time) voltage to the tool and workpiece. Maximum frequency of function generator is 80 MHz. A vibratory unit of piezoactuator is also connected to 230 V main power supply through amplifier module (–20 V to 130 V) for feed control in the nanometer range (resolution 0.12 nm) having maximum stroke 67 μm i.e. translation in Z direction.

Micro tool movement is actuated by a precision screw nut mechanism. The main screw nut mechanism is rotated with the help of stepper motor control unit. Movement of the stepper motor is controlled through a microprocessor unit based on the Intel 8085 processor.

Continuous monitoring of the EMM processes is done by a digital storage oscilloscope make: Tektronix, model: TBS1062. This also acts as data collection system which is inevitable for the control of inter-electrode gap (IEG) during EMM. Narrow IEG $<10$ μm improves the accuracy. After setting of initial gap of 10 μm between micro tool and workpiece within electrolyte medium, machining voltage is applied and automatic feed system is followed through running programs. Various data such as peak to peak voltage, Faradaic voltage, mean voltage, pulse width, and duty ratio are observed during micromachining operation. Short circuit is detected by observing the nature of pulse shape.

## 4.9 FABRICATION OF MICRO TOOL

Micro tool fabrication is very much critical for the developing micromachining technology. Micro tools up to φ 100 μm are fabricated by reverse electro discharge machining (EDM) and wire electro discharge grinding. Nowadays, micro tool of below 10 μm diameter is also being fabricated by ECM at low cost. The tungsten rod has some advantages as micro tool in EMM such as high rigidity and inertness with electrolyte. Both DC current and pulsed current have been used for fabrication of micro tool. Strong NaOH or KOH solution is used for etching of tungsten wire. Tungsten wire is dipped in electrolyte as anode and nickel rod as cathode to a particular depth side by side, maintaining a desired inter-electrode gap. Anodic tungsten rod is rotated for uniform dissolution of metal. Applied voltage and rotational speed have effect on the shape of developed micro tool [39]. Another method of fabrication of micro tool is shown in Figure 4.22. Tungsten rod of φ 300 μm is vertically immersed in the 2 M KOH electrolyte inside cathode ring of inner diameter 8 mm. 20 minutes of etching at around 5 V applied DC voltage develops diameter of φ 6 μm from φ 300 μm tungsten rods [40]. Tungsten rod develops to conical shape as shown in Figure 4.23 b in the range of less than 5 V while it becomes reverse conical shape as shown in Figure 4.23 c in the range of more than 5 V due to formation of diffusion layer around electrode surface. Straight micro tool shape as shown in Figure 4.23a is developed around 5 V. To reduce the time of fabrication of micro tool, longitudinal vibration is applied to anodic tungsten rod [41]. Stainless steel ring of 1600 μm diameter is used as cathode and φ 300 μm straight tungsten rod as anode is immersed centrally to a certain depth for the electrochemical machining with NaOH electrolyte while vibration is applied to the anodic tungsten rod. Cylindrical micro tool can be fabricated within 1 minute to 5 minutes depending on applied voltage, proper amplitude of vibration, and frequency of vibration. Thus, vibration of the tungsten rod plays an important role during the fabrication of micro tool due to hydrodynamic effects on the bubble behavior.

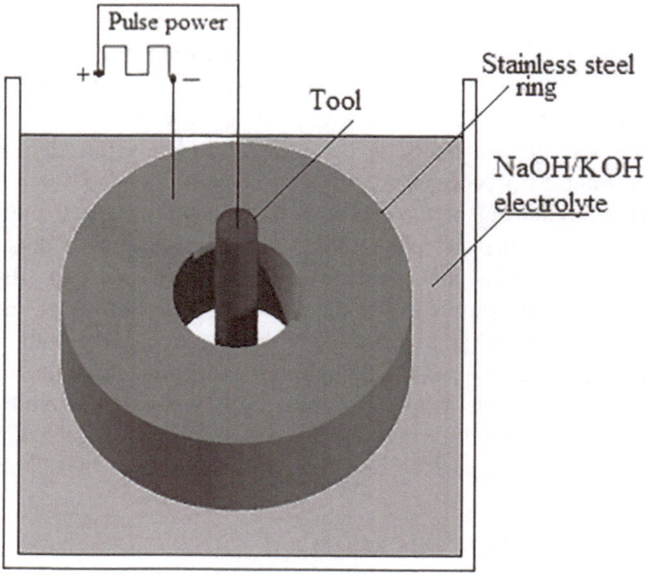


FIGURE 4.22 Schematic fabrication of micro tool in ECM.

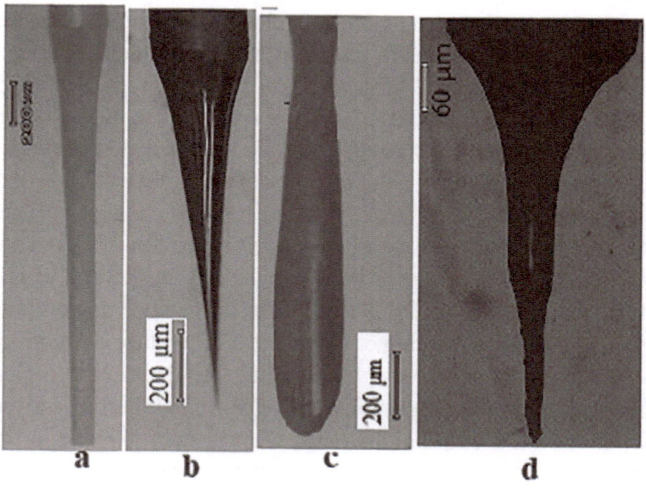


FIGURE 4.23 Various shaped micro tool. a. straight b. Conical c. Reverse conical d. Two stepped.

To fabricate disk shaped micro tool, top & bottom surface of stainless steel ring is insulated. The current is allowed to flow through the sidewall of the hole only. The anodic tool is located at the center of a hole through insulated stainless steel ring. The disk shaped micro tool is used to machine micro features like cylindrical hole with reduced taper angle, taper free microgroove, and 3D microstructure by EMM.

## 4.10 APPLICATION OF MICRO-ELECTROCHEMICAL MACHINING

Different micro features have applications in MEMS, micro engineering, integrated circuits in microelectronics, microfluidic systems, printed circuit board, etc. Some of the typical applications of EMM are as follows. Micro fuel injection nozzles improve atomized spray of fuel so that oxygen of air quickly reacts with fuel and thereby improves efficiency. Micro holes are done on turbine blades by EMM for cooling purpose. Micro-profiles and microchannels are widely used in micro fluidics, aerodynamics, fuel cell technology, and biomedical sectors. Micro holes in artificial body parts made of titanium help in quick assimilation with bone and muscle of human body.

## 4.11 CONCLUSIONS

This chapter deals with fundamentals of EMM to recent developments. Developed MRR equations for single metal as well as alloy will be of very much useful for selection of tool feed rate. Prediction of shape evolution from tool shape has been discussed. Ultra-short pulsed voltage of square, triangular, and sinusoidal voltage waveform of nanosecond pulse width improves accuracy of EMM. Electrolyte influences resistance to current flow, MRR, sludge formation, surface finish, and accuracy of machining. Scientific selection of electrolyte for a particular metal is still a major area of research. Recent developments in the fabrication of micro tools by ECM will further enhance EMM of superalloy for their wide use in aircraft engine parts, automobile, and medical fields due to high strength, good mechanical properties, and corrosion resistance. Very low IEG produces quality machined features. Highly reliable and satisfactory methods of controlling IEG are still a research area.

## REFERENCES

[1] Bhattacharyya, B., Sorkhel, S.K. (1999). Investigation for controlled electrochemical machining through response surface methodology-based approach. *J. Mater. Process. Technol.* 86 (1–3): 200–207.

[2] Bhattacharyya, B., Munda, J., Malapati, M. (2004). Advancement in electrochemical micro-machining. *Int. J. Mach. Tools Manuf.* 44 (15): 1577–1589.

[3] Kim, U.S., Park, J.W. (2019). High-quality surface finishing of industrial three-dimensional metal additive manufacturing using electrochemical polishing. *Int. J. Precision Eng. Manuf.-Green Technol.* 6: 11–21.

[4] Taniguchi, N. (1983). Current status in, and future trends of, ultraprecision machining and ultrafine materials processing. *CIRP Ann. - Manuf. Technol.* 32 (2): 573–582.

[5] Corbett, J., McKeown, R.A., Peggs, G.N., Whatmore, R. (2000). Nanotechnology: International developments and emerging products. *CIRP Ann. - Manuf. Technol.* 49(2): 523–545.

[6] Landolt, D., Chauvy, P.F., Zinger, O. (2003). Electrochemical micro machining, polishing and surface structuring of metals: fundamental aspects and new developments. *Electrochimica Acta* 48: 3185–3201.

[7] Rajurkar, K.P., Zhu, D., McGeough, J.A., Kozak, J.A., Silva, D. (1999). New developments in electrochemical machining. *Ann. CIRP* 48 (2): 567–579.

[8] Datta, M. (1998). Micro fabrication by electrochemical metal removal. *IBM J. Research Develop.* 42 (5): 655–669.

[9] Schuster, R., Kirchner, V., Allongue, P., Ertl, G. (2000). Electrochemical micromachining. *Science* 289 (5476): 98–101.

[10] Jain, V.K. (2002). Advanced machining *Process. Allied Publishars Pvt Ltd, New Delhi.*

[11] Mithu, M.A.H., Fantoni, G., Ciampi, J., Santochi, M. (2012). On how tool geometry, applied frequency and machining parameters influence electrochemical microdrilling. *CIRP J. Manuf. Sci. Technol.* 5: 202–213.

[12] Ghoshal, B., Bhattacharyya, B. (2014). Shape control in micro borehole generation by EMM with the assistance of vibration of tool. *Precision Eng.* 38: 127–137.

[13] Bard, A.J., Faulkner, L.R. (2001). *Electrochemical methods.* 2nd ed, JohnWiley & Sons 200.

[14] Pattavanitch, J., Hinduja, S., Atkinson, J. (2010). Modeling of the electrochemical machining process by the boundary element method. *CIRP Ann. – Manuf. Technol.* 59: 243–246.

[15] Purcar, M., Bortels, L., Bossche, B.V., Deconinck, J. (2004). 3 D electrochemical machining computer simulations. *J. Mater. Process. Technol.* 149: 472–478.

[16] Wang, L., Wang, Q., Hao, X., Ding, Y., Lu, B. (2010). Finite element simulation and experimental study on the through-mask electrochemical micromachining (EMM) process. *Int. J. Adv. Manuf. Technol.* 51: 155–162.

[17] Ghoshal, B., Bhattacharyya, B. (2013). Micro electrochemical sinking and milling method for generation of micro features. *J. Eng. Manuf.* 227 (11): 1651–1663.

[18] Ghoshal, B., Bhattacharyya, B. (2015). Investigation on profile of microchannel generated by electrochemical micromachining. *J. Mater. Process. Technol.* 222: 410–421.

[19] Lee, E., Shin, T., Kim, B., Baek, S. (2010). Investigation of short pulse electrochemical machining for groove process on Ni-Ti shape memory alloy. *Int. J. Precision Eng. Manuf.* 11(1): 113–118.

[20] Bhattacharyya, B., Doloi, B., Sridhar, P.S. (2001). Electrochemical micromachining: New possibilities for micro-manufacturing. *J. Mater. Process. Technol.* 113 (1–3): 301–305.

[21] Ahn, S.H., Ryu, S.H., Choi, D.K., Chu, C.N. (2004). Electro-chemical micro drilling using ultra short pulses. *Precis. Eng.* 28 (2): 129–134.

[22] Kock, M., Kirchner, V., Schuster, R. (2003). Electrochemical micromachining with ultrashort voltage pulses- a versatile method with lithographical precision. *Electrochimica Acta* 48: 3213–3219.

[23] Sharma, V., Patel, D., Agrawal, V., Jain, V.K., Ramkumar, J. (2021). Investigations into machining accuracy and quality in wire electrochemical micromachining under sinusoidal and triangular voltage pulse condition. *J. Manuf. Processes* 62: 348–367.

[24] Schultze, J.W., Lohrengel, M.M. (2000). Stability, reactivity and breakdown of passive films: Problems of recent and future research. *Electrochemical Acta* 45: 2499–2513.

[25] Macdonald, D.D. (1999). Passivity-The key to our metals-based civilization. *Pure Appl. Chemistry* 71 (6): 951–978.

[26] Datta, M. (1993). Anodic dissolution of metals at high rates. *IBM J. Research Develop.* 37(2): 207–226.

[27] Choi, S.H., Kim, B.H., Shin, H.S., Chang, D.K., Chu, C.N. (2013). Analysis of the electrochemical behaviors of WC – Co alloy for micro ECM. *J. Mater. Process. Tech.* 213 (4): 621–630

[28] Ayyappan, S., Sivakumar, K. (2014). Investigation of electrochemical machining characteristics of 20MnCr5 alloy steel using potassium dichromate mixed aqueous NaCl electrolyte and optimization of process parameters. *J. Eng. Manuf.* 229 (11): DOI: 10.1177/0954405414542136.

[29] Bilgi, D.S., Jain, V.K., Shekhar, R., Mehrotra, S. (2004). Electrochemical deep hole drilling in super alloy for turbine application. *J. Mater. Process. Technol.* 149 (1): 445–452.

[30] Thangamani, G., Thangaraj, M., Moiduddin, K., Mian, S.H., Alkhalefah, H., Umer, U. (2021). Performance analysis of electrochemical micro machining of Titanium (Ti-6Al-4V) alloy under different electrolytes concentrations. *Metals* 11: 247 10.3390/met11020247.

[31] Ghoshal, B., Bhattacharyya, B. (2015). Generation of microfeatures on stainless steel by electrochemical micromachining. *Int. J. Adv. Manuf. Technology* 76: 39–50.

[32] Leese, R.J., Ivanov, A. (2016). Electrochemical micromachining: An introduction. *Adv. Mech. Eng.* 8 (1): 1–13.

[33] Kozak, J., Rajurkar, K.P., Makkar, Y. (2004). Study of pulse electrochemical micromachining. *J. Manuf. Processes* 2001: 7–14.

[34] Fan, Z.W., Hourng, L.W., Lin, M.Y. (2012). Experimental investigation on the influence of electrochemical micro-drilling by short pulsed voltage. *Int. J. Adv. Manuf. Technol.* 61: 957–966.

[35] Jain, N.K., Jain, V.K. (2007). Optimization of electro-chemical machining process parameters using genetic algorithms. *Mach. Sci. Technol.* 11 (2): 235–258.

[36] Silva, A.K.M.D., Altena, H.S.J., McGeough, J.A. (2000). Precision ECM by process characteristic modelling. *CIRP Ann. - Manuf. Technol.* 49 (1): 151–155.

[37] Li, D., Zhu, D., Li, H. (2011). Microstructure of electrochemical micromachining using inert metal mask. *Int. J. Adv. Manuf. Technol.* 55: 189–194.

[38] Ghoshal, B., Bhattacharyya, B. (2015). Vibration assisted electrochemical micromachining of high aspect ratio micro features. *Precision Eng.* 42: 231–241.

[39] Fan, Z.W., Hourng, L.W. (2009). The analysis and investigation on the microelectrode fabrication by electrochemical machining. *Int. J. Mach. Tools Manuf.* 49: 659–666.

[40] Wang, M.H., Zhu, D., Peng, W. (2008). Experimental research on electrochemical micromachining, advanced design and manufacture to gain a competitive edge. chapter 6: 775–783, DOI: 10.1007/978-1-8-4800-2418-78.

[41] Ghoshal, B., Bhattacharyya, B. (2013). Influence of vibration on micro-tool fabrication by electrochemical machining. *Int. J. Mach. Tools Manuf.* 64: 49–59.

# 5 Pulsed Fiber Laser Processing of A4 Stainless Steel

*Debasish Panigrahi, Sweta Rout, and S.K. Patel*

## 5.1 INTRODUCTION

In the current age of globalization and technological breakthroughs, smaller in scale but more effective products are highly in demand ranging from the domestic to large industrial sectors. The growing demand for new products with new functionality necessitates extensive research and the improvement of state of the art in manufacturing technology around the globe. The conventional ways of processing different engineering materials with high elasticity and fragility are brutal and suffer from extended machining time and cost ineffectiveness. This is because of the complexity of the machining process. These factors lead to high tolerances with poor machining surfaces, which are incompatible with the demanding quality control standards that are required at this time. On the other hand, unconventional machining approaches with disparate forms of energy as input, such as mechanical, chemical, and thermal, are more suitable to address the difficulties that need to be overcome to achieve a good surface finish with astonishingly good machining quality [1].

The laser has established itself as one of the best machining options among unconventional families, as it takes the least time to complete the process, which is very flexible and cost-effective. Since the laser is a force-free revamping approach, its applicability may be witnessed in several articles while producing a variety of features across many different materials [2]. Consequently, it has shown its supremacy over traditional or unconventional machining procedures while fabricating micro-macro scale structures on high-performance metallic alloys [3]. Fiber lasers have shown significant and constant growth across various sectors due to recent breakthroughs in laser system technology.

### 5.1.1 Stainless Steel and Its Families

The crystal structure can classify stainless steel (SS) into five groups based on their composition. The usual mechanical properties of each family are different


DOI: 10.1201/9781003364948-5

from one another. In addition, each family tends to exhibit the same characteristics in terms of its resistance or susceptibility to specific types of corrosion. Nevertheless, even within the same family, there is the potential for various compositions. Each group, thus, is suitable for use in a wide variety of corrosion conditions.

#### 5.1.1.1 Ferritic Stainless Steel (FSS)

Pure iron and chromium make up the essential kind of SS. Since chromium acts as a ferrite stabilizer, increasing its concentration results in a more robust ferritic structure. The body-centered cubic (BCC) structure of ferrite gives magnetic properties with a reasonably high yield strength but limited ductility and work hardenability. The solubility of such interstitial elements as nitrogen (N) and carbon (C) in ferrite is incredibly low. In FSS grades, a slight temperature shift can cause the behavior to change from ductile to brittle. This transition from ductile to brittle takes place at room temperature with higher nitrogen and carbon contents, particularly at higher levels of chromium (Cr). Due to this risk, FSS grades were not widely used until argon-oxygen decarburization (AOD) was developed. The FSS family used to only include AISI type 446 for oxidation-resistant uses, and AISI types 430 and 434 for corrosion-prone uses, including automotive trim, until now [4]. However, welding or heat exposure easily exposed these grades to intergranular corrosion, further restricting their use.

The activity of C and N could be further diminished with AOD and the addition of stabilizers. However, the stabilizers, which are highly reactive substances, precipitate the remaining interstitials. Stabilizers and AOD could further dampen carbon and nitrogen's reactivity. However, the remaining interstitials are precipitated by the stabilizers, which are very reactive chemicals. AISI type 444 and the higher-alloyed ferritic grades are examples of the newer generation of FSS. Both types of FSS contain meager stabilizing elements or interstitial elements, which improve their resistance to corrosion and weldability. Grades with exceptionally high levels of Cr and molybdenum (Mo) can be produced by manipulating the interstitial elements. Compared to the first generation of FSS, these grades are tougher and more weldable at low sufficient C levels [5]. Their lackluster durability, however, typically prevents them from being used for sheet applications. Chloride stress-corrosion cracking (SCC) is an uncommon occurrence in FSS, and these steels are sometimes even impervious to this type of damage. These metals are commonly considered for use in heat transfer systems. Similarly, the low-alloyed ferritic AISI 409 has flourished due to the same characteristics and benefits [6]. There has been a rise in the use of this grade for exhaust systems in automobiles and other applications subject to moderate to severe environmental exposure.

#### 5.1.1.2 Austenitic Stainless Steel (ASS)

The adverse effects of nitrogen and carbon in ferrite can be eliminated by switching to an austenite crystal structure with a face-centered cubic (FCC) symmetry. Various elements, mainly nickel (Ni), manganese (Mn), and N, are used as austenite stabilizers to effect this transformation. ASS is a nonmagnetic material known for its strong ductility, rapid work hardening rates, outstanding toughness, and low yield strength. The ASS grades, specially AISI type 304, are the most prevalent because of their desirable mechanical qualities and the simplicity with which they may be fabricated [7]. Nitrogen is added to alloys to stabilize the ASS grades with up to 6% Mo, which increases their resistance to corrosion in chloride surroundings. High contents of chromium-enriched ASS are formulated for high-temperature applications, whereas ASS having enriched Ni contents is used in inorganic acid conditions. Welding or extended exposure to heat can make ASS more vulnerable to intergranular corrosion. Due to this prolonged exposure to heat, chromium carbides form at grain boundaries, and the Cr levels in the surrounding areas are depleted.

Low-carbon L-grades (C: 0.03%) and stabilized grades (AISI 347 and 321), which have an additional amount of carbide-stabilizing elements (niobium and titanium), can postpone or prevent sensitization. Standard ASS grades, like AISI 304 and 316, are particularly vulnerable to chloride SSC. There is a vulnerability in all ASS; nonetheless, specific high-Ni, high-Mo grades are adequate concerning stress corrosion attacks in most engineering fields [5]. High-alloy SS are used where regular or modified 300-series austenitic grades can't withstand corrosion. Cb-3 (N08020) illustrates a high-alloy SS created to increase resistance to sulfuric acid. This alloy has seen extensive use in a wide range of chemical and adjacent industries. The addition of niobium in this alloy reduces the likelihood of weld sensitization (intergranular corrosion), and the alloy's high-Ni concentration (32.5–35%) makes it resistant to chloride stress-corrosion cracking. To counteract the corrosive effects of sulfuric acid, a Mo concentration of (2–3%) is added, and a copper concentration of (3–4%) is often added for crevice and pitting corrosion. The high-Ni content grades and the 6% Mo super ASS are other notable examples of high-alloy stainless steels that are useful for heat-resistant applications [6].

#### 5.1.1.3 Martensitic Stainless Steel (MSS)

The austenite obtained at high temperatures from a low-chromium, and a high-carbon alloy, may be transformed into the BCC tetragonal martensite by rapid cooling. These high-strength, brittle martensite crystals may be tempered to achieve the same desirable high-strength/adequate-toughness profiles with plain carbon and low-alloy steels. Due to chromium's ferrite-stabilizing properties, there is a trade-off between corrosion resistance and total Cr concentration in MSS grades. The corrosion and toughness resistance of MSS has been greatly enhanced in recent years by adding Ni, N, and Mo at somewhat reduced C levels [5].

#### 5.1.1.4 Duplex Stainless Steel (DSS)

It's possible to think of these as Mo-Cr FSS to which austenite stabilizers have been added to the point where the ratio of ferrite to austenite is roughly equal at room temperature. High levels of Cr and Mo in these grades provide

FSS with exceptional resistance to corrosion and give ASS their desirable mechanical attributes. The strength of the DSS grades with roughly equal proportions of austenite and ferrite is higher than any phase present alone, and they have exceptional toughness. This phase equilibrium was initially attained in first-generation DSS grades like AISI type 329 by the inclusion of nickel. These initial DSS grades have better characteristics in the annealed state [7]. Still, their corrosion resistance is sometimes greatly diminished due to the dispersion of Mo and Cr between the two phases as they are re-formed after welding. Second-generation DSS grades benefit from N addition because it helps to speed up the process of restoring phase balance and reduces Mo and Cr dispersion without annealing. The newer DSS grades are a perfect example of this, as they are both highly durable and resistant to corrosion and SCC and economically viable to produce even in heavier configurations [5].

#### 5.1.1.5 Precipitation-Hardening SS

High-temperature aging may be used to strengthen these Cr-Ni alloys. The crystal structures of these grades can vary from austenitic and semi-austenitic to martensitic. High-temperature austenite conditioning transforms semi-austenitic structures from formable austenite to tougher martensite. Cold treatment is used to promote transformation in some grades. Copper and aluminum, when added together, generate intermetallic precipitates that strengthen the material as it ages. These grades can be easily created in the solution-annealed condition because their properties are very close to those of the austenitic grades. Since the strengthening can be lost by over-aging the precipitates, the precipitation-hardened grades must not be subjected to further exposure to elevated temperatures from welding or the environment. The corrosion resistance of these grades is generally on par with that of Cr-Ni grades [5].

### 5.1.2 Effects of Alloying Elements vs. Corrosion Behavior: Ferrite and Austenite Grades

Stainless steel's corrosion resistance varies greatly depending on the amount of Cr present and the absence or presence of roughly 10–15 additional alloying elements. Many different grades have shown effectiveness over the years in many environments, from the comparatively innocuous open air to the exceedingly hostile chemically prone product streams of the chemical processing industries. SS are vulnerable to a wide variety of localized corrosive effects, such as:

- Sensitization and intergranular corrosion.
- Stress-corrosion cracking.
- Crevice corrosion.
- Pitting corrosion.

#### 5.1.2.1 Pitting Corrosion

It's a targeted assault that can penetrate SS with nearly little overall mass loss. It's related to a breakdown in the passive film that's only visible first hand. Mechanical flaws, such as inclusions and surface degradation or localized chemical breakdown of the film, may also cause this corrosion. The most prevalent agent for initiating pitting is chloride. The chemical environment around a pit is much more hostile than the surrounding area after the pit has developed into a crevice. Cr and Mo are the primary determinants of the passive film's stability regarding resistance to pitting.

##### *5.1.2.1.1 Austenitic Stainless Steels*

Both the principal alloying components of ASS, Cr and Ni, improve resistance to pitting. In addition, Mo and nitrogen are other crucial alloying components that boost pitting resistance. The highest Mo-containing alloys, 254SMO, and AL-6X, showed the most extraordinary pitting temperature [5]. However, the advent of sigma and chi phase precipitation, which embrittle the alloys and decrease pitting resilience, restricts the amount of Mo that may be added to ASS. As a result, the ability of Mo additives to boost pitting resistance is limited. Adding nitrogen to ASS without Mo may increase the metal's resistance to pitting, but this benefit can be greatly boosted when Mo is also present. These advantages result from a chain reaction that begins with Cr enrichment underneath the passive film, continues with the production of a nitride layer at the film-metal interfaces, and culminates with the formation of a ferrous molybdate layer in the film's periphery. Alloys in the 200 series with Mn and super austenitic with 6% Mo take advantage of nitrogen and molybdenum's synergistic properties [8].

##### *5.1.2.1.2 Ferritic Stainless Steels*

Ferritic stainless steel's pitting resistance varies depending on whether it is talking about the common grades (430 and 434) or the proprietary super-ferritic varieties (29–4C, Sea-cure, and Monit). When exposed to chloride solutions, super-ferritic grades 29–4 show exceptional resistance to pitting, while 400 grades show only moderate resistance. The pitting capability in the exquisite direction is significantly increased by Cr and 4% Mo; hence the combination of these alloying elements should provide a good pitting-resistant structure in super ferritic. In high-purity, inclusion-free ferritic SS, increasing the Cr concentration from 18% to 28% causes the passive films to become thinner and the Cr enrichment inside the films to increase significantly. For the lower-Cr FSS, the effects of different alloying elements have been investigated; vanadium, tungsten, and silicon are advantageous for both the 13% and 24% Cr ferritic [6]. The synergistic interaction takes place when the Cr and Mo combine. The enhancement in pitting capability in the noble direction brought on by each addition of Mo is, therefore, noticeably more significant, even if Mo additions improve both the 13% Cr and 18% Cr ferritic at the higher Cr level. It has also been established that adding titanium is advantageous for a specific level of C plus N and that the value of titanium rises with increasing Mo concentration. The results also show that alloys with low C plus N levels have worthier pitting potentials than alloys with more significant C plus N values [5].

##### *5.1.2.1.3 Duplex Stainless Steels*

Typically, ferrite and austenite make up half of their composition. This way, DSS benefits from the same alloying components that enhance the pitting resistance of ASS and FSS. Pitting resistance comes from Cr, Mo, and Ni. In contrast, Ni balances the austenite and ferrite percentages to 50:50. The Cr and Mo percentages are limited (respectively) to 22–26% and 2–5% because these values prevent the embrittling sigma phase from forming [7]. Pitting resistance can also be improved by including elements like copper, tungsten, and silicon in an alloy. But at high N levels, it has no effect at all.

##### *5.1.2.1.4 Martensitic Stainless Steels and Precipitation-Hardening Stainless Steels*

When exposed to a chloride solution in water, MSS shows pitting potentials much less superior to type 304 stainless steel. They are more resistant to pitting in humid environments when their Cr concentration is higher, and Mo alloying improves their resistance. Compared to type 304 (18% Cr), the pitting potentials of higher Cr martensitic precipitation-hardening SS, such as Custom 450 (15%) and 17-4PH (16%Cr), are slightly less noble [5].

#### 5.1.2.2 Crevice Corrosion

It is probably the most damaging type of pitting. Oxygen can be blocked from reaching an area of interest when a crack, gap, or crevice is present. It can happen anywhere, including metal-to-metal joints, gaskets, fouling, or deposits. It is challenging to eliminate crevices in real life. When it comes to protecting against crevice corrosion, higher Mo and, in particular, higher Cr grades are the way to go. There is a critical temperature for each given environment; thus, there is a critical crevice temperature (CCT). Each grade has a unique corrosion environment, so the temperature required to prevent crevices must be determined individually [9].

##### *5.1.2.2.1 Austenitic Stainless Steels*

Nickel and chromium, the two most common alloying elements, do the same thing for crevice corrosion that they do for pitting resistance. Crevice corrosion resistance can also be improved by including Mo, silicon, and copper. When it comes to increasing resistance to corrosion in crevices, molybdenum is the alloying element that contributes the most [8].

##### *5.1.2.2.2 Ferritic Stainless Steels*

Adding more Cr and Mo to this steel makes it more resistant to crevice corrosion. Compared to type 430 (17.5% Cr), type 446 (with 26% Cr) has better resistance to crevice corrosion in saltwater because of the added benefit of a higher Cr concentration. Research shows that the temperature at which crevice attack is detected increases with rising Mo content from 0% to 5%, assuming a constant Cr level of 25% [6].

##### *5.1.2.2.3 Duplex Stainless Steels*

When the Mo and N levels in DSS are raised, the CCT also rises dramatically. The favorable effect of N has been seen, lending credence to the theory that nitrogen increases Cr retention in the austenite phase and increases its resistance to crevice corrosion. Crevice corrosion resistance in DSS can be enhanced by adding tungsten at a concentration of 0.5% to 1.0%.

#### 5.1.2.3 Sensitization and Intergranular Corrosion

Sensitization, or carbide precipitation at grain boundaries, can occur when ASS are heated over an extended period in the range of 425–870°C (approx.). Carbide precipitation is proportional to the time spent at temperature. Cr carbides precipitating at grain boundaries remove Cr from the surrounding material. Intergranular corrosion, the dissolving of the Low-Cr layer or envelope around each grain, occurs in SS when the precipitation is relatively continuous due to the depletion. Sensitization reduces resistance to the crevice, pitting corrosion, and SSC [9]. One other way to avoid sensitization is to use stabilized steels. These SS include both titanium and niobium. These elements rapidly form carbides with carbon, which keeps the Cr in solution even after prolonged exposure to sensitizing temperatures. Type 304 L is resistant to sensitization over short welding exposure but becomes sensitive after prolonged exposure.

#### 5.1.2.4 Stress-Corrosion Cracking

This corrosion occurs when an environment, a vulnerable alloy, and prolonged tensile stress merge to induce a crack in the metal. SS is highly vulnerable to SSC in chloride environments, and elevated temperatures and the presence of oxygen exacerbate this vulnerability. FSS and DSS have higher immunity and resistance to SCC [9]. There is a degree of susceptibility in all ASS grades, but mainly in types 304 and 316. The same holds for most MSS and precipitation-hardening steels. It isn't easy to spot stress corrosion as it's happening, but once it starts, it can cause catastrophic failures in pressurized machinery very quickly.

##### *5.1.2.4.1 Austenitic Stainless Steels*

The impact of alloying additives on the chloride SCC-resistant properties of SS has been studied in great depth. Though many alloying additives may be deleterious to chloride SCC, others can be classified as advantageous and changeable. The valuable metals are Ni, cadmium, zinc, silicon, beryllium, and copper. It must be stressed that nickel's positive effect only applies to ASS. Nickel is deleterious to chloride SCC susceptibility even when present in trace amounts in FSS. Chloride SCC resistance would require Ni levels of at least 50%. The chloride SCC isn't typically seen at or above 42% Ni concentration. Hence this finding runs counter to the vast majority of industrial experience. Chloride SCC resistance is a common requirement, and alloys 825 (42% Ni), G (44% Ni), and

20Cb3 (33% Ni) are frequently specified [6]. According to recent assessments using pre-cracked fracture-mechanics specimens, magnesium chloride is thought to be in a harsher environment than sodium chloride.

They imply that 30–40 % nickel is needed to prevent SCC in sodium chloride. Cadmium and zinc additions are also considered beneficial. Adding 0.2% Cd to type 304 prevented it from breaking in a vapor test that easily cracked type 304, while adding 1% Zn to a Fe-20Cr-15Ni alloy increased its SCC resistance by a factor of five. Due to this, we must treat as provisional the idea that these elements are helpful. Chloride SCC resistance is improved by silicon alloying [9]. Both beryllium and copper are valuable as alloying additives; copper has a more subtle effect. Both carbon and Cr have negligible impacts on chloride SCC resistance [5]. At the same time, boron, aluminum, and cobalt appear deleterious in low concentrations but advantageous in high concentrations; tin and manganese seem to have no effect in specific ranges and beneficial or counterproductive effects in other ranges. Mo's impact appears to be context-specific. According to the results obtained with magnesium chloride, the minimal resistance to cracking occurs at around 1.5% Mo and then increases as more Mo is added. Testing with sodium chloride reveals no such trend; cracking resistance increases with Mo concentration.

The early studies of alloying additives on ASS chloride SCC resistance used conventionally melted materials with significant impurities. In this context, ASS with a high degree of purity has excellent resistance to SCC in magnesium chloride solutions at high temperatures. For instance, high-purity alloys can be found in Fe-16Cr-20Ni and Fe-18Cr-14Ni. The latter was quoted with 1 ppm of metallic impurities and 10 ppm of non-metal impurities (nitrogen, oxygen, phosphorus, and sulfur). Researchers have found that by reducing the Ni content of Fe-18Cr-10Ni SS to 0.003 wt.%, the material becomes highly resistant to chloride SCC in a magnesium chloride solution at boiling 154°C. Some have hypothesized that the high corrosion potentials of these pure alloys put them beyond the range of potentials at which SCC may take place, making them resistant to corrosion [8].

#### *5.1.2.4.2 Ferritic Stainless Steels*

Chloride SS resistance in FSS can be reduced by adding copper, nickel, molybdenum, cobalt, ruthenium, or carbon to the alloy. In the alloy or the form of sulfur-containing gases in the atmosphere, sulfur may also contribute to SCC in chloride. Fe-18Cr-2Mo alloy, both annealed and welded, is vulnerable to SCC when exposed to magnesium chloride conditions due to high quantities of copper and Ni. These effects may become technologically relevant since Mo is an alloying addition, and copper, Ni, and cobalt may be present if the scrap is employed in the melting charge. The carbon content of 28.5Cr-4.0Mo SS was found to cause SCC when increased from 20 to 171ppm in a magnesium chloride solution at 155°C [6]. Sulfur is also hazardous, as evidenced by the cracking of type 430F (0.15 wt %) in marine environments.

#### *5.1.2.4.3 Duplex Stainless Steels*

Duplex stainless steel's chloride SCC has not been the subject of any reported systematic research on the effects of single-alloying-element additions. However, there are substantial differences in chloride SCC resistance between the various commercial grades. Chloride SCC resistance can be improved by increasing the total amount of Cr, Ni, Mo, and N.

### 5.1.2.5 Oxidation Resistance

Chromium is pivotal in stainless steel since it boosts the metal's resistance to oxidation. Selective oxidation of Cr at the metal surface, leading to Cr depletion at the metal-oxide interface, is required to create the protective $Cr_2O_3$ scale. Bulk Cr concentration must be high enough not to drop below predefined minimum levels in this interface region to preserve or stabilize the $Cr_2O_3$ scale. The bulk Cr content of commercial SS must be comparable to that of type 310 (for example, 25% Cr) to ensure good oxidation resistance at temperatures near 1000°C. The research demonstrates that Ni has a favorable effect, as it affects the adherence of the protective scale, slows down the rate of cation diffusion in $Cr_2O_3$ scales, and prevents the dissolution of the scale. Ni likewise stabilizes the austenitic structure of FCC steel. Austenitic material is superior to ferrite in strength and creep resistance because of its unique crystal structure. The addition of a tiny quantity of silicon (0.5%) is thought to positively affect the oxidation resistance of Fe-26% Cr ferritic alloys. There is solid evidence that silicon enhances the oxidation resistance of commercial ASS. When comparing type 302 (maximum Si content of 1%) to type 302B (maximum Si content of 3%), the difference in oxidation resistance is clear. In terms of oxidation resistance, however, the high-Cr type 310 outperforms it. When the Si content of ASS is increased to 3.5%, the material exhibits oxidation rates similar to type 310. When aluminum is added to SS, delta ferrite and the brittle sigma phase occur, which improves the steel's oxidation resistance but may decrease its cold formability [6]. Therefore, aluminum additions to enhance oxidation resistance should be considered carefully in the context of all other metallurgical criteria. Therefore, proprietary SS and higher alloys that incorporate aluminum to boost oxidation resistance typically have either low-Cr levels. In addition, FSS, like Fe-22Cr-5Al, can become embrittled when subjected to high temperatures for an extended period (1000°C) because the grain coarsens [5].

## 5.1.3 Concepts of Fiber Optics: Fundamentals of Laser Generation

Fiber lasers have come a long way from their inception in 1961, owing to innovations like Bragg gratings, diode-fiber coupling devices, low-loss doped fibers, high-power diode lasers, and many more. Putting a fiber amplifier within a cavity providing optical feedback results in a basic fiber laser. Fiber lasers reflect light through an optical system, which may excite atoms to absorb and emit light at specific wavelengths. The atomic nucleus's outermost orbit electrons may ascend to a higher energy level by absorbing more power. Pumping the diodes excites molecules with lesser energy to higher excited states, producing laser light. The term "spontaneous emission" refers to the process by which the laser is produced. In contrast, the term "stimulated emission" refers to the process by which the laser's output increases along the resonator cavity's axis. Light travels in mode within the fiber. Silica glasses are used to make pure fibers with a core, cladding, and protective covering. While the lasing mode travels through the core, the pump beam travels through the cladding [10]. Light impinging on the fiber core-cladding interface must be at angles larger than the critical angle and undergo total internal reflection. The outer layer coating confines the pump light for high-power core pumping. Modes are the optical fiber's core size (where light travels) [11]. A few examples of fibers and major characteristics for the propagation of light into the fiber are shown in Figures 5.1 and 5.2, respectively.

Based on the modes a fiber laser utilizes, it can be of two types, as mentioned in Figure 5.3. Single-mode lasers often offer superior beam quality and efficiency. The acceptance of light increases with the numerical aperture, which controls the effective angle of light. The core and cladding of fiberglass have different refractive indices. Additionally, the outer layer coating turns low-brightness pump light into a brighter laser beam. In most situations, that fiber gain medium is a rare earth ion-doped fiber such as thulium ($Tm^{3+}$), erbium ($Er^{3+}$), neodymium ($Nd^{3+}$), and ytterbium ($Yb^{3+}$). Among these elements, ytterbium ($Yb^{3+}$) is the most popular due to its high quantum effectiveness (95%) and minimal quantum flaws.

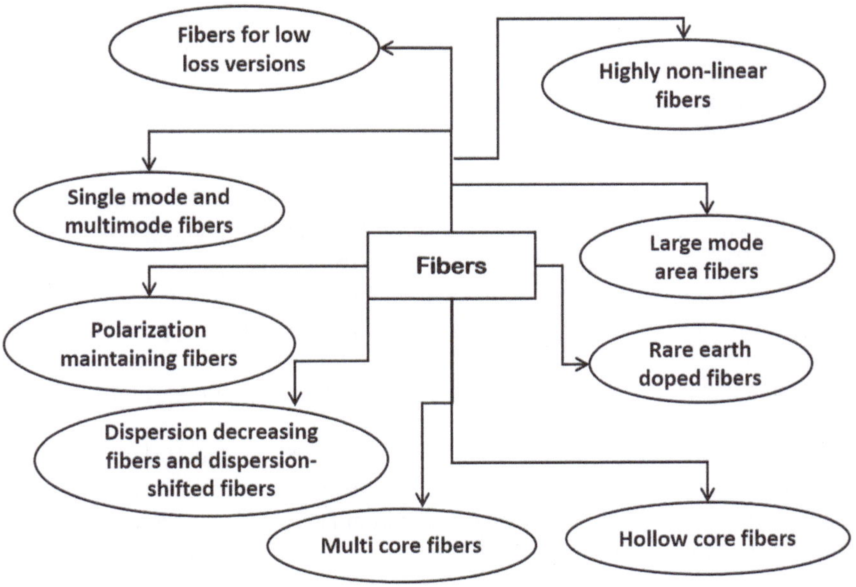


**FIGURE 5.1** Examples of fibers.

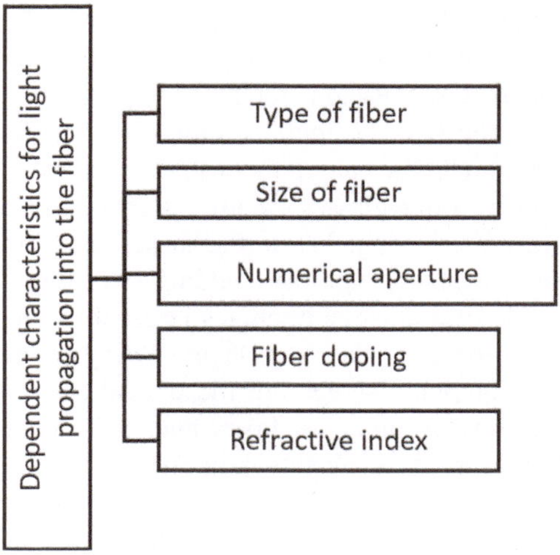


**FIGURE 5.2** Major characteristics for light propagation into the fiber.

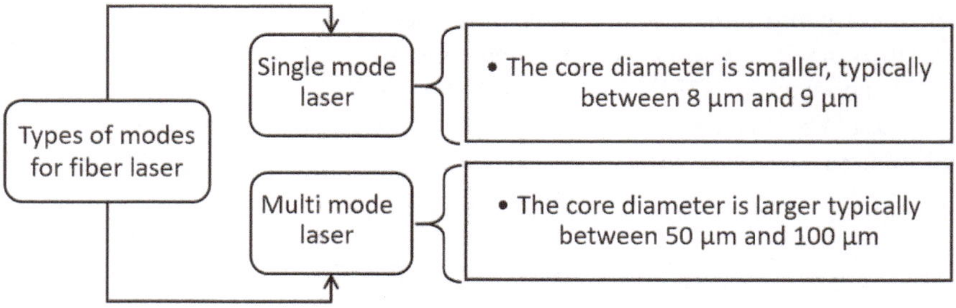


**FIGURE 5.3** Types of modes for fiber laser.

$Yb^{3+}$ doped optical fibers also influence photon absorption and refractive index. Phosphate glasses provide a better option than silica glasses due to their increased Yb3+ emission cross-section and laser efficiency. The fiber receives the energy through the pumping process and stores it for later use. A fiber laser is "pumped" by one or more fiber-coupled laser diodes operating in the 950–980 nm wavelength range. Therefore, diode-pumped fiber lasers are predominating. Through the action of pumping, just a small amount of energy is sent to the front mirror [11].

Depending on the purpose of the fiber laser and the needed output power, the number of laser diodes may vary. The fiber cavity resonator is constructed to improve pumping and emission synchronization [12]. The ytterbium-doped fiber is sliced to a high-reflector fiber Bragg grating at the pump end of the cavity, creating the fundamental structure of a fiber laser. Mirrors in fiber Bragg gratings may be tuned to reflect just a specific wavelength of light. However, a Bragg grating having moderate reflectivity is fitted at the output end. Consequently, the light that propagates along the fiber causes light to remain with the fiber. Doped fibers absorb light according to their length. Long fibers provide a high surface-to-volume ratio, which helps to remove excess amount of heat. Therefore, to construct a high-power fiber laser, it is best to use diodes with a big numerical aperture and a large output fiber diameter [13]. One of the methods used in fiber lasers is "Q switch operation." Introducing the "master oscillator with power amplifier" (MOPA) architecture allows for the independent regulation of the average power, pulse repetition rate, and pulse width. The amount of pulse energy delivered by Q-switched fiber laser systems is not reliant on the fiber's overall length. During the laser creation and propagation down the length of the fiber, it is possible to witness two distinct types of nonlinear effects: Stimulated Brillouin scattering and Raman scattering. The fiber's length may be longer, and the diameter of the core can be made larger, both of which may help prevent the occurrence of these phenomena [14]. In contrast to the quality of the single-mode beam, this may significantly decrease the quantity of laser light being emitted. The quality of the beam may be improved by slightly bending the fiber, which can also affect the amount of light that is accepted by the fiber. In addition to this, there is the possibility of using a residual pump stripper that has a high refractive index at the end of the fiber to compensate for the loss of heat energy.

### 5.1.4 Types of Fiber Laser

It is possible to categorize the types of fiber laser into a variety of subtypes based on the modes of the method of pumping, output power, and propagation, including the following:

#### 5.1.4.1 High-Power Fiber Laser

The early fiber lasers could only generate a few milliwatts, but quick improvements have led to high-power fiber lasers and amplifiers that can output hundreds or even several kilowatts from a single fiber. This likely results from a very high surface-to-volume ratio, which prevents the guiding effect, and excessive heating, which prevents thermo-optical difficulties even under circumstances of severe heating. Together, these two factors make it possible for this potential to exist. Many high-performance fiber lasers are really fiber laser systems, incorporating crucial components like a fiber amplifier or beam combining methods. This combination is a collection of methods for increasing the output power of laser sources by integrating the results of many individual devices. Laser light with a single-mode core may have diffraction-limited beam quality. Despite its lower output power, the fiber laser output may be orders of magnitude brighter than the pump light. Double-clad fiber lasers may transform brightness. A Yb-based fiber laser pumped with efficient high-power diode lasers can have a 50% electrical-to-optical power conversion efficiency. Because of high optical intensities and minimal pump absorption in double-clad fibers with a sizeable cladding ratio to the core area, the core area must be big for high powers. Laser-material processing uses high-power fiber lasers. Laser welding and cutting on metals and other industrial materials are examples of this kind of laser type [15].

#### 5.1.4.2 Upconversion Fiber Laser

Upconversion lasers often have to function on somewhat "tough" laser transitions, which need large pump intensities. The fiber laser idea is best appropriate for implementing upconversion lasers because of this requirement. Fiber lasers can sustain high pump intensities over extended lengths, simplifying operating low-gain transitions. These lasers often have a longer laser wavelength than the pump wavelength. It is a necessary consequence of all pumping techniques in which a single pump photon produces a single laser photon. Upconversion pumping techniques, however, employ multiple pump photons to excite a laser ion to a state where the laser photon energy is greater than that of the pump source with the shorter wavelength. In the case of infrared-pumped visible lasers, the labeled upconversion laser can be employed when the pumping operation requires a greater number of photons than the laser wavelength. Early upconversion lasers required bulk crystals to be cooled to extremely low temperatures, rendering them impractical. Most upconversion lasers employ glass fibers to sustain high pump intensities over extended distances to attain the laser threshold under challenging circumstances. Due to strong multi-phonon transitions and short metastable level lifetimes, silica glass is seldom acceptable. Thus, delicate and costly heavy-metal fluoride glasses like ZBLAN are commonly needed. Thulium-doped fibers generate blue light, whereas ytterbium-doped fibers generate red, orange, and green light [16].

#### 5.1.4.3 Narrow-Linewidth Fiber Laser

Single-frequency, single-mode fiber lasers have a tiny linewidth of a few kilohertz or even under 1 kHz. Even though a longer laser resonator may reduce phase noise and linewidth, one must maintain it short of ensuring long-term steady single-frequency operation without excessive temperature consistency. Distributed Bragg reflector lasers (DBR) fiber lasers with narrow-bandwidth fiber Bragg gratings pick a single resonator mode. Single-frequency fiber lasers have output powers up to 1 W, although most have a few milliwatts to tens of milliwatts. Distributed-feedback lasers (DFB) include the laser resonator in a fiber Bragg grating with a phase shift. In this case, the resonator is relatively short, which may impose a limit on the output power and the linewidth, but the functioning at a single frequency is relatively steady [17].

#### 5.1.4.4 Q-switched Fiber Laser

Q switching method modulates the laser resonator's optical cavity losses and Q factor to produce intense short pulses. This technique is used to generate high-peak-power and energy nanosecond pulses using solid-state bulk lasers. While combining this method in fiber lasers may generate pulses between 10 and 100 nanoseconds using passive or active Q switching. Since all-fiber configurations often cannot be implemented with large mode area fibers and efficient Q switches, the maximum amount of pulse energy that can be achieved with such a configuration is quite constrained. Because of the much higher laser gain, the specifics of Q switching a fiber laser are often qualitatively distinct from those of a bulk laser and more sophisticated. Often one gets a periodic sub-structure comprising several sharp spikes, and there is the potential to generate Q-switched pulses with a period much shorter than the resonator's round-trip duration [18,19].

#### 5.1.4.5 Mode-Locked Fiber Laser

More complicated resonators are used in mode-locked fiber lasers, which can produce picosecond or femtosecond pulses. The laser resonator can have a saturable absorber or an active modulator. An artificial saturable absorber might be produced via nonlinear polarization rotation or a fiber loop mirror. Components used in mode-locked fiber lasers come from the field of telecommunications; these parts have been rigorously tested for long-term dependability and are reasonably priced. Fiber coupling and compatibility with telecom networks are features of the laser output. Immense output powers may be achieved using double-clad fibers [20].

#### 5.1.4.6 Raman Fiber Laser

Fiber Raman lasers are a subcategory of fiber lasers that operate by using the Raman gain related to the fiber's nonlinearity. The Raman conversions may be accomplished in multiple phases using numerous stacked pairings of fiber Bragg gratings. These span the gap of hundreds of nanometers between the pump's and the output's wavelength. Raman fiber lasers, for instance, may be pumped in the range of one micron and create light at 1.4 microns, which is necessary for boosting $Yb^{3+}$ doped fiber amplifiers [21].

### 5.1.5 Fiber Lasers over Conventional Lasers: Benefits and Capabilities

A laser requires a pumping source, gain medium, and optical resonator. Laser diodes and fiber lasers have similar pumping sources, but their optical resonator and gain medium designs are distinct. This innovative arrangement provides seven key advantages.

#### 5.1.5.1 Highly Efficient Gain Medium

Fiber lasers use optical fibers doped with rare earth metal ions like erbium (Er3+), praseodymium (Pr3+), thulium (Tm3+), neodymium (Nd3+), or ytterbium (Yb3+). These active ions take in the vast majority of the pump light and are thus stimulated to generate photons of a specific frequency. Moreover, fiber's flexibility allows more extensive gain lengths than other laser kinds; therefore, the optical gain is high.

#### 5.1.5.2 Implementing Fiber Bragg Gratings for Smart Feedback Loop

Fiber lasers use fiber Bragg gratings—a sequence of glass fibers with varying refractive indices fusion-spliced periodically—for optical feedback instead of dielectric mirrors. The fiber laser's optical cavity is a periodic structure that reflects the laser beam at a specific wavelength. Thus, the optical reservoir is inside the gain medium for a fiber laser, directing photons to the work material inside the fibers. This occurrence helps to stabilize the setup and eliminates complications with beam delivery mirrors and laser beam coupling with beam delivery fiber.

#### 5.1.5.3 Robust Optical Cavity

Fiber lasers are not equivalent to lasers that have optical fibers, a popular misconception. Optical fibers are used exclusively for beam delivery in fiber-coupled diode lasers, which do not employ stimulated emission. Therefore, even when optical fibers are combined with laser systems, the resulting response may not possess all the advantages of using a fiber laser. The one-of-a-kind integrated optical cavity, which utilizes coiled fiber as the gain medium, results in an optical cavity that is both sturdy and reliable.

#### 5.1.5.4 Compact Design

One of fiber lasers' primary benefits is their design's space-saving nature. Fiber lasers have a lower footprint at equivalent output powers than conventional lasers. This is because optical fibers can be bent and coiled. Moreover, optical fibers' versatility allows for additional customization of the optical route, providing designers with more opportunities to build for particular conditions.

#### 5.1.5.5 Reduction of Operating Costs and High Wall Plug Efficiency

The ratio of a laser's electrical power efficiency to its optical power efficiency is termed device's wall plug efficiency. Due to the very efficient pump source and extraction from the gain medium, fiber lasers can attain 25–30% high wall plug efficiency. Fiber lasers may have optical fibers many kilometers long and can achieve significant pumping light gain since the gain medium is thin and flexible. Also, since optical fibers have a large ratio of surface area to volume, the heat generated by fiber lasers can be easily dispersed. Therefore, fiber lasers may operate efficiently at high levels without requiring complex chillers [11].

Consequently, this entails lower running costs and necessitates a smaller electrical supply and cooling needs, up to 300 W, for a small laser head design. In addition, manufacturing this laser requires fewer mechanical components than conventional solid-state lasers. It inevitably makes fiber lasers more competitively priced than other kW-class lasers, such as $CO_2$ lasers, which are employed in various flat-bed cutting applications.

#### 5.1.5.6 Superior Quality Beam

Typically, laser beam quality is considered to measure just how narrowly the beams can be concentrated, and an $M^2$ factor measures it; for optimal beam quality, it should be equal to 1. Moreover, the quality of a laser beam is also influenced by its wavelength, beam waist diameter, and far-field divergence. Due to the long and thin gain medium, high-quality laser beams are generated.

Commercially accessible fiber lasers provide high-quality beams of approximately a few kW. The critical properties of high-quality photons are their ability to be concentrated on a tiny region of the object, resulting in a higher density of power and a shorter process time. Fiber lasers have a lengthier focal length and higher depth of focus than other laser systems [13]. A fixed focal length and focus spot size may reduce beam diameter. Therefore, high beam intensity allows for a broader dimensional space, making the process easily optimized with the optical factors. Single-mode fibers provide the highest beam performance in fiber lasers, enabling multiple applications. This model shields lenses from debris and gases.

#### 5.1.5.7 High Reliability and Low Maintenance

Fiber lasers provide a high degree of dependability and require no essential maintenance. Additionally, since the optical route is encased in protective cladding layers, the laser beam is much less likely to be affected by disturbances from the outside environment. Because of this, fiber lasers often show remarkable stability even while functioning under high temperatures and vibrations. The pump sources in fiber lasers on the market today have a mean time before failure (MTBF) of 100,000 hours, equivalent to over 11 years of use [13]. It is impossible for any manufacturing industry to substitute laser diodes, owing to the system's cost and time efficiency. The minimal amount of regular maintenance required by the diode-pumped fiber laser system is another fascinating aspect that puts a feather in the cap of this system. This characteristic is principally attributable to the fact that the resonator is made entirely of fiber. As a result, neither beam delivery optics nor resonator alignment is necessary for its operation. In addition, the cooling system requirements for operating a fiber laser system are either very low or nearly nonexistent compared to those for operating a Nd-YAG laser system.

#### 5.1.5.8 Difference between Conventional and Fiber Lasers

Although fiber lasers provide a plethora of benefits, it is essential to have a firm grasp of how they compete with other available options. As we can see in the following section, the supremacy of fiber lasers depends on the application domain, and their competitors may be more appropriate in some conditions.

#### 5.1.5.9 Bulk Laser vs. Fiber Laser

The gain mediums like doped glass or bulk crystal are utilized in bulk lasers, such as the Ti sapphire laser and Nd-YAG laser. Fiber lasers have replaced Nd-YAG lasers in micro-machining applications such as micro-cutting stents and thin sheets of ferrous and non-metals in terms of cutting speed, microcrack length, and cutting edge quality. Fiber laser cutting of thick polycrystalline silicon (silicon wafer) also outperforms Nd-YAG laser. Fiber lasers may interact effectively with a broad spectrum of metals due to their essential wavelengths of 1–2 μm [15]. However, bulk lasers are still helpful for material processing. In the 700–1000 nm wavelength range, no fiber laser can replace the customizable Ti sapphire. Bulk lasers are better for material processing since they have greater peak power. Most notably, fiber laser systems may need unique components and complicated operation methodologies, and their more considerable expense may be economically disadvantageous [13].

#### 5.1.5.10 $CO_2$ Laser vs. Fiber Laser

Although both kinds of lasers are excellent at cutting various materials, the primary focus of their respective functions is different. On the one hand, $CO_2$ lasers are excellent in cutting instruments for non-metallic materials such as plastics and other similar substances. Because of their relatively superb beam quality and high efficiency, they are the kind of laser employed the most in the industrial sectors. Fiber lasers, on the other hand, have made considerable strides in cutting metal sheets (primarily SS) in recent years. It is mainly owing to their high cut speed, which is often two to three times quicker than $CO_2$ lasers when operating at equivalent power levels. For mass manufacturing, fiber lasers are better for thinner metals, whereas $CO_2$ lasers are faster and better for thicker metals. $CO_2$ lasers may be better at cutting thicker materials. However, it has been found that cutting thick materials by fiber laser can be more efficient than $CO_2$ laser [22,23]. Their multi-beam fiber laser machining equipment produced burr-free cuts in 1–2 mm AISI 304 SS at various cutting speeds. According to research studies, fiber lasers surpassed $CO_2$ lasers for cutting copper and magnesium alloys [11].

### 5.1.6 Components of the Fiber Laser System

In this part, we study a basic fiber laser system that operates in TEM00 mode and has a power of 50 W and a wavelength of 1064 nm. The components that make up a typical diode-pumped fiber laser micro-machining system are as follows: (a) a power supply unit to laser head; (c) a collimator to beam bender unit; (e) a beam delivery unit to focusing lens; (f) an assist gas supply unit; and (g) galvanometric scanners. The power supply unit and the laser head work together to produce the fiber laser that is contained inside the laser head. The beam delivery system of a fiber laser uses the remaining components of the subsystems that make up the device. The micro-machining setup using a fiber laser is shown in Figure 5.4, which features a diagrammatic representation of the process.

#### 5.1.6.1 Power Supply Unit for Laser Head

An isolation transformer having a backup power source of some kVa make up the power supply unit (UPS). These subsystems of the power supply unit are connected. The isolation transformer and the uninterruptible power supply are linked to the primary power supply unit. The transformer (isolation) is also connected to the UPS. The power supply unit is responsible for pumping the laser diodes that are included inside the laser head of the device. A fiber doped with $Yb^{3+}$ is a component of the diode-pumped fiber laser system, along with a fiber Bragg grating and several diodes. For pumping, the current system makes

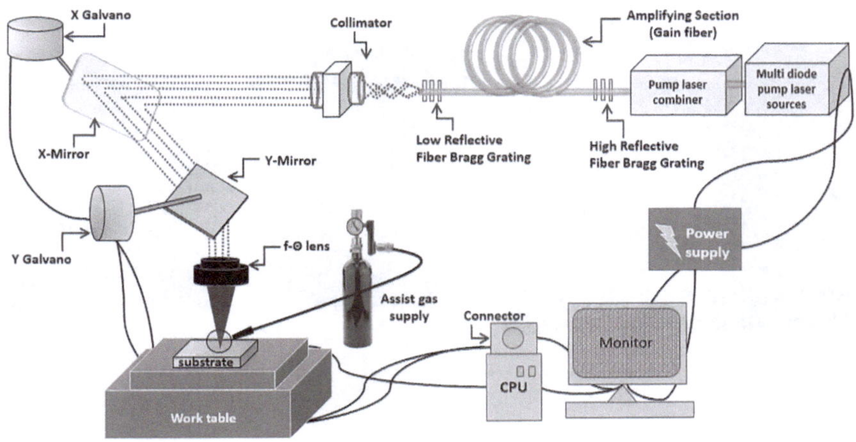


**FIGURE 5.4** Machining setup.

use of multiple diodes as a multi-diode pumped source. The overall output power of the system is used in the pump laser combiner to determine the optimal number of laser diodes to use. In order to generate the laser, fiber Bragg gratings must be cut into two halves and inserted into the ends of the fiber optics. Depending on the amount of output power provided by the fiber laser, the total length of the fiber may vary.

### 5.1.6.2 Collimator to Beam Bender Unit

After completing the laser creation process within the optical fiber, the laser will be transmitted into a collimator. At each of the collimator's two ends, two prisms have been inserted. This collimator functions as a beam expander, meaning the laser may travel the necessary distance to reach its target. Then, at an angle of 45° from the horizontal, a beam bender with a reflectivity of 100% comes after the collimator. It allows the laser to be directed in an orthogonal direction to the focusing lens.

### 5.1.6.3 Beam Delivery Unit to Focusing Lens

In the end, the laser passes through an f-θ lens of 160 mm in focal length. A nozzle shields this lens to protect it from dust and other types of contamination. The 34 μm is the diameter of the spot produced by the fiber laser beam after going through the focusing lens. This lens is set in the lens holder of the laser head (shown in Figure 5.5). It is necessary to ensure that the focusing lens's laser beam projected onto the surface is accurate and properly positioned. If the center of the lens does not align with the center of the beam, the portion of the beam that travels through the lens won't be uniform. It will cause the laser beam to have low energy, resulting in less efficient micro-machining. In addition, the beam has to be concentrated on the surface to get the best possible results from the micro-machining processes. The material starts to melt faster after the

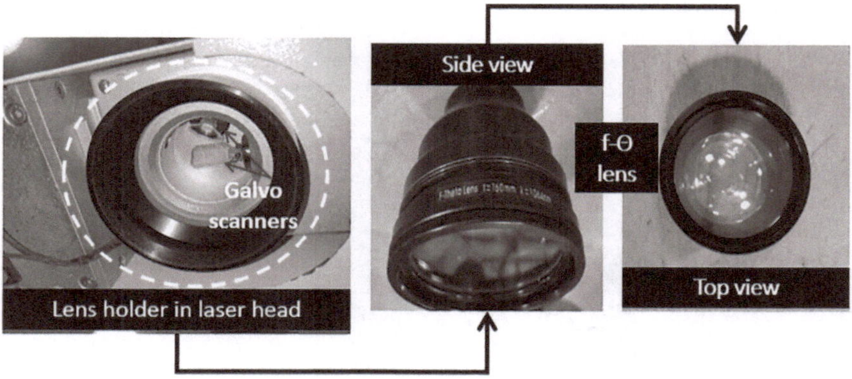


**FIGURE 5.5** Image of the laser head and f-θ lens.

laser beam has been correctly focused, which finally leads it to evaporate. Because of this, the focal point location of the laser beam has to be adjusted so that it is in the appropriate spot. The experimental setup includes a CNC-controlled arrangement, which aims to aid the exact adjustment of the focus point by controlling the lens's movement along the Z-axis. It was done so that the experiment could be carried out successfully.

#### 5.1.6.4 Assist Gas Supply Unit

Assisted gas streams remove molten material from the ablated surface to partly prevent re-solidification in the micro-machining zone. If there is to be any compressed air flow at all, then the supply line has to go through a moisture separator first, and then it should be linked to a pressure regulating valve afterward. Consequently, a stream flow of dry, pressurized air is produced and directed into the zone of laser micro-machining.

#### 5.1.6.5 Galvanometer Scanners

Galvanometer scanners hold mechanisms responsible for directing the laser beam's path. They are made up of two galvanometers with a pair of miniature mirrors that are very reflective and put on top of them. The galvanometers enable rapid and exact movement of the mirrors, enabling the laser beam to be directed to precise locations on the material being cut or marked. Controlling the motion of the axis may be accomplished with the help of servomotors. It is often linked to the computer system, which is the source of the axis movements that are carried out by the laser head. Suppose there is to be any movement along the X-, Y-, or Z-axis. In that case, the computer system will first send a command to the servo interfacing unit, which will then be followed by the acceptance of a command for the relevant servo motor to move in the specified directions. Depending on what the fiber laser system is used for, it can be controlled by either a motor system based on CNC or a motor system based on galvanometers. On the other hand, Galvanometer-type laser systems only need the laser head to move in the X, Y, and Z directions. Because of this, the scanning speed can be breakneck, sometimes up to some thousands per second.

## 5.2 INTRODUCTION OF FIBER LASER TECHNOLOGY IN THE FIELD OF SUBTRACTIVE MANUFACTURING

### 5.2.1 Laser and SS Interaction: Ablation Characteristics and Physical Phenomena

When the laser's high-density beam falls on the work material's surface, the interface zone is surrounded by various physical processes, including refraction, reflection, absorption, dispersion, and propagation. In laser-material processing, the absorption rate at the irradiated region of the laser-matter interface is dependent not only on the laser's wavelength but also on several other elements, such as incidence angle, surface roughness, and temperature at the interaction zone. The absorption phenomena' likelihood increases with the laser's intensity and shorter wavelength. These are the sole mechanisms responsible for the rapid motion of electrons and nucleon lattices at the laser-material interface during the multi-photon agitation process for a shorter interaction period. When this laser radiation strikes a material, its opto-electronic field imposes tremendous forces, inducing a vibrational movement in the electrons of the nuclei-lattice of the work material. As a result, the free and bound electrons' kinetic and excitation energies are amplified by the absorbed laser energy. Hence, the lattice of vibrating nuclei absorbs all the excess energy of the colliding electrons, and part of the incident lasers is transferred to the lattice, producing heat. The intensity gradually decreases as the laser beam penetrates further into the substrate. Sub-physical phenomena like heating, melting, and vaporization also surround the physical phenomenon that occurs in the interaction zone, which depends on the temperature difference and the characteristics of the work material. By means of conduction, convection, and radiation, heat is transferred to the material; at sufficiently high laser fluence, a phase transformation occurs [24]. Fiber laser micromachining relies on the interaction of the laser materials surface to remove minute amounts of material off the surface. It is well understood that the shorter the laser's operating wavelength, the greater the photon energy. The fiber laser processing of materials can be further classified into two different processes, i.e., photothermal and photochemical [25]. Metals, ceramics, and alloys are the most common materials that undergo photothermal changes. Fast heating, melting, and perhaps partial evaporation of the heated volume are the hallmarks of photothermal processes. It is possible to machine a wide range of engineering materials using a fiber laser ranging from nanosecond to femtosecond.

The machining quality varies with the duration of the laser's engagement with the material. Four distinct processes may be identified depending on the temporal regime of the laser-material interaction. Firstly, the absorption mechanism begins with the electron absorbing the photon, and then the series of subsequent physical events vary depending on the time scale. Secondly, phase transitions occur through the modalities of heat transfer by causing heating-melting evaporation. This evaporation causes the generation of recoil pressure (RP) in the melt pool [24]. Thirdly, the ejection mechanism takes place in which molten material is evacuated with the help of rebound force, i.e., RP. Then the evolved gas is converted into a plasma plume upon contact with the intense laser, which shields the machining zone by obstructing the path of the laser. When this plasma, which is made up of atoms, molecules, and metal globules, collides with the laser, its constituent particles are shattered and fall as spatter beyond the machining zone. Lastly, it goes for the long-range change in morphology in the machined zone. It induces thermal distortion in the lattice structure, due to which defects occur. The discussed phenomena can be understood by referring to Figure 5.6. However, from each of the four processes, absorption of the photon by an electron occurs first, and the subsequent series of physical actions vary according to the time scale. When a laser is used for nanosecond ablation, the material is heated to its melting point and vaporization temperature in a single process. During the contact period, energy is dissipated due to heat being conducted into the solid [26]. The harmful heat-affected zone (HAZ) caused by nanosecond pulse ablation is mitigated to some extent by picosecond ablation. However, picosecond ablation isn't without its drawbacks, such as significant evaporation and the creation of the melted zone inside the material. Femtosecond ablation solves this critical issue since there is no time for heat exchange to the lattice throughout the sequence of procedures involved. Thus, there is no HAZ, and the machining properties are fine-tuned [11].

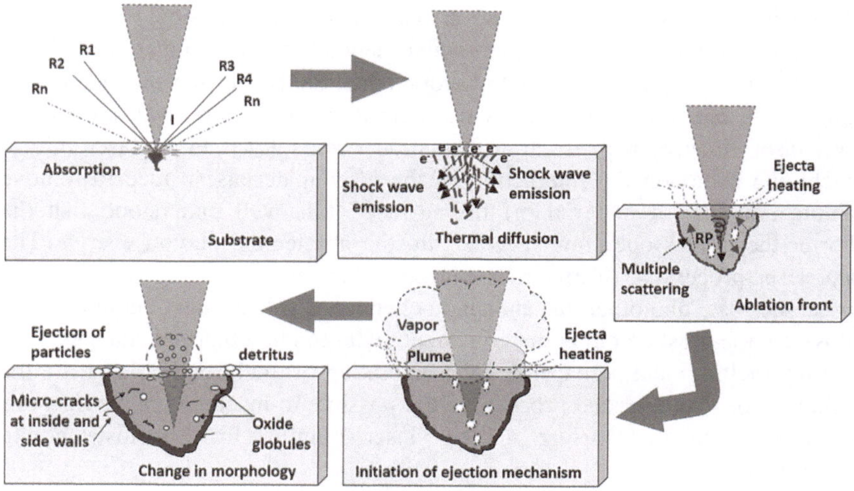


**FIGURE 5.6** Basic phenomena take place during laser-material interaction.

### 5.2.2 Research Proceedings in Fiber Laser Processing of Metallic Substrates

During the past decade, fiber lasers have become the most adaptable and quickly expanding laser technology. They operate in the near-infrared (NIR) spectra, offer several benefits over conventional lasers, and hold the potential to pave the way for brand-new micro-machining applications. While micro-machining with a fiber laser, the photothermal process is the primary determinant for working with metals and alloys. Pulsed fiber laser micro-machining (PFLMP) of a wide variety of engineering materials requires careful consideration of many process parameters, including pulse width, scanning speed, beam focal position, pulse frequency, assisting gas pressure, number of repetitions, laser power, substrate thickness, and composition. The process performance is to be considered when performing PFLMP for volumetric aspects, dimensional aspects (geometry width, depth, taper, diameter, roundness), and surface characteristics (surface morphology and roughness). Furthermore, the micro-machining zone is to be metallurgically evaluated to analyze defects like micro-cracks, heat-affected zone (HAZ), resolidified layer, striation, dross, and recast layer. The author in [27] studied fiber laser surface texturing of AISI D2 steel under an inert environment (argon). They investigated the correlation between laser parameters (energy density, velocity, strategy) and resultant surface micro-texture linewidth and characteristics. While texturing tool steel, it was discovered that using argon gas aids in achieving less HAZ and somewhat reduced surface roughness. Trepanning drilling of Nimonic alloy using compressed air as an aid gas by using a 20 kW Quasi-CW fiber laser. To drill a 0.75 mm hole in a thick plate by Quasi-CW fiber laser, the authors advise a speed of 500 mm/min with low peak, high power, and frequency, resulting in suitable quality holes [28]. The author in [29] found the optimal process parameters during fiber laser micro-drilling of AISI 303 material, resulting in a notable increase in performance in terms of low HAZ and surface roughness under ideal cutting conditions. On ultra-thin foils of aluminum AA5182, the author in [30], developed a micro-scale laser scoring method utilizing a nanosecond pulsed fiber laser. Furthermore, comparing 70 W and 120 W laser types reported here provides crucial insights into laser functioning. A 120 W average power single-mode laser is preferable to process materials quickly and accurately. Millisecond lasers may manufacture holes of acceptable quality in an Al-metal matrix composite with SiC as reinforcement, as discovered [31].

According to the authors' findings, drilling Al-MMC at a higher speed with less energy and peak power is required to achieve optimal hole characteristics compared to other metals and alloys. In addition, compared to the other gases employed in this research, the hole profile of argon-assisted machining can have a small recast layer thickness and stands out as the clear winner. To further understand the fundamental characteristics of melt-evacuation and micro-hole creation during millisecond pulse quasi-CW fiber laser drilling of Nimonic 263, the author in [31] once again did experimental research supported by finite element modeling. The authors in [32] undertake an experiment to compare the effectiveness of fiber laser drilling methods in drilling holes in Inconel 718 super alloy with regards to MRR, specific energy consumption (SEC), and hole taper. After comparing other options, single-pulse drilling was shown to be the most cost-effective and efficient method for MRR and SEC. Other researchers have adopted new parameters over controllable ones to enhance the process and productivity. So, while drilling a hole in Monel k-500 using a fiber laser, the author in [33,34], found that the sawing angle was more critical than scanning speed, pulse repetition rate, duty cycle, and average power in reducing the amount of variation in the hole. In [35], the author used two methods, i.e., preheating the workpiece with a waiting period and a step-like increasing cutting speed, to boost the cutting performance at the outset and compared the results with constant speed cutting. The author performs the cutting operation on a 60-mm thick SS plate using a 6-kW fiber laser. One cutting approach that was easy to implement, alignment-insensitive, and produced satisfactory results was "step-like cutting speed increase." The "step-like cutting speed increase" enhanced the overall cutting performance of thick SS plates compared to the constant speed cutting. An incremental cutting speed resulted in "a step-like cutting speed increase" that was 2.4% faster than that achieved with constant cutting speed. While processing A4SS with a pulsed fiber laser, the authors in [36,37], experimented by adding a unique linear orbit-in-orbit laser beam position method with the existing major parameters to enhance the productivity and dimensional aspects with good surface traits. By adjusting the process variables and by tuning the fiber laser's scan modes, this novel approach was studied in both (active and inactive gas) and (dry and wet) conditions to determine its potential for industrial use. Using a 30 W Q-switched fiber laser, the author milled a 2 mm 2024 aluminum alloy sheet to make a square pocket at varying scan speeds, hatching distances, pulse energies, and repetitions. The roughness (Ra), MRR, and depth were also assessed as an output response [38].

## 5.3 EXPERIMENTAL STUDIES ON PFLMP OF A4SS

In this chapter, a comprehensive discussion has been provided on the details of the setup of the fiber laser micro-machine, as well as the experimental strategy and outcomes.

### 5.3.1 Experimental Setup and Material

The optical fiber, the rare earth dopant ions, the mirrors, the pump sources, and the fiber coupler are the essential components of a fiber laser. The way in which a laser interacts with the material of an optical fiber is contingent on several aspects, including the laser source (primarily in terms of its wavelength and emission regime), as well as the properties of the material itself. The primary component of an optical communication system is an optical fiber, a cylinder-shaped dielectric waveguide composed of low-loss materials such as silica glass.

Doping the material with rare earth elements, primarily ytterbium ($Yb^{3+}$), enhances the material's optical characteristics. Compared to other rare earth elements, $Yb^{3+}$ possesses several benefits, including a longer upper-state lifetime, a minor quantum defect (which leads to a lower thermal load per unit of pump power), and the absence of excited state absorption. These benefits are since $Yb^{3+}$ has these characteristics. Semiconductor saturable absorbers, multilayer dielectric mirrors, and Bragg gratings are examples of the several kinds of mirrors that may be used in fiber lasers to generate a laser cavity. Pumping is accomplished primarily via using semiconductor diode lasers in most instances.

In the present research, all the experiments were performed through multi-diodes pumped solid-state (MDPSS) Yb: YAG fiber laser system by considering average power (Pavg) of 50 W (make: M/s Sahajanand Laser Technology, India) with $M^2 \approx 1.4$. A two-dimensional scanning galvanometer (X-galvo and Y-galvo) focused the beam onto the substrate via a "flat field lens" in this laser system (i.e., focal length of 160 mm) [37]. Figure 5.4 illustrates the present research's PFLMP setup. The laser head travels in software-programmed directions and directs a highly focused beam to the workpiece. The work uses A4SS with a square form (30 × 30) and 1 mm thickness. According to EDS report, the work substrate's major chemical elements are iron (61.84%), chromium (17.35%), and nickel (10.18%) [38].

### 5.3.2 Research Methodology

Due to the uncertainty of the correlation between both laser processing variables and the machining performance, only four process parameters—transverse speed (TS), pulse frequency (PF), power (P), and fill spacing (FS)—were considered for dry (D) machining. Furthermore, an additional factor, such as assist gas pressure (AP), was considered for active (A) and inactive (IA) machining conditions. These factors were considered to assess how fiber laser micro-machining process factors influence distinct responses. For the current PFLMP, the responses were categorized under three aspects: Dimensional, volumetric, and surface attribute. Micro-hole entry diameter (EnD) and taperness of the hole (TN) are dimensional aspects, material removal rate (MAR) is volumetric, and HAZ and width of resolidified material evacuated from the zone of machining (EMW) are surface characteristics. After the PFLMP, the specimens were provided a 15 min cleaning with the ultrasonic cleaner to remove any residual dust particles, followed by being sectioned through a wire electro-discharge machining process (EDM). For the present PFLMP, the MDPSS fiber laser operating variables, such as P (60–100%), PF (34–66 kHz), TS (800–1,000 mm/s), FS (30–70 μm), and AP (0.8–3.2 $kgf/cm^2$), were considered. These variables were evaluated in the context of the D, A (i.e., compressed), and IA (i.e., argon) environments. For the reliability of the study, a few test runs were conducted just before the final set of experiments, which was carried out as per the desired DOE plan. After the scrutinization of the trials' results, various parametric values and the associated operating windows were chosen.

A layer-by-layer removal strategy was used during the PFLMP process to fabricate a micro-hole feature. The procedure was carried out several times to sustain this mechanism at a specific focal position and a predetermined number of scans. In current PFLMP, the laser beam is moved over the surface of the substrate along particular scanning tracks in a multi-scan pattern to remove the materials from the target zone. Initially, a track was defined to conduct the laser scans over the surface of the substrate in accordance with the filling procedure. The detailed processing during machining and post-processing can be understood from Figure 5.7. The laser beam is used like a "tool" to remove the material layer by layer. First, a desired circular geometric pattern was defined in the CAD environment. The fill spacing strategy was implemented through the controlling window, which machined the inner regions restricted by the circle. The multiple laser beam moves along the line after line up under scan combined scan tracks of 135° and 45° to completely cover the area that is to be machined. The distance between the two consecutive laser line tracks is defined by fill spacing (FS) [24].

### 5.3.3 Result and Discussion

The current research work aims to scrutinize the technological aspects of PFLMP. The crucial operating variables like P, PF, TS, FS, and AP were considered. Further, it was necessary to discern the inevitable interaction among the process windows and their consequences against the imperative outcomes like D-MAR, A-MAR, IA-MAR, D-TN, IA-TN, A-TN, D-EnD, A-EnD, IA-EnD, D-EMW, A-EMW, IA-EMW, D-HAZ, A-HAZ, and IA-HAZ. This section holds supportive plots like the parametric effects plots against responses and the OM and SEM report to justify those interpretations.

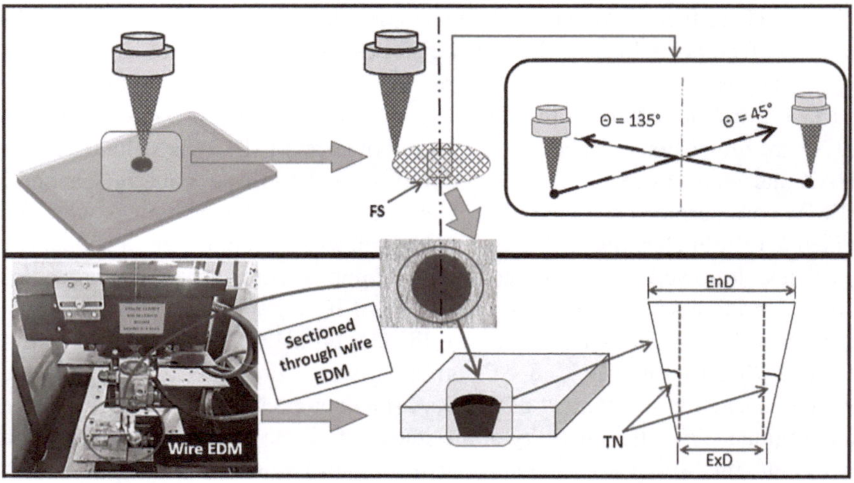


**FIGURE 5.7** Processing of PFLMPed holes to study the cross-sectional view.

### 5.3.3.1 Volumetric Response

During the process of nanosecond pulsed laser ablation, the intense laser beam strikes the targeted area of the material. This raises the concentrated temperature in the interface region to a level higher than the threshold energy of the material, which permits MAR. Nevertheless, in laser interaction mode, the volume of ablated materials and the amount of thermal damage are dependent solely on the irradiance (I) and fluence (LF) levels during laser processing, as seen in PFLMP. This is the case since both factors are proportional to the amount of energy delivered by the laser. P and PF levels were further able to be successfully employed as major operational parameters in order to explore their effects on MAR. These levels were found to be a function of I, pulse energy (PE), and LF. In addition, by considering that PE represents the single-pulse release of energy. It can be inferred that PE, LF, and I are directly proportional to the P levels and inversely linked to the PF factor [37]. P and PF were chosen as the input variables in this study instead of PE, LF, and I due to their superiority. As P increases, the laser beam becomes so intense that energy is transferred from the laser source to A4 SS (interface to the atomic levels) at the region of contact, a process known as interatomic heat transfer. P values for the defined target area (A) determine the threshold energy. Thus, as PE grows with P, increasing energy and heat flux (H). H levels enhance atom mobility, which increases collision and diffusion [39]. Since PE raises recoil pressure (RP), RP raises MAR, as shown in Figure 5.8(a). Thus, RP escalation and cutback are highly vulnerable to laser processing parameters (P, PF) that were successfully used as primary operational factors to examine their effects on MAR [40]. In all three environments with rising P levels, the laser's PE rises, which raises the localized temperature at the interface region (laser source-A4SS) beyond the material's threshold energy, causing interatomic heat transfer.

Additionally, there is an increase in the thermal gradient inside the molten pool, which improves the Marangoni convection. Higher melted liquid components are "pulled out" towards the borders of the molten pool, which supports a robust MAR in all the settings. It takes place due to the fact that an increase in the PE results in an increase in the RP, and an increase in the RP results in a high MAR [38]. However, the opposite impact was seen when PF was being considered; MAR decreased before 50kHz [evident in Figure 5.8(b)] but then abruptly improved when the value of PF was increased. The explanation for this is that as PF rises, PE decreases. As a consequence, the laser's capacity to generate heat is diminished, and the quantity of liquid or molten material that may be expelled is cut down [37]. In addition, a decrease in the PE results in a decrease in the RP. As the PF rises, the laser's pulse cycle will eventually reach its shortest possible length. On and off time are the two components that make up a laser pulse cycle. Because the pulse length is fixed in this endeavor, decreasing the pulse period will reduce the duration of the off-pulse [41]. As a result, a higher PF will decrease the cooling effect during the off-pulse time. Though the accretions of heat between laser pulses are more significant at higher PF [42]; So decreasing PE and RP mayn't dominate the process.

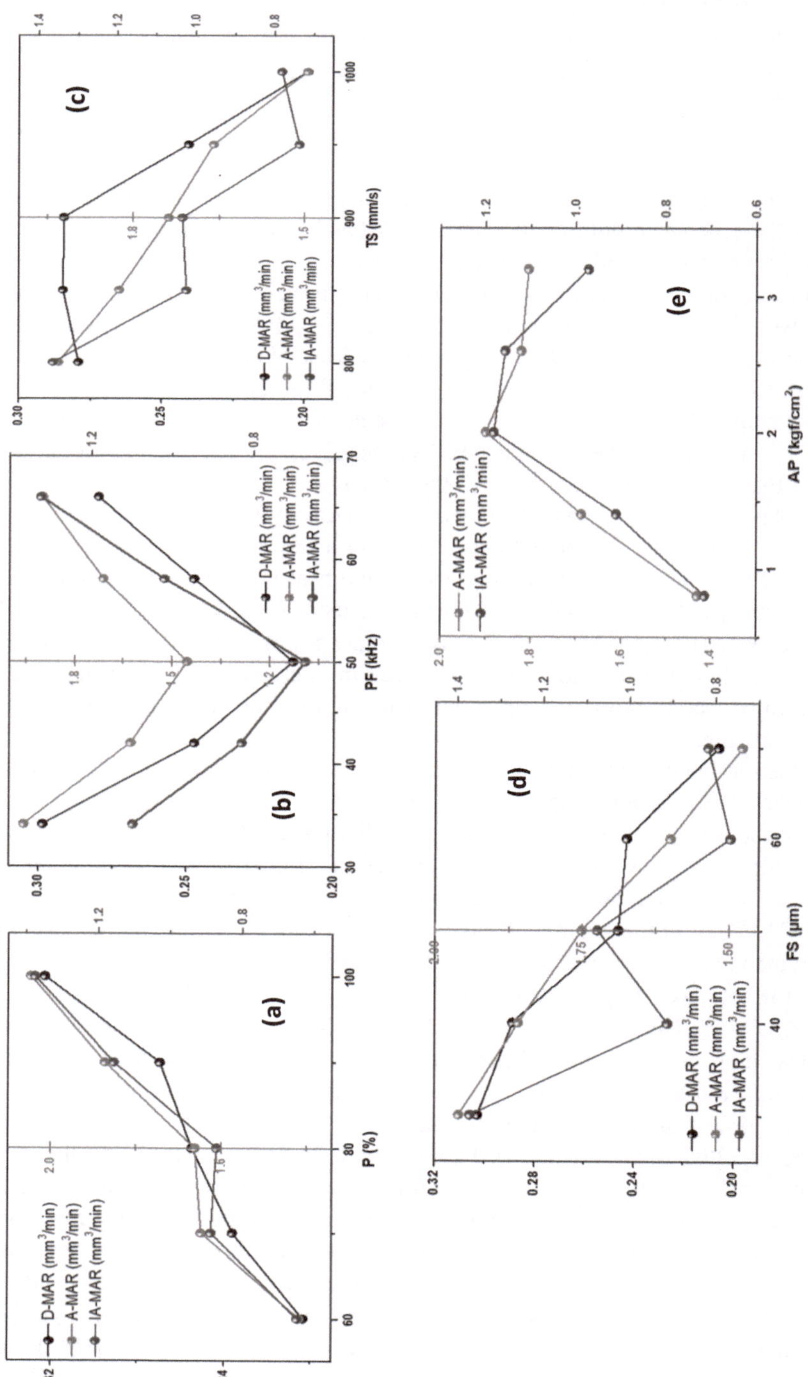


FIGURE 5.8 Parameters vs. MAR.

From Figure 5.8, it can be clear that the MAR values rise when the TS and FS values drop [Figure 5.8(c-d)]. Irradiation occurs irregularly and occasionally during contact with PFLMP. The irradiation rate depends on the movement of the laser footprint; when the footprint falls between two successive laser pulses, it does not ablate the same area of the material. However, the two pulses generate considerable irradiation in the overlapping area [43]. The laser's efficiency rises when the TS levels fall due to the increased laser-A4SS contact time. Moreover, the ablation zone's temperature gradients decrease as the overlapping zone decreases, thus lowering the MAR value [37]. As shown in Figure 5.8(e), the AP and the composition of the assist gases significantly impacted the PFLMP efficiency. In the case of gas environments, particularly in the A-environment, the metallurgical reaction, heat accumulation, and laser irradiation all combined to influence the PFLMP efficiency. The A4SS and the assist gas reaction occurred at high temperatures when the gas was switched from IA to A. This caused a rise in MAR. Compressed air is a very efficient and active gas that contains around 20% oxygen. Therefore, the heat generated by the exothermic reaction in the machining zone contributed to the evaporation of the PFLMPed material [44]. The gas thrust force generated by the AP moves in the direction of the machining area, where it interacts with the material that is being removed. The evaporation of machined material becomes better as the AP is more [36]. When the conditions are dry to IA, MAR rises, notably at A one.

#### 5.3.3.2 Dimensional Aspect

In this present sub-section, the influence of five process parameters on the fabricated hole geometry has been explained in detail.

According to the findings, P directly connects to PE, as discussed in section 5.3.3.1. PF is inversely linked, and its effects may be seen throughout the EnD [Figure 5.9(a-b)]. As PE significantly influences depth, the ablation depth is increased, and the distance that needs to be ablated also increases. This results in lower irradiation of absorption, which in turn leads to a smaller ablated surface in the exit area of the profile [26]. This is because as the PF increases, the ablation process on the bottom side of the hole slows down, which correlates with a reduction in the PE, as stated in [36,37], respectively. The uneven melting or vaporization of the work material, partial melt ejection, and thermal degradation around the perimeter are the causes of the growth of the EnD while at the same time reducing the exit area [45]. The value of TN, on the other hand, decreases with increasing and decreasing levels of P and PF [Figure 5.10(a-b)]; this happens because the bottom surface receives efficient thermal energy within the specified amount of time for interaction, and this energy results in regular dimension, which is maintained throughout the hole (from the entry to exit diameter of hole) [36].

PFLMP variables such as FS and TS maintain secondary control of the process based on the target substrate's responsiveness to the main process controller (i.e., P and PF), as they help to regulate the energies at the ablation zone. It is clear from [Figure 5.9(c-d)] that the response value EnD decreases as the amount of TS and FS that is applied increases. The overlapping of the laser footprints diminishes as the TS and FS levels grow, leading to a loss in effective

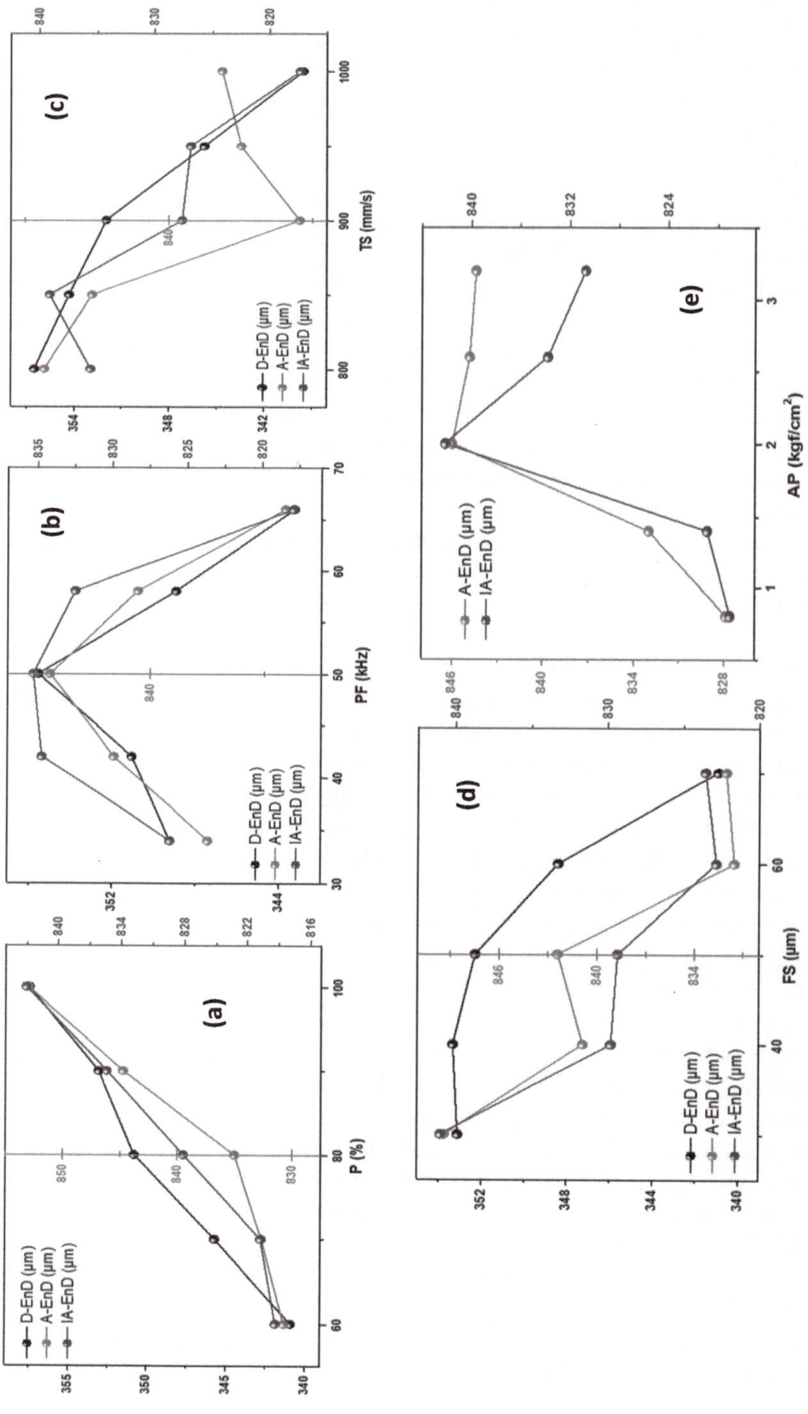


FIGURE 5.9 Parameters vs. EnD.

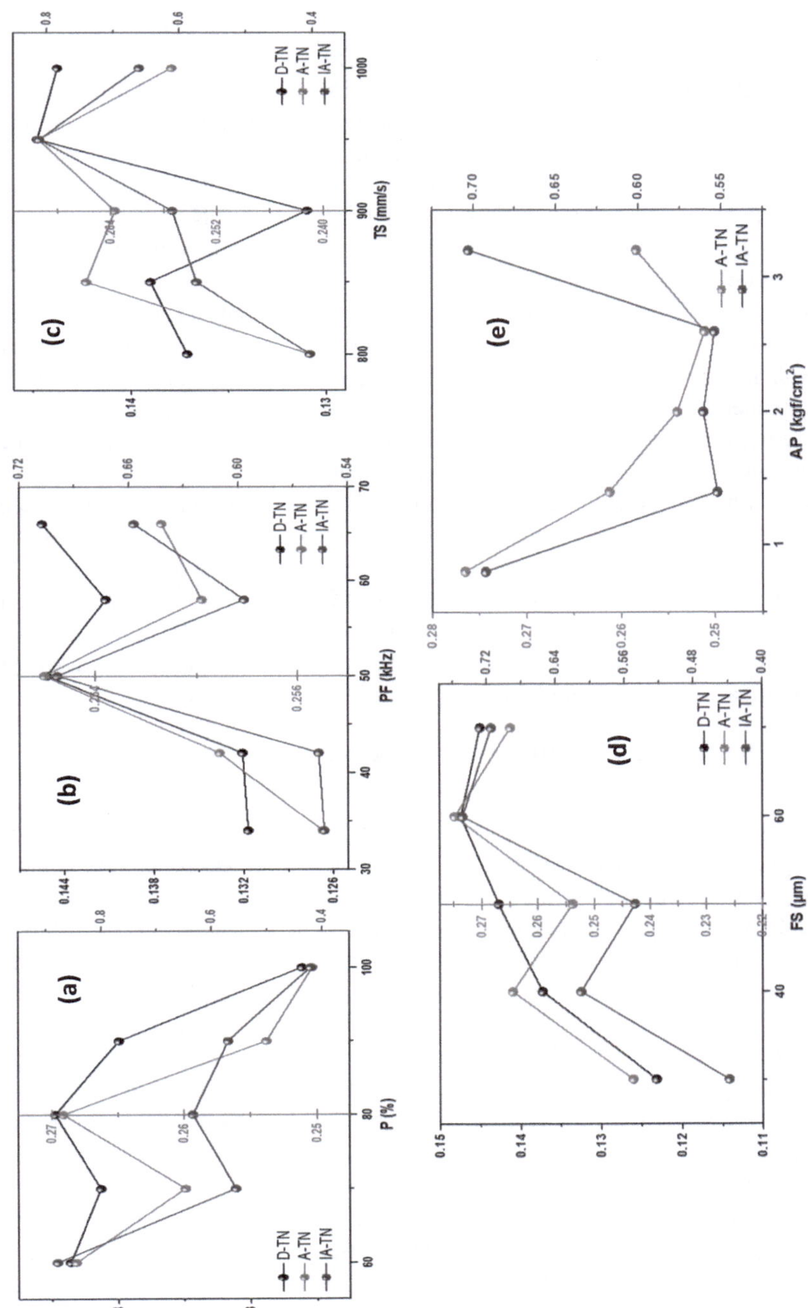


**FIGURE 5.10** Parameters vs. TN.

PE and insufficient thermal energy to be engaged during the prescribed interaction duration. It is because there is less space for the laser footprints to overlap. The laser footprint stays longer at lower TS due to substantial overlapping inside an effective target zone [36]. It results in sufficient melting and promotes a broader EnD. But in the context of TN, the situation is somewhat inverted. As the levels of TS and TD increase, the interaction period between the primary and secondary laser footprints decreases. Consequently, the energy concentration at work substrate debacles, resulting in nonuniformity in the hole entry and exit area, leading to TN formation [Figure 5.10(c-d)]. This TN also rises if the ablated material cannot evacuate quickly enough before solidification and is deposited on the bottom surface of the hole, clogging the existing diameter [24].

Further, the AP is vital in removing these clogged materials. From Figure 5.9(e) and 5.10(e), it can be concluded that AP of 2 kgf/cm$^2$ is recommended for lower TN with maintained EnD. It is likely because the AP supports the ejection of molten material. Still, the continual increase in AP causes the molten material on the hole bottom to be cooled and more sluggish. Therefore, TN will increase if the ablated material cannot leave the hole in a timely manner prior to the onset of solidification. The AP may be the factor that is responsible for causing the molten material to be evacuated; nevertheless, the continual increase in AP causes the molten material on the hole bottom to become cooled and more resistant to movement. As a result of the sluggishness, it collects on the side of the hole that is closest to the exit, which results in a high TN (i.e., observed after AP of 2.6 kgf/cm$^2$). TN's value is higher in the IA environment because the IA environment has a higher cooling impact than the A environment [36].

### 5.3.3.3 Surface Traits with the Photographic Exhibition

#### *5.3.3.3.1 Evacuated Molten Material Width (EMW)*

As discussed in section 5.3.1.1, PFLMP is followed by melting, disintegration, and vaporization. It makes a molten pool in the machined zone, where thermomechanical driving flow happens. Further, it leads to outbursts of liquefied materials at the edges of the holes. At last, the melted material cools down, solidifies, and forms a layer of evacuated molten material (EMM) around the edges. The material evacuation mechanism is well supported by the hypotheses made by [37]. As a result of this endeavor, it was found that the work substrate begins to disintegrate into solutes of carbon, oxygen, silicon, and phosphorus, as well as liquefied elements such as iron, chromium, nickel, manganese, and molybdenum when the temperature of the machining process rises above the melting temperature [37]. At lower levels of TS, FS, and higher levels of P, thermal energy works so well that the material melts quickly. The intense threshold energy and large overlapping pulses in the extended contact zone cause a solid-liquid interface (i.e., melt pool) [38]. Because of this, there will be a greater temperature differential between the top and bottom layers. As a result, there is an increase in EMM due to the escalation in the melt pool's vapor speed and flow velocity [43]. As seen in Figure 5.11, EMW decreases before increasing

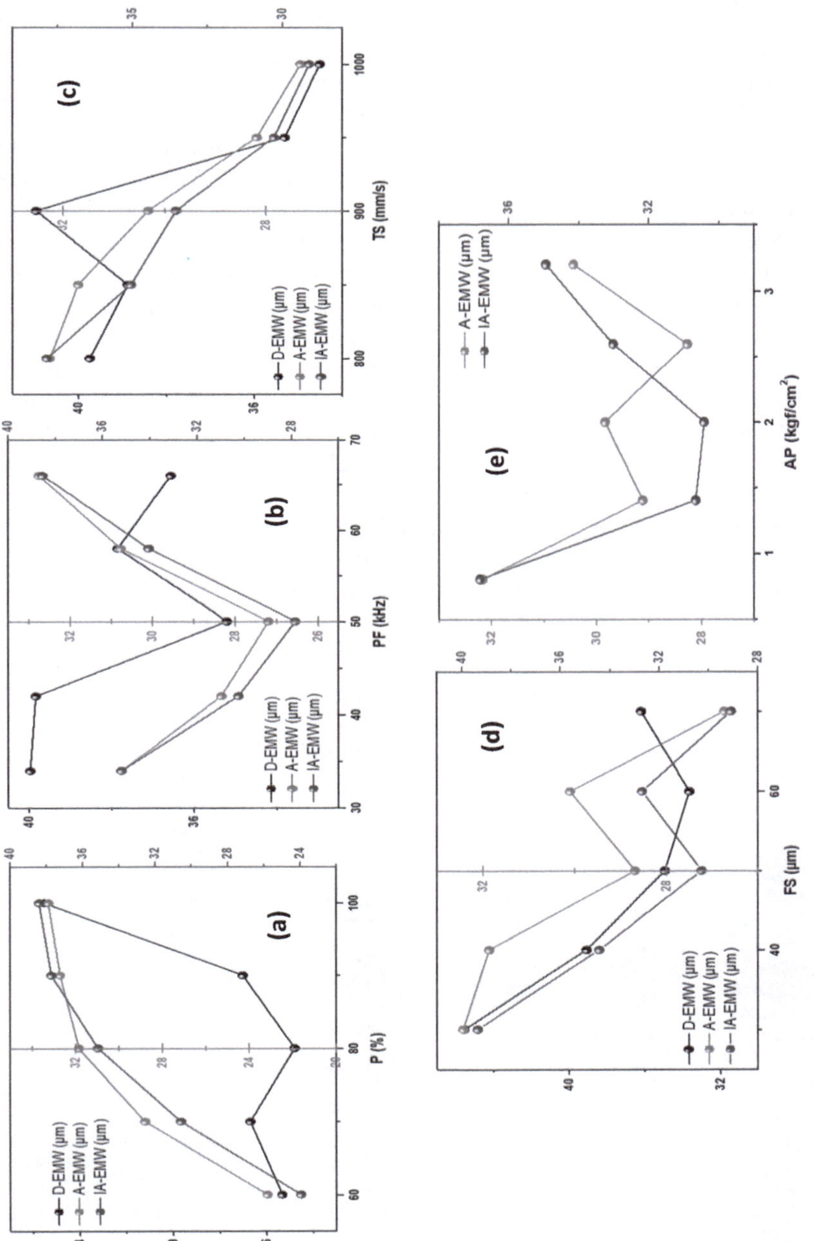


**FIGURE 5.11** Parameters vs. EMW.

in the moderate PF range, which corresponds to 50 kHz. As PF rises, the number of pulses that occur per unit area also rises, leading to an accumulation of PE on the surface [46]. Further, the additional materials are heated, taken from the bottom surface, and deposited around the hole periphery. The disparity in temperature between the molten pool and the surrounding area also helps to boost Marangoni convection [43]. The liquefied materials were further dragged out, which resulted in a good MAR and a more expanding EMW [37].

In the IA environment, the EMW decreases with the increase in AP up to 2 kgf/cm$^2$. From Figure 5.11(e), we can see a sudden rise in EMW after the value rises beyond 2 kgf/cm$^2$. However, this seems to be higher in the A environment after 2.6 kgf/cm$^2$. It occurs due to the high pressure, which causes the molten material to flow out of the hole bed. However, due to the increased cooling effect of the assist gas, the material eventually congeals and settles on the hole's side wall, forming wider EMW, which occurs more frequently in environments with a higher AP of I [36]. For a better understanding, microscopic images of EMW have been shown in Figure 5.12 under different media. It has been observed that the EMW increases slightly in the sequence of A, IA, and D environments. When observing the hole's inner surface, it becomes clear that IA provides comparatively cleaner and smoother surface, although with some micro-cracks, whereas D-PFLMPed shows a few remelted areas and some micro-pores. However, while it is possible to create deeper holes with A-PFLMP, certain oxide flakes, and oxide blankets are the main problems that need to be handled further.

### 5.3.3.3.2 Heat-Affected Zone (HAZ)

The values of the HAZ are influenced by the parameters as well as the thermal diffusivity of the material. The development of HAZ as a result of the accumulation of heat is a natural phenomenon. With reference to Figure 5.13, it has

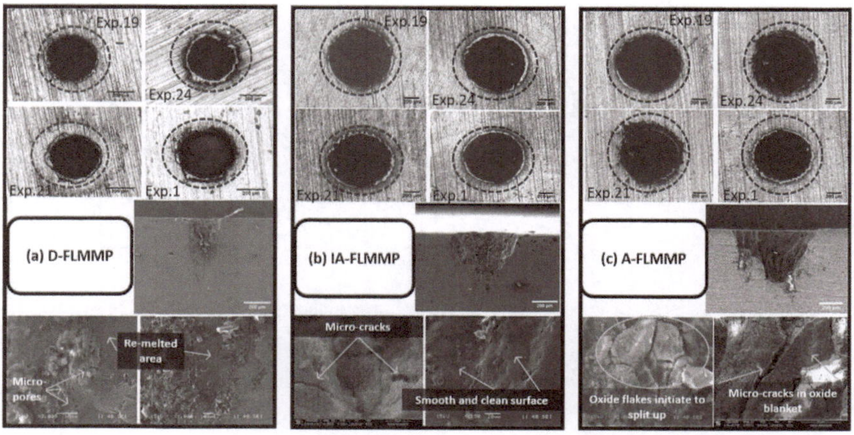


**FIGURE 5.12** Microscopic images of top and cross-sectional view of holes in (a) D, (b) IA, (c) A environment.

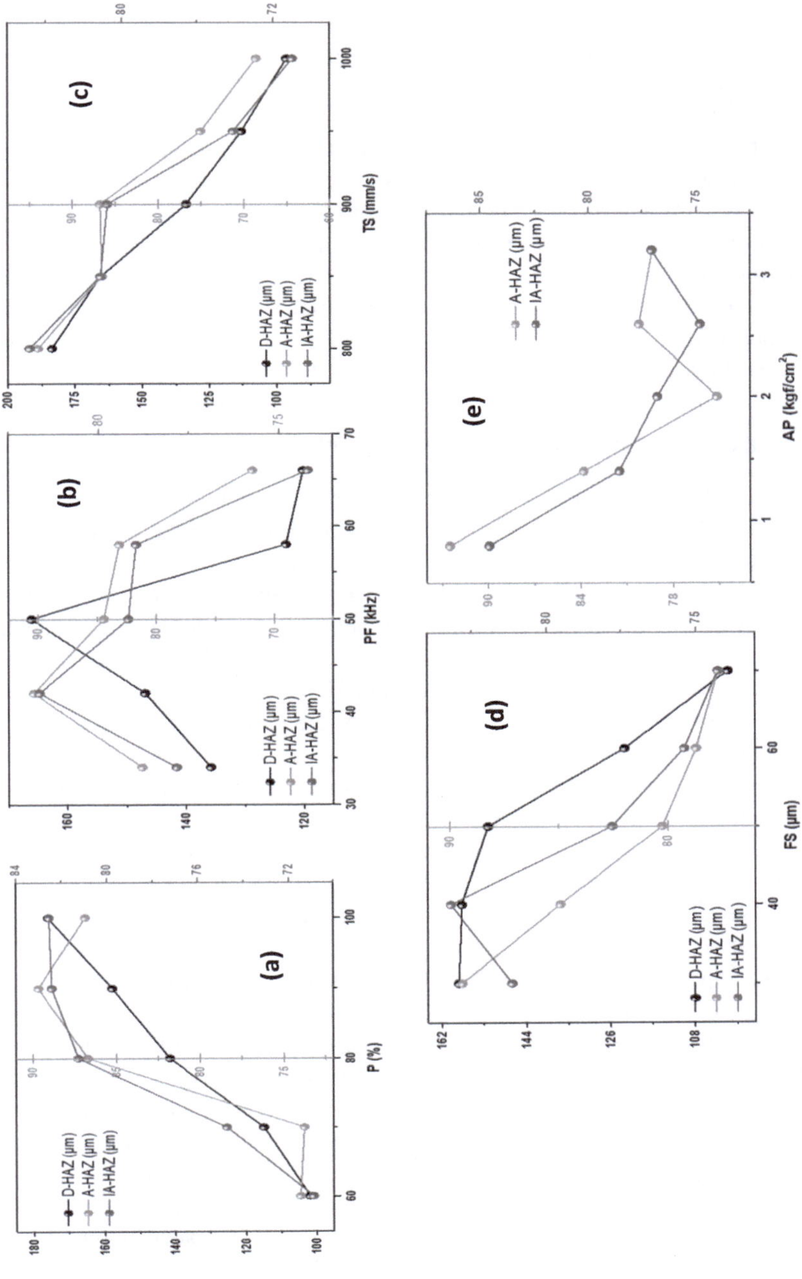


**FIGURE 5.13** Parameters vs. HAZ.

been seen that the HAZ value decreases with an increase in the amounts of PF, TS, and FS. A higher TS level results in shorter contact intervals, while a higher FS level; which results in less overlapping activity [47]. In addition, as the PF levels increase, the A4SS surface experiences shorter pulse overlap durations. This causes the energy to be at its lowest possible level in the spot area, promoting a lower HAZ [24]. On the other hand, the value of HAZ was directly related to operational factors like P levels. The significant increase in P levels significantly induces the temperature field and its genesis, resulting in an easy transfer of heat from the material's target spot to the peripheral surfaces within a short time. It is the cause of the sharp increase in HAZ values that can be seen in contrast to the rise in P values. The significant temperature differences contribute to speeding up the HAZ creation with the appropriate hole [37]. Since the ultimate goal of the AP is to provide a cooling effect to the machining zone, it would appear from Figure 5.13 that the value of HAZ decreases as the AP value rises when the assist gas is switched from A to IA [36].

## 5.4 CONCLUSION AND FUTURE SCOPE

Compared to the present PFLMP, the current research may give effective machining capabilities in D-PFLMP, IA-PFLMP, and A-PFLMP environments. In addition, this research uses an MDPSS fiber laser to create holes in maskless A4SS plates. Surfaces that have been treated with a laser have been subjected to optical and SEM/FESEM analysis in their adopted environments. Parameter interactions and their effects on crucial responses like MAR, EnD, and TN, as well as surface characteristics like EMW and HAZ, have been the subject of research. These possible inferences are derived from the technical interpretations: Concerning environmental effects, it has been discovered that MAR increases in the sequence of A, D, and IA environments. However, the features' surface quality may degrade in the case of D and A. The MAR mainly depends upon the PE, which can be well described as a function of the P and PF. The MAR upsurged with an increased P value and decreased TS and FS values. But the parameter PF is slightly critical to analyze in the case of MAR. There is a sudden hike in MAR when PF exceeds 50 KHz. In the case of EMW, the parametric impacts were found to be relatively similar to MAR except for PF. However, from the surface traits point of view, HAZ seems to be a significant concern in D and A environments compared to IA. However, EMW seems to be meager in the case of A. Moreover, from the dimensional perspective, taperness is a major concern in the case of the IA environment compared to D and A.

The advent of fiber lasers was a game-changer for the laser industry. They have several benefits over competing laser technologies and pave the way for enhanced production efficiency. The combination of an integrated laser cavity and a quite effective gain medium ensures reliable and consistent output capability. Due to their small size, they are beneficial for integration purposes. In spite of this, their supremacy relies on the application context since their output wavelength and tunability are relatively fixed. PFLMP is an innovative technique that allows for more specific hole production in an A4SS metallic substrate. It is impressive to see how far fiber laser technology has come in such a short time from both a scientific and an industrial standpoint. The current study not only simplifies the difficult phenomena of processing A4SS metallic alloys but also makes possible the use of fiber lasers in the machining arena. Parameter settings that maximize better machining quality, more productivity, and the lowest possible cost-effectiveness may be determined with further study.

With this, the authors of this chapter believe that the current chapter will be helpful for future readers to grasp the current research proceedings in the field of PFLMP. Some specific parameters are still available that can improve the current process of subtractive manufacturing. Laser is a widely used tool in many industries for processing the required amount of materials for job shops, batch shops, and mass production to fulfill respective applications. Therefore, more research should be attempted to acquire knowledge of new processing factors and their implementations in the fabrication of desired features within a short period. So, the recent development in fiber lasers and the research discussed in this chapter may open up a wide range of possibilities for studying PFLMP in the context of numerous industrial and biomedical applications.

## ACKNOWLEDGMENT

The authors like to express their gratitude to M/S Sahajanand Laser Technology Ltd., Gujarat, India, for providing the fiber laser machine support to carry out the required experimentations for this study.

## REFERENCES

[1] Cardoso, P., Davim, J.P. 2012. A brief review on micromachining of materials, *Reviews on Advanced Materials Science*, 30:98–102.

[2] Matsunawa, A. 1989. Laser processing of materials in Japan, *Materials and Processing Report*, 4, 10.1080/08871949.1989.11752314.

[3] Davim, J.P., 2011. *Lasers in manufacturing*, ISBN: 978-1-848-21369-2.

[4] Umoru, L.E., Afonja, A.A., Ademodi, B. 2008. Corrosion study of AISI 304, AISI 321 and AISI 430 stainless steels in a Tar Sand Digester, *Journal of Minerals and Materials Characterization and Engineering*, 07:291–299.

[5] Davis, J.R. 2001. *Alloying Understanding Basics, ASM International*, ISBN: 978-0-87170-744-4www.asminternational.org.

[6] Schweitzer, P.A. 2017. *Stainless steels, corrosion and corrosion protection handbook, Second Edition*. 10.1201/9781315140384.

[7] Follansbee, P.S. 2015. Structure evolution in austenitic stainless steels—A state variable model assessment, *Materials Sciences and Applications*, 06:457–463.

[8] Gideon, B., Ward, L., Biddle, G. 2008. Duplex stainless steel welds and their susceptibility to intergranular corrosion', *Journal of Minerals and Materials Characterization and Engineering*, 07:247–263.

[9] Takakuwa, O., Soyama, H. 2015. Effect of residual stress on the corrosion behavior of austenitic stainless steel', *Advances in Chemical Engineering and Science*, 05:62–71.

[10] Correa, E.O. Barbosa, R.P., Burschinelli, A.L. *et al.* 2010. Influence of clad metal chemistry on stress corrosion cracking behaviour of stainless steels claddings in chloride solution, *Engineering*, 02:391–396.
[11] Sen, A., Doloi, B., Bhatacharyya, B. 2015. Fiber Laser Micro-machining of Ti-6Al-4V, 10.1007/978-81-322-2352-8.
[12] Wandera, C. 2016. Fiber Lasers in Material Processing, *Fiber Laser*, 10.5772/62014.
[13] Sen, A., Doloi, B., Bhattacharyya, B. 2017. Fiber Laser Micro-machining of Engineering Materials, 227–252.
[14] Gursel, A.T. 2018. Fiber Lasers and Their Medical Applications, *Optical Amplifiers-A Few Different Dimensions*, 10.5772/intechopen.76610.
[15] Schaeffer, R.D. 1959. *Fundamentals of laser micromachining*, Engineering & Technology, 10.1201/b11881
[16] Galvanauskas, A. 2001. Mode-scalable fiber-based chirped pulse amplification systems, *IEEE Journal on Selected Topics in Quantum Electronics*, 7:504–517.
[17] Silversmith, A.J., Lenth, W., Macfarlane, R.M. 1987. Green infrared-pumped erbium upconversion laser, *Applied Physics Letters*, 51:1977–1979.
[18] Fleming, M.W., Mooradian, A. 1981. Spectral characteristics of external-cavity controlled semiconductor lasers, *IEEE Journal of Quantum Electronics*, 17:44–59.
[19] Tang, Y., Xu, J. 2015. A random Q-switched fiber laser, *Scientific Reports*, 5:9338.
[20] Krausz, F. Fermann, M.E., Brabec, T. *et al.* 1992. Femtosecond solid-state lasers, *IEEE Journal of Quantum Electronics*, 28:2097–2122.
[21] Stolen, R.H., Ippen, E.P., Tynes, A.R. 1972. Raman oscillation in glass optical waveguide, *Applied Physics Letters*, 20:62–64.
[22] Olsen, F.O., Hansen, K.S., Nielsen, J.S. 2009. Multibeam fiber laser cutting, *Journal of Laser Applications*, 21:133–138.
[23] Jia, X., Li, Z., Wang, C. *et al.* 2023. Study of the dynamics of material removal processes in combined pulse laser drilling of alumina ceramic, *Optics and Laser Technology*, 160:109053.
[24] Rout, S., Panigrahi, D., Patel, S.K. *et al.* 2021. Microchanneling on bio-inert dental ceramic using dry pulsed laser ablation and liquid supported pulsed laser ablation approach, *Optics and Lasers in Engineering*, 144:106654.
[25] Geough, J.M. 2002. *Micromachining of engineering materials*, ISBN: 9780429153259.
[26] Rout, S., Patel, S.K. 2022. Fabrication of 2.5D feature by fiber laser step down milling approach on Z-A composite intended for bone scaffolding, *Optics and Laser Technology*, 154:108322.
[27] Yang, L., Deng, Z., He, B. *et al.* 2021. An experimental investigation on laser surface texturing of AISI D2 tool steel using nanosecond fiber laser, *Lasers in Manufacturing and Materials Processing*, 8:140–156.
[28] Marimuthu, S., Antar, M., Dunleavey, J. *et al.* 2017. An experimental study on quasi-CW fibre laser drilling of nickel superalloy', *Optics and Laser Technology*, 94:119–127.
[29] Reddy, V.C., Keerthi, T., Nsihkala, T. *et al.* 2021. Analysis and optimization of laser drilling process during machining of AISI 303 material using grey relational analysis approach, *SN Applied Sciences*, 3:335.
[30] Banat, D., Ganguly, S., Meco, S. *et al.* 2020. Application of high power pulsed nanosecond fibre lasers in processing ultra-thin aluminium foils, *Optics and Laser in Engineering*, 129:106075.

[31] Marimuthu, S., Dunleavey, J., Liu, Y. *et al.* 2019. Characteristics of hole formation during laser drilling of SiC reinforced aluminium metal matrix composites, *Journal of Materials Processing Technology*, 271:554–567.
[32] Sarfraz, S., Shehab, E., Salonities, K. *et al.* 2019. Experimental investigation of productivity, specific energy consumption, and hole quality in single-pulse, percussion, and trepanning drilling of in 718 superalloy, *Energies*, 12:24.
[33] Pramanik, D., Kaur, A.S., Sarkar, S. *et al.* 2020. Evaluation of sawing approach of hole quality characteristics in low power fiber laser trepan drilling of monel k-500 superalloy sheet, *Optik*, 224:165642.
[34] Pramanik, D., Roy, N., Kaur, A.S. *et al.* 2022. Experimental investigation of sawing approach of low power fiber laser cutting of titanium alloy using particle swarm optimization technique, *Optics and Laser Technology*, 147:07613.
[35] Seon, S., Shin, J.S., Oh, S.Y. *et al.* 2018. Improvement of cutting performance for thick stainless steel plates by step-like cutting speed increase in high-power fiber laser cutting, *Optics and Laser Technology*, 103, 311–317.
[36] Panigrahi, D., Patel, S.K. 2023. Pulsed fiber laser processing of A4 SS under active and inactive environments: A comparative study, *Optics and Laser Technology*, 158:108921.
[37] Panigrahi, D., Patel, S.K. 2023 Influence of dry and wet environments during fiber laser processing of metallic substrate under a neoteric scan mode, *Infrared Physics and Technology*, 127, 1044500.
[38] Leone, C., Genna, S. Tagliaferri, F. 2020. Multiobjective optimisation of nanosecond fiber laser milling of 2024 T3 aluminium alloy, *Journal of Manufacturing Processes*, 57:288–301.
[39] Panigrahi, D., Rout, S., Tirpathy, S.S. *et al.* 2022. Optimization and simulation of laser beam trepanning approach on DSS superalloy, *Materials Today: Proceedings*, 44:1916–1924.
[40] Zhang, Y., Wang, Y., Zhang, J. *et al.* 2016. Effects of laser repetition rate and fluence on micromachining of TiC ceramic, *Materials and Manufacturing Processes*, 31:832–837.
[41] Feng, S., Zhang, R., Huang, C. *et al.* 2020. An investigation of recast behavior in laser ablation of 4H-silicon carbide wafer, *Materials Science in Semiconductor Processing*, 105:104701.
[42] Semak, V., Matsunawa, A. 1997. The role of recoil pressure in energy balance during laser materials processing, *Journal of Physics D: Applied Physics* 30:2541.
[43] Wang, X., Huang, Y., Xing, Y. *et al.* 2019. Fabrication of micro-channels on Al 2 O 3/TiC ceramics using picosecond laser induced plasma micromachining, *Journal of Manufacturing Processes*, 44:102–112.
[44] Low, D.K.Y., Li, L., Byrd, P.J. 2002. Hydrodynamic physical modeling of laser drilling, *Journal of Manufacturing Science and Engineering, Transactions of the ASME*, 124:852–862.
[45] Saini, S.K., Dubey, A.K., Upadhay B.N. *et al.* 2018. Study of material characteristics in laser trepan drilling of ZTA, *Optics and Laser Technology*, 103:330–339.
[46] Xing, Y., Liu, L., Wu, Z. *et al.* 2018. Fabrication and characterization of micro-channels on Al2O3/TiC ceramic produced by nanosecond laser, 44:23035–23044.
[47] Leone, C., Pape, I., Tagliaferri, F. 2013. Investigation of CFRP laser milling using a 30 W Q-switched Yb:YAG fiber laser: Effect of process parameters on removal mechanisms and HAZ formation, *Composites Part A: Applied Science and Manufacturing*, 55:129–142.

# 6 Machining Strategies for Micromachining of Glass Using Electrochemical Discharge Machining

## *A Review*

*Jinka Ranganayakulu, P.V. Srihari, K. Venkata Rao, Rithwik Shankar Raj, and Mallikarjun Mahajanshetti*

## 6.1 INTRODUCTION

Due to its optical transference, chemical inertness, low thermal expansion, toughness, and capacity to anode bond with silicon wafer, glass is increasingly used in microdevices such as microfluidics, micro reactors, bio-medical devices, and MEMS. Due to the absence of time-consuming processes like preparation, masking, and developing activities, electrochemical discharge machining (ECDM) is becoming more and more important in the fabrication of such glass-based microdevices. Borosilicate glass is utilized due to its capacity to be anodically bonded to silicon [1]. Electrochemical discharge machining (ECDM) is a hybrid version of electrochemical machining (ECM) and electric discharge machining (EDM). ECDM process is also referred with different names by various researchers. Based on electrochemical discharge phenomena, for the first time in 1968 Kurafuji and Suda, referred this process as "electrical discharge drilling' for micro-holes in glass [2]. This process is also referred to by many other names in literature as "Discharge machining of nonconductors" [3], "Electrochemical Arc Machining" (EAM) [4], "Electrochemical Discharge Machining" (ECDM) [5], "Micro Electrochemical Discharge Machining" [6], "Electro Chemical Spark Machining" (ECSM) [7] and "Spark Assisted Chemical Engraving" (SACE) [8].

 DOI: 10.1201/9781003364948-6

In comparison with laser beam machining (LBM), the ECDM process resulted in lower heat affected zones (HAZ) due to reduced temperatures at the machining zone (about one-fourth as that of LBM). Higher machining speeds are also achieved with ECDM as compared to etching (both wet & dry) and ultrasonic drilling processes [9]. In addition, high aspect ratios above 10 can be achieved with smooth and optically transparent surfaces on glass workpiece [10–12]. Despite the advantages, ECDM suffers low machineability issues at higher machining depths.

The researchers have suggested several strategies to enhance machining performance and overcome the low machinability of glass, particularly at deeper machining depths. In this work, the primary focus is on various electrolytes, such as direct and hybrid electrolytes, anode and cathode configurations, tool kinematics, magnetic and ultrasonic assisted and hybrid variants. In addition, the technical features of the custom engineered ECDM experimental setup are also discussed. Finally, the main areas that still require research are also discussed.

## 6.2 ECDM CELL ARRANGEMENT AND CHEMICAL REACTIONS

The basic ECDM cell arrangement, as shown in Figure 6.1, consists of pulse generation and power module arrangement, servo-controlled unit for X, Y, and Z axes for work, and tool feed, respectively. Tool electrode is connected to negative polarity and anode is connected to positive polarity of the DC source; this facilitates the generation of $H_2$ gas at cathode and $O_2$ gas at anode. Tool electrode and anode are immersed inside the electrolyte with predefined inter-electrode distance (IED) (about 30–40 mm) and electrolyte level is kept at 1–2 mm above the workpiece. About 25 microns of initial gap between the tool tip and the workpiece top surface. The size of the anode is kept relatively larger than the tool electrode to develop higher current densities at cathode and facilitates the faster discharge activity at tool tip.

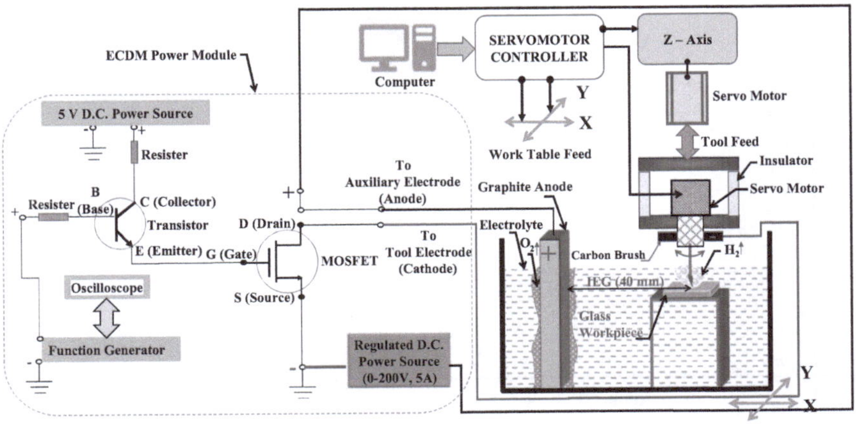


**FIGURE 6.1** Schematic of ECDM cell arrangement.

Electrolysis of water during ECDM follows Equations 6.1 and 6.2 [13]. The chemical reaction of NaOH and KOH electrolytes in ECDM of glass follows Equations 6.3 and 6.4 [14].

At cathode (Tool electrode):

$$2H_2O + 2e^- \rightarrow 2OH^- + H_2\uparrow \tag{6.1}$$

At anodic (Auxiliary electrode):

$$4OH^- \rightarrow 2H_2O + O_2\uparrow + 4e^- \tag{6.2}$$

$$SiO_2 + 2KOH + H_2O \rightarrow Si(OH)_4 + K_2O \tag{6.3}$$

$$SiO_2 + 2NaOH + H_2O \rightarrow Si(OH)_4 + Na_2O \tag{6.4}$$

ECDM is a complex combination of electrochemical machining (ECM) and electric discharge machining (EDM). The ECDM process involves melting and chemical etching of the workpiece due to high electrical energy discharged on the tip of the electrode during electrolysis [14,15].

## 6.3 STAGES OF DISCHARGE ACTIVITY IN ECDM

The discharge activity in ECDM takes place in five stages [16–18], as shown in Figure 6.2.

**Stage - I:** In the region *OE*, since the voltage is less than $V_d$, no current flows between the electrodes, and hence electrolysis is insignificant. This is termed as thermodynamic and overpotential region.

**Stage - II:** There is a linear variation of current in the region *EF* with the increase in the voltage beyond $V_d$. In this region, due to Ohmic effect, the electrolyte

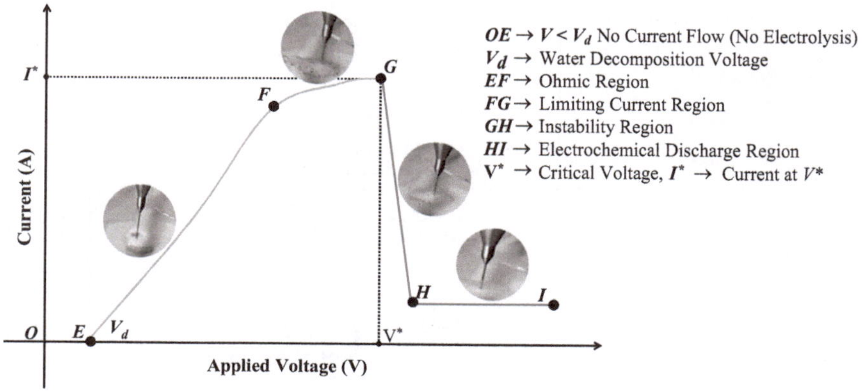


**FIGURE 6.2** Current and voltage characteristics in ECDM.

boils and decomposes water into $H_2\uparrow$ and $O_2\uparrow$ at cathode and anode respectively. The $H_2\uparrow$ gas bubble layer starts to build around the tool tip (cathode).

**Stage - III:** In the region *FG* the current is almost constant, and the $H_2\uparrow$ gas bubbles grow larger in size and form bubble layer around the tool tip. This limiting current ($I_{crit}$) region ends when the applied voltage reaches $V_{crit}$.

**Stage - IV:** In the region *GH,* the electrolyte boils, the electrolysis process loses its stability, and $H_2\uparrow$ bubbles grow and form an insulating gas film layer at the tool tip. This causes increase in resistance between electrolyte and tool tip increases.

**Stage - V:** In the region *HI*, due to the formation of the gas film, the working electrode (tool) is completely isolated from the electrolyte, and breakdown of voltage takes place. Electrochemical discharges take place through the gas film and are used for machining.

In addition, a theoretical model was also developed to study the mechanism of spark generation and validated the dependence of critical voltage on the electrolyte concentration with the experiments. A decrease in critical voltage with increase in the electrolyte concentration was reported with electrolytes KOH, NaOH, KCl, and NaCl correspondingly there was increase in critical current [19]. Also, the discharge in ECDM is considered as an event which is alike to that which happens in arc discharge valves [20].

## 6.4 MATERIAL REMOVAL IN ECDM

Several studies have been conducted regarding the mechanism of material removal in the ECDM process. A theoretical model for MRR was developed and validated with experimental results [21]. In contrast to discharge as breakdown of thin gas film, it was reported discharge as switching process. An increase in MRR was reported with an increase in electrolyte concentration and inductance.

Glass is eroded not only thermally, because of spark generation, but also chemically, as a result of the alkali solution's chemical attack on the glass [15]. A more thorough investigation of the chemical component of the material removal mechanism was conducted and suggested the material removal mechanism as a high temperature etching process [14]. Electrochemical discharge phenomenon similar to that which occurs in arc discharge valves. The spark energy and the approximate order of the hydrogen gas bubbles are then determined by the above proposed model. The material removal rate has been determined by considering the process to be 3D unsteady state heat conduction problem [20]. The model for the ECDM mechanism has been further investigated and analyzed by experimental observations of time-varying current in the circuit [22].

The combined effects of the thermal energy from the spark and chemical etching by the electrolyte are responsible for the MRR in ECDM.

## 6.5 CUSTOM ENGINEERED ECDM EXPERIMENTAL SETUP

The overview of the custom engineered ECDM experimental setup is shown in Figure 6.3 and the front view presented in Figure 6.4 displays the details of the scanning stage, processing cell, and tool with spindle arrangements.

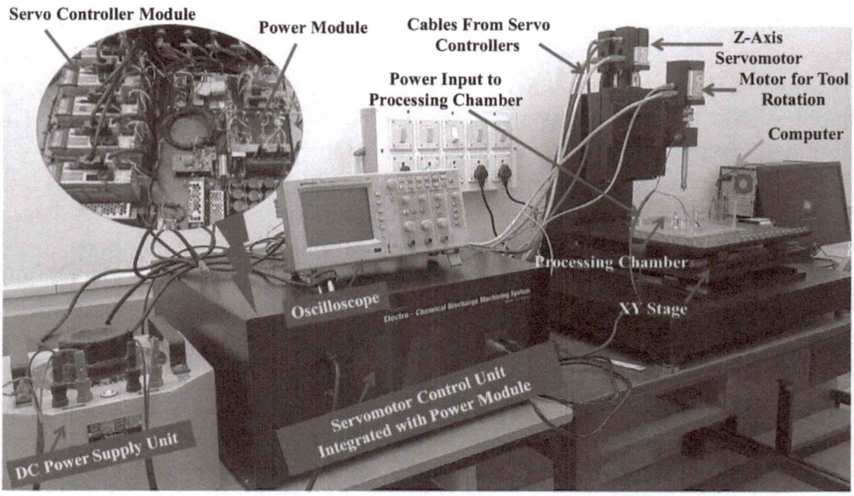


**FIGURE 6.3** Overview of the custom engineered ECDM experimental setup.

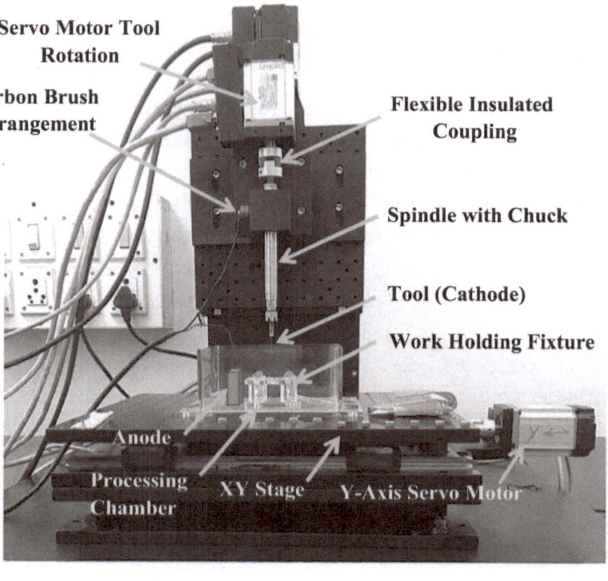


**FIGURE 6.4** Front view of ECDM experimental setup.

The X, Y, and Z moments are controlled by servomotors. It is designed for EDM, ECM, and ECDM operations for machining engineering materials. The tool and table movements are controlled by using the G and M codes of computer numerical control (CNC). The sample Y-shaped microfluidic channel with diameter 0.7mm tungsten carbide drill and 15% wt. KOH electrolyte is presented in Figure 6.5.

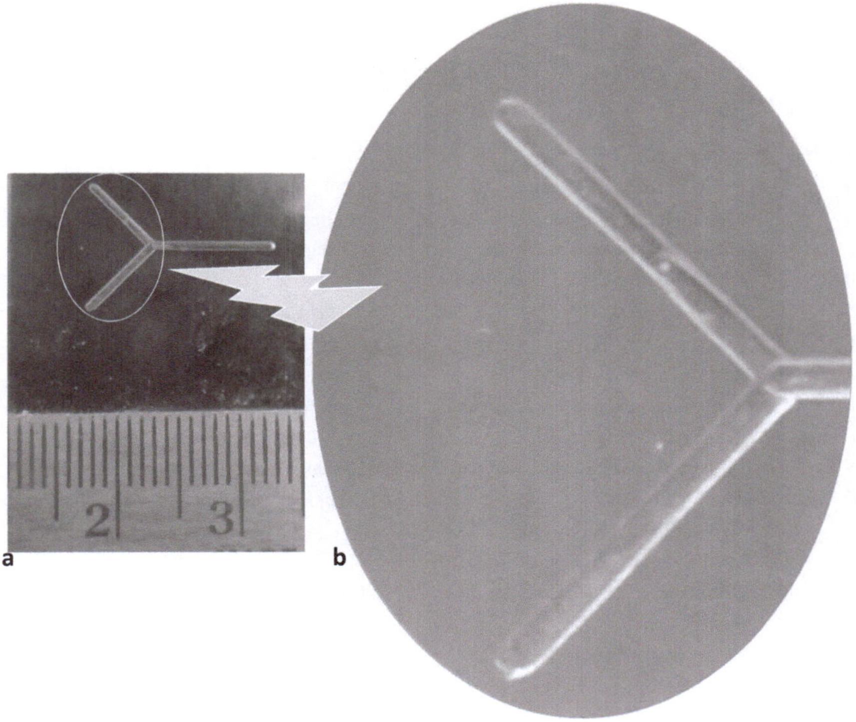


**FIGURE 6.5** Microfluidic Y-channel on glass with diameter 0.7 mm tungsten carbide drill and 15% wt. KOH electrolyte at a) 3.125X and b) 4X magnification.

The detailed technical specifications of the ECDM machine are elaborated in Table 6.1.

## 6.6 OVERVIEW OF ECDM VARIANTS

Figure 6.6 elaborates the overview of ECDM based on electrolyte, tool geometry, tool motion, tool feed, and power supply arrangements. It also includes ECDM process variants assisted with magnetic, ultrasonic, grinding, and traveling wire arrangements.

## 6.7 MACHINING STRATEGIES FOR GLASS MICROMACHINING USING ECDM

For machining non-conducting materials, especially glass workpieces, ECDM is a highly effective method. The ECDM process is capable of milling and slicing glass as well as drilling and trepanning holes in it.

**TABLE 6.1**
**Technical Specifications of ECDM Experimental Setup**

| Sl. No. | Description | Specification |
|---|---|---|
| 1 | Platform Size for XY | 450 mm X 450 mm |
| 2 | Platform Size for Z Stage: | 250 mm X 250 mm |
| 3 | Minimum and Maximum distance between XY stage and Z stage | Minimum Distance 100 mm, Maximum Distance 350 mm |
| 4 | Load Carrying Capacity on Z-axis | 15 kg |
| 5 | Travel Range | X and Y: 100 mm; Z-axis: 100 mm |
| 6 | Resolution | 1 μm in each axis |
| 7 | Position accuracy | 5 μm in each axis |
| 8 | Straightness uncertainty over the total travel range | 5 μm for all axes |
| 9 | Minimum travel Speed | 100 microns/minute in all axes |
| 10 | Motors | Digital AC servo divers for all three axes |
| 11 | Vertical load capacity (load perpendicular to table surface) | 30 kg or more (less than 50 Kg) |
| 12 | Horizontal load capacity | 15kg or more |
| 13 | Power supply | Regulated pulsed DC, 200V and 5A |
| 14 | Processing chamber | Acrylic, 100 mm X 100 mm X 50 mm |
| 15 | Spindle rpm | Maximum 10000 rpm – minimum 10 rpm |
| 16 | Anode | Graphite plate, 50 mm X 50 mm X 5 mm |
| 17 | Tool | Tungsten carbide helical drill bits (dia. 0.1 mm – 0.8 mm) |

Several strategies have been adopted to machine glass workpieces. These strategies include machining with direct (single) and hybrid electrolytes (mixed), anode and cathode configurations, tool kinematics and materials, ultrasonic and magnetic assisted and hybrid variants, and electrolyte circulating methods.

### 6.7.1 Direct and Hybrid Electrolytes

#### 6.7.1.1 Direct Electrolytes

The type and concentration of electrolyte plays a vital role in the machining of glass using the ECDM process. Several studies deal with the effect of electrolyte parameters on the machining process. Among the direct electrolytes majority of research was done with KOH and NaOH electrolytes due to their MRR, surface roughness, machining depth, and stable discharge capabilities. A comparative study was carried out using the KOH and NaOH electrolytes to machine borosilicate glass. Higher MRR was noticed with KOH compared with NaOH electrolytes, due higher mobility, and conductivity of $K^+$ ions than $Na^+$ ions.

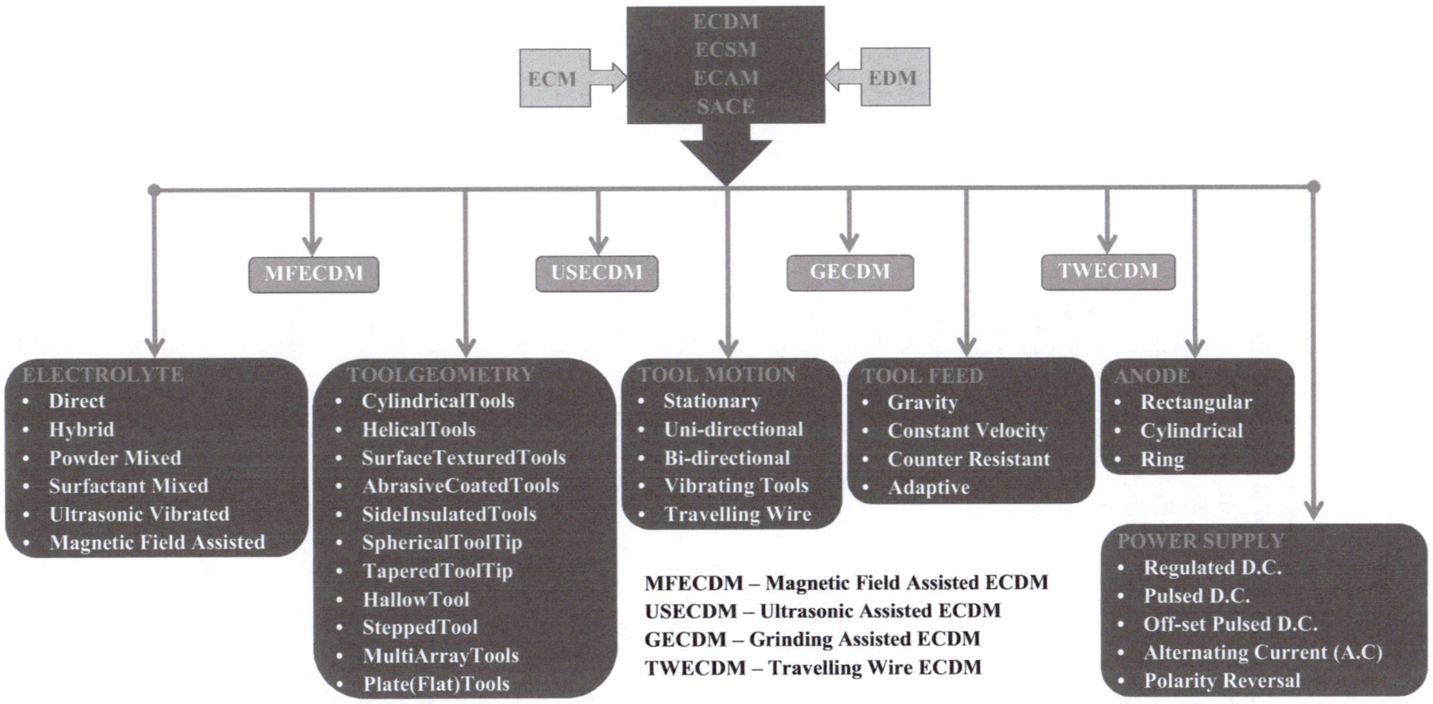


**FIGURE 6.6** Overview of ECDM variants, electrolytes, tool geometry, tool motion, tool feed, and power supply.

However, NaOH electrolyte has improved the surface roughness of machined surface due to low discharge energy applied to the workpiece. Also, reported that the alkaline electrolytes with higher PH value led to higher MRR [14].

Traveling wire ECDM (TWECDM) was implemented to cut optical glass with KOH and NaOH electrolytes. Reported that thermal cracks on the machined surface with KOH due to thermal stress effects with current pulse-off mode compared with NaOH [23]. Machining slits in glass using wire ECDM (WECDM) was carried out and found that the slit expansion was smaller with KOH electrolyte compared with NaOH. It is due to more dense bubble formation with KOH electrolyte and stable discharge [24]. A soda lime grass rod was machined with NaCl as electrolyte. It was found that the applied voltage has a significant influence on the width and depth of the machined grooves and the surface roughness of the machined surface [25]. Micro-structuring of glass was carried out with KOH and NaOH electrolytes. Machined holes in KOH solutions had smaller diameters at the same concentration. The hydrogen bubble density in the KOH solution is higher due to the KOH solution's higher conductivity, which also makes the isolation layer more stable [26]. Higher the concentration of NaOH electrolyte, higher the viscosity, and smoother channels were obtained on the glass workpiece [27]. Comparative study was done with KOH and NaCl electrolytes to machine glass. Higher MRR was noticed with KOH due to high reactivity with glass [28]. The experimental results indicated that NaOH was more effective at machining soda lime glass than $NaNO_3$. With an increase in the applied voltage, an increasing trend was seen in the MRR. When using NaOH electrolyte, fumes were produced that, if the area is not properly ventilated, could endanger the operator's health. Utilizing $NaNO_3$ electrolyte, comparatively, results in the generation of no such dangerous fumes [29]. With a higher NaOH electrolyte concentration, the surface material of borosilicate glass is more frequently vaporized by high-energy sparks, which is a dominant thermal mechanism. However, when the concentration is lower, the hole produced is smaller and has surface wrinkling along the perimeter because chemical etching predominates around the hole [30]. The electrolytes NaOH, KOH, KCl, and NaCl are frequently used in ECDM processes. NaCl and KCl form heavy salt precipitates as a result of chemical etching of the glass substrate, necessitating frequent replacement of the electrolytic solution. NaOH and KOH are preferred as electrolytes in ECDM processes as a result of this [31]. Compared to the NaOH and KOH alkaline electrolytes, $NaNO_3$ had a significantly higher critical voltage [32].

#### 6.7.1.2 Hybrid Electrolytes

Equal parts of KOH and NaOH electrolytes have a higher electrical conductivity than KOH and NaOH alone at the same concentrations. In comparison to KOH and NaOH, this made it easier to fabricate deeper microchannels with sharper sidewalls while maintaining the surface quality of the channel. In comparison to KOH and mixed electrolyte, the tungsten carbide (WC) tool used in NaOH and $NaNO_3$ had more severe wear [28]. When 5% KOH and NaOH are combined, the high conductivity of the hybrid electrolyte (HE) hastens the tendency for gas to form at the cathode and raises MRR [33]. When compared to hybrid electrolyte, lower MRR was obtained with NaOH electrolyte. This is because, for the same electrolyte %wt, HE (NaOH+KOH) has a lower viscosity than the NaOH electrolyte. Because of the lower viscosity, there is enough electrolyte at the tool tip to maintain discharge activity. The lower resistance to current flow with HE also resulted from the higher mobility and conductivity of $K^+$, which increased the discharge energy at the tool tip and consequently increased MRR [34]. A non-Newtonian electrolyte (Polyacrylamide + KOH) significantly weaken the effect of the impact force on the gas film, and thus the gas film was thin and more stable. Compared to KOH electrolyte, non-Newtonian electrolyte resulted in low HAZ and overcut of the microchannels [35].

### 6.7.2 Effect of Abrasive Particles and Powder-Mixed Electrolytes

Using WECDM to machine slits in Pyrex glass, the effects of SiC abrasive particles in KOH and NaOH electrolytes were investigated. The findings demonstrate that as the critical voltage rises, the slit expansion is decreased by the abrasive particles. The particles create an isolating layer around the wire by dispersing the bubble accumulation, raising the critical voltage, and lowering discharge energy. Small cracks and melted zones created by heat erosion are refined with the aid of the abrasive. Additionally, it was discovered that a smaller grit produces a smoother surface [24]. Conductive graphite powder-mixed electrolyte with 30% wt. NaOH, referred to as powder-mixed EDM (PM-EDM) was implemented to machine borosilicate glass. A reduction in micro-cracks on the machined surface of the workpiece was noticed due to controlled spark generation [36]. The surface quality and precision of the micro-slit can be improved by using WECDM of quartz glass under titrated electrolyte with SiC powder added to the KOH electrolyte [37].

### 6.7.3 Ultrasonic-Vibrated Electrolyte

Incorporating ultrasonic vibrations of the electrolyte is one potential strategy to guarantee uniform distribution of the electrolyte and maintain the machining to deeper levels. It was found that the inability to maintain adequate flow of the electrolyte in the small gap between the tool and the machined surface on the workpiece is the main cause of the reduction in spark frequency. To improve the electrolyte's circulation, replenishment, and uniformity, ultrasonic vibrations are employed. It has been demonstrated that using a side-insulated electrode and a pulse-power generator improves the surface finish at the hole entrance and the stability of the gas film. In this instance, the electrode prevented stray electrolysis and limited the workpiece's thermal damage [38]. For the drilling of a glass workpiece, high ultrasonic frequencies (27–28 kHz) with amplitudes up to 3.5 mm were used. The results demonstrated that greater MRR was produced by consecutive discharges during ultrasonic vibration with an amplitude less than

2 m. On the other hand, surface finish and quality improved with ultrasonic vibrations with an amplitude of 2–3.5 m, but MRR decreased [39]. When electrolytes are sonicated, the electrolyte flushing increases, allowing the ECDM process to operate at its maximum capacity and manipulate materials outside of the hydrodynamic regime [40].

### 6.7.4 Surfactant Mixed Electrolytes

The surfactant mixed electrolyte resulted in lower critical voltages and higher current densities. At machining voltage, frequency, and smaller duty ratios, taper of the machined holes was reduced significantly [41]. Sodium dodecyl sulphate (SDS), an anionic surfactant, and cationic surfactant Cetyltrimethylammonium bromide (CTAB), a cationic surfactant, were dissolved in the 25 wt% KOH and NaOH alkaline solutions in order to decrease the thickness of the gas film by altering the physicochemical characteristics of the electrolyte and the wettability of the electrode. According to the experiment's findings, microchannels with higher MRR, lower overcut, and lower HAZ were developed [42].

### 6.7.5 Magnetic Field Assisted ECDM (MFECDM)

In ECDM the machining efficiency reduces with increase in machining depth due to insufficient electrolyte at the tool tip to enhance electrolyte, supply a magnetohydrodynamic (MHD) system was implemented. During ECDM, the developed Lorenz forces due to the magnetic field enhance the electrolyte circulation. Higher machining depths were achieved at lesser time [43]. MHD convection is triggered by the presence of a magnetic field, which in turn speeds up the expulsion of bubbles from the cathodic surface. The obtained results show that for lower electrolyte concentration values and higher machining voltages, the machined surface will be smoother when the magnetic field is applied [44].

The circulation of the electrolyte in the restricted area between the tool (wire) and the workpiece has been improved by traveling wire ECSM (TWECSM) assisted with magnetic field, and MRR has been found to vary from 9.09% to 200% compared to without magnetic field [45]. High-speed tool rotation assisted with magnetic field enhanced electrolyte circulation and improved MRR [46].

### 6.7.6 Tool Parameters

The tool used for machining the glass workpiece plays a very important role in determining the overall MRR, the surface finish and quality, the depth of machining, and other important machining parameters.

#### 6.7.6.1 Tool with Vibration

The machining time for vibration amplitudes of 10 μm was reduced by about 50% when the tool was subjected to low-frequency vibrations (5–30 Hz), but the vibration amplitude higher than 10 μm had no noticeable effect on the MRR [47].

To further improve MRR, tool electrode vibration caused by a piezo actuator was added [48]. When a square waveform voltage is applied to the actuator in the case of the cylindrical rod, the resulting vibration increases the rate of material removal by up to 40% and allows for the drilling of deeper holes in a shorter amount of time [49]. While machining soda lime glass, ultrasonic vibration increased material removal by up to 35% while reducing tool wear by between 3% and 14%. Additionally, there is an inverse correlation between the gas film thickness and the vibration amplitude. The thickness of the gas film is reduced by applying ultrasonic vibration, which also reduces overcut [50]. Ultrasonic vibrated tool resulted in high aspect ratios with low specific energy due to removal of residue from bottom face of the tool [51].

#### 6.7.6.2 Tool Rotary Effects

In tool rotation assisted ECDM, the discharge is focused uniformly across the machined surface and also allows the fresh electrolyte to the machining region [14,52]. A decrease in the groove width was noticed with an increase in the tool rotational speed. The main reason behind this is at higher rotational speeds the gas fil becomes very thin and uniform and causes stable discharge [53]. It was discovered that rotating the tool electrode improved the circularity of the machined hole while having no effect on ECDM parameters like the critical voltage and critical current [12]. The replenishment cycle of the electrolyte under centrifugal force will be accelerated and material removal will be improved with an increase in electrode rotational speed [54]. The fringing effect makes spark generation with cylindrical tools challenging. Spark generation from tapered tools is more reliable, and they can be modeled more precisely [55]. A periodic bi-directional tool rotational method was implemented to improve the machining depth using high-speed helical tool [34].

#### 6.7.6.3 Tools with Surface Texture

Tool surface roughness plays a vital role in the formation of gas film around the tool. On a tool electrode with a rougher surface, larger mean-diameter bubbles form, creating a thicker gas film. This can be attributed to the poor wettability caused by the larger contact angle on surface with greater roughness [56]. Crack free surfaces were produced with textured cylindrical tools compared to smooth tools due to uniform distribution of spark discharge [57].

#### 6.7.6.4 Tools with Abrasive Coating

The MRR with conventional tools compared with conical abrasive coated tools to machine borosilicate glass. Due to improved discharge activity brought on by the availability of electrolyte in the small gap between the tool and the workpiece, coated tools showed higher MRR. Additionally, the machining was also aided by the abrasive coating [58]. A hybrid version of ECDM assisted with polycrystalline diamond micro grinding has improved the surface quality. In addition, the total machining was reduced considerably [59]. Rotating diamond core drill was used to implement GECDD. The MRR is achieved through a combination of thermal melting caused by electric discharges, diamond grit grinding action, and chemical etching action [60].

#### 6.7.6.5 Tools with Insulation

A major problem in ECDM is overcut problem raised due to side sparking. To overcome this, a tungsten tool with side insulation was implemented to machine micro grooves on the glass. Due to side insulation, the side sparking from the tool got eliminated and total discharge activity focussed on the tool tip and thus reduced the overcut [61,62]. A side-insulated electrode and a pulse-power generator were used to increase the gas film formation's stability and the hole entrance's surface quality. With side-insulation stray electrolysis prevented and the entire spark discharge focused on the tool tip [38]. In ECDM micro-hole drilling, a diamond coating side-insulated tool electrode was used to prevent the sidewall discharge. The side-insulated tool electrode achieves a smaller hole diameter and better surface integrity without a heat affected zone at the hole entrance when compared to the traditional tool electrode [63].

#### 6.7.6.6 Tool Geometry

To machine cover glass, various tool configurations, including cylindrical needles and spherical and triangular wedge-shaped tools, were used with robotic tool feeding arrangements [6]. Additionally, the effects of different tool tip shapes—such as flat ends, flat ends with tapered walls, and curved tips with sidewalls—were examined. According to studies, the improved electrolyte availability and stable sparking discharge increase the rate of machining with tapered walls and curved tips [13]. According to the investigations, a tool with a spherical tip produces a profile with a smaller entrance diameter compared to cylindrical tool, and by adjusting the tool feed-rate, the overcut can be significantly decreased [64]. Micro-holes, microchannels, micro slits, and complex three-dimensional features with little side clearance were created on ultra-clear glass using a rotating helical tool [65]. Through holes were machined on a glass substrate using tubular electrodes with a notched shape. Because of the tubular electrode's vertical notches, the electrolyte could easily reach greater depths, making it simple to machine deeper holes [66].

#### 6.7.6.7 Traveling Wire ECDM (TWECDM)

A modern technology TWECDM was developed to slice optical glass. Demonstrated machining slots with NaOH electrolyte have improved the surface finish compared to KOH. In addition, chemical etching with KOH electrolyte has etching is more intense compared to NaOH [23]. A comparative study was made with WECDM, by providing reciprocating mechanism to the wire. The quality of the slit has improved due to additional machining with reciprocating motion of the wire [24]. A theoretical model was developed to study traveling wire ECSM (TWECSM). An increase in MRR was noticed decrease in spark radius and increase in ejection efficiency [67]. TWECSM applied to cut borosilicate glass

[68,69]. With titrated flow electrolyte supply, wire electrochemical discharge machining (WECDM) was used to machine quartz glass [37,70].

#### 6.7.6.8 Multiarray Tools

The performance of single-electrode ECDM was compared with those of multiple electrodes connected electrically in parallel. The divided power caused a circular hole on soda lime glass [71]. With the aid of wire EDM, a stain-free steel multi-tip array tool with tapered tips was created, and it was used to simultaneously drill multiple holes on the glass substrate in order to create through-glass vias for MEMS packaging [72]. To create deeper microchannels on the glass workpiece, a customized SS304 line-array tool was used [73]. Multiarray tools implemented to machine through-glass vias [74].

#### 6.7.6.9 Tool Feed

At present two tool methods are in practice, gravity tool feed and constant velocity tool feed. With gravity tool feed, tool always makes contact with the workpiece and restricts the electrolyte supply to the machining zone. As machining progress, it promotes the hydrodynamic regime and creates tool bending and workpiece breakage problems. Use of constant velocity tool feed ensures a gap between the tool and workpiece, and this allows the electrolyte to reach the machining zone effectively and hence limits the hydrodynamic regime and machining depths can be increased. But the tool feed with constant velocity feed should be smaller than machining speed of ECDM [75]. Recent discussions have focused on the potential for using the current signal as a control input. It has been established that this signal can provide important information [76]. Repeatability of the machining was improved, and tool forces decreased with counter resistant tool feeding compared to gravity tool feeding arrangement [77]. Adaptive tool feed was implemented to minimize tool contact with the workpiece and improved the geometry of the micro-holes [78]. An adaptive tool feed system is used to keep the tool away from the work material and maintain an efficient machining gap between them, in contrast to the gravity tool feed and constant velocity tool feed methods [79].

## 6.8 POWER SOURCE IN ECDM

Among the parameters like applied voltage, tool geometry, type of electrolyte, and inter-electrode gap, the applied voltage has more influence on MRR [80,81].

### 6.8.1 Machining with Pulsed DC Power Supply

Rectified DC voltages are frequently used as the applied voltage in ECDM. The overcut phenomenon appears inevitable at this voltage because the thermal energy of discharge is continuous. Since a pulse voltage offers cooling during the pulse-off period to lessen the thermal heating zone, it is believed to be possible to overcome this shortage. Recently, it has been shown that rectangular pulse voltage can be used to increase the accuracy of micro-drilling [82]. To reduce the HAZ, a series of rectangular voltage pulses, were applied in this study instead of the rectified or full-wave DC voltages. The effect of the frequency and duty ratio of the voltage pulse on the ECDM of Pyrex glass was experimentally investigated. The findings also demonstrate that while the removal rate increases along with the duty ratio, the drilled surface becomes rougher. Consequently, taking into account the reduction in removal rate, a high frequency and a low duty ratio are needed to create a precise hole with little thermal damage. With a smaller diameter tool, it is discovered that the tool wear rate and clearance rise [11]. Experiments were conducted by increasing the duty factor by increasing the spark on time ($t_{on}$) and keeping off time ($t_{off}$) constant. At higher duty factor spark energy transfers to the workpiece for more time and leads to higher MRR [83]. To increase gas film stability, a brand-new pulse voltage configuration called offset pulse voltage has been developed and used in the ECDM process. At $T_{off}$ duration, the offset pulse voltage adds a constant voltage known as the offset voltage to improve the stability of the gas film and to further advance discharge performance. In addition, the experimental findings showed that more stable discharges are produced by an increase in offset voltage at the pulse-off duration as opposed to those produced by a conventional pulse voltage [84]. Due to the periodic pulse-off time introduced by the pulse voltage, thermal damage to the workpiece surface was reduced [38]. The use of an offset pulse voltage enhances machining performance and prevents the phenomenon of tool erosion [85].

### 6.8.2 Machining with Polarity Reversal

The tool polarity has a significant effect on machining performance in glass machining. When the tool electrode is given positive polarity for drilling in glass a spherical hole was obtained. A conical tapered hole was noticed with negative tool polarity due to accumulation of highly concentrated OH radicals of the electrolyte at the entrance of the hole [86]. Straight polarity (SP) and reverse polarities (RP) were applied to machine grooves on silica glass. Using a mixed electrolyte of NaOH 3%+KOH 1% with SP, the machining depth was increased while the surface quality improved. And with a mixed electrolyte of NaOH 1% +KOH 3% with RP, less tool wear was observed [87].

## 6.9 ANODE GEOMETRY

Researchers made attempts with different anode configurations to regulate the current flow between cathode and anode. The implemented shapes are rectangular [11] and cylindrical tungsten rods [25] in lathe type ECDM process. The ring type implemented to reduce the difference in the ion-translation path during machining, a ring-type auxiliary electrode is placed around the wafer (acting as an anode) [88].

## 6.10 CONCLUSION AND SCOPE FOR IMPROVEMENT

In the present chapter, the detailed review of state-of-art of machining strategies for glass machining is presented. Which includes the process parameters affecting the machining, like types of electrolytes, tool parameters, ultrasonic and magnetic field assisted electrolyte circulation, power sources, and anode geometry, are presented. From this work, the following conclusions can be made.

- The technical features of the custom engineered ECDM experimental setup along with the detailed spark discharge phenomena and material removal mechanism are presented.
- Among various electrolytes KOH and NaOH are found suitable for effective machining. Hybrid, powder-mixed, surfactant mixed, and ultrasonic & magnetic field assisted electrolytes have improved the machining performance.
- Tool parameters like tool vibration, rotation, surface texture, abrasive coatings, insulation, geometry, traveling wire, multiarray tools, and tool feed are implemented, and significant improvement is achieved in terms increased MRR, reduced overcut, and HAZ.
- Power sources with pulsed DC supply reduced the chances of overcut and HAZ by allowing electrolyte to the machining zone. Polarity reversal has resulted in a spherical hole in drilling of glass. Straight polarity with mixed electrolyte has improved MRR and tool wear has reduced with reverse polarity.

In addition to the above methods of machining the following aspects are still open for the improvement.

- The electrolyte minimization methods need to be developed towards green manufacturing.
- Active feedback mechanism to maintain a constant gap between the tool and work is still open for research.

## ACKNOWLEDGMENTS

The authors are thankful to Vision Group on Science and Technology, KSTePS, Department of IT, BT and S&T, Karnataka—560-001, for the support through KFIST Level (1) project under the VGST Scheme, No: KSTePS/VGST/GRD-678/KFIST(L1)/2018, 27.08.2018.

## REFERENCES

[1] R. Wüthrich, and V. Fascio, Machining of non-conducting materials using electrochemical discharge phenomenon—an overview, *International Journal of Machine Tools & Manufacture*, 45 (2005), 1095–1108. doi:10.1016/j.ijmachtools.2004.11.011.

[2] H. Kurafuji, and K. Suda, Electrical discharge drilling of glass, *Annals of the CIRP*, 16 (1968), 415–419.

[3] N. H. Cook, G. B. Foote, P. Jordan, and B. N. Kalyani, Experimental studies in electro-machining, *Transactions of ASME, Journal of Engineering for Industry*, (1973), 945–950.

[4] M. Kubota, Proceedings of the International Conference on Production Engineering, Tokyo, (1974), p. 51.

[5] K. Allesu, A. Ghosh, and M. K. Muju, Preliminary qualitative approach of a proposed mechanism of material removal in electrical machining of glass, *European Journal of Mechanical Engineers*, 36 (1992), 202–207.

[6] H. Langen, J.-M. Breguet, H. Bleuler, Ph. Renaud, and T. Masuzawa, Micro electrochemical discharge machining of glass, *International Journal of Electrical Machining*, 3 (1998), 65–69.

[7] S. Tandon, V. K. Jain, P. Kumar, and K. P. Rajurkar, Investigations into machining of composites, *Precision Engineering*, 12 (1990), 227–238.

[8] H. Langen, V. Fascio, R. Wüthrich, and D. Viquerat, Three-dimensional structuring of pyrex glass devices—trajectory control, International Conference of the European Society for Precision Engineering and Nanotechnology (EUSPEN) 2 Eindhoven (2002), 435–438.

[9] L. Hof, and J. Ziki, Micro-hole drilling on glass substrates—A review, *Micromachines*, 8 (53) (2017), 23, Doi:10.3390/mi8020053.

[10] B.-H. Yan, C.-T. Yang, F.-Y. Huang, and Z.-H. Lu, Electrophoretic deposition grinding (EPDG) for improving the precision of microholes drilled via ECDM, *J. Micromech. Microeng*, 17 (2007), 376–383. Doi:10.1088/0960-1317/17/2/025

[11] D. J. Kim, Y. Ahn, S. H. Lee, and Y. K. Kim, Voltage pulse frequency and duty ratio effects in an electrochemical discharge microdrilling process of Pyrex glass, *Int. J. Mach. Tools Manuf*, 46 (2006), 1064–1067. 10.1016/j.ijmachtools.2005.08.011.

[12] S. K. Jui, A. B. Kamaraj, and M. M. Sundaram, High aspect ratio micromachining of glass by electrochemical discharge machining (ECDM). *J. Manuf. Process*, 15 (2013), 460–466. 10.1016/j.jmapro.2013.05.006.

[13] B. Bhattacharyya, B. N. Doloi, and S. K. Sorkhel, Experimental investigations into electrochemical discharge machining (ECDM) of non-conductive ceramic materials, *Journal of Materials Processing Technology*, 95 (1–3) (1999), 145–154, ISSN 0924-0136, 10.1016/S0924-0136(99)00318-0.

[14] C. Yang, S.-S. Ho, and B. Yan, Micro hole machining of borosilicate glass through electrochemical discharge machining (ECDM), *Key Engineering Materials - KEY ENG MAT.*, 196 (2001), 149–166. Doi:10.4028/www.scientific.net/KEM.196.149

[15] V. Fascio, R. Wuthrich, D. Viquerat, and H. Langen, 3D microstructuring of glass using electrochemical discharge machining (ECDM). MHS'99. Proceedings of 1999 International Symposium on Micromechatronics and Human Science (Cat. No.99TH8478), (1999), 179–183. Doi:10.1109/MHS.1999.820003.

[16] V. Fascio, R. Wüthrich, and H. Bleuler, Spark assisted chemical engraving in the light of electrochemistry, *Electrochimica Acta*, 49 (22–23) (2004), 3997–4003, ISSN 0013-4686, 10.1016/j.electacta.2003.12.062.

[17] R. Wüthrich, Ch. Comninellis, and H. Bleuler, Bubble evolution on vertical electrodes under extreme current densities, *Electrochimica Acta*, 50 (25–26) (2005), 5242–5246. ISSN 0013-4686, 10.1016/j.electacta.2004.12.052.

[18] R. Wüthrich, and L. A. Hof, The gas film in spark assisted chemical engraving (SACE)—A key element for micro-machining applications, *International Journal of Machine Tools and Manufacture*, 46 (7–8) (2006), 828–835. ISSN 0890-6955, 10.1016/j.ijmachtools.2005.07.029.

[19] I. Basak, and A. Ghosh, Mechanism of spark generation during electrochemical discharge machining: a theoretical model and experimental verification, *Journal of Material Processing Technology*, 62 (1996), 46–53. 10.1016/0924-0136(95)02202-3.

[20] V. K. Jain, P. M. Dixit, and P. M. Pandey, On the analysis of the electrochemical spark machining process, *International Journal of Machine Tools and Manufacture*, 39 (1) (1999), 165–186. ISSN 0890-6955, 10.1016/S0890-6955(98)00010-8.

[21] I. Basak, and A. Ghosh, Mechanism of material removal in electrochemical discharge machining: a theoretical model and experimental verification, *Journal of Material Processing Technology*, 71 (1997), 350–359. 10.1016/S0924-0136(97)00097-6.

[22] A. Kulkarni, R. Sharan, and G. K. Lal, An experimental study of discharge mechanism in electrochemical discharge machining, *International Journal of Machine Tools and Manufacture*, 42 (2002), 1121–1127.

[23] W. Y. Peng, and Y. S. Liao, Study of electrochemical discharge machining technology for slicing non-conductive brittle materials, *Journal of Materials Processing Technology*, 149 (1–3) (2004), 363–369. ISSN0924-0136, 10.1016/j.jmatprotec.2003.11.054.

[24] C. T. Yang, S. L. Song, B. H. Yan, and F. Y., Huang, Improving machining performance of wire electrochemical discharge machining by adding SiC abrasive to electrolyte, *International Journal of Machine Tools and Manufacture*, 46 (15) (2006), 2044–2050. ISSN 0890-6955, 10.1016/j.ijmachtools.2006.01.006.

[25] K. Furutani, and H. Maeda, Machining glass rod with lathe type electro-chemical discharge machine, *Journal of Micromechanics and Microengineering*, 18 (065006) (2008). Doi: 10.1088/0960-1317/18/6/065006.

[26] X. Cao, B. H. Kim, and C. Chu, Micro-structuring of glass with features less than 100μm by electrochemical discharge machining, *Precision Engineering*, 33 (2009), 459–465. Doi: 10.1016/j.precisioneng.2009.01.001.

[27] J. D. A. Ziki, T. F. Didar, and R. Wüthrich, Micro-texturing channel surfaces on glass with spark assisted chemical engraving, *International Journal of Machine Tools and Manufacture*, 57 (2012), 66–72. ISSN 0890-6955, 10.1016/j.ijmachtools.2012.01.012.

[28] N. Sabahi, and M. R. Razfar, Investigating the effect of mixed alkaline electrolyte (NaOH + KOH) on the improvement of machining efficiency in 2D electrochemical discharge machining (ECDM), *Int J Adv Manuf Technol*, 95 (2018), 643–657. 10.1007/s00170-017-1210-4.

[29] C. Jawalkar, P. Kumar, and A. Sharma, Investigations on performance of ECDM process using NaOH and NaNO3 electrolytes while micro machining soda lime glass, *International Journal of Manufacturing Technology and Management*, 28 (2014), 80–93. 10.1504/IJMTM.2014.064623

[30] K. R. Kolhekar, and M. Sundaram, A study on the effect of electrolyte concentration on surface integrity in micro electrochemical discharge machining, *Procedia CIRP*, 45 (2016), 355–358. ISSN 2212-8271, 10.1016/j.procir.2016.02.146.

[31] S. Saranya, A. Nair, and A. Ravi Sankar, Experimental investigations on the electrical and 2D-machining characteristics of an electrochemical discharge machining (ECDM) process, *Microsyst Technol*, 23 (2017), 1453–1461. 10.1007/s00542-016-3027-8.

[32] Z. Zhang, L. Huang, Y. Jiang et al., A study to explore the properties of electrochemical discharge effect based on pulse power supply, *Int J Adv Manuf Technol*, 85 (2016), 2107–2114. 10.1007/s00170-015-8302-9.

[33] J. Ranganayakulu, and P. V. Srihari, Multi-objective Optimization Using Taguchi's Loss Function-Based Principal Component Analysis in Electrochemical Discharge Machining of Micro-channels on Borosilicate Glass with Direct and Hybrid

Electrolytes. Advances in Manufacturing Processes. *Lecture Notes in Mechanical Engineering*. Springer, Singapore, (2019). 10.1007/978-981-13-1724-8_34

[34] J. Ranganayakulu, P. V. Srihari, and K. V. Rao, A strategy to improve performance in electrochemical discharge machining using periodic bi-directional tool rotation, *Int J Adv Manuf Technol*, 123 (2022), 1459–1476. 10.1007/s00170-022-10227-x.

[35] Z. Zou, Z. Guo, K. Zhang, Y. Xiao, T. Yue, and J. Liu, Electrochemical discharge machining of microchannels in glass using a non-Newtonian fluid electrolyte, *Journal of Materials Processing Technology*, 305 (2022) 117594. ISSN 0924-0136, 10.1016/j.jmatprotec.2022.117594.

[36] M.-S. Han, B.-K. Min, and S. J. Lee, Improvement of surface integrity of electrochemical discharge machining process using powder-mixed electrolyte, *Journal of Materials Processing Technology*, 191 (1–3) (2007), 224–227. ISSN 0924-0136, 10.1016/j.jmatprotec.2007.03.004.

[37] K. Y. Kuo, K. L. Wu, C. K. Yang, and B.-H. Yan, Effect of adding SiC powder on surface quality of quartz glass microslit machined by WECDM, *Int J Adv Manuf Technol*, 78 (2015), 73–83. 10.1007/s00170-014-6602-0.

[38] M.-S. Han, B.-K. Min, and S. Lee, Geometric improvement of electrochemical discharge micro-drilling using an ultrasonic-vibrated electrolyte, *Journal of Micromechanics and Microengineering*, 19 (2009), 065004, p8. Doi: 10.1088/0960-1317/19/6/065004.

[39] M. Rusli, and K. Furutani, Performance of micro-hole drilling by ultrasonic-assisted electro-chemical discharge machining, *Advanced Materials Research*, 445 (2012), 865 - 870. Doi: 10.4028/www.scientific.net/AMR.445.865.

[40] K. V. Bhargav, P. S. Balaji, and R. K. Sahu, Micromachining of borosilicate glass using an electrolyte-sonicated-μ-ECDM system, *Materials and Manufacturing Processes*, 38 (2022), 64 - 77. Doi:10.1080/10426914.2022.2089893.

[41] Y. S. Laio, L. C. Wu, and W. Y. Peng, A study to improve drilling quality of electrochemical discharge machining (ECDM) process, *Procedia CIRP*, 6 (2013), 609–614, ISSN 2212-8271, 10.1016/j.procir.2013.03.105.

[42] N. Sabahi, M. R. Razfar, and M. Hajian, Experimental investigation of surfactant-mixed electrolyte into electrochemical discharge machining (ECDM) process, *Journal of Materials Processing Technology*, 250 (2017), 190–202. ISSN 0924-0136, 10.1016/j.jmatprotec.2017.07.017.

[43] C. Cheng, K. Wu, C. Mai, Y. Hsu, and B. Yan, Magnetic field-assisted electrochemical discharge machining, *Journal of Micromechanics and Microengineering*, 20 (2010), 075019. Doi:10.1088/0960-1317/20/7/075019

[44] M. Hajian, M. R. Razfar, and S. Movahed, An experimental study on the effect of magnetic field orientations and electrolyte concentrations on ECDM milling performance of glass, *Precision Engineering*, 45 (2016), 322–331. ISSN 0141-6359, 10.1016/j.precisioneng.2016.03.009.

[45] N. Rattan, and R. Mulik, Improvement in material removal rate (MRR) using magnetic field in TW-ECSM process, *Materials and Manufacturing Processes*, 32 (2016), 101–107. Doi:10.1080/10426914.2016.1176197.

[46] M. Harugade, S. Waigaonkar, G. Kulkarni, and M. Diering, Experimental investigations of magnetic field-assisted high-speed electrochemical discharge drilling, *Materials and Manufacturing Processes*, (2021), 1–12. Doi: 10.1080/10426914.2021.2016814.

[47] R. Wüthrich, B. Despont, P. Maillard, and H. Bleuler, Improving the material removal rate in spark-assisted chemical engraving (SACE) gravity-feed micro-hole drilling, *Journal of Micromechanics and Microengineering*, 16 N28 (2006). Doi: 10.1088/0960-1317/16/11/N03.

[48] B. Jiang, S. Lan, and J. Ni, Experimental investigation of drilling incorporated electrochemical discharge machining, Proceedings of the ASME 2014 International Manufacturing Science and Engineering Conference collocated with the JSME 2014 International Conference on Materials and Processing and the 42nd North American Manufacturing Research Conference. Volume 2: Processing. Detroit, Michigan, USA. June 9–13, 2014. V002T02A056. ASME. 10.1115/MSEC2014-4054.

[49] M. R. Razfar, A. Behroozfar, and J. Ni, Study of the effects of tool longitudinal oscillation on the machining speed of electrochemical discharge drilling of glass, *Precision Engineering*, 38 (4) (2014), 885–892. ISSN 0141-6359, 10.1016/j.precisioneng.2014.05.004.

[50] S. Elhami, and M. R. Razfar, Study of the current signal and material removal during ultrasonic-assisted electrochemical discharge machining, *The International Journal of Advanced Manufacturing Technology*, 92 (2017), 1591–1599. Doi:10.1007/S00170-017-0224-2.

[51] T. Singh, A. Dvivedi, A. Shanu, and P. Dixit, Experimental investigations of energy channelization behavior in ultrasonic assisted electrochemical discharge machining, *Journal of Materials Processing Technology*, 293 (2021), 117084, ISSN 0924-0136, 10.1016/j.jmatprotec.2021.117084.

[52] N. Gautam, and V. K. Jain, Experimental investigations into ECSD process using various tool kinematics, *Int. J. Mech. Tools and Manufact*, 38 (1998), 15–27.

[53] Z. Zheng, W. Cheng, F. Y. Huang, and B. Yan, 3D microstructuring of Pyrex glass using the electrochemical discharge machining process, *Journal of Micromechanics and Microengineering*, 17 (2007), 960 - 966. Doi: 10.1088/0960-1317/17/5/016.

[54] K. L. Wu, H. M. Lee, and K. H. Chin, Application of electrochemical discharge machining to micro-machining of quartz, *In Advanced Materials Research*, 939 (2014), 161–168. 10.4028/www.scientific.net/amr.939.161.

[55] B. Jiang, S. Lan, J. Ni, and Z. Zhang, Experimental investigation of spark generation in electrochemical discharge machining of non-conducting materials, *Journal of Materials Processing Technology*, 214 (4) 2014, 892–898. ISSN 0924-0136, 10.1016/j.jmatprotec.2013.12.005.

[56] C.-K. Yang, C.-P. Cheng, C.-C. Mai, A. C. Wang, J.-C. Hung, and B.-H. Yan, Effect of surface roughness of tool electrode materials in ECDM performance, *International Journal of Machine Tools and Manufacture*, 50 (12) (2010), 1088–1096. ISSN: 0890-6955, 10.1016/j.ijmachtools.2010.08.006.

[57] M.-S. Han, B.-K. Min, and S. Lee, Micro-electrochemical discharge cutting of glass using a surface-textured tool, *Cirp Journal of Manufacturing Science and Technology*, 4 (2011). Doi: 10.1016/j.cirpj.2011.06.007.

[58] V. K. Jain, S. K. Choudhury, and K. M. Ramesh, On the machining of alumina and glass, *International Journal of Machine Tools and Manufacture*, 42 (11) 2002, 1269–1276. ISSN 0890-6955, 10.1016/S0032-3861(02)00241-0.

[59] X. D. Cao, B. H. Kim, and C. N. Chu, Hybrid micromachining of glass using ECDM and micro grinding, *Int. J. Precis. Eng. Manuf.*, 14 (2013), 5–10. 10.1007/s12541-013-0001-6.

[60] V. G. Ladeesh, and M. Rajan, Experimental Investigation on the Performance of Grinding Assisted Electrochemical Discharge Drilling of Glass. MATEC Web of Conferences, (2016). Doi: 51. 03001. 10.1051/matecconf/20165103001.

[61] M.-S. Han, B.-K. Min, and S. J. Lee, Reduction of overcut in the electrochemical discharge machining process using a side-insulated electrode, 2008 International Conference on Smart Manufacturing Application, Goyangi, Korea (South), (2008), 251–254. doi: 10.1109/ICSMA.2008.4505652.

[62] M.-S. Han, B.-K. Min, and S. Lee, Modeling gas film formation in electrochemical discharge machining processes using a side-insulated electrode, *Journal of Micromechanics and Microengineering*, 18 (2008), 045019, (8 pp). doi:10.1088/0960-1317/18/4/045019.

[63] W. Tang, X. Kang, and W. Zhao, Enhancement of electrochemical discharge machining accuracy and surface integrity using side-insulated tool electrode with diamond coating, *J. Micromech. Microeng.*, 27 (6) (2017). Doi: 10.1088/1361-6439/aa6e94.

[64] S. Sambathkumar, and R. Sankar, Effect of tool shape and tool feed rate on the machined profile of a quartz substrate using an electrochemical discharge machining process, 2nd International Symposium on Physics and Technology of Sensors (ISPTS) (2015), 313–316. Doi: 10.1109/ISPTS.2015.7220137.

[65] Y. Liu, C. Zhang, S. Li, C. Guo, and Z. Wei, Experimental study of micro electrochemical discharge machining of ultra-clear glass with a rotating helical tool, *Processes*, 7(4) (2019), 195. 10.3390/pr7040195.

[66] J. Arab, and P. Dixit, Formation of macro-sized through-holes in glass using notch-shaped tubular electrodes in electrochemical discharge machining, *Journal of Manufacturing Processes*, 78 (2022), 92–106. ISSN: 1526-6125, 10.1016/j.jmapro.2022.03.052.

[67] M. Panda, and V. Yadava, Finite element prediction of material removal rate due to traveling wire electrochemical spark machining. *The International Journal of Advanced Manufacturing Technology*, 45 (2009), 506–520. Doi: 10.1007/s00170-009-1992-0.

[68] B. K. Bhuyan, and V. Yadava, Experimental modeling and multi-objective optimization of traveling wire electrochemical spark machining (TW-ECSM) process, *J Mech Sci Technol*, 27 (2013), 2467–2476. 10.1007/s12206-013-0632-7.

[69] B. Bhuyan, and V. Yadava, Experimental study of traveling wire electrochemical spark machining of borosilicate glass, *Materials and Manufacturing Processes*, 29 (2014). Doi: 10.1080/10426914.2013.852216.

[70] K.-Y. Kuo, K.-L. Wu, C.-K. Yang, and B.-H. Yan, Wire electrochemical discharge machining (WECDM) of quartz glass with titrated electrolyte flow, *International Journal of Machine Tools and Manufacture*, 72 (2013), 50–57. ISSN 0890-6955, 10.1016/j.ijmachtools.2013.06.003.

[71] K. Furutani, H. Shintani, Y. Murase, and S. Arakawa, Performance of electrochemical discharge machining by forced discharge dispersion, *International Journal of Electrical Machining*, 19 (2014), 9–15. Doi:10.2526/ijem.19.9.

[72] J. Arab, P. Adhale, D. K. Mishra, and P. Dixit, Micro array hole formation in glass using electrochemical discharge machining, *Procedia Manufacturing*, 34 (2019), 349–354. ISSN 2351-9789, 10.1016/j.promfg.2019.06.174.

[73] D. K. Mishra, and P. Dixit, Experimental investigation into tool wear behaviour of line-array tool electrode during the electrochemical discharge micromilling process, *Journal of Manufacturing Processes*, 72 (2021), 93–104. ISSN 1526-6125. 10.1016/j.jmapro.2021.10.009.

[74] T. Singh, D. K. Mishra, and P. Dixit, Effect of pulse frequency and duty cycle on electrochemical dissolution behavior of multi-tip array tool electrode for reusability in the ECDM process, *J Appl Electrochem*, 52 (2022), 667–682. 10.1007/s10800-021-01662-x

[75] R. Wüthrich, U. Spaelter, Y. Wu, and H. Bleuler, A systematic characterization method for gravity-feed micro-hole drilling in glass with spark assisted chemical engraving (SACE), *Journal of Micromechanics and Microengineering*, (2006). Doi: 16. 1891. 10.1088/0960-1317/16/9/019.

[76] R. Wüthrich, U. Spaelter, and H. Bleuler, The current signal in spark-assisted chemical engraving (SACE): what does it tell us?, *Journal of Micromechanics and Microengineering*, 16 (2006), 779 - 785. DOI:10.1088/0960-1317/16/4/014.
[77] Y. Xu, J. Chen, B. Jiang, and J. Ni, Investigation of micro-drilling using electrochemical discharge machining with counter resistant feeding, *Journal of Materials Processing Technology*, 257 (2018), 141–147. ISSN 0924-0136, 10.1016/j.jmatprotec.2018.02.023.
[78] V. Rajput, G. Mudimallana, and N. Suri, Geometrical improvement in machined micro-hole using adaptive tool feed system in electrochemical based discharge machining, IOP Conference Series: Materials Science and Engineering, 1168 (2021), 012014, p8. Doi: 10.1088/1757-899X/1168/1/012014.
[79] V. Rajput, M. Goud, and N. M. Suri, Performance analysis of closed-loop electrochemical discharge machining (CLECDM) during micro-drilling and response surface methodology based multi-response parametric optimisation. *Advances in Materials and Processing Technologies*, 8 (2021), 1352 - 1382. DOI:10.1080/2374068X.2020.1860494.
[80] B. R. Sarkar, B. Doloi, and B. Bhattacharyya, Parametric analysis on electrochemical discharge machining of silicon nitride ceramics, *Int. J. Adv. Manuf. Technol*, 28 (2006), 873–881. Doi: 10.1007/s00170-004-2448-1.
[81] M. R. Razfar, J. Ni, A. Behroozfar, and S. Lan, An investigation on electrochemical discharge micro-drilling of glass. In Proceedings of the ASME 2013 International Manufacturing Science and Engineering Conference collocated with the 41st North American Manufacturing Research Conference, Madison, WI, USA, (10–14 June 2013).
[82] Z.-P. Zheng, H.-C. Su, F.-Y. Huang, and Y. Biing Hwa, The tool geometrical shape and pulse-off time of pulse voltage effects in a Pyrex glass electrochemical discharge microdrilling process, *Journal of Micromechanics and Microengineering*, 17 (2007), 265. Doi: 10.1088/0960-1317/17/2/012.
[83] K. L. Bhondwe, V. Yadava, and G. Kathiresan, Finite element prediction of material removal rate due to electro-chemical spark machining, *International Journal of Machine Tools and Manufacture*, 46 (14) (2006), 1699–1706. ISSN 0890-6955, 10.1016/j.ijmachtools.2005.12.005.
[84] Z.-P. Zheng, J.-K. Lin, F.-Y. Huang, and Y. Biing Hwa, Improving the machining efficiency in electrochemical discharge machining (ECDM) microhole drilling by offset pulse voltage. *Journal of Micromechanics and Microengineering*, 18 (2008), 025014, (6pp). Doi: 10.1088/0960-1317/18/2/025014.
[85] K. H. Nguyen, P. A. Lee, and B. H. Kim, Experimental investigation of ECDM for fabricating microstructures of quartz, *Int. J. Precis. Eng. Manuf.*, 16 (2015), 5–12. 10.1007/s12541-015-0001-9.
[86] J. West, and A. Jadhav, ECDM methods for fluidic interfacing through thin glass substrates and the formation of spherical microcavities, *Journal of Micromechanics and Microengineering*, 17 (2007), 403–409. Doi: 10.1088/0960-1317/17/2/028.
[87] B. Mallick, B. R. Sarkar, B. Doloi, and B. Bhattacharyya, Improvement of surface quality and machining depth of µ-ecdm performances using mixed electrolyte at different polarity, *Silicon*, 14 (2022), 8223–8232. 10.1007/s12633-021-01587-2.
[88] Y. S. Liao, and W. Peng, Study of hole-machining on pyrex wafer by electrochemical discharge machining (ECDM), *Materials Science Forum*, 505-507 (2006), 1207 – 1212. Doi:10.4028/www.scientific.net/MSF.505-507.1207.

# 7 Application and Mechanism of the Cavitation Effect in the Preparation of Magnetorheological Polishing Fluid

*Ce Guo, Shengqiang Yang, Xiuhong Li, and Wenhui Li*

## 7.1 INTRODUCTION

Magnetorheological polishing fluid (MRPF) is a solid-liquid hybrid smart material whose mechanical properties can be regulated by an external magnetic field, which has the advantages of being green, environmentally friendly, and intelligently controllable. The special properties of MRPF make it a strategic material for many advanced technologies, such as magnetorheological polishing, magnetic jet polishing, and multi-physical field-assisted liquid jet polishing [1,2]. However, during the preparation of MRPF, the solid-phase particles tend to coalesce, agglomerate, and settle [3]. Thus, the dispersion process is necessary for homogenization of MRPF. The question of how to improve the preparation theory and process of MRPF has become a challenging problem that limits its development.

Traditional mechanical dispersion methods such as the blade or ball mill stirring are time consuming, low precision, and poorly controllable, making it difficult to disperse solid-phase particles uniformly in the base liquid [4]. Ultrasonic dispersion uses the cavitation effect generated by ultrasonic waves in a liquid to prepare suspensions. The cavitation effect can produce large numbers of micro-bubbles that oscillate violently, which are also known as cavitation bubbles. At the same time, physical phenomena such as microjets and shock waves released by cavitation bubbles collapsing in the vicinity of solid-phase particles can have a strong impact and efficient dispersion effect on the solid-phase particle population [5].

 DOI: 10.1201/9781003364948-7

However, research on the ultrasonic dispersion of MRPF is currently not well studied. Cavitation bubbles in MRPF are still being understood and explored.

The characteristics of the cavitation bubble movement can be described by the bubble dynamic model. A number of well-known models have been used to approximate and explain the bubble motion, such as the Rayleigh model, Rayleigh–Plesset model, Keller–Miksis model, and Gilmore model [6–8]. Unfortunately, current bubble dynamic models are primarily applied to aqueous solutions or liquid environments that can be approximated as aqueous solutions and are difficult to use directly to describe a solid-liquid mixed suspension. In recent years, computational fluid dynamics and multiphysics field simulation have been widely used in modeling the acoustic and flow fields of the suspension. However, these numerical calculation methods often ignore the cavitation effects of liquids in engineering applications because of the complexity of bubble models. The search for a suitable model for explaining the cavitation effects in suspension is becoming a focus of ultrasonic dispersion, in particular for MRPF.

Based on classical bubble theory, the bubble dynamics models for MRPF under ultrasonic preparation and mechanical preparation are established, considering the solid-liquid mixing effect of suspension. The bubble motion under different processes is discussed and verified, in order to provide theoretical support for the cavitation effect of the preparation process of MRPF.

## 7.2 PROPERTIES OF MRPF

The general composition of MRPF is shown in Table 7.1. The liquid phase of MRPF is formed by mixing deionized water and propanetriol, which are the main components of MRPF [9]. The solid phase of MRPF is composed of carbonyl iron particles, green silicon carbide, and a small amount of additives. The carbonyl iron particle, as a magnetically sensitive particle, provides MRPF

**TABLE 7.1**
**General Composition of MRPF**

| Composition | | Number | Unit | Volume Fraction $\alpha$/% |
|---|---|---|---|---|
| Liquid phase | Deionized water | 70 | ml | 60.99 |
| | Propanetriol | 1 | ml | |
| Solid phase | Carboxymethylcellulose sodium | 0.33 | g | 5.89 |
| | Sodium hexametaphosphate | 1.35 | g | |
| | Sodium nitrite | 0.2 | g | |
| | Anhydrous sodium carbonate | 0.4 | g | |
| | Nanosilica | 4.32 | g | |
| | Carbonyl iron particle | 230 | g | 25.07 |
| | Green silicon carbide | 30 | g | 8.12 |

with obvious rheological properties and high shear yield strength. The green silicon carbide allows MRPF to have the ability of grinding and polishing. Additives can be seen as surface activators for solid particles, inhibiting agglomeration of solid particles in MRPF and preventing them from settling to some extent. Some of the commonly used additives are carboxymethylcellulose sodium, sodium hexametaphosphate, sodium nitrite, anhydrous sodium carbonate, and nanosilica.

From the composition of MRPF, it is clear that MRPF can be considered as a two-phase suspension consisting of a variety of solid particles and base fluid. When MRPF is prepared, the particles of the solid phase are rapidly and uniformly dispersed in the base fluid under externally applied fluid pressure and velocity. When the size of solid-phase particles is in the range of microns or nanometers, the solid-phase particles and the liquid medium can move at the same speed. In this case, the solid-liquid two-phase flow can be regarded as a strong coupling effect, and the solid-liquid two-phase flow motion can be approximated as a unidirectional flow motion. Thus, the equation of continuity of the mixture is introduced to describe the MRPF [10]:

$$\frac{\partial \rho}{\partial t} + \nabla(\rho u) = 0 \tag{7.1}$$

$$\rho = \sum_k \alpha_k \rho_k \tag{7.2}$$

Where $u$ is the flow rate of the mixture, $k$ is the type of phase, $\alpha_k$ is the volume fraction of each phase, and $\rho_k$ is the density of each phase.

The percentage of volume of tiny gas nuclei in the base fluid can be negligible when the MRPF is not affected by the ultrasonic cavitation effect.

Thus, the density $\rho$ of MRPF can be expressed as:

$$\rho = \alpha_p \rho_p + \alpha_l \rho_l \tag{7.3}$$

Where $\alpha_\mathrm{p}$ and $\alpha_\mathrm{l}$ are the volume fraction of the solid particles and the base liquid, respectively, and $\rho_\mathrm{p}$ and $\rho_\mathrm{l}$ are the densities of the solid particles and the base liquid, respectively.

In addition, the viscosity of MRPF is linked to the type of solid particles, the particle geometry, the particle concentration, and the ratio of the base fluid. However, it is the volume fraction of the solid phase that determines the viscosity of MRPF.

For the diluted suspension, Einstein gives the following equation to explain the viscosity [11]:

$$\mu = \mu_0(1 + k_0 \alpha_p) \tag{7.4}$$

Where $\mu_0$ is the viscosity of the base fluid and $k_0$ is the constant of the shape of the particles, for spherical particles $k_0 = 2.5$.

Equation (7.4) is used for the volume fraction of the solid phase up to 2% and cannot meet the process requirements of MRPF. On this basis, Vand proposed another viscosity equation for a high volume fraction of the solid phase [12], as follows:

$$\mu = \mu_0 e^{k_0 \alpha_p / (1 - a' \alpha_p)} \tag{7.5}$$

Where $a'$ is the second constant of the shape of the particle, for spherical particles, $a' = 39/64 = 0.609$. Taking into account the viscosity change in the preparation of MRPF, the Vand viscosity equation is introduced to describe the viscosity of MRPF.

## 7.3 CAVITATION BUBBLE DYNAMICS MODEL OF MRPF

MRPF usually uses an external magnetic field for the polishing operation, and it can be aggregated into a chain structure along the magnetic induction lines under the magnetic field. In this case, MRPF is a highly viscous semi-solid substance. However, during the preparation of MRPF, an additional magnetic field is not needed, where the solid phase in MRPF should be uniformly dispersed in the base fluid and behave as a Newtonian fluid. Thus, the cavitation bubble dynamics model discussed in the study is applicable to MRPF under the conditions of a nonmagnetic field.

The assumptions for the bubble of MRPF during ultrasonic dispersion are as follows: 1) the bubble always remains spherical in motion and the center of the bubble is fixed; 2) the gas inside the bubble is ideal and its isothermal and adiabatic processes are considered; 3) the solid-liquid mixing ratio is taken into account and the liquid is incompressible; and 4) heat exchanges, interface phase changes, and chemical reactions are not considered.

The dynamic behavior of the bubble in MRPF can be likened to the system of a spring oscillator. The gas inside the bubble corresponds to the spring, and the MRPF outside the bubble corresponds to a mass. Then the increase in internal energy of the gas inside the bubble $\mathrm{d}U$ is equal to the work done by the bubble in the MRPF $\mathrm{d}W$, that is,

$$\mathrm{d}U = \mathrm{d}W \tag{7.6}$$

The internal energy of the gas inside the bubble is produced by the change in bubble radius and can be expressed as follows [13]:

$$U = \int_R^{\infty} \frac{1}{2} \rho \left( \frac{R^2}{r^2} \frac{dR}{dt} \right)^2 4\pi r^2 dr \tag{7.7}$$

where $R$ is the instantaneous radius of the bubble and $r$ is the distance from the center of the bubble.

The work done on the bubble by the surrounding liquid is as follows:

$$W = -\int_{R_0}^{R} p \cdot 4\pi R^2 dR \tag{7.8}$$

$$p = p_g + p_v - \frac{2\sigma}{R} - 4\mu\frac{\dot{R}}{R} - p_\infty \tag{7.9}$$

where $p_g$ is the pressure inside the bubble, $p_v$ is the saturated vapor pressure inside the bubble, $\sigma$ is the surface tension of the liquid, and $p_\infty$ is the far-field liquid pressure.

The pressure $p_g$ within the gas bubble of the MRPF is expressed as follows [14]:

$$p_g = \begin{cases} \left(p_0 + \frac{2\sigma}{R_0} - p_v\right)\left(\frac{R_0}{R}\right)^3, & R_0 \le R \le R_{max} \\ \left(p_0 + \frac{2\sigma}{R_0} - p_v\right)\left(\frac{R_0^3}{R^3 - a^3}\right)^\gamma, & R_{min} \le R \le R_0 \end{cases} \tag{7.10}$$

where $R_0$ is the initial radius of the bubble, $R_{min}$ is the minimum radius of the bubble, $R_{max}$ is the maximum radius of the bubble, $\gamma$ is the gas adiabatic index, and $a$ is the radius of van der Waals ($R_0/a = 8.54$).

In the process of ultrasonic dispersion of MRPF, the far-field fluid pressure is expressed as follows:

$$p_\infty = p_0 - p_a \sin(2\pi f t) \tag{7.11}$$

where, $p_0$ is the static pressure of MRPF, $p_a$ is the pressure amplitude of the ultrasonic wave, $f$ is the ultrasonic frequency, and $t$ is the time.

However, for mechanical dispersion of MRPF, the rotational fluid movements are considered. When the fluid rotates around the axis of a roller, the liquid will generate dynamic pressure due to centrifugal forces. Thus, the far-field pressure is expressed as:

$$p_\infty = p_0 + \rho a_r r_r = p_0 + 2\rho(\pi n r_r)^2 \tag{7.12}$$

Where $a_r$ is the average acceleration of the rotating fluid, $r_r$ is the radius of the roller, and n is the speed of the roller.

Combining equations (7.3) and (7.5) and taking equations (7.7) to (7.11) in equation (7.6), the bubble dynamic model for the ultrasonic dispersion of MRPF is obtained as follows:

$$R\frac{d^2R}{dt^2} + \frac{3}{2}\left(\frac{dR}{dt}\right)^2 = \frac{1}{\rho}(p_g + p_v - \frac{2\sigma}{R} - \frac{4\mu}{R}\frac{dR}{dt} - p_0 + p_a \sin(2\pi ft)) + \frac{R}{\rho c}\frac{d}{dt}(p_g + p_a \sin(2\pi ft)) \tag{7.13}$$

Similarly, associating equations (7.3) and (7.5) and bringing equations (7.7) to (7.10) and (7.12) into equation (7.6), the bubble dynamic model for the mechanical dispersion of MRPF is obtained as follows:

$$R\frac{d^2R}{dt^2} + \frac{3}{2}\left(\frac{dR}{dt}\right)^2 = \frac{1}{\rho}(p_g + p_v - \frac{2\sigma}{R} - \frac{4\mu}{R}\frac{dR}{dt} - p_0 + \rho d\,(\pi n)^2) + \frac{R}{\rho c}\frac{dp_g}{dt} \tag{7.14}$$

## 7.4 BUBBLE MOTION CHARACTERISTICS OF MRPF UNDER DIFFERENT PROCESSES

The numerical simulation of equations (7.13) and (7.14) was complicated using the fourth-order Runge–Kutta method and MATLAB tools. The initial conditions are as follows: For $t = 0$, $R = R_0$, and $dR/dt = 0$. The calculated parameters for MRPF can be seen in Table 7.1, where $\rho_{\text{Carbonyl iron particle}} = 7.89 \times 103$ kg/m$^3$, $\rho_{\text{green silicon carbide}} = 3.17 \times 103$ kg/m$^3$, $\rho_{\text{additive}} \approx 1.22 \times 103$ kg/m$^3$, $\rho_l = 1.00 \times 103$ kg/m$^3$, and the remaining parameters are as follows: $c = 1{,}481$ m/s, $\gamma = 4/3$, $p_v \approx 0$ Pa, $p_0 = 1.013 \times 105$ Pa, $\sigma = 7.28 \times 10^{-2}$ N/m, and $\mu_0 = 1.005 \times 10^{-3}$ Pa·s. Furthermore, for ultrasonic dispersion, the ultrasonic amplitude $p_a = 0.3 \times 106$ Pa and the ultrasonic frequency $f = 20$ kHz, while for mechanical dispersion, the radius of the roller $r_r = 100$ mm and the maximum stirring speed of the roller $n = 350$ r/ min.

Figure 7.1 shows the bubble motion characteristics of MRPF under ultrasonic dispersion, where the initial radius of the bubble $R_0$ is taken as 10 μm, 20 μm, 50 μm, and 100 μm. To reveal the effect of the solid-phase composition of the MRPF on the cavitation effect, the bubble motion characteristics are discussed in two cases, the liquid phase and the solid-liquid phase. As can be seen from Figure 7.1, the bubble of MRPF exhibits the dynamic process of growth, expansion, compression, collapse, and rebound under one acoustic cycle, but the solid phase in the base fluid has an obvious effect on the cavitation bubble motion. The addition of the solid phase to the base fluid significantly reduces the amplitude of the bubble expansion and prolongs the time of bubble collapse. This indicates that the addition of solid particles to MRPF significantly weakens the original cavitation effect of the liquid phase.

Comparing Figures 7.1(a) to (d), it can be seen that as the initial radius of the bubble increases, the amplitude of the bubble expansion is weaker, the time of the bubble collapse is longer, and the degree of the bubble rebound is also reduced. It demonstrates that the larger bubbles in MRPF are not conducive to the generation and development of the cavitation effect. From the current awareness of the ultrasonic wave, the bubbles are subjected to the alternating

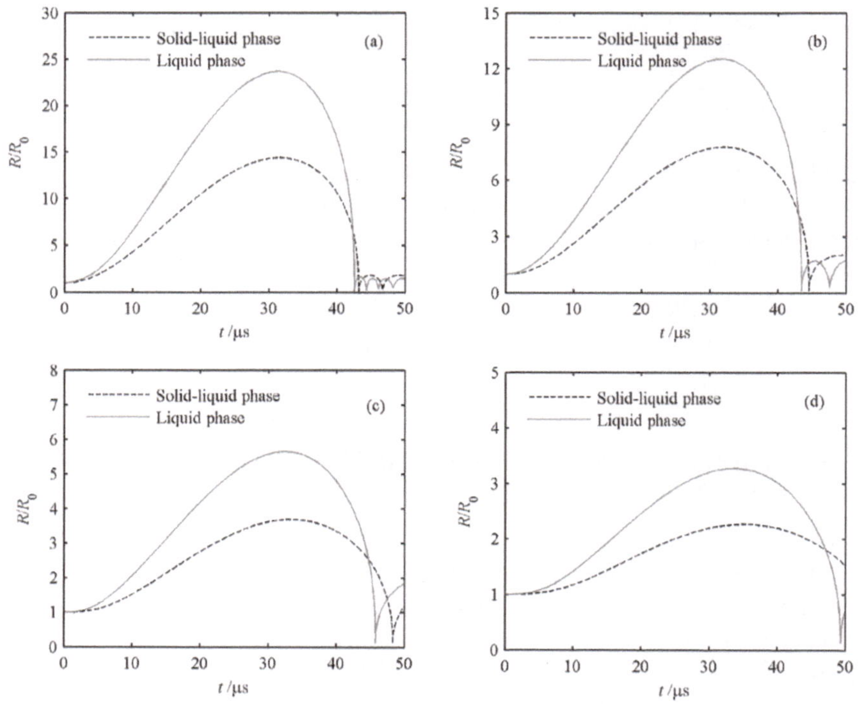


**FIGURE 7.1** Bubble motion characteristics of MRPF under ultrasonic dispersion. (a) $R_0$ = 10 μm; (b) $R_0$ = 20 μm; (c) $R_0$ = 50 μm; (d) $R_0$ = 100 μm.

action of the negative and positive pressure of the ultrasonic wave during an acoustic cycle. The bubbles can grow under negative pressure and shrink under positive pressure. However, the larger bubble starts to shrink in the positive pressure before it grows to the maximum size in the negative pressure, and then its cavitation effect is less intense.

Figure 7.2 shows the bubble motion characteristics of MRPF under mechanical dispersion, where the initial radius of the bubble $R_0$ is taken as 10 μm, 20 μm, 50 μm, and 100 μm. From Figure 7.2, it can be seen that the bubble of MRPF produced by the mechanical dispersion does not undergo expansion but is directly compressed and then undergoes several bounces. With increasing time, the amplitude of the bubble bounce becomes smaller. Meanwhile, solid particles in the base fluid have an obvious influence on bubble movement. With the addition of solid particles to the base fluid, the amplitude of the bubble rebound is increased, and the rebound time is extended. It indicates that, for the mechanical dispersion, the solid phase has the ability to enhance the original cavitation effect of the liquid phase.

Comparing Figures 7.2(a) to (d), it can be found that with the increase of the initial radius of the bubbles in MRPF, the amplitude of the bubble bounce varies

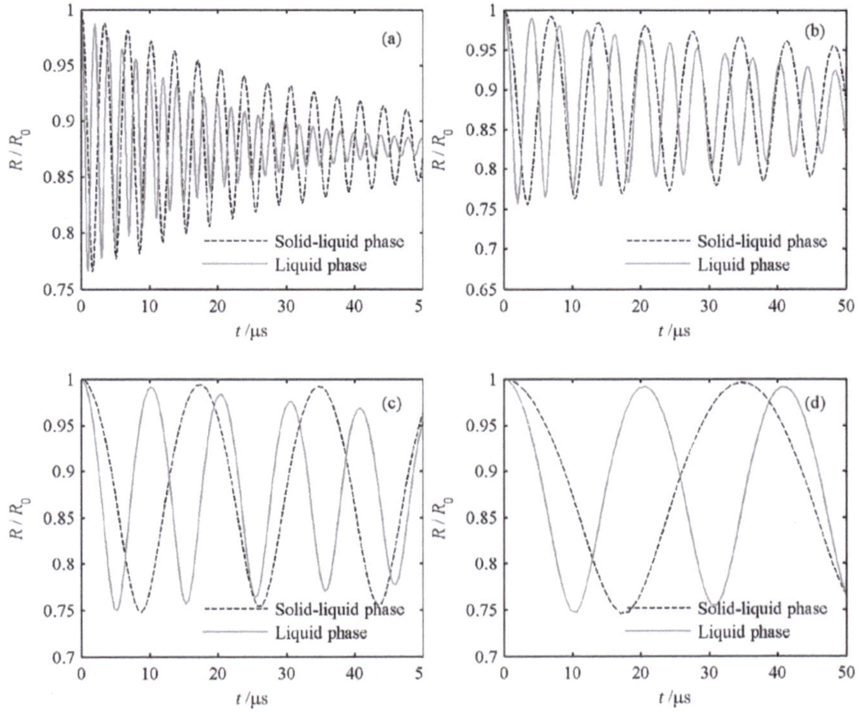


**FIGURE 7.2** Bubble motion characteristics of MRPF under mechanical dispersion. (a) $R_0$ = 10 μm; (b) $R_0$ = 20 μm; (c) $R_0$ = 50 μm; (d) $R_0$ = 100 μm.

very little, but the time of the bubble bounce is significantly longer, which also indicates that the larger bubbles are not conducive to the improvement of cavitation effect.

In order to further reveal the difference in intensity of cavitation effect under ultrasonic and mechanical dispersion, the maximum pressure inside the bubble and maximum velocity of the bubble wall in the MRPF under the two processes are given in Figures 7.3 and 7.4, respectively. As can be seen from Figure 7.3, the maximum pressure produced by the bubble of different initial radii is up to the order of $10^9$ Pa. As the initial radius of the bubble increases, the maximum pressure inside the bubble gradually decreases. For ultrasonic dispersion, the maximum pressure inside the bubble decreases significantly when solid particles are added to the base liquid. However, for mechanical dispersion, the maximum pressure inside the bubble increases gradually as solid particles are added to the base liquid. It is mainly related to the fact that the bubble produced by ultrasonic dispersion can undergo the dynamic processes of expansion, compression, collapse, and rebound, whereas the bubble produced by mechanical dispersion only has the behavior of compression and rebound. A comparison of Figures 7.3(a) and (b) shows that the maximum pressure inside the bubble produced by

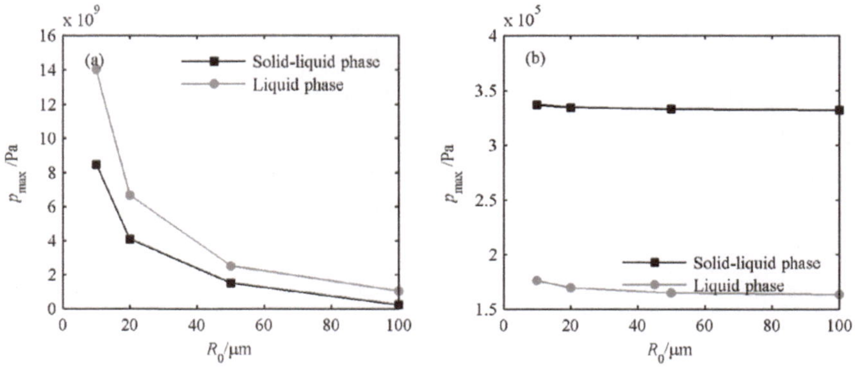


**FIGURE 7.3** Maximum pressure inside the bubble of MRPF. (a) Ultrasonic dispersion; (b) Mechanical dispersion.

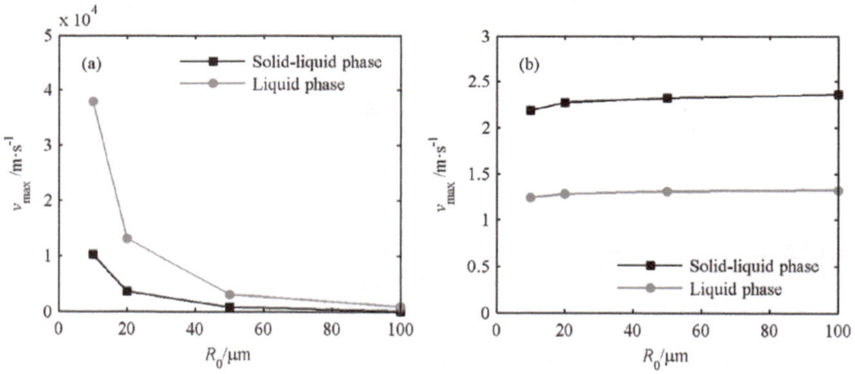


**FIGURE 7.4** Maximum velocity of the bubble wall of MRPF. (a) Ultrasonic dispersion; (b) Mechanical dispersion.

ultrasonic dispersion of the MRPF is $10^4$ orders of magnitude higher than that of mechanical dispersion. It indicates that MRPF has a more intense cavitation effect under ultrasonic action.

Figure 7.4 presents the maximum velocity of the bubble wall of MRPF under both processes, and the variation is similar to that of Figure 7.3. The ultrasonic dispersion of MRPF produces a velocity of the bubble wall $10^4$ orders of magnitude higher than that of the mechanical dispersion. It again proves that the ultrasonic dispersion of MRPF has a stronger cavitation effect than that of the mechanical preparation. Furthermore, for bubbles of different initial radii, the velocity of the bubble wall under the ultrasonic wave is of the order of $10^4$. The velocity of the bubble wall is more likely to be greater than the speed of sound in the base fluid ($c$ = 1481m/s) [15]. Thus, the bubble can produce a high-speed microjet effect near the solid particles, which will result in a strong impact and dispersion effect on the solid particles. The phenomenon is also responsible for one of the microscopic reasons for the superiority of ultrasonic dispersion over mechanical dispersion.

## 7.5 SEDIMENTATION AND DISPERSION OF MRPF

Sedimentation of MRPF is an important indicator to evaluate its performance and can indirectly reflect the dispersion of the particles [16]. Figure 7.5 gives a comparison of the experimental results of the sedimentation rate of MRPF under ultrasonic and mechanical dispersion. The MRPF was prepared by choosing the composition and specifications in Table 7.1, and the additional parameters of both processes were as follows: For ultrasonic dispersion, the ultrasonic power was 500 W, the ultrasonic frequency was 20 kHz, and the ultrasonic time was 30 s; for mechanical dispersion, the roller stir speed was 200 r/min, and the stirring time was 3 h. The MRPF was observed after settling for 0.5~4 h.

In Figure 7.5, it can be observed that the sedimentation rates of MRPF increase with the extension of the observation time. However, the sedimentation rate of MRPF under ultrasonic dispersion is much lower than that of the mechanical dispersion during the same observation time. This phenomenon can also be found in the theoretical analysis in Figures 7.1 to 7.4, the dramatic cavitation effect of MRPF can occur in the process of ultrasonic dispersion compared to the mechanical dispersion. The ultrasound wave causes the cavitation bubble of MRPF to undergo the dynamic process of growth, expansion, compression, collapse, and rebound and also generates up to $10^4$ orders of magnitude of velocity of the bubble wall and $10^9$ orders of magnitude of pressure inside the bubble. Whatever, the bubble near the solid particles is also prone to generate high-speed microjets, thus resulting in a strong impact and dispersion

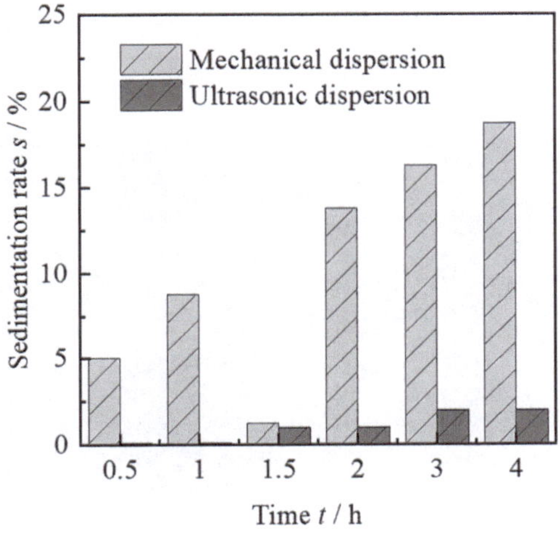


**FIGURE 7.5** Sedimentation rate of MRPF.

effect on the solid particles of the MRPF. On the contrary, under the mechanical dispersion process, the bubble in MRPF just undergoes a forced damping motion and the velocity of the bubble wall and pressure inside the bubble may be relatively low. It has a confined impact effect on the solid particle of MRPF and then leads to a significant increase in the sedimentation rate of MRPF. The experimental results for MRPF are in general in agreement with the results of the analysis of the dynamic characteristics of the bubble in the MRPF.

The microscopic dispersion of MRPF from Table 7.1 is shown in Figure 7.6. The microstructure of MRPF was recorded using a DYS-107 trinocular biomicroscope. The magnification of the image is 40x. In Figure 7.6, the microstructure of MRPF is divided into black area and white area. The black area represents the various solid particles of the solid phase, while the white area represents the liquid phase. As can be seen in Figure 7.6, solid particles of the MRPF under ultrasonic dispersion can be uniformly dispersed in the base solution, while the mechanical dispersion tends to agglomerate and behave unevenly. It again demonstrates the effectiveness of the ultrasonic dispersion process.

The process of MRPF is of great importance to the industrial application of technologies such as magnetorheological polishing. Figure 7.7 gives several examples of applications of MRPF in the field of magnetorheological polishing. The material in Figure 7.7(a) and (b) is 6061 aluminum and that of Figure 7.7(c) and (d) is 440C stainless steel, respectively. As can be observed in Figures 7.7(a) to (d), the surface of the specimen before polishing has rust stains, more pits, and edge burrs. However, the surface of the machined specimen is bright, free of pits and scratches, and the edge burrs have been effectively removed. The results of such tests are on the one hand related to the processing parameters of magnetorheological polishing such as current, processing distance, abrasive type, particle size, processing time, and speed. On the other hand, it is closely related to the properties of MRPF. From the analysis of the study, it is clear that the cavitation effect of MRPF has a significant impact on the performance of MRPF, and that such an effect will directly affect the technological maturity of

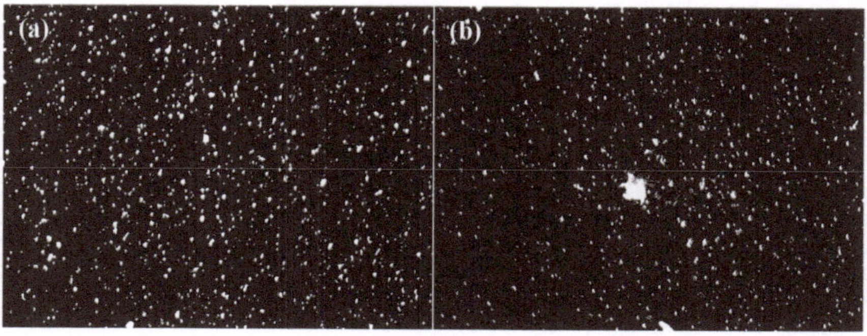


**FIGURE 7.6** Microscopic dispersion of the MRPF. (a) Ultrasonic dispersion; (b) Mechanical dispersion.

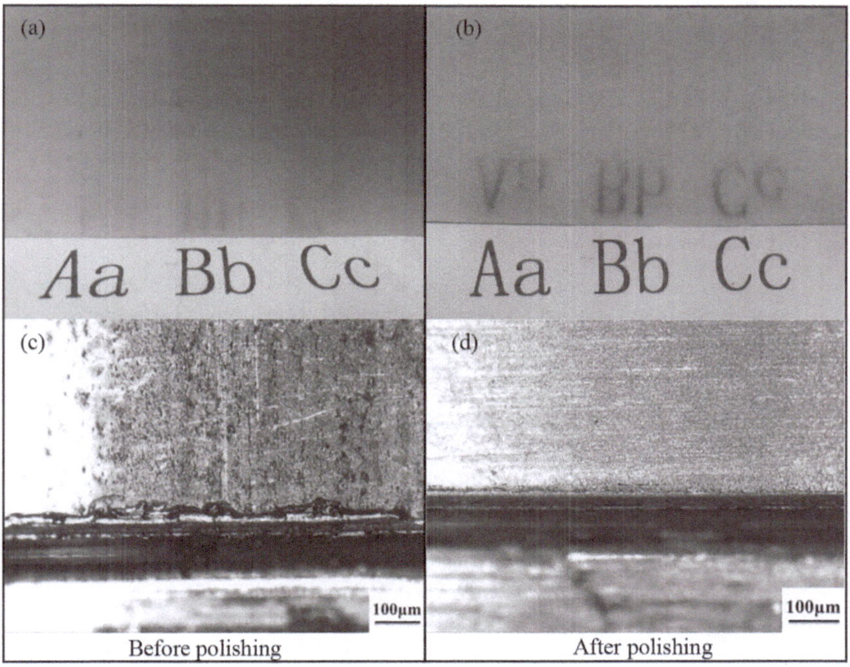


**FIGURE 7.7** Examples of applications of MRPF. (a) and (b) 6061 aluminum; (c) and (d) 440C stainless steel.

technologies such as magnetorheological polishing in industry. However, the control and exploitation of the cavitation effect may not be easy to achieve, especially for the MRPF, and it deserves to be explored in depth subsequently.

## 7.6 CONCLUSION

The mixture continuity equation and the Vand viscosity equation were used to construct the bubble dynamic model of MRPF during ultrasonic and mechanical dispersion. Numerical simulation of the bubble dynamic model revealed that the bubble in MRPF reacted more vigorously under the ultrasonic dispersion compared to the mechanical dispersion. Sedimentation and dispersion behaviors of MRPF were considered to be superior under ultrasonic dispersion than under mechanical dispersion.

Furthermore, the pressure inside the bubble and the wall velocity of the bubble produced by the ultrasonic dispersion of MRPF are $10^4$ orders of magnitude higher than those of the mechanical dispersion, and MRPF is more likely to produce microjets under the ultrasonic dispersion, which is one of the microscopic features leading to the superiority of the ultrasonic dispersion over the mechanical dispersion.

## ACKNOWLEDGMENTS

This work was supported by the National Natural Science Foundation of China (Nos. 51975399 and 52075362), the project funded by the Postdoctoral Research Foundation of China (No. 2020M670712), and the National Science Foundation of Shanxi Province of China (No. 202203021221067 and 201901D211016).

## REFERENCES

[1] Dorosi, A. H., Ghatee, M., Norouzi, M., 2020, Preparation and characterization of water-based magnetorheological fluid using wormlike surfactant micelles. *Journal of Magnetism and Magnetic Materials* 498:166193.
[2] Barman, A., and Das, M., 2018, Nano-finishing of bio-titanium alloy to generate different surface morphologies by changing magnetorheological polishing fluid compositions. *Precision Engineering* 51:145–152.
[3] Zhang, P., Dong, Y. Z., Choi, H. J., et al., 2020, Reciprocating magnetorheological polishing method for borosilicate glass surface smoothness. *Journal of Industrial and Engineering Chemistry* 84:243–251.
[4] Sumitomo, S., Koizumi, H., Uddin, M. A., et al., 2018, Comparison of dispersion behavior of agglomerated particles in liquid between ultrasonic irradiation and mechanical stirring. *Ultrasonics Sonochemistry* 40:822–831.
[5] Vinatoru, M., Mason, T. J., Calinescu, I., 2017, Ultrasonically assisted extraction (UAE) and microwave assisted extraction (MAE) of functional compounds from plant materials. *Trends in Analytical Chemistry* 97:159–178.
[6] Mahdi, M., Ebrahimi, R., Shams, M., 2011, Numerical analysis of the effects of radiation heat transfer and ionization energy loss on the cavitation Bubbles dynamics. *Physics Letters A*, 375(24):2348–2361.
[7] Behnia, S., Sojahrood, A. J., Soltanpoor, W., et al., 2009, Towards classification of the bifurcation structure of a spherical cavitation bubble. *Ultrasonics* 49(8):605–610.
[8] Guo, C., Liu, J., Li, X. H., et al., 2021, Effect of cavitation bubble on the dispersion of magnetorheological polishing fluid under ultrasonic preparation. *Ultrasonics Sonochemistry* 79:105782.
[9] Sarawathamma, K., Jha, S., Rao, P. V., 2017, Rheological behaviour of magnetorheological polishing fluid for Si polishing. *Materials Today Proceedings* 4(2):1478–1491.
[10] Peters, A., Sagar, H., Lantermann, U., et al., 2015, Numerical modelling and prediction of cavitation erosion. *Wear* (338-339):189–201.
[11] Einstein, A., 1906, Eine neue Bestimmung der Molekuldimensionen. *Annals of Physics* 19:289–306.
[12] Vand, V., 1948, Viscosity of solutions and suspensions. *The Journal of Physical and Colloid Chemistry* 52:277.
[13] Guo, C., and Zhu, X. J., 2018, Effect of ultrasound on dynamics characteristic of the cavitation bubble in grinding fluids during honing process. *Ultrasonics* 84:13–24.
[14] Ruuth, S. J., Putterman, S., Merriman, B., 2002, Molecular dynamics simulation of the response of a gas to a spherical piston: Implications for sonoluminescence. *Physical Review E Statal Nonlinear & Soft Matter Physics* 66(3):036310.
[15] Vignoli, L. L., Barros, A. L. F. D., Thomé, R. C. A., et al., 2013, Modeling the dynamics of single-bubble sonoluminescence. *European Journal of Physics* 34(3):679–688.

[16] Bica, I., Liu, Y. D., Choi, H. J., 2013, Physical characteristics of magnetorheological suspensions and their applications, *Journal of Industrial and Engineering Chemistry* 19:394–406.

## NOMENCLATURE

| | |
|---|---|
| $a$ | radius of van der Waals (m) |
| $a_r$ | average acceleration of the rotating fluid (m/s$^2$) |
| $a$ | second constant of the shape of the particle |
| $c$ | speed of sound in the base fluid (m/s) |
| $f$ | ultrasonic frequency (Hz) |
| $k$ | type of phase |
| $k_0$ | constant of the shape of the particles |
| $n$ | speed of the roller (r/min) |
| $p_a$ | pressure amplitude of the ultrasonic wave (Pa) |
| $p_g$ | pressure inside the bubble (Pa) |
| $p_0$ | ambient static pressure (Pa) |
| $p_v$ | saturated vapor pressure inside the bubble (Pa) |
| $p$ | far-field liquid pressure (Pa) |
| $r$ | distance from the center of the bubble (m) |
| $r_r$ | radius of the roller (m) |
| $R$ | instantaneous radius of the bubble (m) |
| $R_{max}$ | maximum radius of the bubble (m) |
| $R_{min}$ | minimum radius of the bubble (m) |
| $R_0$ | initial radius of the bubble (m) |
| $s$ | sedimentation rate (%) |
| $t$ | time (s) |
| $u$ | flow rate of the mixture (m/s) |
| $U$ | internal energy of the gas inside the bubble (J) |
| $W$ | work done by the bubble (J) |

## *GREEK LETTERS*

$\alpha_l$ volume fraction of the base liquid (%)

$\alpha_k$ volume fraction of each phase (%)

$\alpha_p$ volume fraction of the solid particles (%)

*Gamma* gas adiabatic index

*Sigma* surface tension of the liquid(N /m)

*row* density of the liquid ($kg/m^3$)

$\rho_{Carbonyl\ iron\ particle}$ density of the carbonyl iron particle ($kg/m^3$)

$\rho_{green\ silicon\ carbide}$ density of the green silicon carbide ($kg/m^3$)

$\rho_{additive}$ density of the average of various additives ($kg/m^3$)

$\rho_k$ density of each phase ($kg/m^3$)

*Meu* liquid viscosity (Pa·s)

$\mu_0$ viscosity of the base fluid (Pa·s)

# 8 Surface Coating

*K. Algadah and A. Alaboodi*

## 8.1 INTRODUCTION

Recently, additive manufacturing has improved the performance of numerous industries. This includes metallic parts and polymeric components [1]. The size of structure that is less than 100 μm is classified as a microstructure. In aerospace and automotive industries, thermal spray coating is widely utilized. In thermal spray process, ceramics, alloys, and cermet are common micro-materials used [2]. Plasma spray coating was well-known coating process in the last 25 years [3]. Modifying the microstructure of the coating surface would enhance its mechanical properties. The resultant morphology of the structure and the crystallization behavior influenced the mechanical properties of a nano structural TiCN coated by means of plasma spray [4]. The effect of mixing two gases such as Ar and $H_2$ for the plasma spray process of micro $Al_2O_3$ particles has stimulated the particles to melt and impinged on the substrate significantly faster [5]. In terms of cavitation erosion, solid particles erosion slurry erosion, and mechanical properties, high velocity air fuel (HVAF) is much better than high velocity oxygen fuel (HVOF) in the aforementioned properties [6]. HVOF is considered a highly effective technique in industry. Flame spray coating was the first thermal method had been employed. It started in 1909 [7]. Laser treatment is implemented to enhance the surface composition which decreases the porosity of the coating [8]. It is used for parts repair, fast manufacturing, and coating [9]. The higher the temperature of the substrate, the faster the deposition time for the powder particles [10]. Extreme high speed laser cladding (EHLA) was introduced in 2013 to overcome some conventional laser cladding disadvantages such as low surface preparation rate [11]. In addition, in 1960s, physical vapor deposition (PVD) was first emerged [12]. Sputtering was the first vacuum coating process. It was used with induction coils from 1852–1920s [13]. Pratt and Whitney aerospace company was the first who used electron beam physical vapor deposition (EBPVD) back in 1970s [14]. According to Delimi et al.'s experiment, a carbon steel coated with organosilicon layer 2.5 μm in thickness by plasma enhanced vapor deposition improved its corrosion resistance [15]. Electroless plating of titanium alloy substrate was coated with Ni-B and Ni-B-W and it enhanced its hardness and wear resistance [16]. Surface roughness of Ni-Co coating was enhanced by tungsten disulfide

DOI: 10.1201/9781003364948-8

($WS_2$) nanoparticles [17]. Ling et al. found a difference in the coating thickness of a composite plate of Al-Mg after it was coated by micro arc oxidation (MAO) process [18]. They found that Al plate had higher thickness than the Mg plate. Moreover, plasma immersion ion implantation was used for surface coating and device manufacturing [19]. It was used with CrN coating to produce high-density crystal defects which improved the hardness and maintained its toughness [19].

This chapter demonstrates many surface coating methods which are used for micro additive manufacturing. Thermal spray processes, cladding processes, physical vapor deposition processes, chemical vapor deposition processes, electrodeposition processes, and ion implantation processes are elaborated.

## 8.2 THERMAL SPRAY COATING

### 8.2.1 Plasma Spray

Plasma spray coating is one common type of thermal spray coating. It is used to form coating with superior microstructure properties. It is cheap, robots, and replicable process. It forms a micro ceramic layer that covers metal and composite substrates. The principle of this coating is that the coating powder is injected towards the plasma jet.

Because of the high temperature of the plasma jet which ranges from 10000–30000 °C, the particles melt swiftly [19]. They will eventually hit the substrate to construct a coating layer. The direct current flows through the anode and cathode to ignite the plasma gas. Many gases could be used in this process such as nitrogen, hydrogen, helium, or the mixture of them [20]. Figure 8.1 shows the plasma spray coating process.

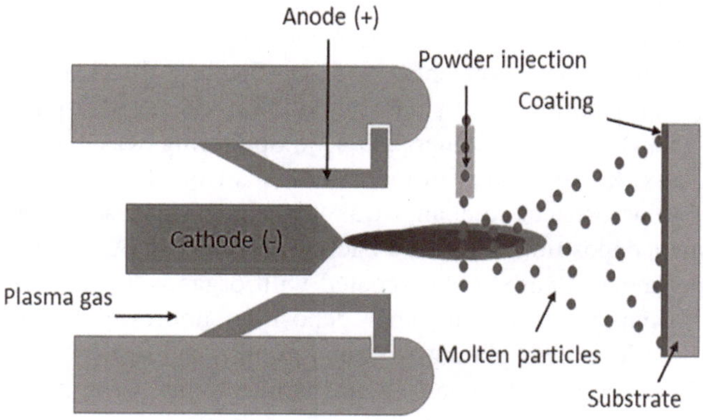


**FIGURE 8.1** Plasma spray coating process.

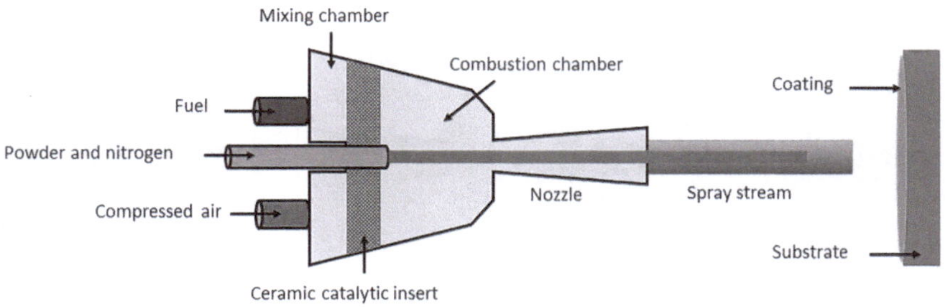


**FIGURE 8.2** HVAF coating process.

### 8.2.2 High Velocity Air Fuel

High velocity air fuel (HVAF) coating process uses fuel, compressed air, powder, and nitrogen to perform the coating. The coating process is illustrated in Figure 8.2. HVAF provides very excellent coating surface with less porosity, oxidation, and decarburization under low processing temperature [3]. Many types of gausses fuel could be injected, for example, natural gas, propylene, and propane [6]. The fuel, compressed air, powder, and nitrogen are mixed in a mixing chamber. Next, they go through an insert ceramic catalytic to the combustion chamber. The gas flow has supersonic speed resulted from the divergent nozzle. The nozzle hastens the particles to melt. The flame temperature is 1900–2000 °C [21]. This process could be used for composite coating. For example, Ni-P reinforced with c-BN has been coated by means of HVAF and then it gave better wear resistance to the coating layer [22].

### 8.2.3 High Velocity Oxygen Fuel

High velocity oxygen fuel (HVOF) is another thermal coating technique. HVOF spraying affords less oxidized and denser coating in comparison to plasma spraying [2]. Its working principle is quite similar to HVAF but it differs in some aspects. A difference between HVOF and HVAF that is HVOF uses oxygen to burn the fuel whereas HVAF uses dry air [21]. Furthermore, the flame temperature in HVOF is higher. It ranges between 2700–3100 °C [21]. The coating process is depicted schematically in Figure 8.3. Oxygen gas and fuel mix together to burn and create flame. Acetylene, propane, propylene, and kerosene (liquid) are fuels that could be used [23]. However, low quality fuels create high-temperature corrosion [24]. The size of powder plays a big role in deposition. The recommended particle size is within the range of 10–40 μm [23]. Big particle will not melt easily whereas the small particles will diminish before the deposition on the substrate occurs. Particles are injected into the direction of the flame. They move toward the substrate to form a coating layer. The gas velocity exceeds 2000 m/s [2]. It is more than Mach number speed which is equal to 343.29 m/s.

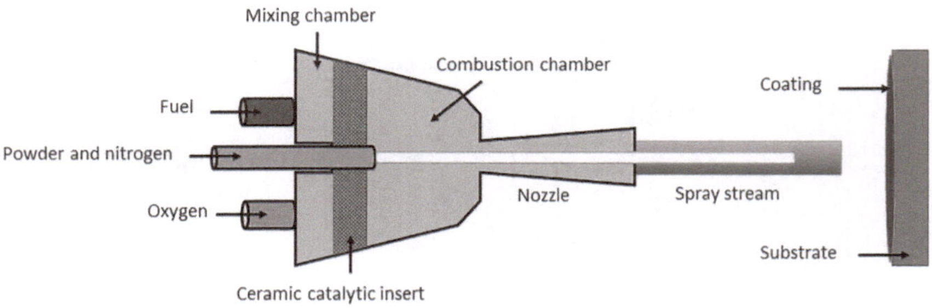


**FIGURE 8.3** HVOF coating process.

### 8.2.4 Cold Spray

In the cold spray (CS) process, mixing of the coating powder and the gas happens before the converging-diverging nozzle. Nitrogen or helium is used as carrier gas.

As it is shown in Figure 8.4, gas is heated before it enters the nozzle using an external heater. The gas flow temperature reaches 1200 °C [25]. The nozzle accelerates the powder particles velocity. It ranges between 300–1200 m/s [26]. Particle size ranges from 5 to 75 μm [25]. The particles consequently impact the substrate to form a thin coating surface. CS processes are classified as low pressure cold spray and high pressure cold spray [26]. The pressure ranges between $680 \times 10^{-3}$–$6.9 \times 10^{-6}$ Pa [25]. Furthermore, it is used for metal, alloys, and composite metal-metal coatings. The metals are Ta, Ti, and Ni, the alloys are Hastelloy and Inconel, and the composite metal-metal is Cu-W [27]. CS coating provides higher hardness value and minimizes the wear by up to two-thirds when it is compared with plasma spray coating having the same coating materials [28].

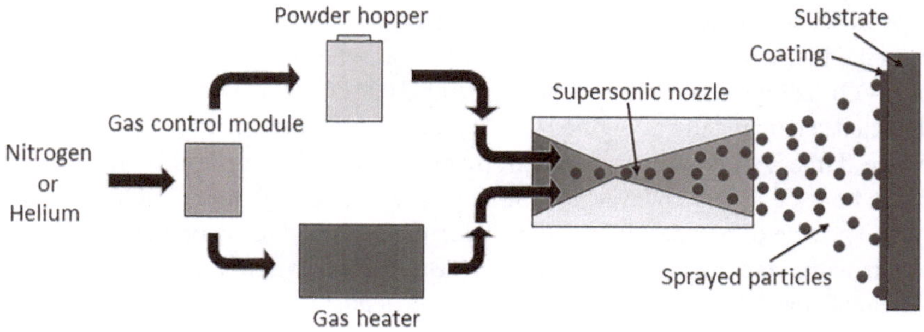


**FIGURE 8.4** CS coating process.

### 8.2.5 Flame Spray

Flame spray coating melts powder particles by combustion flame. It is a cheap process when it is compared with HVOF [29]. The process is illustrated in Figure 8.5. Fuel and oxygen mixtures are used for combustion. The process temperature is 3000 °C approximately [7]. The powder particles are injected before the nozzle. The nozzle speeds up the particle's velocity and lets them diffuse. Particle velocity increases to 100 m/s [7]. The molten particles form a thin layer on the substrate. Table 8.1 summarizes the parameters of thermal spray coating processes.

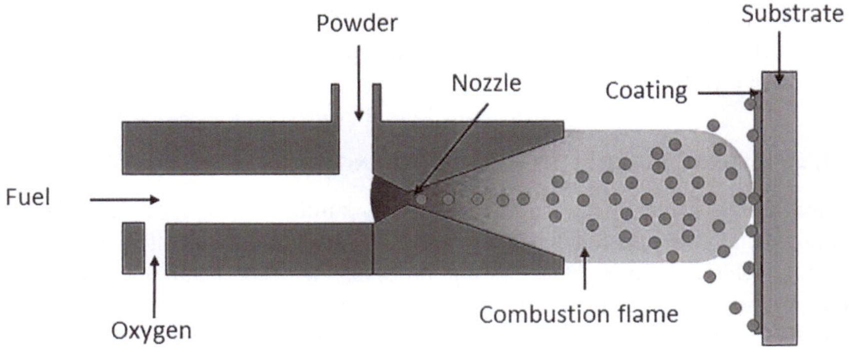


**FIGURE 8.5** Flame spray coating process.

**TABLE 8.1**
**Thermal Spray Coating Process Parameters**

| Coating Process | Process Temperature (°C) | Coating Materials | Gas Mixture |
|---|---|---|---|
| Plasma spray | 10000–30000 | Powder | Nitrogen<br>Hydrogen<br>Helium |
| HVAF | 1900–2000 | Powder | Natural gas<br>Propylene<br>Propane<br>Nitrogen |
| HVOF | 2700–3100 | Powder | Oxygen<br>Nitrogen<br>Fuel |
| Cold spray | 1200 | Powder | Helium |
| Flame spray | 3000 | Powder | Oxygen<br>Fuel |

## 8.3 CLADDING

### 8.3.1 Laser Cladding

Coating layer with micro size can be obtained using laser cladding. There are many different methods of laser cladding. The most popular method used is coaxial powder [9]. The one that is exposed in Figure 8.6 is a coaxial powder process. The powder particles are injected into two paths. Laser beam melts the powder particles forming a molten pool. When the molten pool becomes cold it constructs a cladding layer. Functionally graded materials can be made using laser cladding with the ease of controlling the feed rate of powders and the ability of mixing multiple powder types [30]. It is a versatile method to form multilayer coating from dissimilar compositions [31]. Nevertheless, it is a very expensive coating process compared to HVOF and flame spray [29].

### 8.3.2 Extreme High Speed Laser Cladding

Extreme high speed laser cladding process is similar to the conventional laser cladding that was explained in section 8.3.1, but it is implemented under high moving velocity. During the coating process, the laser beam moves in one direction with a preselected velocity. The coating layer can be formed with velocity of 3.3 m/s [32]. The speed influences the coating thickness. The microstructure of the coating that is formed by high speed laser cladding is smaller and denser [33]. The micro hardness, wear, and corrosion that are obtained due to this process are much better than the results obtained from the classical laser cladding process [33].

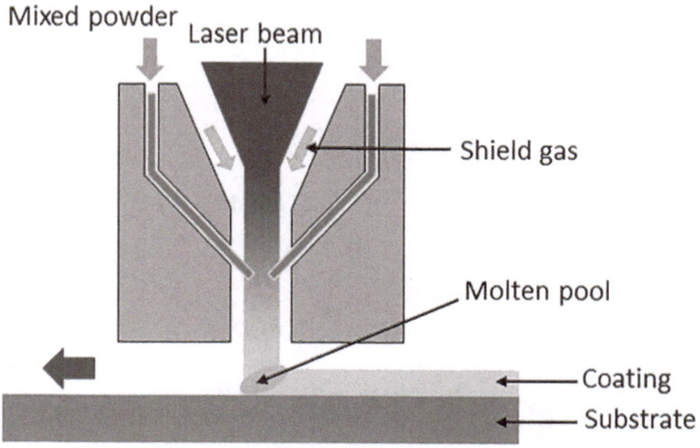


**FIGURE 8.6** Laser cladding coating process.

## 8.4 PHYSICAL VAPOR DEPOSITION

### 8.4.1 Thermal Evaporation

Surfaces can be coated using thermal evaporation system. Coating process occurs inside a vacuum chamber. Powder particles are heated in a crucible. They evaporate and raise towards the substrate making a thin film. The materials experience condensation and evaporation. It transforms from condensation to evaporation when the materials diffuse. It then returns to condensation while the coating is being shaped [34]. This process is used for coating as well as thin films manufacturing. However, it is not appropriate to be used for complex shaped coating such as membranes [12]. The thermal evaporation process is simplified and shown in Figure 8.7.

### 8.4.2 Sputtering

Sputtering is another physical vapor deposition which is employed in micro coating. It allows mixing of multiple materials [12]. Figure 8.8 illustrates sputtering coating process. It uses an electric current to control the powder particles' movement. It enables plasma ions to impact the target and go backward to deposit on the substrate. Nitrogen plasma or argon is used [35]. The gas excites the powder inside the processing chamber. The positive terminal of a direct current is connecting to the substrate. The negative terminal is connected to the target plate. Parameters such as the distance between the sputter target and the substrate as well as the gas velocity affect coating morphology [12].

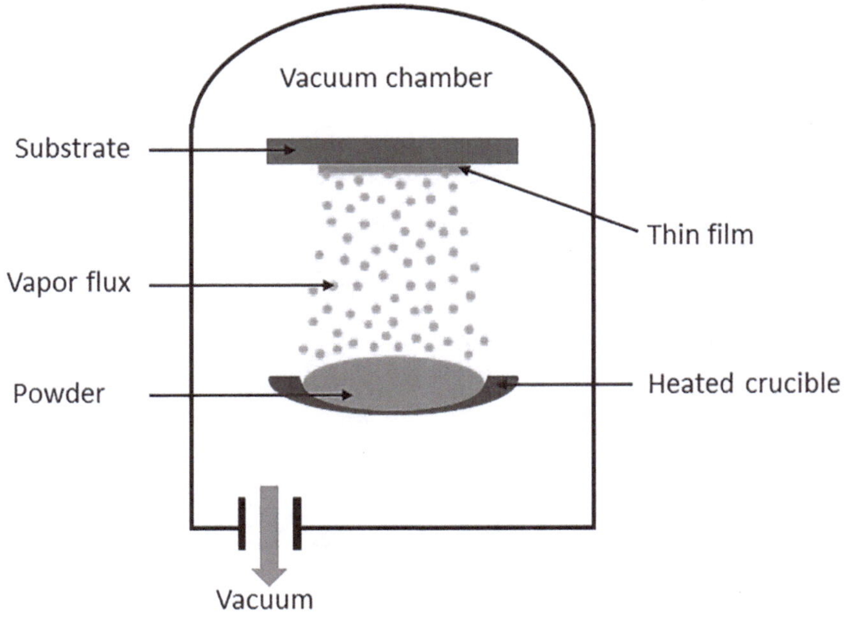


**FIGURE 8.7** Thermal evaporation coating process.

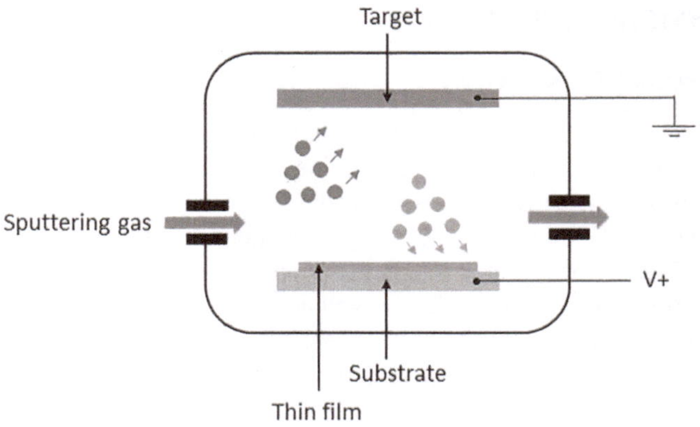


**FIGURE 8.8** Sputtering coating process.

### 8.4.3 Ion Plating

Ion plating coating process is similar to the thermal evaporation process. However, an electric current is used to excite the powder particles. Deposition of powder is carried out under vacuum pressure. The vacuum pressure is $4 \times 10^{-3}$ Pa [36]. The substrate is connected to ground. The positive terminal of the electrical source is connected to the plate. The distance between the substrate and the target is 200 mm [37]. The vapor flux initiates. The flux forms a thin film on the substrate. Deposition time is 90 min [37]. The ion plating process is presented in Figure 8.9.

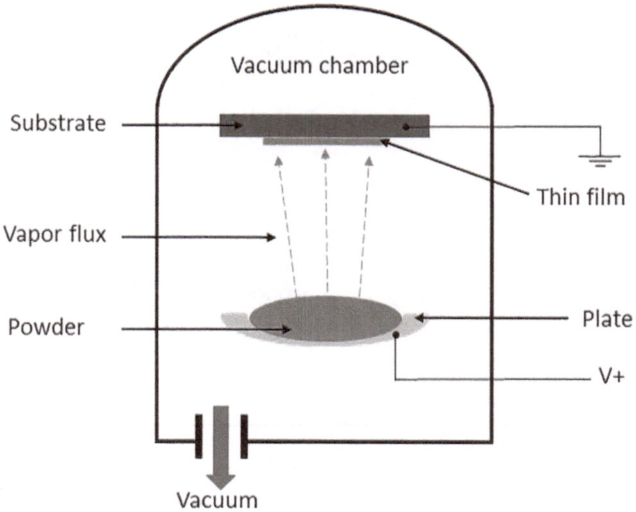


**FIGURE 8.9** Ion plating coating process.

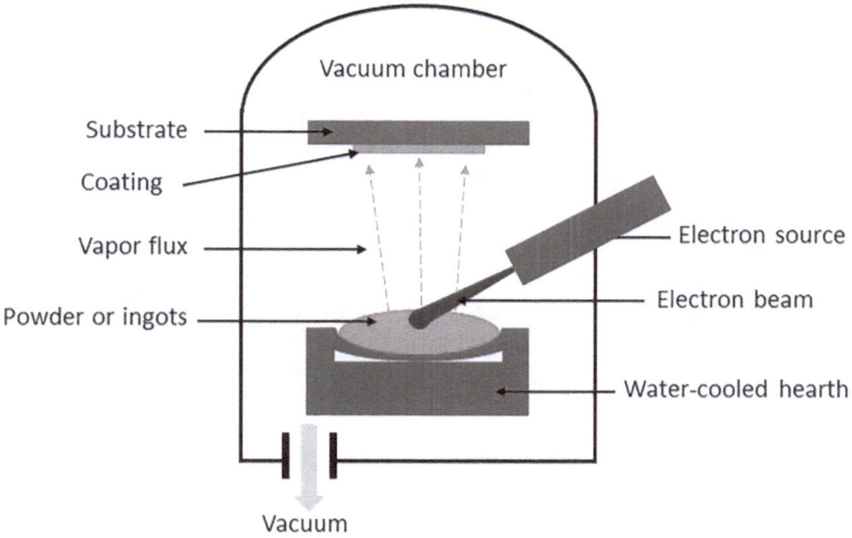


**FIGURE 8.10** Electron beam physical vapor deposition coating process.

### 8.4.4 Electron Beam Physical Vapor Deposition

Electron beam physical vapor deposition (EBPVD) is a process in which the coating materials are melted by a high energy electron beam and then deposited on a substrate under vacuum condition [38]. It is an expensive method because of its low deposition rate [14]. Figure 8.10 shows the EBPVD process. Vacuum pressure is used inside the reaction chamber. Electron beam is directed to the powder particles. As a result, vaporization occurs. The types of vapors are ions, molecules, atoms, and cluster of atoms [39]. Thus, vapor flux forms a coating layer on the substrate. EBPVD also can offer a single crystal columnar microstructure which has superior mechanical adhesion, high-temperature resistance, and high strain tolerance [14]. The vapor pressure and the deposition rate have a proportional relationship with the coating component [39]. Table 8.2 demonstrates the physical vapor deposition processes parameters.

**TABLE 8.2**
**Physical Vapor Deposition Process Parameters**

| Coating Process | Coating Materials | Gas | Supplement | Vacuum Pressure |
|---|---|---|---|---|
| Thermal evaporation | Powder | None | None | Yes |
| Sputtering | Powder | Nitrogen Argon | Electrical source | None |
| Ion plating | Powder | None | Electrical source | None |
| Electron beam physical vapor deposition | Powder | None | Electron beam | None |

## 8.5 CHEMICAL VAPOR DEPOSITION

### 8.5.1 Conventional

In chemical vapor deposition (CVD), the coating process occurs under chemical reaction. It is widely used in microfabrication to deposit materials [40]. It is unlike the PVD process. In CVD, coating materials are in gaseous form whereas the coating materials of PVD are in solid form. It elevates the purity of crystals of the semiconductors. CVD process is characterized by substrate heating, physical characteristics of vapor, and operating conditions [40]. The CVD process is illustrated in Figure 8.11. A gas carrier enters the reaction chamber to stimulate the precursor. The coating stream moves towards the substrate to form a thin film. The process temperature is 500–1100°C [36]. The gas exits the chamber from another side.

### 8.5.2 Plasma Enhanced Chemical Vapor Deposition

Plasma enhanced chemical vapor deposition (PECVD) is a chemical coating process activated by plasma. Figure 8.12 exposes the plasma enhanced chemical vapor deposition coating process. It is a beneficial method for producing dielectric thin films in micro level [41]. It has a low proceeding temperature up to 350°C [42]. The carrier gas enters the chamber to fill it. Radio frequency (RF) power supply is used to energize the upper electrode. The bottom electrode is connected to the ground. Plasma beam initiates between the two electrodes. By means of plasma, gases hit the substrate to constitute a coating layer. Table 8.3 illustrates the chemical vapor deposition processes parameters.

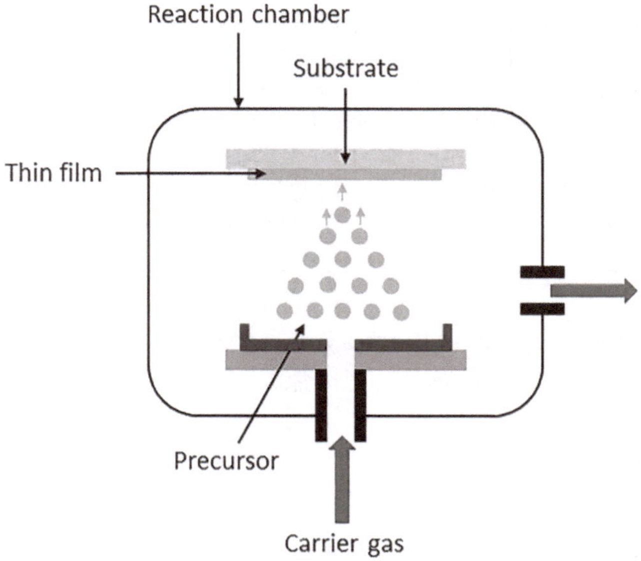


**FIGURE 8.11** Chemical vapor deposition coating process.

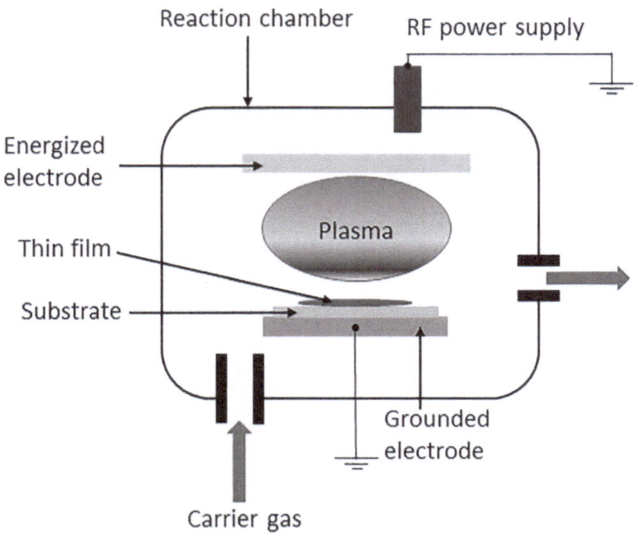


**FIGURE 8.12** Plasma enhanced chemical vapor deposition coating process.

**TABLE 8.3**
**Chemical Vapor Deposition Process Parameters**

| Coating Process | Process Temperature (°C) | Coating Materials | Supplement |
|---|---|---|---|
| Conventional | 500–1100 | Gas | None |
| Plasma enhanced chemical vapor deposition | 350 | Gas | Plasma beam<br>Electrodes |

## 8.6 ELECTRODEPOSITION

### 8.6.1 Conventional

Electrodeposition or electroplating is an electrochemical method that is used for coating materials in micro size. It uses an electric current to deposit ionic species on electrolyte [43]. The materials that are used for coating are connected to anode. The surface which will be coated is connected to cathode. To deposit mix of metals or alloys enough voltage is required [44]. Ions move towards the cathode to form a coating layer. Figure 8.13 shows the electrodeposition coating process.

### 8.6.2 Electroless Plating

Electroless plating process is a fully chemical coating process. Figure 8.14 depicts the electroless plating process. There is no current involved in this

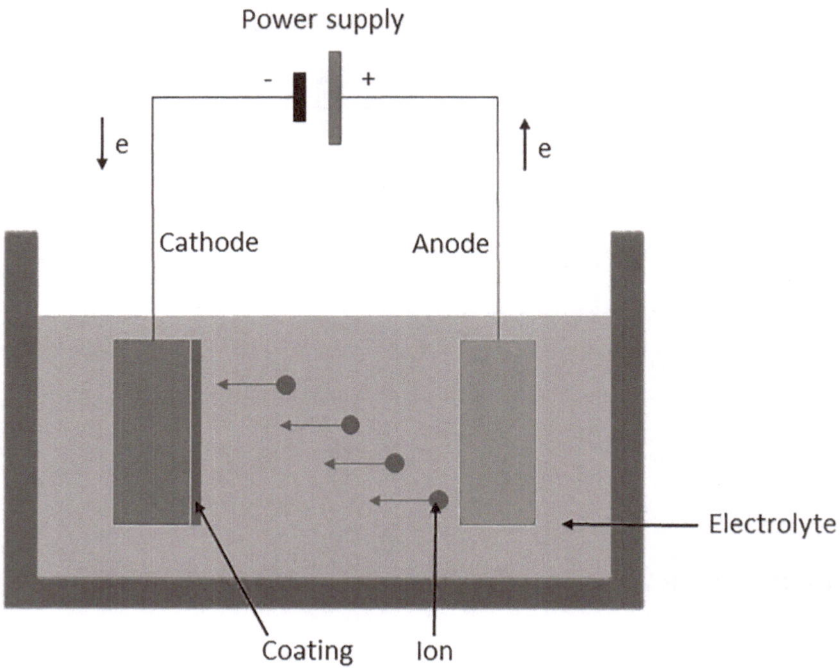


**FIGURE 8.13** Electrodeposition coating process.

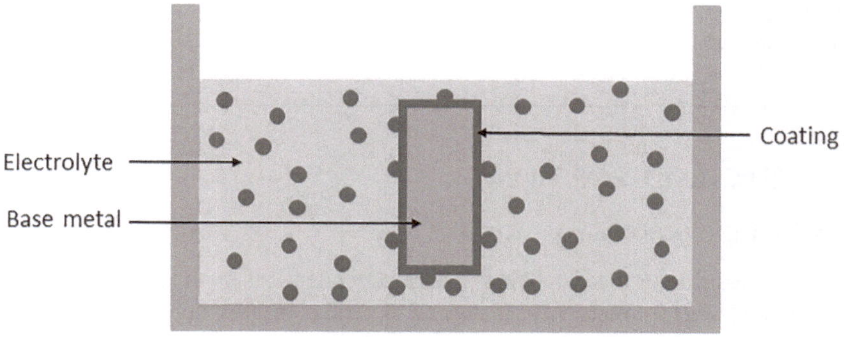


**FIGURE 8.14** Electroless plating process.

process. It relies on aqueous solutions. Oxides and salts are used to coat metallic and nonmetallic surfaces [45]. The deposition takes place by way of the reaction of the substrate creating its composition, a reagent, or a metal ion [46]. Ions go to the cathode to form a coating layer upon the coated part.

### 8.6.3 Electrocomposite Coating

For composite materials coating, electroplating is utilized. Reinforcement particles can be mixed into metallic matrix using undivided cell through electroplating [47]. The substrate should be well cleaned before coating. Distilled water, sonicating the sample to be coated in anhydrous ethanol, and hydrochloric acid are all used for cleaning [17]. It is as same as the process in Figure 8.13.

### 8.6.4 Micro Arc Oxidation

Miro arc oxidation (MAO) which is also called plasma electrolytic oxidation (PEO) is used to form ceramic coatings on metal surfaces. It offers higher hardness, more corrosion resistance, and a uniform thickness [18]. It is used for valve metals made from aluminum, titanium, or magnesium [48]. A positive current is connected to the sample. The ground is connected to the metallic container. A propeller is dipped inside the container to mix the electrolyte. The electrolyte is cooled using an external cooling system to maintain its temperature; therefore, the coating properties will not be compromised. The arc ions will be attracted to form a thin coat on the substrate. Figure 8.15 shows the micro arc oxidation coating process.

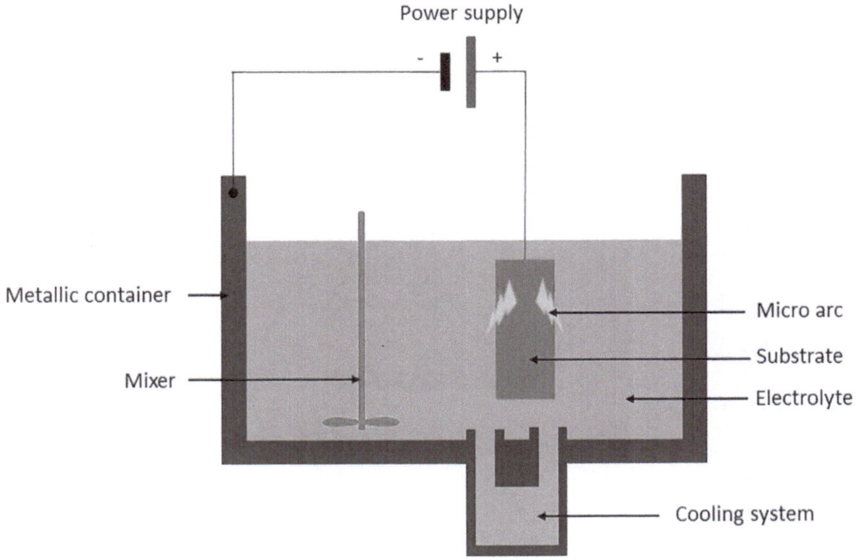


**FIGURE 8.15** Micro arc oxidation coating process.

## 8.7 ION IMPLANTATION

### 8.7.1 Conventional

Ion implantation is a coating process that uses ions to deposit powders and carrier gas. Figure 8.16 shows the ion implantation coating process. Vacuum pressure is applied to the process. The gas is fed through an ion source. This allows the gas to be deposited by the ion beam. Ions' impact energy on the substrate is 10–200 keV [49]. Hence, a coating layer is formed on the substrate. This process is used for micro coating. It can go less than micro level to coat less than 100 nm size [49].

### 8.7.2 Plasma Immersion

Plasma immersion coating process incorporates plasma beam to induce ions for gases deposition. This process is carried out under vacuum pressure. Vacuum pressure is $2 \times 10^{-3}$ Pa [50]. Pulse generator is connected to the substrate. The accelerating voltage is 40 kV and the current is 3 mA [50]. A gas such as nitrogen enters the process chamber across the plasma beam. Ions stream assists the deposition. Consequently, Ions are accelerated to impinge the substrate to form a coating layer [51]. The immersion time is 60 min [50]. Figure 8.17 shows the plasma immersion coating process.

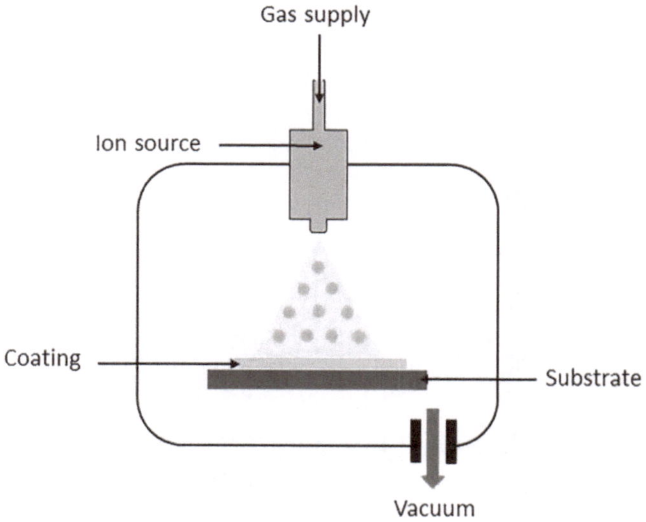


**FIGURE 8.16** Ion implantation coating process.

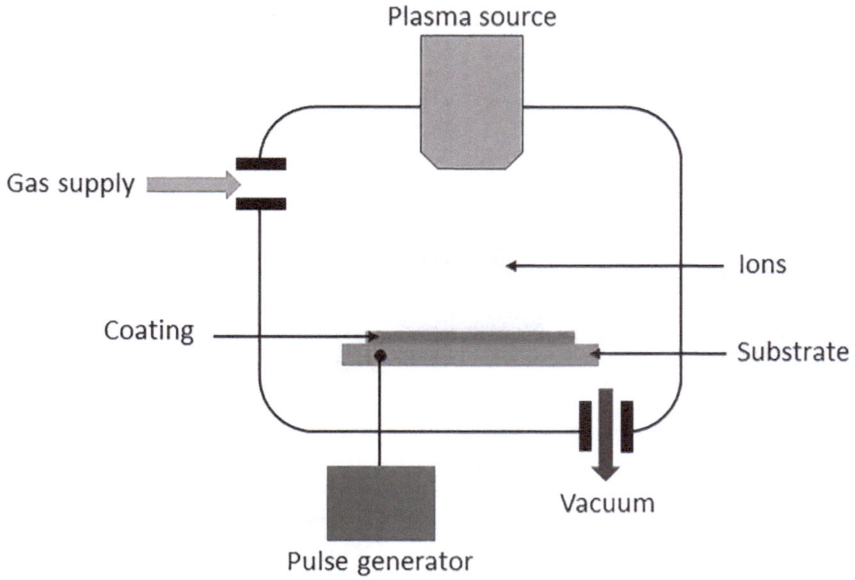


**FIGURE 8.17** Plasma immersion coating process.

## 8.8 CONCLUSION

Micro additive manufacturing requires special surface coating techniques. Plasma spray coating process, high velocity air fuel coating process, high velocity oxygen fuel coating process, cold spray coating process, and flame spray were illustrated. Moreover, cladding process and extreme high speed laser cladding process were highlighted. Thermal evaporation process, sputtering, ion plating, and electron beam physical vapor deposition process were discussed. Conventional chemical vapor deposition process and plasma enhanced chemical vapor deposition process were explained. Furthermore, conventional electrodeposition process, electroless plating process, electrocomposite coating process, and micro arc oxidation were demonstrated. Finally, conventional ion implantation process and plasma immersion were described. These above-stated surface coating processes provide superior coating properties for additive manufacturing.

## REFERENCES

[1] R. F. Vaz, A. Garfias, V. Albaladejo, J. Sanchez, and I. G. Cano, "A review of advances in cold spray additive manufacturing," *Coatings*, vol. 13, no. 2, p. 267, 2023, doi: 10.3390/coatings13020267.

[2] G.-J. Yang, X.-K. Suo, and G.-R. Li, "Introduction to Advanced Micro-Nano Coating Materials and Thermal Spray," *Advanced Nanomaterials and Coatings by Thermal Spray*. Elsevier, pp. 1–11, 2019. doi: 10.1016/b978-0-12-813870-0.00001-2.

[3] Joshi and Nylen, "Advanced coatings by thermal spray processes," *Technologies*, vol. 7, no. 4, p. 79, 2019, doi: 10.3390/technologies7040079.

[4] Y. Qin *et al.*, "Study on the crystallization behaviors and mechanical properties of reactive plasma sprayed TiCN coatings," *Ceram. Int.*, vol. 48, no. 20, pp. 30490–30498, 2022, doi: 10.1016/j.ceramint.2022.06.329.

[5] T. Zhu, M. Baeva, H. Testrich, T. Kewitz, and R. Foest, "Effect of a spatially fluctuating heating of particles in a plasma spray process," *Plasma Chem. Plasma Process.*, vol. 43, no. 1, pp. 1–24, 2022, doi: 10.1007/s11090-022-10290-y.

[6] Y. Y. Avcu, M. Güney, and E. Avcu, "High-velocity air fuel coatings for steel for erosion-resistant applications," *J. Electrochem. Sci. Eng.*, 2022, doi: 10.5599/jese.1369.

[7] D. Tejero-Martin, M. Rezvani Rad, A. McDonald, and T. Hussain, "Beyond traditional coatings: a review on thermal-sprayed functional and smart coatings," *J. Therm. Spray Technol.*, vol. 28, no. 4, pp. 598–644, 2019, doi: 10.1007/s11666-019-00857-1.

[8] N. Al Harbi, K. Y. Benyounis, L. Looney, and J. Stokes, "Laser Surface Modification of Ceramic Coating Materials," *Encyclopedia of Smart Materials*. Elsevier, pp. 445–461, 2018. doi: 10.1016/b978-0-12-803581-8.11386-4.

[9] L. Zhu *et al.*, "Recent research and development status of laser cladding: A review," *Opt. Laser Technol.*, vol. 138, p. 106915, 2021, doi: 10.1016/j.optlastec.2021.106915.

[10] Z. Khalkhali and J. P. Rothstein, "Characterization of the cold spray deposition of a wide variety of polymeric powders," *Surf. Coatings Technol.*, vol. 383, p. 125251, 2020, doi: 10.1016/j.surfcoat.2019.125251.

[11] Y. Ren, L. Li, Y. Zhou, and S. Wang, "In situ synthesized VC reinforced Fe-based coating by using extreme high-speed laser cladding," *Mater. Lett.*, vol. 315, p. 131962, 2022, doi: 10.1016/j.matlet.2022.131962.

[12] M. O. Mavukkandy, S. A. McBride, D. M. Warsinger, N. Dizge, S. W. Hasan, and H. A. Arafat, "Thin film deposition techniques for polymeric membranes– A review," *J. Memb. Sci.*, vol. 610, p. 118258, 2020, doi: 10.1016/j.memsci.2020.118258.

[13] D. M. Mattox, "Physical Sputtering and Sputter Deposition," *The Foundations of Vacuum Coating Technology*. Elsevier, pp. 87–149, 2018. doi: 10.1016/b978-0-12-813084-1.00004-2.

[14] A. Avci, A. A. Eker, and B. Eker, "Microstructure and Oxidation Behavior of Atmospheric Plasma-Sprayed Thermal Barrier Coatings," *Exergetic, Energetic and Environmental Dimensions*. Elsevier, pp. 793–814, 2018. doi: 10.1016/b978-0-12-813734-5.00045-7.

[15] A. Delimi *et al.*, "Corrosion protection performance of silicon-based coatings on carbon steel in NaCl solution: a theoretical and experimental assessment of the effect of plasma-enhanced chemical vapor deposition pretreatment," *RSC Adv.*, vol. 12, no. 24, pp. 15601–15612, May 2022, doi: 10.1039/d1ra08848c.

[16] F. Zhao *et al.*, "Mechanical and tribological properties of Ni-B and Ni-B-W coatings prepared by electroless plating," *Lubricants*, vol. 11, no. 2, p. 42, 2023, doi: 10.3390/lubricants11020042.

[17] M. Wang *et al.*, "Pulse electrodeposited super-hydrophobic Ni-Co/WS2 nanocomposite coatings with enhanced corrosion-resistance," *Coatings*, vol. 12, no. 12, p. 1897, 2022, doi: 10.3390/coatings12121897.

[18] K. Ling *et al.*, "Growth characteristics and corrosion resistance of micro-arc oxidation coating on Al–Mg composite plate," *Vacuum*, vol. 195, p. 110640, 2022, doi: 10.1016/j.vacuum.2021.110640.

[19] V. Chauhan, P. Singh, and R. Kumar, "Ion Beam-Induced Modifications in ZnO Nanostructures and Potential Applications," *Nanostructured Zinc Oxide*. Elsevier, pp. 117–155, 2021. doi: 10.1016/b978-0-12-818900-9.00011-5.
[20] E. Bakan, D. E. Mack, G. Mauer, R. Vaßen, J. Lamon, and N. P. Padture, "High-Temperature Materials for Power Generation in Gas Turbines," *Advanced Ceramics for Energy Conversion and Storage*. Elsevier, pp. 3–62, 2020. doi: 10.1016/b978-0-08-102726-4.00001-6.
[21] A. Garfias Bulnes, V. Albaladejo Fuentes, I. Garcia Cano, and S. Dosta, "Understanding the influence of high velocity thermal spray techniques on the properties of different anti-wear wc-based coatings," *Coatings*, vol. 10, no. 12, p. 1157, 2020, doi: 10.3390/coatings10121157.
[22] S. Mathiyalagan *et al.*, "High velocity air fuel (HVAF) spraying of nickel phosphorus-coated cubic-boron nitride powders for realizing high-performance tribological coatings," *J. Mater. Res. Technol.*, vol. 18, pp. 59–74, 2022, doi: 10.1016/j.jmrt.2022.02.058.
[23] M. S. El-Eskandarany, "Utilization of Ball-Milled Powders for Surface Protective Coating," *Mechanical Alloying*. Elsevier, pp. 309–334, 2020. doi: 10.1016/b978-0-12-818180-5.00012-1.
[24] Y. Ozgurluk, "Investigation of oxidation and hot corrosion behavior of molybdenum coatings produced by high-velocity oxy-fuel coating method," *Surf. Coatings Technol.*, vol. 444, p. 128641, 2022, doi: 10.1016/j.surfcoat.2022.128641.
[25] V. Champagne, N. Matthews, and V. Champagne, "Introduction to Supersonic Particle Deposition," *Aircraft Sustainment and Repair*. Elsevier, pp. 799–844, 2018. doi: 10.1016/b978-0-08-100540-8.00014-5.
[26] H. Singh, M. Kumar, and R. Singh, "An overview of various applications of cold spray coating process," *Mater. Today Proc.*, vol. 56, pp. 2826–2830, 2022, doi: 10.1016/j.matpr.2021.10.160.
[27] M. R. Dorfman, "Thermal Spray Coatings," *Handbook of Environmental Degradation of Materials*. Elsevier, pp. 469–488, 2018. doi: 10.1016/b978-0-323-52472-8.00023-x.
[28] Q. Cao, G. Huang, L. Ma, and L. Xing, "Comparison of a cold-sprayed and plasma-sprayed Fe25Cr20Mo1Si amorphous alloy coatings on 40Cr substrates," *Mater. Corros.*, vol. 71, no. 11, pp. 1872–1884, 2020, doi: 10.1002/maco.202011558.
[29] A. B. Elshalakany, T. A. Osman, W. Hoziefa, A. V. Escuder, and V. Amigó, "Comparative study between high-velocity oxygen fuel and flame spraying using MCrAlY coats on a 304 stainless steel substrate," *J. Mater. Res. Technol.*, vol. 8, no. 5, pp. 4253–4263, 2019, doi: 10.1016/j.jmrt.2019.07.035.
[30] J. Mazumder, "Laser-Aided Direct Metal Deposition of Metals and Alloys," *Laser Additive Manuf.* Elsevier, pp. 21–53, 2017. doi: 10.1016/b978-0-08-100433-3.00001-4.
[31] A. Sylvester Bolokang and M. Ntsoaki Mathabathe, "Laser Cladding—A Modern Joining Technique," *Advanced Welding of Deforming*. Elsevier, pp. 291–319, 2021. doi: 10.1016/b978-0-12-822049-8.00011-6.
[32] T. Li *et al.*, "Extreme high-speed laser material deposition (EHLA) of AISI 4340 steel," *Coatings*, vol. 9, no. 12, p. 778, 2019, doi: 10.3390/coatings9120778.
[33] W. Yuan, R. Li, Z. Chen, J. Gu, and Y. Tian, "A comparative study on microstructure and properties of traditional laser cladding and high-speed laser cladding of Ni45 alloy coatings," *Surf. Coatings Technol.*, vol. 405, p. 126582, 2021, doi: 10.1016/j.surfcoat.2020.126582.

[34] A. V. Rane, K. Kanny, V. K. Abitha, and S. Thomas, "Methods for Synthesis of Nanoparticles and Fabrication of Nanocomposites," *Synthesis of Inorganic Nanomaterials*. Elsevier, pp. 121–139, 2018. doi: 10.1016/b978-0-08-101975-7.00005-1.

[35] M. S. Zafar, I. Farooq, M. Awais, S. Najeeb, Z. Khurshid, and S. Zohaib, "Bioactive Surface Coatings for Enhancing Osseointegration of Dental Implants," *Biomedical, Therapeutic and Clinical Applications of Bioactive Glasses*. Elsevier, pp. 313–329, 2019. doi: 10.1016/b978-0-08-102196-5.00011-2.

[36] B. A. Bhanvase, V. B. Pawade, S. H. Sonawane, and A. B. Pandit, "Nanomaterials for Wastewater Treatment: Concluding Remarks," *Handbook of Nanomaterials for Wastewater Treatment*. Elsevier, pp. 1125–1157, 2021. doi: 10.1016/b978-0-12-821496-1.00034-9.

[37] Wei, Dong, Qu, Ma, and Shen, "Effect of deposition parameters on micro-structure of the Ti-Mg immiscible alloy thin film deposited by multi-arc ion plating," *Metals (Basel)*, vol. 9, no. 11, p. 1229, 2019, doi: 10.3390/met9111229.

[38] Y. Xu, X. Hu, F. Xu, and K. Li, "Rare earth silicate environmental barrier coatings: Present status and prospective," *Ceram. Int.*, vol. 43, no. 8, pp. 5847–5855, 2017, doi: 10.1016/j.ceramint.2017.01.153.

[39] S. Bose, *High Temperature Coatings*. Butterworth-Heinemann, 2017.

[40] M. Mittal, S. Sardar, and A. Jana, "Nanofabrication Techniques for Semiconductor Chemical Sensors," *Handbook of Nanomaterials for Sensing Applications*. Elsevier, pp. 119–137, 2021. doi: 10.1016/b978-0-12-820783-3.00023-3.

[41] R. Satpathy and V. Pamuru, "Making of Crystalline Silicon Solar Cells," *Solar PV Power*. Elsevier, pp. 71–134, 2021. doi: 10.1016/b978-0-12-817626-9.00004-6.

[42] K. Gupta, N. K. Jain, and R. Laubscher, "Advances in Gear Manufacturing," *Advanced Gear Manufacturing and Finishing*. Elsevier, pp. 67–125, 2017. doi: 10.1016/b978-0-12-804460-5.00004-3.

[43] S. Saha and S. Das, "Nanomaterials in Thin-Film Form for New-Generation Energy Storage Device Applications," *Chemical Solution Synthesis for Materials Design and Thin Film Device Applications*. Elsevier, pp. 561–583, 2021. doi: 10.1016/b978-0-12-819718-9.00017-0.

[44] M. Bellardita, A. Di Paola, S. Yurdakal, and L. Palmisano, "Preparation of Catalysts and Photocatalysts Used for Similar Processes," *Heterogeneous Photocatalysis*. Elsevier, pp. 25–56, 2019. doi: 10.1016/b978-0-444-64015-4.00002-x.

[45] S. S. Djokić, "Fundamentals of Electroless Deposition," *Encyclopedia of Interfacial Chemistry*. Elsevier, pp. 161–173, 2018. doi: 10.1016/b978-0-12-409547-2.11701-9.

[46] M. Roy, "Protective Hard Coatings for Tribological Applications," *Materials Under Extreme Conditions*. Elsevier, pp. 259–292, 2017. doi: 10.1016/b978-0-12-801300-7.00008-5.

[47] A. Lelevic and F. C. Walsh, "Electrodeposition of Ni P composite coatings: A review," *Surf. Coatings Technol.*, vol. 378, p. 124803, 2019, doi: 10.1016/j.surfcoat.2019.07.027.

[48] W. Wang, S. Feng, Z. Li, Z. Chen, and T. Zhao, "Microstructure and properties of micro-arc oxidation ceramic films on AerMet100 steel," *J. Mater. Res. Technol.*, vol. 9, no. 3, pp. 6014–6027, 2020, doi: 10.1016/j.jmrt.2020.04.005.

[49] M. M. Verdian, "3.13 Finishing and Post-Treatment of Thermal Spray Coatings," *Comprehensive Materials Finishing*. Elsevier, pp. 191–206, 2017. doi: 10.1016/b978-0-12-803581-8.09200-6.

[50] S. Guo *et al.*, "Effects of carbon and nitrogen plasma immersion ion implantation on bioactivity of zirconia," *RSC Adv.*, vol. 10, no. 59, pp. 35917–35929, Sep. 2020, doi: 10.1039/d0ra05853j.
[51] Y. Chen, C. Xu, C. Wang, M. M. M. Bilek, and X. Cheng, "An effective method to optimise plasma immersion ion implantation: Sensitivity analysis and design based on low-density polyethylene," *Plasma Process. Polym.*, vol. 19, no. 6, p. 2100199, 2022, doi: 10.1002/ppap.202100199.

# 9 Direct Ink Writing

*Muhammad Abas, Ziaullah Jan, and Khalid Rahman*

## 9.1 INTRODUCTION

Direct ink writing (DIW) is a commonly used additive manufacturing technology widely used for patterning and selective deposition of different types of materials. DIW technology allows the direct printing of materials on variety of flat as well as conformal surfaces and is capable to form single and multiple layers with high resolution [1]. This method has been widely studied and developed and has emerged as a potential approach to manufacture the microstructures such as antennas, transistors, capacitors, and different kinds of sensors and wiring. Existing fabrication techniques i.e. etching use masks for the replication of CAD images to define the patterns of the layers, which involves the additional tooling and manufacturing steps thus increasing the chances of human error and increase in manufacturing cost and time. DIW provides a simpler, cost-efficient, and environmentally friendly alternative to traditional electronic fabrication methods, which often involve complex and time-consuming processes [2]. There are various DIW technologies available, with extrusion and droplet-based technologies being the most well-known. Although both methods are developed for selective deposition of conductive traces, their specific goals may differ based on the intended application [3].

## 9.2 EXTRUSION-BASED DIW

Extrusion-based direct ink writing is a method of fabricating electronic components and devices by directly depositing material in a controlled manner to form the desired structures. It's an additive manufacturing technique that enables the production of complex structures with high accuracy and precision and has found applications in a wide range of areas, including electronics, biomedicine, and materials science [4].

In extrusion-based direct ink writing, a material, typically a paste or a liquid, is loaded into a syringe or a similar device and dispensed through a small nozzle onto a substrate. The material is then cured, sintered, or otherwise processed to form a solid structure. By adjusting the temperature, speed, and pressure of the material as it is dispensed, the process can be controlled, thus allowing the precise control over the shape and size of the structures that are produced [5,6].

 DOI: 10.1201/9781003364948-9

The main advantage of extrusion-based direct ink writing is its versatility and flexibility, as it can be used to produce a wide variety of structures, including electronics components, sensors, and actuators [2]. It also enables the production of structures with complex shapes and features, such as multi-level interconnects and 3D architectures. Additionally, it has the potential to reduce the cost and complexity of the manufacturing process, compared to traditional methods such as photolithography or etching [7].

### 9.2.1 Robocasting

Robocasting is an extrusion-based direct ink write (DIW) system, in which object is created layer by layer via extruding continuous filament of ink through nozzle. The basic components of the system are syringe, nozzle, X-Y stage, and CAD/CAM software. The physical model to be generated are the first model in CAD/CAM software. The CAD/CAM software uses the STL file which slices the object into a number of layers having similar thickness. To fabricate an object, ink stored in syringe barrel is extruded continuously through the nozzle. Position of nozzle is controlled by a robotic system guided through a program developed in CAD/CAM software for object. When the first layer is patterned X-Y stage moves down in the Z direction to deposit second layer. This process is repeated until 3D structure is obtained [8]. The schematic of robocasting is shown in Figure 9.1. Robocasting is well suited for 3D printing of composites and ceramic materials [9].

There are four categories of robocasting depending on the techniques of lashing fluid out of the nozzle. They are time pressure, positive displacement, rotary screw, and shutoff valve.

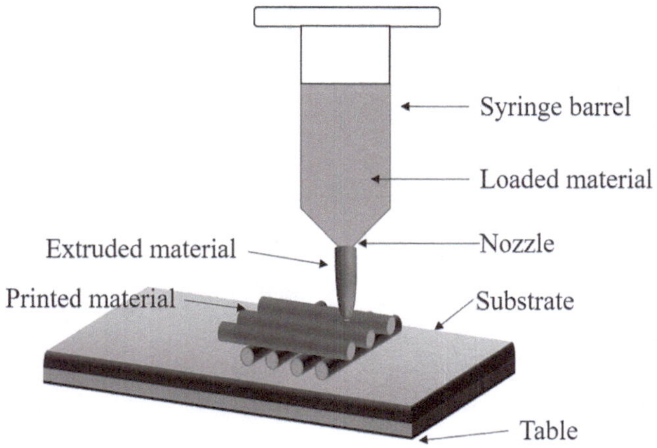


**FIGURE 9.1** Schematic of robocasting process.

In time pressure dispensing, air pressure is applied at the top of syringe barrel to force the material out of the nozzle. The applied pressure depends on the amount of liquid to be dispensed and also viscosity of liquid. To control the amount of fluid to be dispensed the air pressure is regulated as well as its duration [10]. This method is mostly adopted due to its ease of use, and can even dispense small amount of liquid. However, a major disadvantage is that the air volume in the syringe barrel, as well as the viscosity of the material, can impact the dispensing process [11]. The schematic of time pressure dispensing is shown in Figure 9.2a.

Unlike time pressure, the shutoff valve approach contains a shutoff valve at nozzle end. During dispensing of material, the valve is opened. The flow of dispensing material can be controlled by controlling the valve opening and closing. This method is best suited for low viscosity fluids to prevent dripping from the nozzle. However, a drawback of this approach is the reduction of the valve's lifespan due to frequent opening and closing, as well as the potential for clogging. As a result, this method can be costly and time-consuming [12]. The schematic of the shutoff valve is shown in Figure 9.2b.

Rotary screw contains the screw in syringe barrel as shown in Figure 9.2c. The fluid is forced down by rapid rotation of screw to move material out of the nozzle. Compare to time pressure and shutoff valve it is better and improved method. Viscosity of the fluid affects the dispensing of the material [13].

The positive displacement method involves inserting a piston or plunger into the syringe barrel, as shown in Figures 9.2d and 9.2e. With the movement of piston, the fluid is forced out at the nozzle end. The linear motion of the piston can be controlled by air pressure/plunger. Due to the continuous movement of the piston large amount of material can be dispensed. This method is not suitable for dispensing small amount of fluid due to effect of fluid compressibility [14]. Compare to previous three approaches, it provides an accurate flow rate because amount of fluid dispensed depends on movement of piston [15].

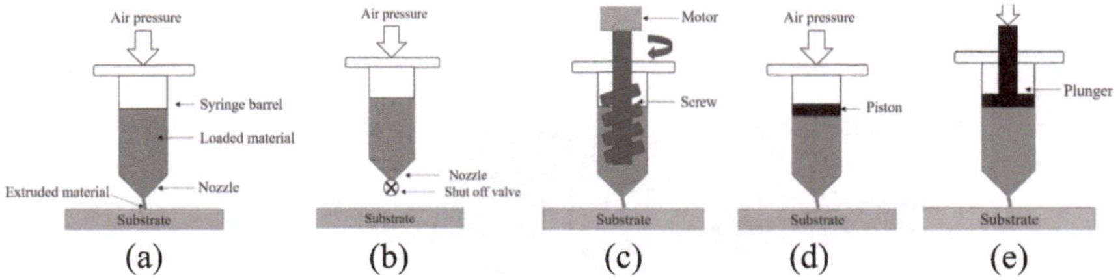


**FIGURE 9.2** Schematic of (a) time pressure (b) shutoff valve (c) rotary screw (d) with piston and (e) with plunger.

### 9.2.2 Fused Deposition Modeling

Like robocasting it also generates object layer by layer using CAD/CAM software files. The starting material is in the form of solid filament, which is melted by the heater to extrude it through a nozzle onto a substrate placed on X-Y stage. The extruded material is solidified rapidly upon deposition. Motion of stage is controlled through the computer; when the first layer is deposited, stage moves up in the Z direction by thickness of layer. The process replicates until part is fully fabricated. The solid filament may be in the form of wax or polymers such as polyethylene, polyimide, polycystic acid, and rubber [16,17]. The schematic of fused deposition is shown in Figure 9.3.

FDM is commonly used in the rapid prototyping and small-scale production of electronic components, such as enclosures and housings for circuit boards, connectors, and other similar parts. The process is known for its affordability, accuracy, and speed, which makes it a popular choice for small-scale electronic manufacturing [18].

It is important to note that FDM is not suitable for all types of electronic components and applications, as the mechanical properties of the finished parts can be limited by the properties of the thermoplastic material used. In addition,

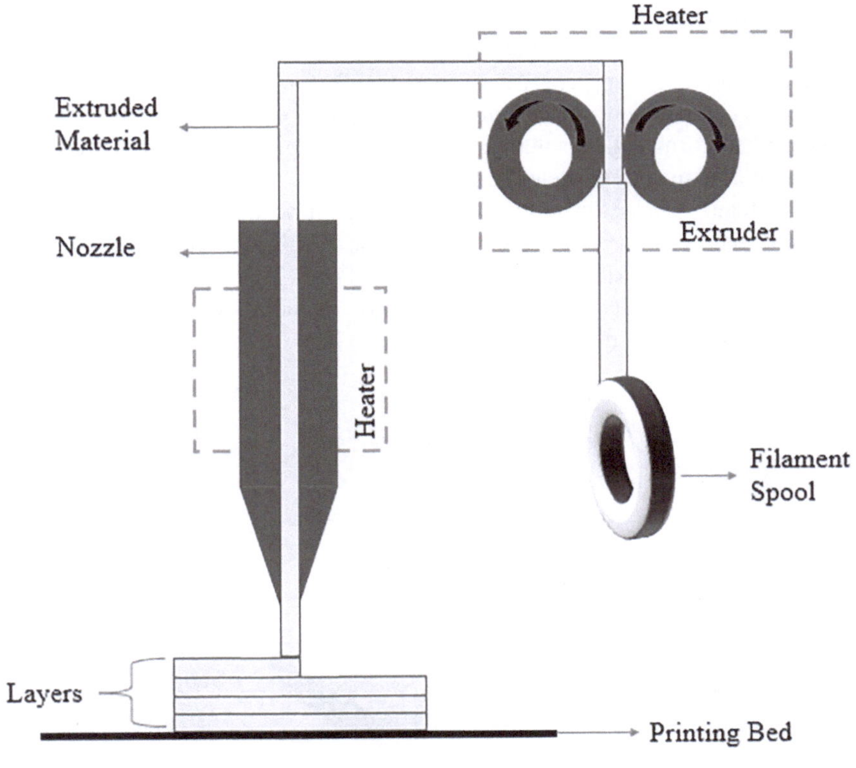


**FIGURE 9.3** Schematic of FDM.

the surface finish of the parts produced by FDM may not be ideal for certain applications, such as those requiring a high level of electrical conductivity. For these types of applications, alternative fabrication methods, such as injection molding or metal stamping, may be more appropriate.

## 9.3 DROPLET-BASED DIW TECHNOLOGIES

### 9.3.1 Ink Jetting

In droplet-based technologies, materials are dispensed onto a substrate in the form of small droplets. Some inkjet systems use a small orifice and piezoelectric diaphragm for dispensing material in the liquid form. As shown in Figure 9.4 expansion of diaphragm applies positive pressure on the fluid and expels it through an orifice. While in some system ink is heated to generate bubbles. Pressure is delivered by this bubble to force the droplet through an orifice. One of the major drawbacks of the thermal inkjet system is rapid volatility of the material, which results in high concentrations of particle causing blockage at orifice [19,20].

Ink jetting system also uses metal or carbon-based ink. The viscosity should be mediocre. Comparing to an extrusion-based system, for the same tip diameter of the orifice, blockage occurs at low concentration and also it is more sensitive to particle loading. Appropriate dispensing ink for inkjet system is metallo-organic decomposition (MOD) ink. It has good conductivity and also the capability to form high-resolution feature size ranging from 20μm to 150μm because it contains no particles [21].

Advantages of inkjet printing for electronics fabrication include its ability to produce small and intricate patterns, low cost, high-speed production, and the ability to print on flexible and non-planar surfaces. This makes it a promising technology for a wide range of applications, including wearable electronics, smart packaging, and flexible displays [22].

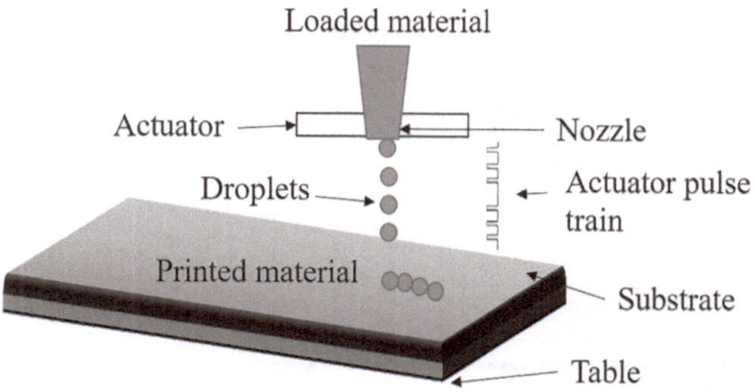


**FIGURE 9.4** Schematic of inkjet deposition.

However, there are also some limitations to consider, such as the difficulty in achieving high conductivity with inkjet-printed materials and ensuring long-term stability and reliability of the printed devices [23]. Additionally, the properties of the inkjet-printed materials may vary depending on the type of ink and the printing process used, so careful selection and optimization of materials and processes are critical to achieve desired results [24].

### 9.3.2 Aerosol Jet

Aerosol Jet technology is a type of 3D printing that uses an aerosol spray of material to create objects layer by layer (Figure 9.5). The process starts by transforming a material into an aerosol mist, which is then directed towards a substrate, where it solidifies to form a layer. This process is repeated to build up the 3D structure [25].

One of the key benefits of Aerosol Jet printing is its versatility, as it can be used to print with a wide range of materials, including polymers, ceramics, metals, and even biological materials like cells and proteins. This makes it suitable for a wide range of applications, including the creation of electronic components, medical implants, and other complex structures [26]. Another advantage of Aerosol Jet printing is its speed, as it can produce parts much faster than traditional 3D printing methods like Fused Deposition Modeling or Stereolithography. This makes it suitable for high-volume production applications [27]. The process can also be used to print on a variety of substrates, including flexible and curved surfaces, making it a good choice for the creation of 3D objects that require a high degree of flexibility or conformability [28]. This method replaces the technologies like photolithography and screen printing [29]. In comparison to inkjet printing, it is more robust because it process variety of ink material on a wide range of substrate. For dispensing it supports a variety of materials such as insulators, conductive polymers, nanoparticle inks, adhesives, and screen printing pastes [30]. The pattern generated for the electronic feature through this technique has smaller resolution having a line thickness of 10nm to 5μm and width ranging from 10 to 150 μm [31].

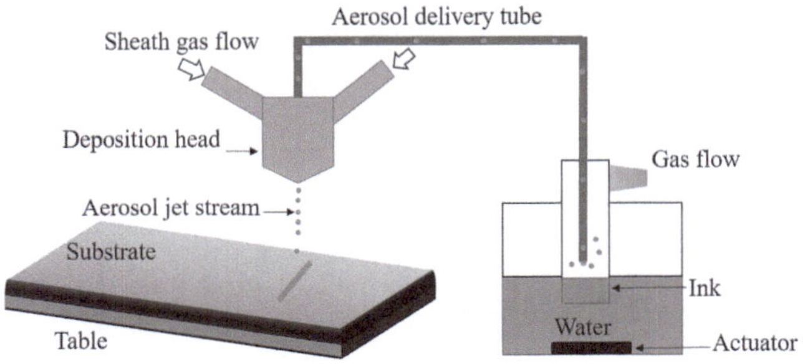


**FIGURE 9.5** Aerosol jet process.

Despite its benefits, Aerosol Jet printing has some limitations. One of the main limitations is the requirement for specialized equipment, which can be expensive and challenging to operate. The process is also limited by the resolution, which can be affected by factors such as the size of the nozzle and the viscosity of the material. Additionally, the surface finish of the printed parts may not be as smooth as parts produced using other 3D printing technologies [32,33].

## 9.4 LASER-BASED DIRECT WRITE

Laser-based direct write system uses laser energy to deposit material to develop 2D/3D structure. The well-known technologies are laser, chemical vapor deposition (LCVD), and Matrix-assisted pulsed-laser evaporation direct write (MAPLE DW).

### 9.4.1 Laser Chemical Vapor Deposition (LCVD)

Laser Chemical Vapor Deposition (LCVD) is a chemical vapor deposition (CVD) method that uses a laser to supply energy to the reaction. In LCVD, a laser is used to generate a high-temperature plasma, which is then used to activate a chemical reaction between the reactant gases and the substrate, leading to the growth of thin films. The laser energy can be used to control the temperature of the substrate and the gas phase, thereby enabling the control of the quality and composition of the deposited film as shown in Figure 9.6 [34].

LCVD has several advantages over traditional CVD methods, including the ability to produce high-quality films at lower temperatures and the ability to deposit films with complex structures, such as 3D geometries and superlattices. LCVD is widely used in the production of electronic and optical devices, including high-temperature superconductors, semiconductors, and optoelectronic materials [35].

Despite its advantages, LCVD also has some limitations that need to be considered when deciding whether it is the appropriate choice for a given application. LCVD involves the use of multiple process parameters, such as laser

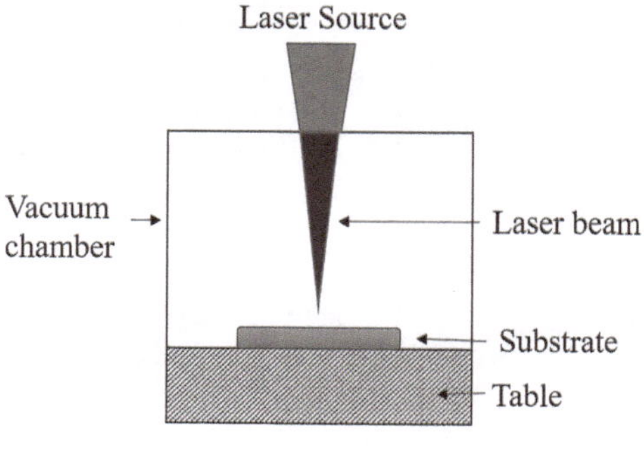


**FIGURE 9.6** Schematic of LCVD.

energy, gas flow rates, substrate temperature, and substrate orientation, that must be carefully controlled to achieve the desired film quality [35]. This complexity can increase the cost and difficulty of implementing LCVD compared to other deposition methods. LCVD is often limited by the quality and uniformity of the deposited films. Factors such as non-uniform laser energy delivery and non-uniform substrate heating can lead to film quality issues such as cracking, porosity, and non-uniform thickness [36]. LCVD is often limited to certain types of substrates, such as silicon, that are compatible with the laser energy used in the process. This can limit the range of applications for which LCVD is suitable [37]. Compared to other deposition methods, LCVD may be relatively slow, particularly when dealing with large substrates. This limitation may restrict its usefulness in high-volume manufacturing applications [38]. LCVD can generate harmful byproducts, such as toxic fumes, that must be carefully controlled to ensure worker safety and minimize environmental impact [39].

### 9.4.2 Matrix-Assisted Pulsed-Laser Evaporation Direct Write (MAPLE DW)

The method was initially developed to fabricate the electronic devices such as resistors, interconnects, and capacitors [40]. It generates patterns rapidly with the help of CAD/CAM system. The schematic is shown in Figure 9.7.

In a laser absorbent matrix, the sensitive material is mixed to form ink and then coated on ribbon. The ribbon is transparent to laser irradiation. Through the back side of the ribbon, the energized laser is focused to evaporate the layer of matrix near support interface. The ink is released in evaporation, which is propelled uniformly and gently away from support. In this way, passive electronic devices such as capacitors, resistors, and interconnects are produced with a line width of 10 μm [41].

MAPLE DW allows for precise control over the size, composition, and arrangement of deposited nanoparticles, making it an attractive method for the production of high-quality, uniform thin films and nanostructures [42]. It can be

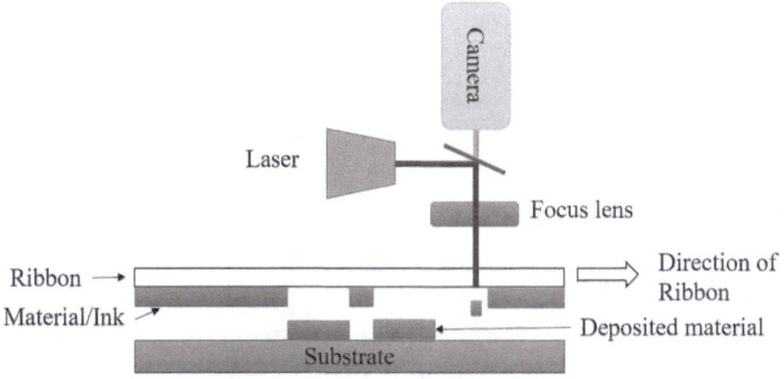


**FIGURE 9.7** Schematic of MAPLE DW.

used to produce a wide range of materials, including metals, semiconductors, insulators, and organic compounds, making it a versatile method for the fabrication of advanced materials [43]. It can be used to deposit materials over large areas, making it useful for the production of large-scale devices and structures. It is well suited for the deposition of materials onto complex or sensitive substrates, such as biological samples, making it an attractive method for the production of bio-compatible materials and devices [44]. It allows for high-resolution patterning of materials, making it useful for the production of nanoscale devices and structures [45].

MAPLE DW can be a relatively expensive process, due to the cost of the laser and the matrix materials. In addition, the process can be time-consuming and requires specialized equipment and expertise to perform [46]. It is typically limited to materials that are compatible with the matrix material used for the process. This restricts the range of materials that can be used for the process. The selection of ribbon or matrix material can significantly affect the quality of the resulting structures. In addition, the matrix material may not be suitable for all applications and may need to be changed for different materials or processes [47]. The resolution of MAPLE DW is limited by the size of the laser pulse and the laser spot size. It is also affected by the matrix material and its properties, which can affect the final size and shape of the deposited material. It can be difficult to precisely control the thickness of the deposited film in MAPLE DW, which can affect the overall quality of the resulting structures [46].

## 9.5 INKS FOR DIRECT INK WRITING

Major types of inks studied in DIW are sol-gel, nanoparticle inks, and polyelectrolyte inks.

### 9.5.1 Nanoparticle Ink

The Nanoparticle ink is made by suspending nanoparticles, such as metallic or semiconducting particles, in a solvent, which can be dispensed through a printhead, such as an inkjet printhead, to form patterns and structures on a substrate. The most common metals are Copper (Cu), Aluminum (Al), Silver (Ag), and Gold (Au). Cu and Al loaded ink are available commercially, but not used widely because of their rapid oxidation in the presence of reducing agent and air [48]. On the other hand, the most common metals used are silver and gold because of their higher conductivity and resistance to oxidation. These metals used in the form of powder having a particle size of Nanoscale and mixed with organic solvent as additives to form ink [49]. The most common additives used and their functions in ink are: Polyvinylpyrrolidone (PVP) and surfynol (Surfactant) to prevent silver particle from agglomerating during the fabrication process and also to improve ink wettability; different types of alcohol e.g. methyl alcohol and ethanol as carrier medium and reducing agent [50]; and polyethyleneimine (PEI) to sustain ink consistency as well as to promote adhesion

[9]. The concentration level of metal particle is important in ink rheology. Higher concentration of metal particle increased viscosity of ink, therefore resulting in higher conductivity. While low concentration makes ink less viscous and is therefore less conductive. Concentration of additives should also be maintained to improve electrical performance [51]. The ink must also be compatible with substrate and its curing temperature for good printing and to achieve minimum resolution. Depending on the type of substrate used curing temperature of ink varies. For flexible polymeric substrate, the curing temperature of ink must be less than 150°C [52]. Metal loaded inks are available easily due to their application in electronic devices such as sensors, RFID tags, solar cells, PCBs and OLEDs, etc. One of the major drawbacks is that they are expensive [53]. For better conductivity metal content should be higher in ink rheology, but it results in clogging of material at nozzle tip during dispensing. The metal content should be between 20% and 70% by weight in ink solution for good performance and to achieve fine patterns [54].

However, there are also some challenges associated with nanoparticle ink that need to be addressed. For example, the stability and shelf life of the ink can be affected by factors such as particle size, concentration, and the presence of impurities. The uniformity and reproducibility of the printed structures can also be influenced by factors such as the viscosity and stability of the ink, and the print conditions, such as the nozzle size, print speed, and substrate temperature [53].

Despite these challenges, nanoparticle ink is a rapidly developing field, with the potential for new advances and innovations in the creation of electronic devices and components, such as flexible and stretchable electronics, sensors, and displays.

### 9.5.2 Sol-Gel

Sol-gel is a process that involves the synthesis of materials as a colloidal suspension, or sol, which is then transformed into a solid gel through a process of chemical or physical crosslinking. The sol-gel process is a versatile and flexible method that can be used to produce a wide variety of materials, including ceramics, glasses, metals, and polymers, with controlled composition, structure, and properties. The sol-gel process starts with the preparation of a sol, which is typically made by mixing a precursor, such as a metal or metal oxide, with a solvent and a gelling agent. The mixture is then subjected to various treatments, such as heating, aging, or the addition of a catalyst, to promote the formation of a gel. The gel is then dried and subjected to further processing, such as sintering or calcination, to remove the solvent and produce a solid material [55].

The most common metal oxides used in sol jet are Ti, Zr, Sn, Zn, Pb, Hf, Sr, and Ni, and non-transition metals such as Al, Si, and Ba are most commonly used. The most common additives used are PVP, Polyvinyl alcohol (PVA), ethanol, glycerin, ethylene glycol, ammonia hydroxide, and distilled/deionized water. The advantage of Sol-gel over Nanoparticle is that it can be prepared easily, is low cost, and requires low sintering temperature. It has a wide application in biosensors, optics, electronic devices, and medicine [56,57].

Sol-gel materials, such as silicon dioxide and aluminum oxide, have been used as gate insulators in metal-oxide-semiconductor field-effect transistors (MOSFETs) due to their high dielectric constant, low trap density, and thermal stability [58]. Tin oxide and indium tin oxide (ITO), have been used as transparent conductors in electronic devices due to their high transparency in the visible region and good electrical conductivity [59]. Lead zirconate titanate (PZT) has been used as a piezoelectric material in sensors, actuators, and other electronic devices due to its high piezoelectric coefficient and good mechanical stability [60]. Iron oxide has been used as a magnetic material in magnetic data storage devices and other electronic applications due to its magnetic properties and good stability [61].

One of the key features of the sol-gel process is its ability to produce materials with uniform and fine microstructures, which is due to the colloidal nature of the sol. This allows for the synthesis of materials with tailored and controlled properties, such as size, shape, composition, and porosity, that can be optimized for specific applications. Another advantage of the sol-gel process is its low processing temperatures, compared to traditional materials synthesis methods such as melting or high-temperature solid-state reactions. This makes it possible to synthesize materials that are sensitive to high temperatures or that have low melting points. It also enables the synthesis of materials that are not thermodynamically stable at high temperatures, such as certain metals or alloys [55].

Sol-gel technology can also be used to incorporate nanoparticles, such as metallic or semiconducting particles, into the material, leading to the creation of hybrid materials with unique properties and functionalities. This can be useful for applications that require a combination of different properties, such as catalysts that need both high surface area and good electrical conductivity [62].

In addition to its advantages, the sol-gel process also has some limitations that need to be considered. For example, the process can be time-consuming, especially for large-scale production, and it may be difficult to obtain consistent and reproducible results. There may also be challenges in controlling the morphology and size distribution of the particles in the sol, and in preventing the formation of unwanted byproducts or impurities [63]. Despite these challenges, sol-gel technology remains an important and rapidly developing field, with a wide range of applications and the potential for further advancements and innovations.

### 9.5.3 Polyelectrolyte Ink

Polyelectrolyte inks are suspensions of charged polymers in a solvent that can be used for various printing and coating applications. The composition of a polyelectrolyte ink can vary depending on the desired properties and application, but typically includes the following components: Polyelectrolyte, solvent, dispersant, and additives. Polyelectrolyte ink is a charged polymer, such as a polycation or polyanion that imparts the ink with its electrical conductivity and other desired properties. The solvent serves as the medium for the polyelectrolyte and is chosen to balance viscosity, solubility, and other physical properties. Common solvents include water, alcohols, and organic solvents. A dispersant is used to prevent agglomeration and improve the stability of the polyelectrolyte suspension. This can be a surfactant, a small charged molecule, or another type of stabilizing agent. other additives, such as stabilizers, viscosifiers, or pH adjusters, may be included in the ink composition to tailor its properties for a specific application [64,65]. The composition of a polyelectrolyte ink can be carefully adjusted to optimize its properties for a specific application, such as conductivity, viscosity, stability, and printability. For example, in applications requiring high conductivity, the concentration of the polyelectrolyte and the type of solvent used may be adjusted to improve conductivity. In applications requiring high stability, the type and concentration of the dispersant may be adjusted to improve the stability of the ink [66].

The most common conductive polymers are polyanilines, poly (thiophene)s (PT), poly(3,4-ethylenedioxythiophene) (PEDOT), poly(pyrrole)s (PPY), and Poly(p-phenylene vinylene) (PPV). Common additives used for synthesis of ink are ethanol, PVP, and PEG. They have wide applications in chemical sensors, OLED'S, Organic solar cells, and displays [67,68].

One of the benefits of using polyelectrolyte ink for printing electronic devices is that it allows for the creation of thin, flexible, and stretchable films that can conform to various substrates, such as plastic or paper. This is because the charged polymers in the ink form strong electrostatic interactions with the substrate, resulting in a stable and uniform film. This property makes polyelectrolyte ink well suited for the creation of wearable and portable electronics, as well as electronic components that can be easily integrated into various products and applications [53,69].

Another advantage of polyelectrolyte ink is its ability to create multilayer structures, by printing different inks with different compositions and properties in a sequential manner. For example, a layer of anionic polyelectrolyte ink can be followed by a layer of cationic polyelectrolyte ink, resulting in a bilayer structure with different electrical properties. This multilayer capability enables the creation of complex and functional devices, such as sensors and displays, that can perform multiple functions [70].

However, there are also some challenges associated with polyelectrolyte ink that need to be addressed. For example, the stability and shelf life of the ink can be affected by factors such as temperature, humidity, and the presence of impurities. The uniformity and reproducibility of the printed structures can also be influenced by factors such as the viscosity and stability of the ink, and the print conditions, such as the nozzle size, print speed, and substrate temperature [71,72].

Overall, polyelectrolyte ink is a promising technology for the creation of flexible and stretchable electronics, offering new opportunities for the development of innovative products and applications in areas such as biomedicine, wearable technology, and the Internet of Things.

## 9.6 STUDIES ON APPLICATION OF DIRECT INK WRITING (DIW)

Peng et al. [5] have proposed a hybrid 3D printing system combining top-down digital light processing (DLP) and DIW printing. The combination of two printing technologies, namely DLP and DIW, enables the high-quality and efficient printing of intricate designs using a variety of materials. DLP is ideal for printing matrix materials with complex shapes as it provides high speed and resolution, while DIW allows for the printing of functional materials like liquid crystal elastomers and conductive silver inks. This system allows for the creation of functional composites with tunable properties and multiple functions, useful for a variety of applications such as soft robotics, electronics, and biomedical devices. Poulin et al. [73] developed an electrically conductive ink using carbon particles in shellac for robocasting. The shellac binder is environmentally friendly and biodegradable, and the ink demonstrates impressive flexibility and stability. The ink has been used to print proximity and deformation sensors, showcasing its versatility and significance in sustainable printed electronics. Fekiri et al. [74] fabricated a high-sensitivity piezoresistive sensor using DIW and a composite material consisting of multi-walled carbon nanotubes dispersed in polydimethylsiloxane (MWCNT-PDMS). The MWCNT-PDMS composite was prepared for 3D printing using extrusion, and the DIW system was specifically designed to print functional materials with different viscosities. The resulting sensor demonstrated good printability and showed excellent characteristics, including, high sensitivity, low hysteresis and non-linearity error, and a fast response to loading and loading cycles. Poulin et al. [75] prepared a water-activated, single-use paper battery with low ecological impact. The battery was made using stencil printing and can be shaped in various forms. It has a potential of 1.2 V and was demonstrated to be suitable for low-power electronics and IoT systems through a proof of concept with a two-cell battery powering an alarm clock. Hassan et al. [76] in their study developed a flexible impact wave propagation sensor using DIW technology. They combined barium titanate and polydimethylsiloxane to improve flexibility and uniformity. The DIW technique was used to add multi-walled carbon nanotube-based electrodes to the sensor, and its impact wave was analyzed by measuring output voltage at different locations. The custom-designed sensor was used to determine the particle-wave velocity of stainless steel and low-density polyethylene, which matched theoretical values. Votzke et al. [77] fabricated connections for flexible electronics using metal paste extrusion. LED oscillator circuit and strain sensors were printed to observe the flexible properties of the paste. A modified 3D printer containing a syringe having the liquid metal paste was used. The fabricated sensors showed stable conductivity of metal paste on sufficiently large strain cycles and large deformations. Chen et al. [78] fabricated flexible electronics circuits using Direct Ink write technique. To make the conductive ink, a blend of ethylene glycol, alcohol, glycerol, and deionized water was used, which contained carbon black. High gloss photographic paper being a flexible material was used as a substrate for printing the electronic circuit. Results showed that the reduced graphene oxide and graphene polypyrrole combination was successful. A significant increase in conductivity and reduction in viscosity is observed. Anelli et al. [79] studied the creation of symmetrical cells using robocasting and inkjet printing. The electrodes were made from robocasted LSM-YSZ slurries and the dense electrolyte was produced through inkjet printing with water-based YSZ ink. The process was found to be feasible and reproducible, with an impressive ASR value of approximately 2.1 Ω $cm^2$ at 750°C. these findings highlight the potential of using robocasting and inkjet printing for fabrication of multi-material energy devices. In the study of Bernard et al. [80] a hybrid additive machine that combines stereolithography and robocasting was developed to produce ceramic/metal parts. The machine produced high-temperature co-fired ceramic components, with dielectric alumina produced through stereolithography and a conductive molybdenum network achieved through robocasting. The co-debinding and co-sintering cycles were optimized for improved performance. Abas et al. [81] in their study fabricated flexible sensor using screen printing carbon paste printed based on direct ink writing. The results showed that the sensor is stable in performance and economical. In another study, they modified the same carbon paste using polyvinyl alcohol for flexible circuit fabrication [82].

Kim et al. [83] used particle-based inks to fabricate microscale 3D thermoelectric structures through direct writing. They achieved high figures of merit and power density in the final structures by controlling particle size and surface oxidation and creating high viscoelastic colloidal inks for 3D printing. A superhydrophobic electrode made of fluoropolymer and carbon nanotubes has been developed by Yang et al. [84] using DIW. This electrode is chemically inert and can be patterned on flexible surfaces, providing a current and power output of 2 mA and 0.12 W. It can also be used in self-powered touch sensing electronic skin. Wang et al. [85] used direct ink writing with auxiliary electric potential to create polyvinylidene fluoride (PVDF) thin films with added graphene, resulting in higher crystallinity and a 67.30% increase in beta-phase content. The sensitivity of the composite film was also found to be significantly higher than that of pure PVDF. Tang et al. [86] developed a 3D-printed anti-swelling polyelectrolyte evaporator using k-carrageenan, poly (diallyldimethylammonium chloride), and carbon nanotubes. The evaporator was constructed using non-covalent supramolecular interactions, was intrinsically antibacterial, and showed remarkable swelling resistance and stable solar thermal evaporation. Sabadin et al. [87] used direct ink writing to create light-emitting devices using two dichroic dyes embedded in a liquid crystal oligomer. Despite modest order, the devices showed positive edge emission ratios and have potential for use in security features, spectral converters, and separators in agriculture. Li et al. [88] created a composite electrode from reduced graphene oxide and Super-P using Direct Ink Writing. The tests showed that the best performance was achieved with a 2:1 mass ratio of GO/Super-P, and the electrode has high conductivity, a large specific area, and a porous structure, making it a potential alternative to traditional graphite felt in vanadium redox flow batteries.

According to Nassar and Dahiya [89], the use of copper-based filaments for 3D printing of electronics is growing because it allows for complex systems to be made with simpler processes. Their study shows that 3D circuits with both planar and vertical interconnections can be created, with resistance of tracks influenced by printing parameters and orientation. Further, the study leads to the development of higher-density FDM 3D-printed circuits and the demonstration of digital data transmission by a 3D-printed circuit. The study of Junpha et al. [90] investigates the effectiveness of 3D-printed electrodes for use as sensors in electronic tongue analysis. These electrodes were fabricated using the combination of Polylactic acid (PLA) and carbon nanotube (CNT), as well as CNT/copper (Cu) and CNT/zinc oxide (ZnO) composites. The study found that the addition of Cu and ZnO nanostructures enhanced the electrical conductivity and thermoelectric properties of the material, leading to distinct electrochemical response to specific chemicals. Overall, the findings suggest that the CNT/PLA-based electrodes fabricated using 3D-printed technology hold promise for use in electronic tongue applications. Georgopoulou et al. [91] compare the performance of a commercial conductive thermoplastic polyurethane (TPU) and a styrenic block copolymer (TPS) as FDM filaments for 3D printing. The TPU could only be extruded through a 0.5mm nozzle and showed plastic deformation at higher elongation. The TPS showed lagging of the resistive value at low strain due to material relaxation. The TPS material produced a reproducible signal, while the TPU material was unsuitable for use as a sensor. In another study, Georgopoulou and Clemens [92] used styrene-based thermoplastic elastomers (TPS) to 3D print soft gripper structures and elastic strips with an integrated piezoresistive sensor. The soft substrate and the integrated sensing element were found to affect the sensitivity of the sensor. The soft compliant gripper structure was successfully printed and could detect positions, spool sizes, and gripper openings/closures. The most sensitive performance was found with the TPS of a lower shore hardness (25 A). The study shows that FDM printers can print potential soft robotic structures with integrated sensor structures. Hassan et al. [93] in their study used electrohydrodynamic (EHD) jet printing to make a strain sensor with high resolution, flexibility, and transparency. PDMS/xylene elastomer was used for fabrication, which resulted in a stable and long-lasting sensing structure with the capability to withstand stretching and bending. The printed mesh structure had a high resolution and sensitivity, resulting in a significant increase in a guage factor. Study highlights the potential of EHD technology in producing high-resolution microchannels, 3D printing, and electronic devices.

## 9.7 FUTURE POTENTIAL

Direct ink writing, which involves dispensing a material, such as a polymer, through a printhead, to form patterns and structures on a substrate, is a rapidly developing field with a wide range of potential applications in the creation of electronic devices and components. However, there are several challenges that need to be addressed in order to advance the technology and increase its impact in the industry. Some of the future challenges in direct ink writing are:

1. Ink stability and shelf life: One of the challenges in direct ink writing is ensuring the stability and shelf life of the ink. Factors such as temperature, humidity, and the presence of impurities can affect the stability and performance of the ink, leading to issues with uniformity and reproducibility of the printed structures.
2. Material compatibility: Another challenge is finding materials that are compatible with the direct ink writing process and that have the desired properties, such as conductivity, transparency, and mechanical strength. This requires a better understanding of the interactions between the materials and the substrate, as well as the ability to control the processing conditions, such as the nozzle size, print speed, and substrate temperature.
3. High-resolution printing: Direct ink writing is often limited by the resolution that can be achieved with the printhead and the printing process. Improving the resolution of the printed structures is important for creating detailed and complex devices, such as sensors and displays.
4. Printing speed: Another challenge is increasing the printing speed, while maintaining the quality and uniformity of the printed structures. This requires a better understanding of the printing process and the development of optimized printing conditions and materials.
5. Multifunctionality: Another challenge is creating electronic devices and components with multiple functions, such as sensing, energy harvesting, and communication. This requires the integration of multiple materials and functionalities into a single device, which can be challenging due to the compatibility and integration issues.

Despite these challenges, direct ink writing is a rapidly developing field, with the potential for new advances and innovations in the creation of electronic devices and components, such as flexible and stretchable electronics, sensors, and displays. By addressing these challenges, direct ink writing has the potential to become a powerful tool for the creation of advanced and functional electronic devices in the future.

## REFERENCES

[1] Bui, H. M.; Großmann, P. F.; Berger, A.; Seidel, A.; Tonigold, M.; Szesni, N.; Fischer, R.; Rieger, B.; Hinrichsen, O. Comparison of Direct Ink Writing and Binder Jetting for Additive Manufacturing of Pt/Al2O3 Catalysts for the Dehydrogenation of Perhydro-Dibenzyltoluene. *Chem. Eng. J.*, 2023, 141361.

[2] Saadi, M.; Maguire, A.; Pottackal, N. T.; Thakur, M. S. H.; Ikram, M. M.; Hart, A. J.; Ajayan, P. M.; Rahman, M. M. Direct Ink Writing: A 3D Printing Technology for Diverse Materials. *Adv. Mater.*, 2022, *34* (28), 2108855.

[3] Hou, Z.; Lu, H.; Li, Y.; Yang, L.; Gao, Y. Direct Ink Writing of Materials for Electronics-Related Applications: A Mini Review. *Front. Mater.*, 2021, *8*, 647229.
[4] Rocha, V. G.; Saiz, E.; Tirichenko, I. S.; García-Tuñón, E. Direct Ink Writing Advances in Multi-Material Structures for a Sustainable Future. *J. Mater. Chem. A*, 2020, *8* (31), 15646–15657.
[5] Peng, X.; Kuang, X.; Roach, D. J.; Wang, Y.; Hamel, C. M.; Lu, C.; Qi, H. J. Integrating Digital Light Processing with Direct Ink Writing for Hybrid 3D Printing of Functional Structures and Devices. *Addit. Manuf.*, 2021, *40*, 101911.
[6] Abas, M.; Naeem, K.; Habib, T.; Khan, I.; Farooq, U.; Khalid, Q. S.; Rahman, K. Development of Prediction Model for Conductive Pattern Lines Generated Through Positive Displacement Microdispensing System Using Artificial Neural Network. *Arab. J. Sci. Eng.*, 2020. 10.1007/s13369-020-05103-3.
[7] Rau, D. A.; Forgiarini, M.; Williams, C. B. Hybridizing Direct Ink Write and Mask-Projection Vat Photopolymerization to Enable Additive Manufacturing of High Viscosity Photopolymer Resins. *Addit. Manuf.*, 2021, *42*, 101996.
[8] Peng, E.; Zhang, D.; Ding, J. Ceramic Robocasting: Recent Achievements, Potential, and Future Developments. *Adv. Mater.*, 2018, *30* (47), 1802404.
[9] Lamnini, S.; Elsayed, H.; Lakhdar, Y.; Baino, F.; Smeacetto, F.; Bernardo, E. Robocasting of Advanced Ceramics: Ink Optimization and Protocol to Predict the Printing Parameters-A Review. *Heliyon*, 2022, e10651.
[10] Liu, S.; Xu, D.; Li, Y.-F.; Shen, F.; Zhang, D.-P. Nanoliter Fluid Dispensing Based on Microscopic Vision and Laser Range Sensor. *IEEE Trans. Ind. Electron.*, 2016, *64* (2), 1292–1302.
[11] Lee, J. M.; Yeong, W. Y. A Preliminary Model of Time-Pressure Dispensing System for Bioprinting Based on Printing and Material Parameters: This Paper Reports a Method to Predict and Control the Width of Hydrogel Filament for Bioprinting Applications. *Virtual Phys. Prototyp.*, 2015, *10* (1), 3–8.
[12] Lindahl, J.; Hassen, A.; Romberg, S.; Hedger, B.; Hedger Jr, P.; Walch, M.; Deluca, T.; Morrison, W.; Kim, P.; Roschli, A. *Large-Scale Additive Manufacturing with Reactive Polymers*; Oak Ridge National Lab.(ORNL), Oak Ridge, TN (United States), 2018.
[13] Tan, H. W.; Choong, Y. Y. C.; Kuo, C. N.; Low, H. Y.; Chua, C. K. 3D Printed Electronics: Processes, Materials and Future Trends. *Prog. Mater. Sci.*, 2022, 100945.
[14] Keesey, R.; LeSuer, R.; Schrier, J. Sidekick: A Low-Cost Open-Source 3D-Printed Liquid Dispensing Robot. *HardwareX*, 2022, *12*, e00319.
[15] Fathi, S.; Lei, I. M.; Cao, Y.; Huang, Y. Y. S. Microcapillary Cell Extrusion Deposition with Picolitre Dispensing Resolution. *Bio-Design Manuf.*, 2022, 1–11.
[16] Sharma, A.; Rai, A. Fused Deposition Modelling (FDM) Based 3D & 4D Printing: A State of Art Review. *Mater. Today Proc.*, 2022, *62*, 367–372. 10.1016/j.matpr.2022.03.679.
[17] Abas, M.; Habib, T.; Noor, S.; Khan, K. M. Comparative Study of I-Optimal Design and Definitive Screening Design for Developing Prediction Models and Optimization of Average Surface Roughness of PLA Printed Parts Using Fused Deposition Modeling. *Int. J. Adv. Manuf. Technol.*, 2022. 10.1007/s00170-022-10784-1.
[18] Manzo, E.; Downey, N.; Cheetham, P.; Pamidi, S. Fabrication and Characterization of Additively Manufactured Electrical Insulation System Components Using SLA and FDM Printers. In *2022 IEEE Electrical Insulation Conference (EIC)*; IEEE, 2022; pp 69–72.

[19] Zhou, H.; Song, Y. Fabrication of Electronics by Electrohydrodynamic Jet Printing. *Adv. Electron. Mater.*, 2022, *8* (11), 2200728.
[20] Farooq, U.; Khan, I.; Ahmad, S.; Abas, M.; Zaib Khan, M. A.; Rahman, K. Fabrication of PEDOT: PSS Conductive Patterns on Photo Paper Substrate through Electro-Hydrodynamic Jet Printing Process. *Int. J. Light. Mater. Manuf.*, 2019, *2* (4). 10.1016/j.ijlmm.2019.06.002.
[21] Šakalys, R.; Mohammadlou, B. S.; Raghavendra, R. Fabrication of Multi-Material Electronic Components Applying Non-Contact Printing Technologies: A Review. *Results Eng.*, 2022, 100578.
[22] Beedasy, V.; Smith, P. J. Printed Electronics as Prepared by Inkjet Printing. *Materials (Basel)*, 2020, *13* (3), 704.
[23] Nachimuthu, M.; P.K., R. Inkjet Four-Dimensional Printing of Shape Memory Polymers: A Review. *Rapid Prototyp. J.*, 2022, *ahead-of-p* (ahead-of-print). 10.1108/RPJ-08-2021-0198.
[24] Kaçar, R.; Serin, R. B.; Uçar, E.; Ülkü, A. A Review of High-End Display Technologies Focusing on Inkjet Printed Manufacturing. *Mater. Today Commun.*, 2023, *35*, 105534. 10.1016/j.mtcomm.2023.105534.
[25] Werum, K.; Mueller, E.; Keck, J.; Jaeger, J.; Horter, T.; Glaeser, K.; Buschkamp, S.; Barth, M.; Eberhardt, W.; Zimmermann, A. Aerosol Jet Printing and Interconnection Technologies on Additive Manufactured Substrates. *J. Manuf. Mater. Process.*, 2022, *6* (5), 119.
[26] Ćatić, N.; Wells, L.; Al Nahas, K.; Smith, M.; Jing, Q.; Keyser, U. F.; Cama, J.; Kar-Narayan, S. Aerosol-Jet Printing Facilitates the Rapid Prototyping of Microfluidic Devices with Versatile Geometries and Precise Channel Functionalization. *Appl. Mater. Today*, 2020, *19*, 100618.
[27] Agrawaal, H.; Thompson, J. E. Additive Manufacturing (3D Printing) for Analytical Chemistry. *Talanta Open*, 2021, *3*, 100036.
[28] Jing, Q.; Pace, A.; Ives, L.; Husmann, A.; Ćatić, N.; Khanduja, V.; Cama, J.; Kar-Narayan, S. Aerosol-Jet-Printed, Conformable Microfluidic Force Sensors. *Cell Reports Phys. Sci.*, 2021, *2* (4), 100386.
[29] Lall, P.; Goyal, K.; Kothari, N.; Leever, B.; Miller, S. Effect of Process Parameters on Aerosol Jet Printing of Multi-Layer Circuitry. In *International Electronic Packaging Technical Conference and Exhibition*; American Society of Mechanical Engineers, 2019; Vol. 59322, p V001T04A005.
[30] Secor, E. B. Principles of Aerosol Jet Printing. *Flex. Print. Electron.*, 2018, *3* (3), 35002.
[31] Smith, M.; Choi, Y. S.; Boughey, C.; Kar-Narayan, S. Controlling and Assessing the Quality of Aerosol Jet Printed Features for Large Area and Flexible Electronics. *Flex. Print. Electron.*, 2017, *2* (1), 15004.
[32] Zhang, H.; Choi, J. P.; Moon, S. K.; Ngo, T. H. A Knowledge Transfer Framework to Support Rapid Process Modeling in Aerosol Jet Printing. *Adv. Eng. Informatics*, 2021, *48*, 101264.
[33] Fujimoto, K. T.; Watkins, J. K.; Phero, T.; Litteken, D.; Tsai, K.; Bingham, T.; Ranganatha, K. L.; Johnson, B. C.; Deng, Z.; Jaques, B. Aerosol Jet Printed Capacitive Strain Gauge for Soft Structural Materials. *npj Flex. Electron.*, 2020, *4* (1), 32.
[34] Lisha, F.; Fan, L.; Guolong, W.; Kovalenko, V. S.; Jianhua, Y. Research Progress of Laser-Assisted Chemical Vapor Deposition. *Opto-Electronic Eng.*, 2022, *49* (2), 210331–210333.
[35] Odusanya, A.; Rahaman, I.; Sarkar, P. K.; Zkria, A.; Ghosh, K.; Haque, A. Laser-Assisted Growth of Carbon-Based Materials by Chemical Vapor Deposition. *C*, 2022, *8* (2), 24.

[36] Sun, L.; Yuan, G.; Gao, L.; Yang, J.; Chhowalla, M.; Gharahcheshmeh, M. H.; Gleason, K. K.; Choi, Y. S.; Hong, B. H.; Liu, Z. Chemical Vapour Deposition. *Nat. Rev. Methods Prim.*, 2021, *1* (1), 5.

[37] Łysień, M.; Witczak, Ł.; Wiatrowska, A.; Fiączyk, K.; Gadzalińska, J.; Schneider, L.; Stręk, W.; Karpiński, M.; Kosior, Ł.; Granek, F. High-Resolution Deposition of Conductive and Insulating Materials at Micrometer Scale on Complex Substrates. *Sci. Rep.*, 2022, *12* (1), 9327.

[38] Arnault, J.-C.; Saada, S.; Ralchenko, V. Chemical Vapor Deposition Single-crystal Diamond: A Review. *Phys. status solidi (RRL)–Rapid Res. Lett.*, 2022, *16* (1), 2100354.

[39] Saleh, T. A. Large-Scale Production of Nanomaterials and Adsorbents. In *Interface Science and Technology*; Elsevier, 2022; Vol. 34, pp 167–197.

[40] Enea, N.; Ion, V.; Viespe, C.; Constantinoiu, I.; Buiu, O.; Romanitan, C.; Scarisoreanu, N. D. Laser Processed Hybrid Lead-Free Thin Films for SAW Sensors. *Materials (Basel)*, 2022, *15* (23), 8452.

[41] Saksena, J.; Sklare, S. C.; Phamduy, T. B.; Huang, Y.; Chrisey, D. B. The Power of CAD/CAM Laser Bioprinting at the Single-Cell Level: Evolution of Printing. In *3D Bioprinting and Nanotechnology in Tissue Engineering and Regenerative Medicine*; Elsevier, 2022; pp 93–121.

[42] Socol, M.; Preda, N.; Socol, G. Organic Thin Films Deposited by Matrix-Assisted Pulsed Laser Evaporation (MAPLE) for Photovoltaic Cell Applications: A Review. *Coatings*, 2021, *11* (11), 1368.

[43] Alfe, M.; Minopoli, G.; Tartaglia, M.; Gargiulo, V.; Caruso, U.; Pepe, G. P.; Ausanio, G. Coating of Flexible PDMS Substrates through Matrix-Assisted Pulsed Laser Evaporation (MAPLE) with a New-Concept Biocompatible Graphenic Material. *Nanomaterials*, 2022, *12* (20), 3663.

[44] Calin, B. S.; Paun, I. A. A Review on Stimuli-Actuated 3D Micro/Nanostructures for Tissue Engineering and the Potential of Laser-Direct Writing via Two-Photon Polymerization for Structure Fabrication. *Int. J. Mol. Sci.*, 2022, *23* (22), 14270.

[45] Grumezescu, V.; Grumezescu, A. M.; Ficai, A.; Negut, I.; Vasile, B. Ştefan; Gălăţeanu, B.; Hudiţă, A. Composite Coatings for Osteoblast Growth Attachment Fabricated by Matrix-Assisted Pulsed Laser Evaporation. *Polymers (Basel)*, 2022, *14* (14), 2934.

[46] Craciun, C.; Andrei, F.; Bonciu, A.; Brajnicov, S.; Tozar, T.; Filipescu, M.; Palla-Papavlu, A.; Dinescu, M. Nitrites Detection with Sensors Processed via Matrix-Assisted Pulsed Laser Evaporation. *Nanomaterials*, 2022, *12* (7), 1138.

[47] Dumitrescu, L. N.; Ionita, E.-R.; Birjega, R.; Lazea-Stoyanova, A.; Ionita, M.-D.; Epurescu, G.; Banici, A.-M.; Brajnicov, S.; Andrei, F.; Matei, A. Kaolinite Thin Films Grown by Pulsed Laser Deposition and Matrix Assisted Pulsed Laser Evaporation. *Nanomaterials*, 2022, *12* (3), 546.

[48] Tan, H. W.; An, J.; Chua, C. K.; Tran, T. Metallic Nanoparticle Inks for 3D Printing of Electronics. *Adv. Electron. Mater.*, 2019, *5* (5), 1800831.

[49] Muldoon, K.; Song, Y.; Ahmad, Z.; Chen, X.; Chang, M.-W. High Precision 3D Printing for Micro to Nano Scale Biomedical and Electronic Devices. *Micromachines*, 2022, *13* (4), 642.

[50] Pajor-Świerzy, A.; Szczepanowicz, K.; Kamyshny, A.; Magdassi, S. Metallic Core-Shell Nanoparticles for Conductive Coatings and Printing. *Adv. Colloid Interface Sci.*, 2022, *299*, 102578.

[51] Potts, S.-J.; Korochkina, T.; Holder, A.; Jewell, E.; Phillips, C.; Claypole, T. The Influence of Carbon Morphologies and Concentrations on the Rheology and Electrical Performance of Screen-Printed Carbon Pastes. *J. Mater. Sci.*, 2022, *57* (4), 2650–2666.

[52] Chandrasekaran, S.; Jayakumar, A.; Velu, R. A Comprehensive Review on Printed Electronics: A Technology Drift towards a Sustainable Future. *Nanomaterials*, 2022, *12* (23), 4251.
[53] Rao, C. H.; Avinash, K.; Varaprasad, B.; Goel, S. A Review on Printed Electronics with Digital 3D Printing: Fabrication Techniques, Materials, Challenges and Future Opportunities. *J. Electron. Mater.*, 2022, *51* (6), 2747–2765.
[54] Pandya, K. S.; Shindalkar, S. S.; Kandasubramanian, B. Breakthrough to the Pragmatic Evolution of Direct Ink Writing: Progression, Challenges, and Future. *Prog. Addit. Manuf.*, 2023, 1–26.
[55] Baskaran, K.; Ali, M.; Gingrich, K.; Porter, D. L.; Chong, S.; Riley, B. J.; Peak, C. W.; Naleway, S.; Zharov, I.; Carlson, K. Sol-Gel Derived Silica: A Review of Polymer-Tailored Properties for Energy and Environmental Applications. *Microporous Mesoporous Mater.*, 2022, 111874.
[56] Benrezgua, E.; Deghfel, B.; Abdelhalim, Z.; Basirun, W. J.; Amari, R.; Boukhari, A.; Yaakob, M. K.; Kheawhom, S.; Mohamad, A. A. Synthesis and Properties of Copper Doped Zinc Oxide Thin Films by Sol-Gel, Spin Coating and Dipping: A Characterization Review. *J. Mol. Struct.*, 2022, 133639.
[57] Zanurin, A.; Johari, N. A.; Alias, J.; Ayu, H. M.; Redzuan, N.; Izman, S. Research Progress of Sol-Gel Ceramic Coating: A Review. *Mater. Today Proc.*, 2022, *48*, 1849–1854.
[58] Sinha, S.; Pal, T. A Comprehensive Review of FET-based PH Sensors: Materials, Fabrication Technologies, and Modeling. *Electrochem. Sci. Adv.*, 2022, *2* (5), e2100147.
[59] Villarreal, C. C.; Sandoval, J. I.; Ramnani, P.; Terse-Thakoor, T.; Vi, D.; Mulchandani, A. Graphene Compared to Fluorine-Doped Tin Oxide as Transparent Conductor in ZnO Dye-Sensitized Solar Cells. *J. Environ. Chem. Eng.*, 2022, *10* (3), 107551.
[60] Jo, E.-B.; Lee, Y.-A.; Cho, Y.-A.; Günther, P. A.; Gebhardt, S. E.; Neubert, H.; Kim, H.-S. The 0-3 Lead Zirconate-Titanate (PZT)/Polyvinyl-Butyral (PVB) Composite for Tactile Sensing. *Sensors*, 2023, *23* (3), 1649.
[61] Tanveer, M.; Nisa, I.; Nabi, G.; Hussain, M. K.; Khalid, S.; Qadeer, M. A. Sol-Gel Extended Hydrothermal Pathway for Novel Cd-Zn Co-Doped Mg-Ferrite Nano-Structures and a Systematic Study of Structural, Optical and Magnetic Properties. *J. Magn. Magn. Mater.*, 2022, *553*, 169245.
[62] Yang, H.; Zhu, M.; Li, Y. Sol–Gel Research in China: A Brief History and Recent Research Trends in Synthesis of Sol–Gel Derived Materials and Their Applications. *J. Sol-Gel Sci. Technol.*, 2022, 1–16.
[63] Ostovan, A.; Arabi, M.; Wang, Y.; Li, J.; Li, B.; Wang, X.; Chen, L. Greenificated Molecularly Imprinted Materials for Advanced Applications. *Adv. Mater.*, 2022, *34* (42), 2203154.
[64] Li, Z.; Liu, X.; Wang, G.; Li, B.; Chen, H.; Li, H.; Zhao, Y. Photoresponsive Supramolecular Coordination Polyelectrolyte as Smart Anticounterfeiting Inks. *Nat. Commun.*, 2021, *12* (1), 1363.
[65] Li, S.; Lu, Z.; Zhang, H.; Ai, Z.; Ran, Y.; Li, Y.; Deng, X.; Li, D. Rheological Behavior of Multi-Sized SiC Inks Containing Polyelectrolyte Complexes Specifically for Direct Ink Writing. *J. Eur. Ceram. Soc.*, 2022, *42* (12), 4810–4816.
[66] Honaryar, H.; Amirfattahi, S.; Niroobakhsh, Z. Associative Liquid-In-Liquid 3D Printing Techniques for Freeform Fabrication of Soft Matter. *Small*, 2023, 2206524.

[67] Abel, S. B.; Frontera, E.; Acevedo, D.; Barbero, C. A. Functionalization of Conductive Polymers through Covalent Postmodification. *Polymers (Basel)*, 2023, *15* (1), 205.
[68] Onggar, T.; Kruppke, I.; Cherif, C. Techniques and Processes for the Realization of Electrically Conducting Textile Materials from Intrinsically Conducting Polymers and Their Application Potential. *Polymers (Basel)*, 2020, *12* (12), 2867.
[69] Tran, V. Van; Lee, S.; Lee, D.; Le, T.-H. Recent Developments and Implementations of Conductive Polymer-Based Flexible Devices in Sensing Applications. *Polymers (Basel)*, 2022, *14* (18), 3730.
[70] Wang, C.; Park, M. J.; Gonzales, R. R.; Phuntsho, S.; Matsuyama, H.; Drioli, E.; Shon, H. K. Novel Organic Solvent Nanofiltration Membrane Based on Inkjet Printing-Assisted Layer-by-Layer Assembly. *J. Memb. Sci.*, 2022, *655*, 120582.
[71] Taghizadeh, M.; Taghizadeh, A.; Yazdi, M. K.; Zarrintaj, P.; Stadler, F. J.; Ramsey, J. D.; Habibzadeh, S.; Rad, S. H.; Naderi, G.; Saeb, M. R. Chitosan-Based Inks for 3D Printing and Bioprinting. *Green Chem.*, 2022, *24* (1), 62–101.
[72] Tong, C.; Tong, C. Conductive Materials for Printed Flexible Electronics. *Adv. Mater. Print. Flex. Electron.*, 2022, 119–157.
[73] Poulin, A.; Aeby, X.; Siqueira, G.; Nyström, G. Versatile Carbon-Loaded Shellac Ink for Disposable Printed Electronics. *Sci. Rep.*, 2021, *11* (1), 23784.
[74] Fekiri, C.; Kim, H. C.; Lee, I. H. 3D-Printable Carbon Nanotubes-Based Composite for Flexible Piezoresistive Sensors. *Materials*. 2020. 10.3390/ma13235482.
[75] Poulin, A.; Aeby, X.; Nyström, G. Water Activated Disposable Paper Battery. *Sci. Rep.*, 2022, *12* (1), 11919. 10.1038/s41598-022-15900-5.
[76] Hassan, M. S.; Zaman, S.; Rodriguez, A.; Molina, L.; Dominguez, C. E.; Morgan, R.; Bernardin, J.; Lin, Y. Direct Ink Write 3D Printing of Wave Propagation Sensor. *Flex. Print. Electron.*, 2022, *7* (4), 45011.
[77] Votzke, C.; Daalkhaijav, U.; Mengüç, Y.; Johnston, M. L. 3D-Printed Liquid Metal Interconnects for Stretchable Electronics. *IEEE Sens. J.*, 2019, *19* (10), 3832–3840.
[78] Chen, Y.; Zhou, L.; Wei, J.; Mei, C.; Jiang, S.; Pan, M.; Xu, C. Direct Ink Writing of Flexible Electronics on Paper Substrate with Graphene/Polypyrrole/Carbon Black Ink. *J. Electron. Mater.*, 2019, *48*, 3157–3168.
[79] Anelli, S.; Rosa, M.; Baiutti, F.; Torrell, M.; Esposito, V.; Tarancón, A. Hybrid-3D Printing of Symmetric Solid Oxide Cells by Inkjet Printing and Robocasting. *Addit. Manuf.*, 2022, *51*, 102636. 10.1016/j.addma.2022.102636.
[80] Bernard, M.; Pateloup, V.; Passerieux, D.; Cros, D.; Madrangeas, V.; Chartier, T. Feasibility of Manufacturing of Al2O3–Mo HTCC by Hybrid Additive Process. *Ceram. Int.*, 2022, *48* (11), 14993–15005. 10.1016/j.ceramint.2022.01.354.
[81] Abas, M.; Rahman, K. Fabrication of Flex Sensors through Direct Ink Write Technique and Its Electrical Characterization. *Appl. Phys. A Mater. Sci. Process.*, 2016, *122* (11). 10.1007/s00339-016-0507-8.
[82] Abas, M.; Salman, Q.; Khan, A. M.; Rahman, K. Direct Ink Writing of Flexible Electronic Circuits and Their Characterization. *J. Brazilian Soc. Mech. Sci. Eng.*, 2019, *41* (12). 10.1007/s40430-019-2066-3.
[83] Kim, F.; Yang, S. E.; Ju, H.; Choo, S.; Lee, J.; Kim, G.; Jung, S.; Kim, S.; Cha, C.; Kim, K. T.; et al. Direct Ink Writing of Three-Dimensional Thermoelectric Microarchitectures. *Nat. Electron.*, 2021, *4* (8), 579–587. 10.1038/s41928-021-00622-9.

[84] Yang, G.; Wu, H.; Li, Y.; Wang, D.; Song, Y.; Zhou, Y.; Hao, J.; Zi, Y.; Wang, Z.; Zhou, G. Direct Ink Writing of Fluoropolymer/CNT-Based Superhydrophobic and Corrosion-Resistant Electrodes for Droplet Energy Harvesters and Self-Powered Electronic Skins. *Nano Energy*, 2021, *86*, 106095. 10.1016/j.nanoen.2021.106095.

[85] Wang, A.; Chen, C.; Liao, L.; Qian, J.; Yuan, F.-G.; Zhang, N. Enhanced β-Phase in Direct Ink Writing PVDF Thin Films by Intercalation of Graphene. *J. Inorg. Organomet. Polym. Mater.*, 2020, *30* (5), 1497–1502. 10.1007/s10904-019-01310-0.

[86] Tang, S.; Lu, X.; Geng, P.; Chen, D.; Shi, Y.; Su, J.; Zhou, Y.; Su, B.; Wen, S. A Non-Covalent Supramolecular Dual-Network Polyelectrolyte Evaporator Based on Direct-Ink-Writing for Stable Solar Thermal Evaporation. *Mater. Adv.*, 2023, *4* (1), 223–230.

[87] Sabadin, M.; Sol, J. A. H. P.; Debije, M. G. Direct Ink Writing of Anisotropic Luminescent Materials. *Crystals*, 2022. 10.3390/cryst12111642.

[88] Li, Q.; Dong, Q.; Wang, J.; Xue, Z.; Li, J.; Yu, M.; Zhang, T.; Wan, Y.; Sun, H. Direct Ink Writing (DIW) of Graphene Aerogel Composite Electrode for Vanadium Redox Flow Battery. *J. Power Sources*, 2022, *542*, 231810. 10.1016/j.jpowsour.2022.231810.

[89] Nassar, H.; Dahiya, R. Fused Deposition Modeling-Based 3D-Printed Electrical Interconnects and Circuits. *Adv. Intell. Syst.*, 2021, *3* (12), 2100102. 10.1002/aisy.202100102.

[90] Junpha, J.; Wisitsoraat, A.; Prathumwan, R.; Chaengsawang, W.; Khomungkhun, K.; Subannajui, K. Electronic Tongue and Cyclic Voltammetric Sensors Based on Carbon Nanotube/Polylactic Composites Fabricated by Fused Deposition Modelling 3D Printing. *Mater. Sci. Eng. C*, 2020, *117*, 111319. 10.1016/j.msec.2020.111319.

[91] Georgopoulou, A.; Sebastian, T.; Clemens, F. Thermoplastic Elastomer Composite Filaments for Strain Sensing Applications Extruded with a Fused Deposition Modelling 3D Printer. *Flex. Print. Electron.*, 2020, *5* (3), 35002. 10.1088/2058-8585/ab9a22.

[92] Georgopoulou, A.; Clemens, F. Pellet-Based Fused Deposition Modeling for the Development of Soft Compliant Robotic Grippers with Integrated Sensing Elements. *Flex. Print. Electron.*, 2022, *7* (2), 25010. 10.1088/2058-8585/ac6f34.

[93] Hassan, R. U.; Khalil, S. M.; Khan, S. A.; Ali, S.; Moon, J.; Cho, D.-H.; Byun, D. High-Resolution, Transparent, and Flexible Printing of Polydimethylsiloxane via Electrohydrodynamic Jet Printing for Conductive Electronic Device Applications. *Polymers*, 2022. 10.3390/polym14204373.

# 10 Ultrasonic Molding

*Marcel Janer, Maria Dolors Riera, and Xavier Plantà*

## 10.1 ULTRASONIC MOLDING TECHNOLOGY: ELEMENTS AND MAIN STEPS

### 10.1.1 General Overview

Ultrasonic molding (USM) is a relatively new technology specifically designed for the production of mini and micro plastic parts. This technology was first developed in 2002 [1], using high-power ultrasound to melt a small quantity of polymer and shape it inside a mold. Since then, numerous experiments have been conducted by a number of authors in an effort to investigate fresh processing options and get beyond some of micro-injection molding's ($\mu$IM) restrictions.

$\mu$IM is a well-established technology for low-cost mass production and has been able to successfully process parts with sub-micron or even nano sizes. However, for some applications wherein the molded parts display cross-scale features, $\mu$IM could be challenging in terms of replication fidelity, material utilization, and energy consumption.

On the other hand, combining high-power ultrasound with injection molding has been shown to improve the replication of micro features and make melt filling easier [2]. Additionally, as has been shown in consolidated procedures like ultrasonic welding or ultrasonic hot embossing, using high-power ultrasound to process polymers can produce incredibly repetitive outcomes with low energy-consumption ratios. As a result, power ultrasonics are used in injection molding operations in an effort to improve their performance in difficult applications while also lowering their energy usage. High-power ultrasound applications involve the use of low ultrasonic frequencies (typically from 16kHz to 100kHz) and high amplitudes of vibration.

This technology has been applied since the 1950s to process polymers and is extensively used in common industrial applications such as ultrasonic welding [3] or ultrasonic cutting [4], among several others [5]. One of the distinctive elements of this technology is the *sonotrode* (also known as ultrasonic horn), which is the element that delivers the ultrasonic vibration to the processed material.

The main advantages of USM technology over the conventional process are:

- **Shorter time of residence at high temperatures:** Due to the rapidity of the process, the polymer is subjected to high temperatures within a very short period of time [6], which allows its injection without degradation of the polymer.


DOI: 10.1201/9781003364948-10

- **Lower temperature needed to inject the material:** The material can flow at a lower temperature as the application of ultrasonic waves to a polymer reduces its viscosity due to physical and chemical factors [7].
- **Lower pressure of injection:** The better fluidity of the material makes USM technology capable of injecting materials at pressures much lower than in conventional micro-injection. Masato et al. injected a polypropylene specimen using USM at a pressure three times lower than in conventional injection molding [8]. This should allow the USM technology to process materials with geometries beyond the scope of conventional injection methods.
- **Energy savings:** As each pellet in the polymer is heated locally, the energy needed to increase the temperature of the material is expected to be much lesser than that used in conventional machines [9].
- **Material savings:** The only material melted is the one injected. That means that there is no remaining material in the plasticizing chamber, and it is even possible to change the material without wasting material. In comparison with conventional injection molding, the USM can save 40% to 70% of the material [2].

The principle on which the USM is based is the transmission of the oscillatory energy from a sonotrode to a thermoplastic polymer with which it is in contact. The mechanical energy is then transformed into thermal energy, which heats and melts the polymeric pellets. In the second stage, the melt is introduced into a mold wherein it solidifies. These steps, along with the main elements involved in the process will be described in this section. A more detailed explanation of this technology and its advances can be found in reference [10].

USM technology combines: An acoustic unit (similar to the ones used in other high-power ultrasound applications), a molding unit adapted from the injection molding technology, and a feeding unit usually designed and fabricated exclusively for this technology (see Figure 10.1). As the acoustic will be described in detail in section 10.2, the other two elements are explained in this section.

#### 10.1.1.1 Feeding Unit

The feeding unit is used to automate the dosage of material in USM technology. As the metering of injection material in USM is done at the feeding stage, the same amount of material needs to be dosed each time. The feeder is composed of a dosing apparatus that includes a hopper, which pours pellets into a vibrating chute (see Figure 10.2). The vibration aligns and moves the pellets until they drop into the mold. A laser is used to count the pellets at the end of the vibrating chute. The dosing can also be done manually, by counting the pellets, or by using a balance, which is more accurate due to the weight dispersion of some polymer pellets.

#### 10.1.1.2 Molding Unit

Molds used in USM are adapted from conventional micro-injection technology. In USM, the molds are placed horizontally and divided into two parts: A moving part

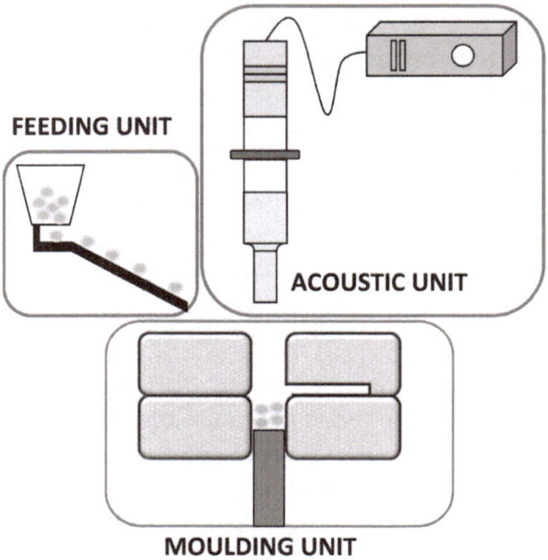


**FIGURE 10.1** Main units of the USM equipment: An acoustic unit, a feeding unit and a molding unit.

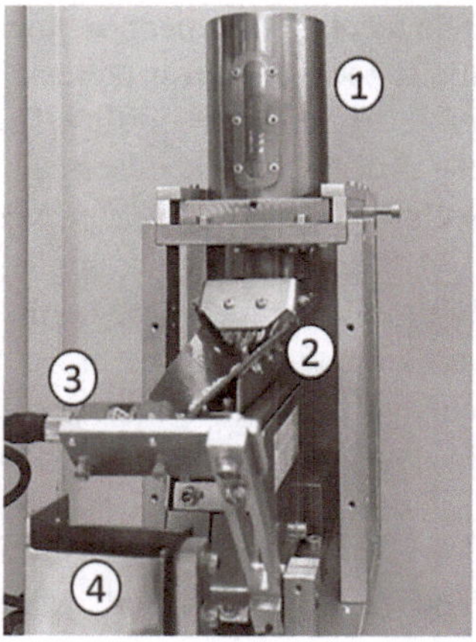


**FIGURE 10.2** Image of a feeder for USM process used in Sonorus®, with a hopper(1), a vibrating chute (2), a laser to count the pellets (3), and a case connected to the mold (4).

(superior) and a fixed part (inferior). Electrical heaters are usually included, along with an ejection unit to demold parts. In addition, ultrasonic molds also include a plasticizing chamber where the material is deposited, and holes to introduce the sonotrode and the plunger (a piston used to push the material into the mold). The main components of a typical USM mold are presented in Figure 10.3.

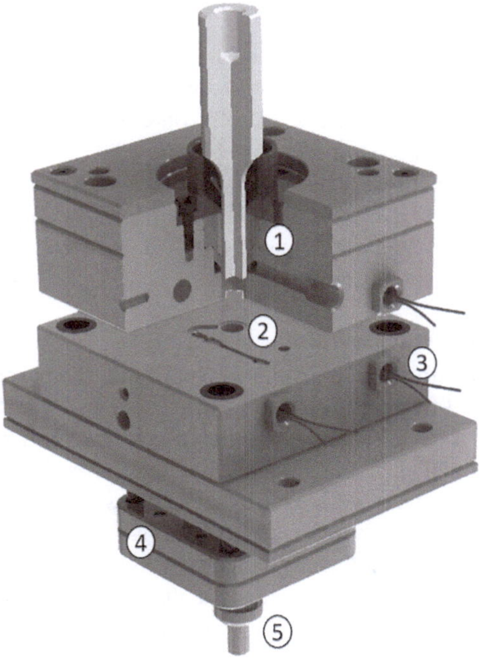


**FIGURE 10.3** Render image of a USM mold; main elements are: Hole and sleeves for the sonotrode (1), plasticization chamber and mold cavity (2), heaters (3), ejection unit (4), plunger and sleeves (5).

### 10.1.2 High Power Ultrasound

Ultrasonic waves are vibratory oscillations at frequencies above the hearing range of a person (typically 16 kHz). These waves may be considered as acoustic waves, which are compression waves that propagate in gases and several liquids, or as elastic waves, which are propagated in solids and depend on the elastic properties of the medium [11].

Ultrasonic applications can be divided into two main categories: Low-power ultrasound and high-power ultrasound. Applications involving low-power ultrasound are used for measurement and control, whereas high-power ultrasonics are used for processing. Currently, power ultrasonics are attracting increased attention in industrial applications, as they might constitute flexible green alternatives for energy efficient processes [12].

High-power ultrasound waves have been applied since the 1950s to process polymers in multiple fields, such as extrusion [13], injection molding [14], welding, and recently, injection molding [9,15–19].

#### 10.1.2.1 Acoustic Unit

Industrial high-power ultrasound waves are created by an *acoustic unit*, which encompasses the main elements needed to generate and deliver the mechanical vibrations at ultrasonic frequency: An ultrasonic generator, an ultrasonic transducer, a booster, and a sonotrode (see Figure 10.4). High power ultrasonics means larger piezoelectrics; so, it involves lower ultrasonic frequencies (typically between 16kHz to 100 kHz). To be able to deliver high power with minimum losses, all the elements must be tuned to work near the resonance frequency of the ultrasonic transducer.

##### *10.1.2.1.1 Ultrasonic Generator*

The ultrasonic generator is responsible for supplying the proper electrical signal to the transducer and for controlling the entire process. To transmit power to the transducer in the most efficient way, the output waveform of an ultrasonic generator is sinusoidal, and the frequency delivered is fixed at a very narrow range [20]. Typical powers delivered by generators range from a few hundred watts to 4 kW. Most commercial generators are equipped with phase-locked loop-frequency systems, which automatically match the generator frequency and phase with the ultrasonic transducer using a VCO (voltage-controlled oscillator) [21]. Some of the newest generators can also provide real-time power consumption, amplitude, and frequency through analogue outputs.

##### *10.1.2.1.2 Ultrasonic Transducer*

Two types of transducers (also known as converters) are most widely applied across industrial applications, that is, magnetostrictive and piezoelectric transducers [12]. Particularly, when using high-power ultrasound equipment in polymer applications, transducers are usually piezoelectric and built following a Langevin design. This type of converter, also called *sandwich transducer*, was invented by physicist Paul Langevin in the 1920s for the development of SONAR. However, his scope of use has expanded since then, due to its high-power ratio and high energy conversion efficiency [22]. Langevin transducers are usually made of two or four lead zirconate titanate (PZT) ceramics sandwiched between two resonator metal blocks [23]. Usually, steel is used as a back-mass,

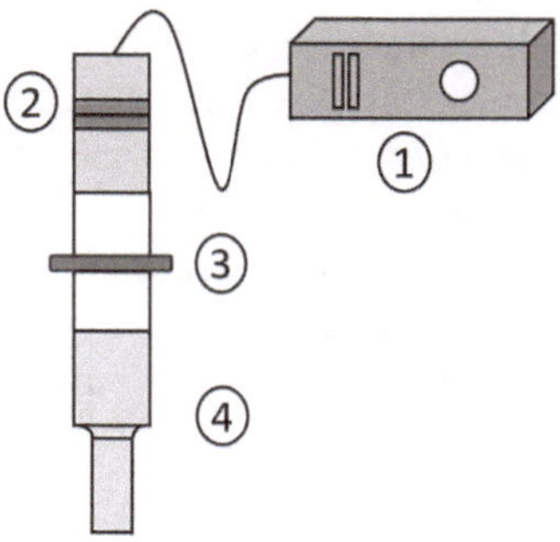


**FIGURE 10.4** Sketch of the main elements of an acoustic unit: 1) ultrasonic generator, 2) ultrasonic transducer, 3) booster, 4) sonotrode.

and aluminum is used as a front-mass. The sandwich is clamped using a bolt, to force the system to resonate according to the entire structure. Since the piezo-electric elements are placed in the middle of the transducer, near the vibrating node, the long displacements are supported by the metal blocks, and it is possible to get high power without breaking the ceramics [24].

##### 10.1.2.1.3 Booster

A booster is a mechanical element designed to amplify or reduce the vibration of the transducer and can also be used as a mounting point for the acoustic assembly. The gain of the booster is related to the mass of the booster at each side of its nodal point and is usually displayed as a ratio between input amplitude and output amplitude. Typical gain values for commercial boosters range from 1 to 2.5. In some devices, such as handheld equipment, boosters may not be necessary.

##### 10.1.2.1.4 Sonotrode

A sonotrode (also called *ultrasonic horn*) is an element that delivers the mechanical energy into the material. This element must be carefully designed to match the resonance frequency of the ultrasonic transducer attached. Typical sonotrode geometries for ultrasonic welding are axisymmetrical with different profiles, depending on the amplitude desired and stress supported by the material. Typical geometries include stepped, exponential, and catenoidal section profiles [25]. Usually, stepped sonotrodes can achieve higher gains (difference between input and output amplitudes) but have higher stress. On the other hand, not all materials are suitable for manufacturing sonotrodes, as they may have to bear ultra-high cycle fatigue. This means that a material with good acoustical properties and low internal damping is needed. Besides, complex geometries require material easy to machine. Thus, the established criteria for material selection constitute a balance among cost, affordability, machinability, good fatigue behavior, and good acoustic properties. Accordingly, some of the typical materials used for sonotrode fabrication are listed in Table 10.1.

**Stepped sonotrode design:** USM technology uses half-wave stepped sonotrodes. However, in practice real stepped sonotrodes have a radius between both cylinders and they no longer measure half wavelength. Thus, to design a sonotrode to resonate at a certain frequency, the total length must be carefully calculated. As a first guess, there are several analytical equations that can be used to estimate the geometry based on empirical data. However, currently, most sonotrodes are designed using finite element methods, through a simple modal analysis study [26]. Finite element methods can also be used to ensure that the gain of the sonotrode and the stress supported by the material are correct.

Once the sonotrode is designed and machined, a tuning step must be performed to adjust the entire acoustic system to the desired frequency. This step is executed, partly to account for possible differences between the real material properties and the given ones, and partly to account for the influence of the different parts of the

**TABLE 10.1**
**Typical Materials for Sonotrode Fabrication**

| Material | Alloy | Benefits | Drawbacks |
|---|---|---|---|
| Titanium | Ti6Al4V Ti7Al4Mo | Good fatigue strength Good surface hardness | Expensive Difficult to machine |
| Aluminum | Al7075 | Easy to machine Low cost | Poor surface hardness |
| Steel | – | Good wear resistance | Low fatigue strength |
| Niobium | – | Erosion resistance during cavitation | Very expensive |
| Syalon | – | Erosion resistance during cavitation Temperature resistance | Very expensive Very difficult to machine |

acoustic unit that have not been simulated (i.e., booster, transducer). For this step, an impedance measuring equipment must be used to match the resonance frequency of the acoustic unit with the one delivered by the ultrasonic generator. Although professional impedance analyzers can be prohibitive in terms of their cost, currently there exist some cheap impedance analyzers, specifically designed for ultrasonic equipment [27]. The analysis of the impedance curve provides important information about the overall performance of the acoustic unit, such as the resonance frequencies and the quality factor. For further information about the design of customized high-power sonotrodes, one can consult reference [23].

### 10.1.3 Ultrasonic Molding Steps

The main steps in a typical USM application are as follows: Feeding the material, heating the polymer pellets until they melt, filling the mold with the melt, packing the material, and cooling it until it can be de-molded (see Figure 10.5). These steps are similar to the ones used in conventional injection molding. However, the coupling of the steps and the high speed of the heating step pose a challenge to the complete understanding of the process.

#### 10.1.3.1 Material Feeding

The first step required for the USM process is to count and place the material in the plasticization chamber. The dosage of the material can be set manually or by using an automated feeder system. The variability of the volume of dosing associated with the dispersion of the volume of the pellets is thought to be one of the main problems affecting the stability of the process. Although the raw material for USM process are usually pellets, some studies have also been carried out using powder [9,17] and irregular-shaped polymers [28].

#### 10.1.3.2 Polymer Heating

Once the polymer is placed in the plasticizing chamber, the upward advance of the plunger compresses the pellets against themselves, against the walls of the

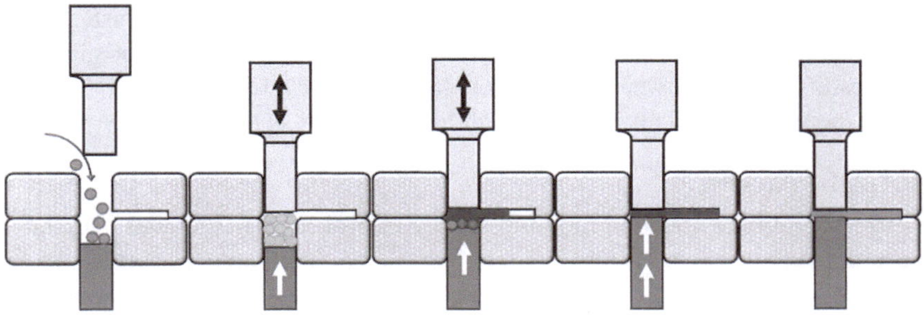


**FIGURE 10.5** Main steps in USM.

chamber and against the sonotrode that is already vibrating. In the first stage, the particles get rearranged and the friction developed in the areas of contact among the particles and with the tools is the basic mechanism responsible for the heating of the material. When the applied stress reaches a certain value, the particles begin to deform, and their contact areas increase progressively. Once the particles are compacted and a complete contact is established between the sonotrode and the polymer, the high frequency vibration intensely increases the temperature of the material closest to the sonotrode until it melts. Conversely, the material that is in contact with the plunger, continues to be in a solid state since its heating rate is much lower than the one developed in the region directly in contact with the sonotrode.

Most of the studies published in extant literature have applied ultrasonic energy in a continuous manner during the heating step. However, Sánchez-Sánchez et al. [28] considered the mode of activation of the wave as a new parameter. They evaluated the possibility of using an intermittent methodology in the application of the ultrasound energy during the heating stage. In continuous ultrasound mode, the ultrasonic wave is continuously applied throughout the process, while in intermittent mode, ultrasound is activated only during a specific phase of the plunger path. The objective of this strategy is to have a larger surface area prior to the application of ultrasonic irradiation, thus improving interfacial friction heating.

### 10.1.3.3 Cavity Filling

The filling stage takes place simultaneously along with the heating. Once the first polymer layer has melted, the upward movement of the plunger against the material makes it flow toward the cavity in the mold, while the rest of the polymer is still heating under the action of ultrasonic vibration. The latest USM machines allow the mold filling to be carried out by keeping the force exerted by the plunger constant (similar to an ultrasonic welding press), or by maintaining constant a speed of advancement (as in conventional injection molding machines).

A comparison with the results obtained by injecting with either a constant force or constant velocity, in the study carried out by Montes [29], does not show significant differences between the methodologies. However, velocity driven injection is more stable and allows the use of a force value as a *"switchover point"*.

#### 10.1.3.4 Packing

As in conventional injection molding, the USM process also includes packing as the final step in cavity filling. In this step, a holding force is applied to the polymer to compensate the shrinkage and to ensure the correct geometrical properties of the sample. The transition between the filling and the packing steps is known as the *"switchover point"* and is one of the most critical aspects in conventional injection molding. Various factors can be used as a switchover point in conventional injection molding, although the ones most used are the screw position and the hydraulic pressure [30].

In the case of $\mu$IM, the position of the piston is usually used as a switchover point. However, in the USM process, the position of the plunger cannot be used as a switchover point, because there can be a significant variation in material from shot-to-shot, depending upon the dispersion of volume between pellets. Thus, the factor used as a switchover point across all the works analyzed is the ultrasonic time [18]. However, the newest versions of *Sonorus® 1G* machine allow the value of the force induced by the plunger to be used as a switchover point. In this case, when the plunger reaches the force chosen, the filling step is switched to the packing step. A more detailed explanation of the use of force as a switchover point in USM process, can be found in reference [31].

#### 10.1.3.5 Cooling

During the cooling stage, the polymer injected is cooled down inside the mold, from the melting state to the de-molding temperature. Although the cooling stage is the most time-consuming one among all the injection processes, while manufacturing micro-samples, there is no need for such long time periods due to the small volume of material to be cooled. In any case, the time needed to demold depends on several parameters: The thickness of the sample, the thermal properties of the polymer, the material of the mold, and on the thermal contact conductance (the quality of the contact between the sample and the mold).

## 10.2 MATERIALS AND GEOMETRIES PROCESSED WITH USM

USM technology has already been tested with multiple materials and applications. Although there is still a great deal of work pending, the technology has shown to be able to process a large variety of polymers. Among them, polypropylene (PP) [6], polyoxymethylene (POM) [32], polylactic acid (PLA) [15], polymethyl methacrylate (PMMA), polyphenylsulfone (PPSU) [19], polybutylene succinate (PBS) [17], ultra-high molecular weight polyethylene (UHMWPE) [33], polyamide 12 (PA12) [34], polycaprolactone (PCL) [35], and synthesized polymers, such as poly (nona-methylene azelate) [36], among others [37].

### 10.2.1 Responses Studied in the Literature

In conventional injection molding, a sample can be considered well processed when it has the desired geometry, the adequate mechanical properties, and no signs of defects or impurities. To evaluate the use of the USM process, different responses related to the quality of the sample have to be studied. The following responses are the ones usually considered in the literature:

#### 10.2.1.1 Filling Capability

A completely filled sample is an indispensable requirement for a good molding process. If this requirement is not fulfilled, most of the rest of the requirements are worthless. Thus, the filling capability is a common response used to find the processing window of a material [38]. However, few authors study in depth the repeatability of the process. Although USM process is capable to fill completely figures with commodity polymers as PP or PS, there is a low repeatability of the process for certain high performance materials like PPSU [19]. In general, to obtain completely filled samples, a minimum ultrasonic time and pressure are needed, and most studies find that higher amplitudes improve the filling of the samples [39]. High mold temperatures also influence part filling positively [28].

#### 10.2.1.2 Mechanical Properties

To ensure the applicability of a polymeric sample, mechanical requirements must be fulfilled. The common method to study the mechanical properties of a material is the stress-strain curve obtained from a tensile stress test. Thus, several authors used the tensile strength of the samples [40] and the Young's modulus [33] as responses for their studies. Other authors only studied the mechanical properties in the optimal molding points. In this case, good repeatability of the stress-strain curves are found for PLA [9].

Comparing other processes, Dorf et al. [40] found that the PEEK specimens obtained from USM had similar tensile strengths to the ones obtained by conventional injection molding but presented a 56% higher crosshead extension. On the other hand, in comparison with the compression molding process, greater values of the elastic modulus and tensile strength but lower values of tensile strain were found [33]. Conversely, Masato et al. [8] found that the mechanical properties obtained when processing PP were significantly better than those obtained with conventional micro-injection, although lower repeatability was observed.

#### 10.2.1.3 Molecular Weight

It is known from the literature that ultrasonic energy reduces the molecular chain entanglement and increases the fluidity of a polymer [7]. However, the application of ultrasonic energy can also cause chain breakages and degrade the material [41]. The measurement of the average molecular weight of the polymer is a method to estimate if there have been chain breakages during the process. Sacristan et al. [10] studied the effect of processing conditions on the average molecular weight of PLA samples. Although a general behavior could not be observed, the authors found that the molding pressure plays a significant role in the molecular weight, while there was no clear tendency with the ultrasonic amplitude. Moreover, the average molecular weight was maintained when the samples were processed at optimal conditions.

#### 10.2.1.4 Crystallization

The first USM tests reported that it was possible to process materials with regular crystallization and homogeneous structures [32]. An amorphous outer layer was observed for ultrasonic molded PP and POM specimens. In fact, the detailed chemical analysis performed by Sacristan et al. [10] with PLA suggests that ultrasonic processing increases the amorphous phase content of the samples. Jian et al. [42] showed that for HDPE, under ultrasonic excitation, the molecular chain becomes shorter, the chain structure becomes simpler, and the symmetry stereo-regularity of the molecule is better. According to these authors, these results lead to faster crystallization velocity, lower crystallization temperature, and more uniform grains of the polymer. In comparison with conventional injection, Masato et al. [9] found higher crystallization in USM processed PP.

#### 10.2.1.5 Dimensional Accuracy

Grabalosa et al. [16] studied the average width and thickness of a PA tensile specimen, concluding that thickness accuracy improves with maximum pressures and maximum ultrasonic time. In contrast, width dimensions did not show any trend under different processing conditions. The authors of the article attributed this due to that the width dimension is two times larger than the thickness (it has greater contact with the mold), so it solidifies earlier and reduces the influence of the packing pressure. Heredia et al. [18] also measured the thickness of processed PLA in a thin wall plate shape. Two different process conditions varying ultrasonic time and a pre-compaction of the pellets have been studied, performing 10 trials for each configuration. However, the dispersion of the results makes it difficult to obtain any influence of the process parameters. The authors point out that maybe the location of the thermal resistances are affecting the results (the location is different for both configurations). Similar results were obtained by Ferrer et al. [38] when measuring the thickness of Polystyrene thin plates at different points and analyzing the results using principal components analysis.

### 10.2.2 Processed Geometries

This section shows some impressive geometries processed with USM, courtesy of *Ultrasion®*.

One of the great benefits of USM technology is its ability to process high aspect ratio figures. In this context, thin spirals as the one shown in Figure 10.6 are used usually to evaluate the flow capacity of a material. In this example, a polypropylene was used to fill the figure.

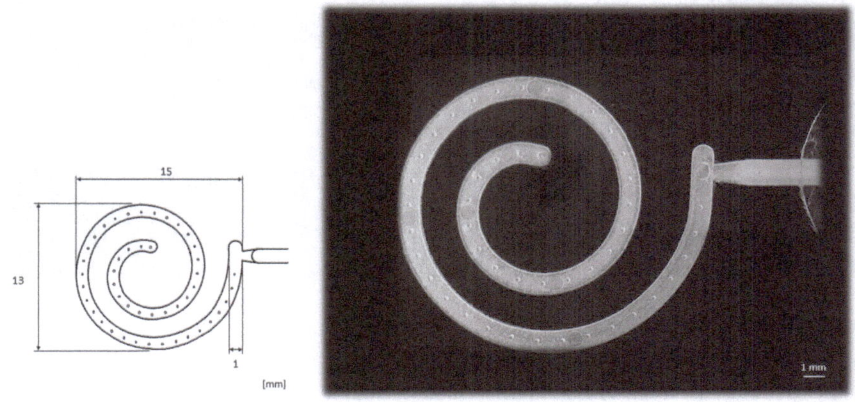


**FIGURE 10.6** 0.2 mm thick polypropylene spiral processed with USM.

The overmolding of miniaturized samples is another application suitable for USM technology, as the polymer is injected at lower pressures than in conventional microinjection molding. In Figure 10.7 overmolded filters placed on a 1 cent coin are displayed.

**FIGURE 10.7** Overmolded polypropylene filters processed with USM placed on a 1 cent coin.

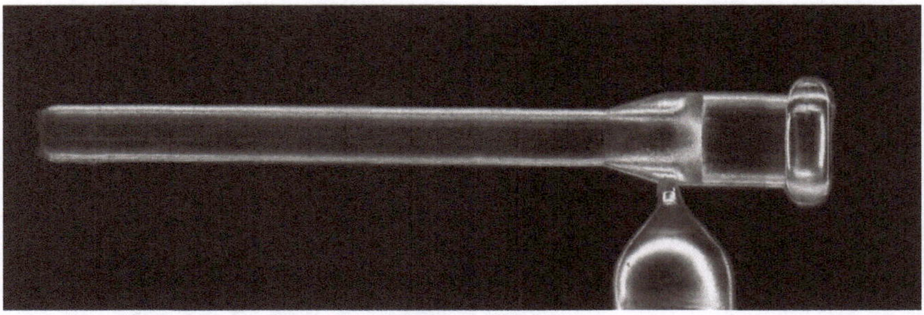

**FIGURE 10.8** Cannula made from cyclic olefin polymer (COP).

Finally, good results are also expected when using USM technology for processing optical materials, as the low injection velocities used in this process will reduce the shear stress of the samples. As an example, Figure 10.8 shows a cannula made of cyclic olefin polymer (COP).

## 10.3 NODAL POINT METHOD, A NEW CONFIGURATION FOR USM

As described previously, ultrasonic molding (USM) technology is an interesting alternative to conventional injection molding and has been applied to process multiple materials for different applications. However, up to now, its use has mainly been limited to research centers and universities due to the lack of repeatability of the technology. It can be said that the performance of standard USM methodology does not meet the requirements from industry standards, preventing its adoption in most of the markets. However, recently, a new configuration for USM has been developed [43] with the aim to solve all the main drawbacks encountered in USM process. This new configuration has been called as *nodal point* due to the use of the resonance vibration properties of the sonotrode involved and will be presented in this section.

### 10.3.1 Identified Drawbacks of USM Standard Methodology

Most of the experimental tests performed with USM in the literature are done using the *standard configuration*. This configuration sonotrode placed in the mold partition level and a plunger to push the material against the sonotrode as it is shown in Figure 10.9. This configuration has a plasticizing chamber where the polymer is melted just above the partition level, and a runner that connects the mold cavity with the plasticizing chamber. However, this chamber is not completely sealed because there is a gap between the mold and the sonotrode, which is necessary to allow the vibration of the sonotrode.

The presence of this gap has two big implications in USM technology. On the one hand, it implies the need of a centering ring around the acoustic

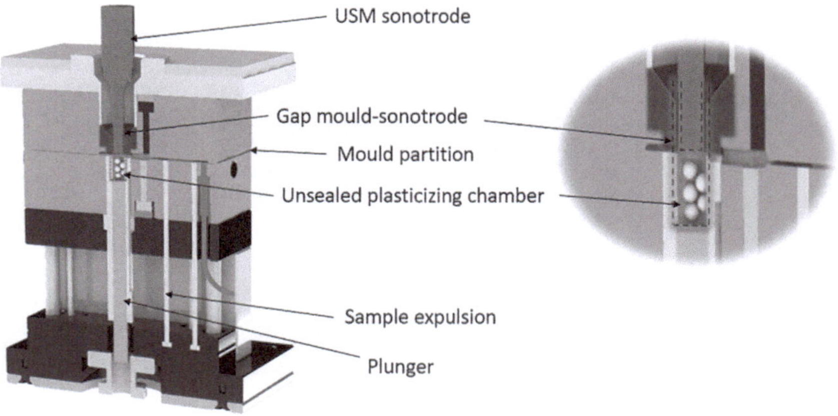


**FIGURE 10.9** Standard USM configuration.

unit to avoid the contact between a moving sonotrode tip and the mold, which would contaminate the sample. To be able to have a correctly centered sonotrode, most of ultrasonic machines used in the experiments reported in the literature, use copper centering rings and place PEEK sleeves between the sonotrode and the copper element to prevent the sonotrode from touching the mold.

On the other hand, even with the centering system, this gap between the mold and sonotrode is believed to be a key element in the instability of the process. The main problems associated with USM technology due to the presence of this gap are listed below:

- **Flash in the material**. The standard configuration allows the leakage of material through the sonotrode-mold gap. This is more critical when the cavity gate has a similar dimension to the sonotrode-mold gap, or when processing polymers with low viscosity, but can appear in almost all the situations. In fact, as showed in Figure 10.10, there are multiple images in the literature where it can be observed flash in the samples processed.
- **Contamination**. If the gap is reduced or the force applied during the filling step is high, the sonotrode will bend and it can eventually touch the centering ring of the mold, leaving copper particles (or, eventually, titanium particles) in the sample. This can be seen in the sample of Figure 10.10.
- **Lack of repeatability**. The sonotrode - mold gap prevents to have a completely sealed plasticization chamber. This causes instability during the filing process, and this seems to be directly related to the poor repetitiveness of the process. This lack of repeatability can be detected in the dispersion of in-mold temperatures Figure 10.10.

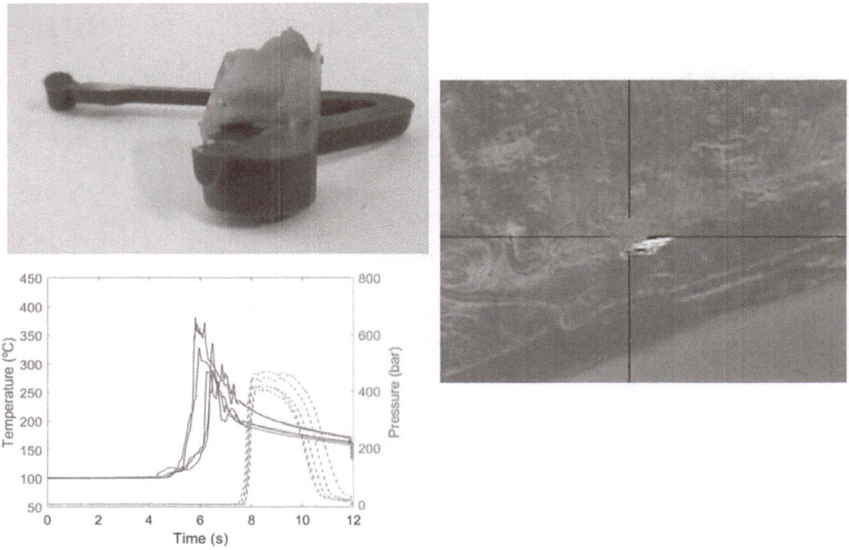


**FIGURE 10.10** Main drawbacks of USM technology with standard configuration: Flash in the processed sample (upper left), contamination inside the sample (right), and lack of repeatability (lower left).

## 10.3.2 Nodal Point Configuration: Main Elements

The new nodal point method improves the performance of USM technology. In particular, the lack of repeatability in filling, in the in-mold measurements, and in the mechanical properties of the samples are the main issues to be addressed from standard USM method. Nodal point method attempts to solve these issues using a new mold configuration and a dedicated methodology [43].

Nodal point ultrasonic molding (NPUSM) configuration changes the placement of the ultrasonic elements in the mold to have a sealed plasticizing chamber. A sealed chamber is expected to improve the performance of the method and prevent the problems associated with the gap between the mold and the sonotrode. Thus, NPUSM configuration places such sonotrode in a low position until its nodal point is in contact with the mold (see Figure 10.11). This new configuration was developed at Eurecat [44].

### 10.3.2.1 Sonotrodes

Conventional half-wave stepped sonotrodes are used in standard USM configurations. However, due to the specific geometric requirements of NPUSM configuration, a customized half-wave stepped sonotrode needs to be designed. This new sonotrode has a flat region around the nodal point to ease the contact with the mold during the process, sealing the plasticizing chamber. Another significant difference of this sonotrode is the conical shape of its lower part, needed to de-mold the material. Typical NPSUM sonotrodes

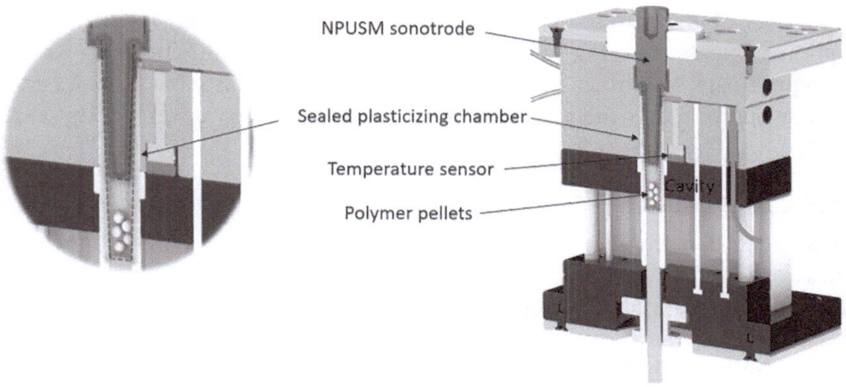


**FIGURE 10.11** Nodal point ultrasonic molding configuration.

are made of Ti6Al4V with a gain (ratio between output and input amplitude) of 4. When working in resonance condition, the amplitude of vibration delivered varies sinusoidally through the length of the sonotrodes, as it is detailed in Figure 10.12). The use of the nodal point (also called nodal plane)

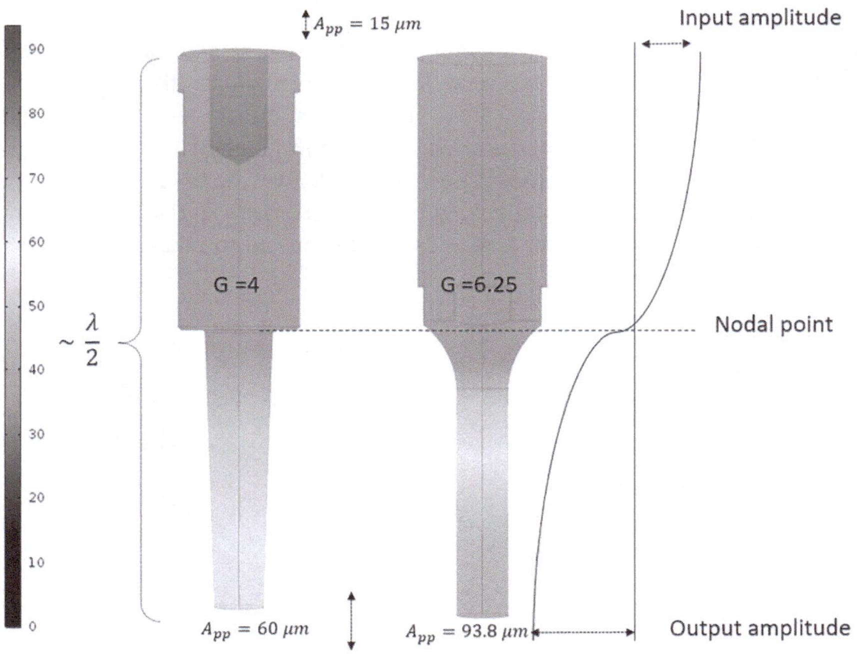


**FIGURE 10.12** Distribution of the amplitude obtained from FEM analysis for a nodal point sonotrode (left) and a standard one (right). $A_{pp}$ is the peak-to-peak amplitude and $G$ is the gain of the sonotrode.

properties of the ultrasonic elements has been used previously in industrial equipment to fix housing assemblies [45]. In this case, this property is used to support the sonotrode in the mold and close the plasticizing chamber. A similar approach for the extrusion process has also been developed by Perez et al. [46].

#### 10.3.2.2 Plasticizing Chamber

This new configuration allows to have a sealed plasticizing chamber. However, this solution comes with a price, i.e., a significant increase of the material of the sprue. Closing the plasticizing chamber at the sonotrode nodal point leaves a hollow cylindrical sprue with a height around 4 centimeters and a similar weight to conventional injection ones. However, its weight is much higher than standard USM and conventional microinjection sprues (see Figure 10.13). This issue can be a major drawback when molding expensive polymers.

#### 10.3.2.3 Plasticizing Chamber Design

As previously discussed, the main benefit of NPUSM configuration is to have a completely sealed plasticizing chamber that allows a better repeatability of the process.

One of the problems that arise when there is an open plasticizing chamber is that the polymer flow may not be continuous. A qualitative study of the influence of the plasticizing chamber in the polymer flow should be done using Moldex3D® software. This software is suited to simulate the conventional injection molding process and cannot account for effects related with USM process. Temperatures and pressures obtained in the simulations won't be valid, nor will be the filling time. However, the simulation can be used to estimate the qualitative influence of the plasticizing chamber geometry in the polymer flow. The continuity of the polymer flow must be guaranteed to obtain the best performance of the method [43].

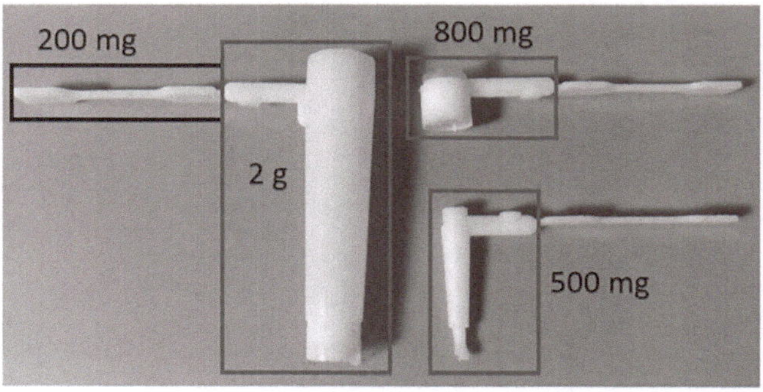


**FIGURE 10.13** Sprue comparison between nodal point method (left), standard USM (upper right) and Babyplast® 6/10 microinjection molding sprue (lower right).

### 10.3.3 Nodal Point Configuration: Validation

In this section, the main improvements obtained by using nodal point configuration are described. The results analyzed are obtained from reference [10], where a comparison between nodal point configuration, standard USM configuration, and conventional injection is done.

#### 10.3.3.1 Molding Equipment

USM experiments described in this section were carried out with the new Sonorus® 2G machine prototype, using a Branson ® DCX 30 kHz generator with its corresponding Langevin transducer. A Babyplast ® 6/10 machine was used for the conventional injection tests in the comparison.

The molds were machined with the same figure cavity for USM technology and conventional injection molding technology. In both cases the figure was a dumbbell shaped sample with dimensions according to the EN ISO 527–2/5B standard and two in-mold sensors are used: A Futaba ® EPSSZL infrared sensor that is able to measure the temperature of the polymer melt up to 430°*C*, and a KISTLER 6157 piezoelectric pressure sensor (0–2000 bar range) (see Figure 10.14).

Two materials were used to validate the *nodal point* configuration for ultrasonic molding method: A polyoxymethylene (POM) Delrin® 500P NC010 and a cyclic olefin polymer (COP) ZEONEX® E483. POM material is commonly used for high-performance engineering components such as small gear wheels, where it needs to fulfil mechanical requirements.

The processing conditions recommended by the manufacturers for the two materials tested are shown in Table 10.2.

#### 10.3.3.2 Filling Stability

Filling stability has been one of the main drawbacks of the USM process, particularly when dealing with high performance polymers [40]. Instead, according to reference [10], the use of NPUSM method improves significantly the filling stability of the process. The comparison of the results from the three methods shows that NPUSM configuration is much more stable than the standard USM one. However, still, conventional injection molding has lower scatter. In Figure 10.15, cavity temperatures and pressures obtained in the best configurations for each method are presented. It can be visually observed that the samples processed with standard USM method have a high scatter in temperature and pressure values.

It is worth noticing that, although the recommended temperature range for POM Delrin® NC10 processing is 210–220°C according to the manufacturer datasheet, when processing this material with standard USM methodology, the material must be heated at least 100°C more to fill completely the sample. In contrast, with nodal point method it is possible to process the material at the temperature recommended in the material datasheet. Such difference on material behavior is probably related with the hesitation effect produced by the gap between the mold and the sonotrode that obstructs the filling of the material, then the material stops and increases its temperature until it can flow again.

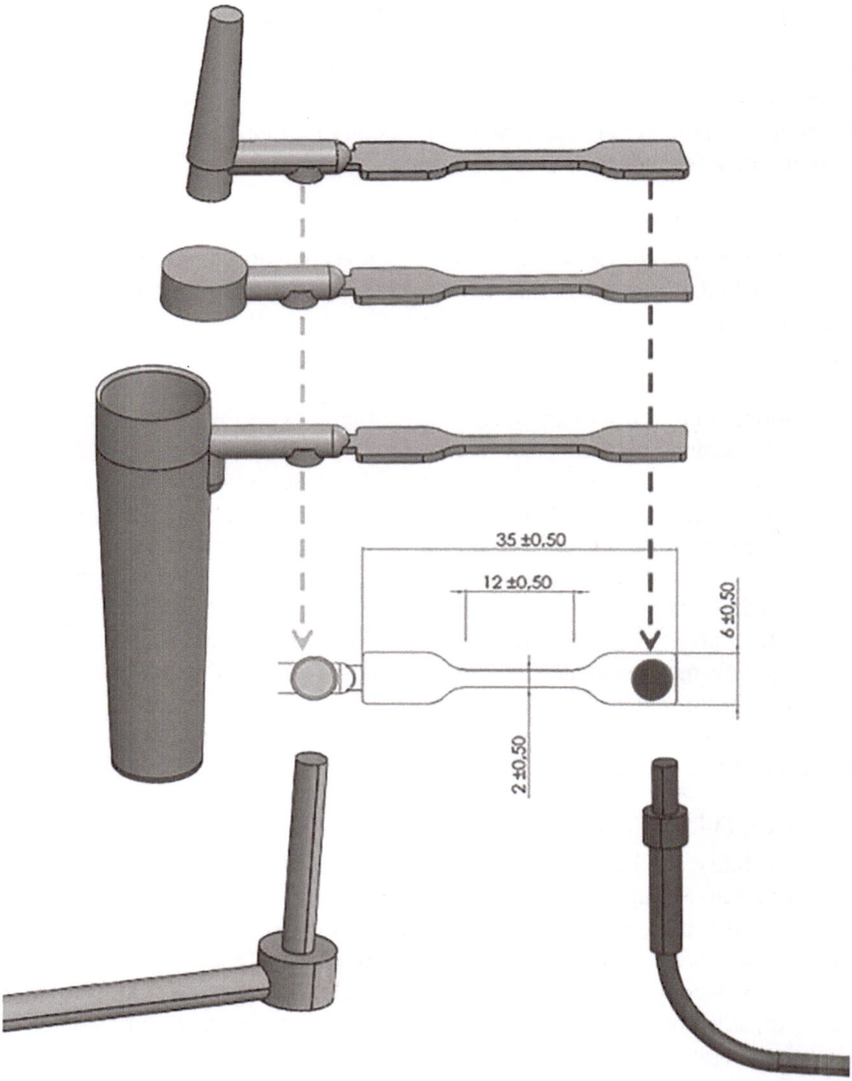


**FIGURE 10.14** Specimens, with their sprue, obtained from: Conventional injection molding (top), USM standard (middle), and NPUSM (bottom). Temperature sensor location is represented in red and pressure sensor in green. Sample dimensions are in millimeters.

Another hypothesis is that the lower distance between the plasticizing chamber and the cavity gate in standard USM is dragging unmelted material and making it difficult to fill the cavity unless the material is much more fluid (higher temperatures). The short distance between the plasticization chamber to the molded part was already identified by Grabalosa et al. [16] as a possible source

**TABLE 10.2**
**Manufacturer Processing Conditions**

| | Material | Melt Temperature | Mold Temperature | Holding Pressure |
|---|---|---|---|---|
| POM | Delrin® 500P NC010 | 210–220°*C* | 80–100°*C* | 800–1000 bar |
| COP | Zeonex® E483 | 265–295°*C* | 95–135°*C* | no specified |

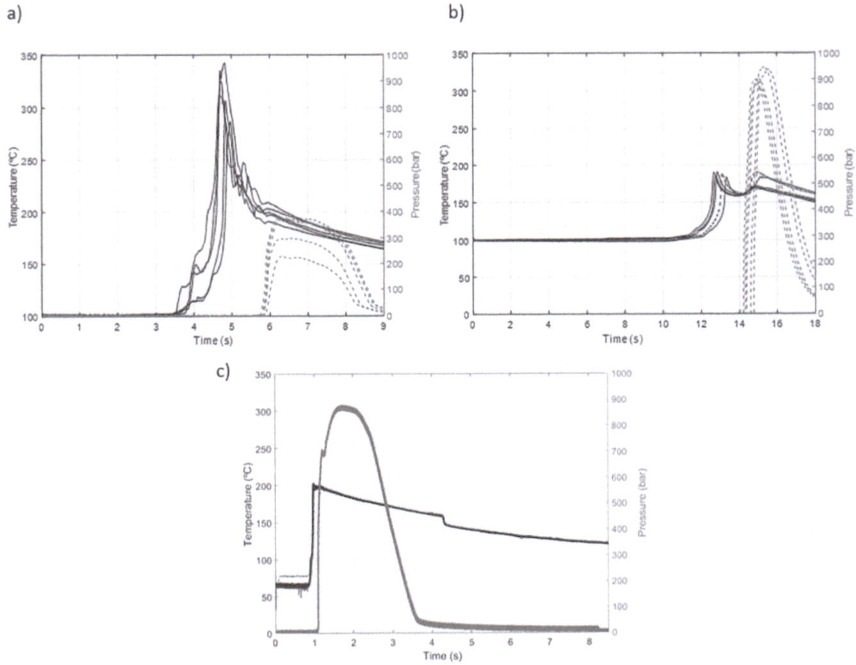


**FIGURE 10.15** In mold temperatures and pressures obtained for a) standard USM (55% amplitude, 4 mm/s velocity), b) Nodal point USM (55% amplitude, 3 mm/s velocity), and c) Babyplast® with optimized process parameters.

for defects at the sample caused by a lack of homogenization. However, in nodal point method, there is a larger distance between the plasticization chamber and the cavity gate, and the material is flowing through a vibrating sonotrode which is expected to help the homogenization of the material.

Similar results are obtained when processing COP in [10]. Standard USM methodology is not able to correctly fill the dogbone sample with this material in a repetitive way. In contrast, using the nodal point method it is possible to process the material at the temperature recommended in the material datasheet in a repetitive way (see Figure 10.16).

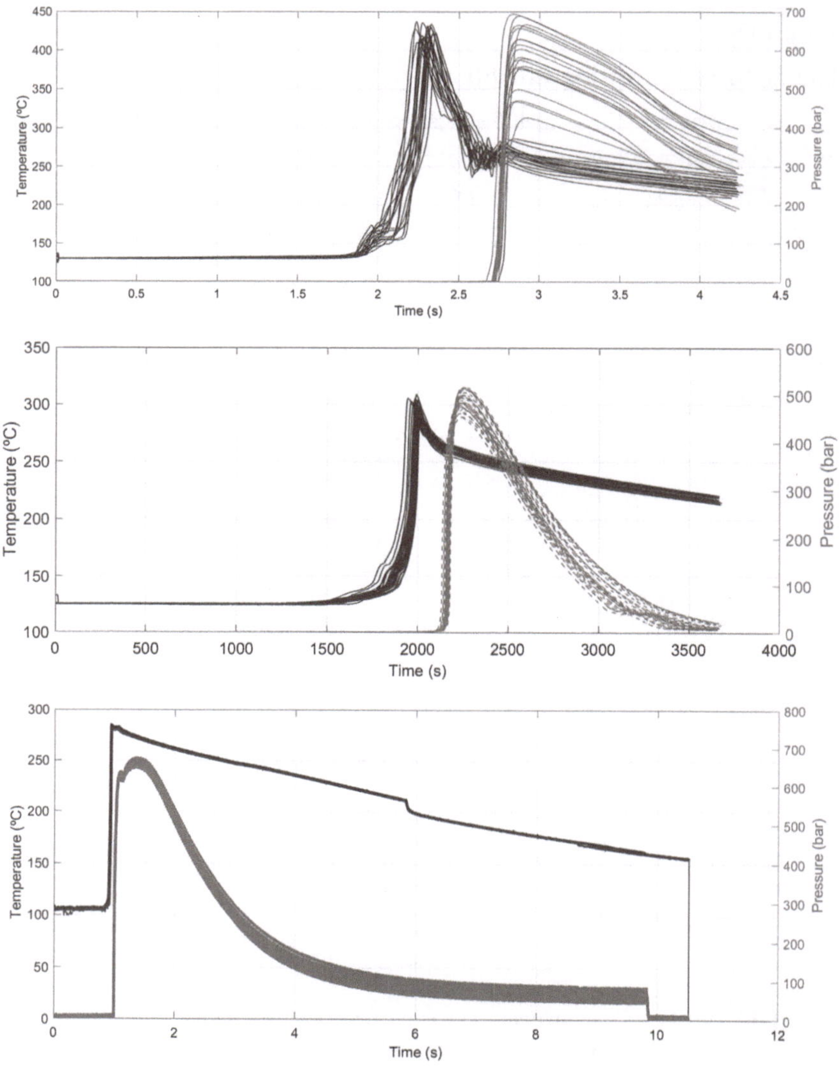


**FIGURE 10.16** In mold temperatures and pressures obtained for (top) standard USM, (middle) NPUSM (90% amplitude, 12 mm/s velocity) and, (bottom) Babyplast® with optimized process parameters for COP samples.

#### 10.3.3.3 Mechanical Properties

The mechanical properties of the materials were determined by means of uniaxial tensile tests to analyze the quality of the parts fabricated by NPUSM and evaluate the capacity of this new NPUSM configuration to manufacture long series of pieces in a stable. Mechanical properties of specimens were evaluated through tensile tests under UNE EN 527 regulation.

**TABLE 10.3**
**POM Tensile Test Results Comparison for the Best Configurations of Each Molding Technology**

| | Manufacturer Datasheet | Standard USM | Nodal Point USM | Babyplast 6/10 |
|---|---|---|---|---|
| (Amplitude, Velocity) | | (55%, 4*mm*/*s*) | (60%, 5*mm*/*s*) | Best parameters |
| Yield stress (MPa) | 70 | 67.5 ± 1 | 72.1 ± 0.23 MPa | 68.2 ± 0.1 |
| Strain at break (%) | 30 | 13.5 ± 4.5 | 26.9 ± 3% | 26.8 ± 0.6 |

In Table 10.3, the mechanical properties of POM specimens processed with the best configuration of each method are presented and compared with the values recommended in the material datasheet. These results show that the samples obtained with NPUSM configuration have the highest yield stress.

For the COP samples, both the samples obtained using NPSUM and conventional injection molding fulfil the mechanical requirements provided by the manufacturer, while none of the configurations used for standard USM provided samples with acceptable mechanical properties. In Table 10.4, the mechanical properties of the specimens prepared with the best configuration of each method are presented and compared with the values recommended in the material datasheet. These results show that the samples obtained using NPUSM configuration have, again, the highest yield stress and strain at yield.

From the experimental results presented in this section, it can be concluded that the use of the nodal point concept in USM technology shows an improvement in the quality of the manufactured parts (as well as in the repeatability of his quality).

**TABLE 10.4**
**COP Tensile Test Results Comparison for the Best Configurations of Each Molding Technology**

| | Manufacturer Datasheet | Standard USM | Nodal Point USM | Babyplast 6/10 |
|---|---|---|---|---|
| (Amplitude, Velocity) | | (100%, 5*mm*/*s*) | (90%, 12*mm*/*s*) | Best parameters |
| Yield stress (MPa) | 73 | 49.8 ± 12.6 | 78.2 ± 0.7 | 74.3 ± 0.1 |
| Strain at yield (%) | 5 | 5.2 ± 1.6 | 7.9 ± 0.1 | 5 ± 0.1 |

Best process parameters of NPUSM allowed the manufacture, under stable conditions, of a large number of tensile specimens, whose physical and mechanical properties (geometry, presence / absence of defects, stress-strain ratio in uniaxial tensile tests, molecular weight, and stress patterns) meet the requirements for this type of components. In addition, the results obtained for NPUSM processing are comparable with those obtained in equal specimens manufactured using the well-established conventional injection molding technology. The ability to process these materials has improved greatly with the use of NPUSM configuration, in comparison to standard USM. The whole comparison between all three methods can be found in reference [10].

## REFERENCES

[1] W. Michaeli, R. Maesing, T. Kampes, S. Hessner, A. Spennemann, and R. Gärtner, "New Plastification Concepts for Micro Injection Moulding," *Microsyst. Technol.*, vol. 8, no. 1, pp. 55–57, 2002, doi: 10.1007/s00542-001-0143-9.

[2] B. Zhao, Y. Qiang, W. Wu, and B. Jiang, *Tuning power ultrasound for enhanced performance of thermoplastic micro-injection molding: Principles, methods, and performances*, vol. 13, no. 17. 2021.

[3] A. Benatar, "Ultrasonic welding of advanced thermoplastic composites," Massachusetts Institute of Technology, 1987.

[4] M. Lucas, J. N. Petzing, A. Cardoni, and L. J. Smith, "Design and Characterisation of Ultrasonic Cutting Tools," *CIRP Ann. - Manuf. Technol.*, vol. 50, no. 1, pp. 149–152, 2001, doi: 10.1016/S0007-8506(07)62092-7.

[5] J. Sackmann *et al.*, "Review on Ultrasonic Fabrication of Polymer Micro Devices," *Ultrasonics*, vol. 56, pp. 189–200, 2015, doi: 10.1016/j.ultras.2014.08.007.

[6] W. Michaeli, T. Kamps, and C. Hopmann, "Manufacturing of Polymer Micro Parts by Ultrasonic Plasticization and Direct Injection," *Microsyst. Technol.*, vol. 17, no. 2, pp. 243–249, Mar. 2011, doi: 10.1007/s00542-011-1236-8.

[7] J. Chen, Y. Chen, H. Li, S.-Y. Y. Lai, and J. Jow, "Physical and Chemical Effects of Ultrasound Vibration on Polymer Melt in Extrusion," *Ultrason. Sonochem.*, vol. 17, no. 1, pp. 66–71, Jan. 2010, doi: 10.1016/j.ultsonch.2009.05.005.

[8] D. Masato, M. Babenko, B. Shriky, T. Gough, G. Lucchetta, and B. Whiteside, "Comparison of Crystallization Characteristics and Mechanical Properties of Polypropylene Processed by Ultrasound and Conventional Micro-injection Molding," *Int. J. Adv. Manuf. Technol.*, vol. 99, no. 1–4, pp. 113–125, 2018, doi: 10.1007/s00170-018-2493-9.

[9] M. Planellas *et al.*, "Micro-molding With Ultrasonic Vibration Energy: New Method to Disperse Nanoclays in Polymer Matrices," *Ultrason. Sonochem.*, vol. 21, no. 4, pp. 1557–1569, 2014, doi: 10.1016/j.ultsonch.2013.12.027.

[10] M. Janer, "Ultrasonic nodal point: a new configuration for ultrasonic moulding. Advances towards the complete industrialisation of the technology," Phd thesis, Universitat Politecnica de Catalunya, 2022.

[11] D. Ensminger and L. J. Bond, *Ultrasonics: fundamentals, technologies, and applications*. CRC press, 2011.

[12] Y. Yao, Y. Pan, and S. Liu, "Power Ultrasound and its Ppplications: A State-of-the-art Review," *Ultrason. Sonochem.*, vol. 62, p. 104722, 2020, doi: 10.1016/j.ultsonch.2019.104722.

[13] A. I. Isayev, R. Kumar, and T. M. Lewis, "Ultrasound Assisted Twin Screw Extrusion of Polymer-nanocomposites Containing Carbon Nanotubes," *Polymer (Guildf)*, vol. 50, no. 1, pp. 250–260, 2009, doi: 10.1016/j.polymer.2008.10.052.
[14] Y. J. Yang, C. C. Huang, S. K. Lin, and J. Tao, "Characteristics Analysis and Mold Design for Ultrasonic-assisted Injection Molding," *J. Polym. Eng.*, vol. 34, no. 7, pp. 673–681, 2014, doi: 10.1515/polyeng-2013-0328.
[15] M. Sacristán, X. Plantá, M. Morell, and J. Puiggalí, "Effects of Ultrasonic Vibration on the Micro-molding Processing of Polylactide," *Ultrason. Sonochem.*, vol. 21, no. 1, pp. 376–386, 2014, doi: 10.1016/j.ultsonch.2013.07.007.
[16] J. Grabalosa, I. Ferrer, A. Elías-Zúñiga, and J. Ciurana, "Influence of Processing Conditions on Manufacturing Polyamide Parts by Ultrasonic Molding," *Mater. Des.*, vol. 98, pp. 20–30, 2016, doi: 10.1016/j.matdes.2016.02.122.
[17] P. Negre, J. Grabalosa, I. Ferrer, J. Ciurana, A. Elías-Zúñiga, and F. Rivillas, "Study of the Ultrasonic Molding Process Parameters for Manufacturing Polypropylene Parts," *Procedia Eng.*, vol. 132, pp. 7–14, 2015, doi: 10.1016/j.proeng.2015.12.460.
[18] U. Heredia, E. Vázquez, I. Ferrer, C. A. Rodríguez, and J. Ciurana, "Feasibility of Manufacturing Low Aspect Ratio Parts of PLA by Ultrasonic Moulding Technology," *Procedia Manuf.*, vol. 13, pp. 251–258, 2017, doi: 10.1016/j.promfg.2017.09.065.
[19] T. Dorf, K. Perkowska, M. Janiszewska, I. Ferrer, and J. Ciurana, "Effect of the Main Process Parameters on the Mechanical Strength of Polyphenylsulfone (PPSU) in Ultrasonic Micro-moulding Process," *Ultrason. Sonochem.*, vol. 46, pp. 46–58, 2018, doi: 10.1016/j.ultsonch.2018.03.024.
[20] W. Kardyś, A. Milewski, P. Kogut, and P. Kluk, "Universal Ultrasonic Generator for Welding," *Acta Phys. Pol. A*, vol. 124, no. 3, pp. 456–458, 2013, doi: 10.12693/APhysPolA.124.456.
[21] A. Ramos-Fernandez, J. A. Gallego-Juarez, and F. Montoya-Vitini, "Automatic System for Dynamic Control of Resonance in High Power and High Q Ultrasonic Transducers," *Ultrasonics*, vol. 23, no. 4, pp. 151–156, 1985, doi: 10.1016/0041-624X(85)90023-X.
[22] S. Ghenna, F. Giraud, C. Giraud-Audine, M. Amberg, and B. Lemaire-Semail, "Vector control applied to a Langevin transducer," *2015 17th Eur. Conf. Power Electron. Appl. EPE-ECCE Eur. 2015*, no. October, 2015, doi: 10.1109/EPE.2015.7309410.
[23] H. Al-Budairi, "Design and analysis of ultrasonic horns operating in longitudinal and torsional vibration," University of Glasgow, 2012.
[24] S. H. Kristensen, "Design, construction and characterization of high power ultrasound sources," no. November, 2009.
[25] M. J. Troughton, *Handbook of plastics joining: A practical guide*. Cambridge: The Welding Institute, 2008.
[26] M. Tadvi, A. Pandey, J. Prajapati, and J. Shah, "Design and Development of Sonotrode for Ultrasonic Drilling," 2016, doi: 10.1115/imece2015-53023.
[27] "Atcp ultrasonics." https://www.atcp-ndt.com/en/products/trz.html (accessed Dec. 22, 2021).
[28] X. Sánchez-Sánchez, M. Hernández-Avila, L. E. Elizalde, O. Martínez, I. Ferrer, and A. Elías-Zuñiga, "Micro Injection Molding Processing of UHMWPE Using Ultrasonic Vibration Energy," *Mater. Des.*, vol. 132, pp. 1–12, 2017, doi: 10.1016/j.matdes.2017.06.055.
[29] D. Montes, "Estudi dels parámetres del procés de micromoldeig per ultrasons mitjançant disseny experimental," Master Thesis, Universitat Politécnica de Catalunya, http://hdl.handle.net/2117/165706, Barcelona, Spain, 2016.

[30] M.-S. S. Huang, "Cavity Pressure Based Grey Prediction of the Filling-to-Packing Switchover Point for Injection Molding," *J. Mater. Process. Technol.*, vol. 183, no. 2–3, pp. 419–424, 2007, doi: 10.1016/j.jmatprotec.2006.10.037.

[31] M. Janer, X. Plantà, and D. Riera, "Ultrasonic Moulding: Current State of the Technology," *Ultrasonics*, vol. 102. Elsevier B.V., Mar. 01, 2020, doi: 10.1016/j.ultras.2019.106038.

[32] W. Michaeli and D. Opfermann, "Ultrasonic Plasticising for Micro Injection Moulding," in *4M 2006 - Second International Conference on Multi-Material Micro Manufacture*, 2006, pp. 345–348, doi: 10.1016/B978-008045263-0/50078-7.

[33] X. Sánchez-Sánchez, A. Elias-Zuñiga, and M. Hernández-Avila, "Processing of Ultra-high Molecular Weight Polyethylene/graphite Composites by Ultrasonic Injection Moulding: Taguchi Optimization," *Ultrason. Sonochem.*, vol. 44, pp. 350–358, 2018, doi: 10.1016/j.ultsonch.2018.02.042.

[34] J. Grabalosa, E. Vázquez, I. Ferrer, A. Elias-Zúñiga, E. Sirera, and F. Rivillas, "Processing of polyamide by ultrasonic molding for medical applications. Preliminar study," in *Proceedings of the 2nd International Conference on Design and Processes for Medical Devices*, 2014, pp. 52–56.

[35] C. Olmo *et al.*, "Preparation of Nanocomposites of Poly ("-caprolactone) and Multi-walled Carbon Nanotubes by Ultrasound Micro-molding. Influence of Nanotubes on Melting and Crystallization," *Polymers (Basel).*, vol. 9, no. 8, pp. 1–18, 2017, doi: 10.3390/polym9080322.

[36] A. Díaz, M. Casas, and J. Puiggalí, "Dispersion of Functionalized Silica Micro- and Nanoparticles into Poly (nonamethylene Azelate) by Ultrasonic Micro-Molding," *Appl. Sci.*, vol. 5, no. 4, pp. 1252–1271, 2015, doi: 10.3390/app5041252.

[37] U. Heredia-Rivera, I. Ferrer, and E. Vázquez, "Ultrasonic Molding Technology: Recent Advances and Potential Applications in the Medical Industry," *Polymers (Basel).*, vol. 11, no. 4, p. 667, 2019, doi: 10.3390/polym11040667.

[38] I. Ferrer, M. Vives-Mestres, A. Manresa, and M. L. Garcia-Romeu, "Replicability of Ultrasonic Molding for Processing Thin-Wall Polystyrene Plates with a Microchannel," *Materials (Basel).*, vol. 11, no. 8, p. 1320, 2018, doi: 10.3390/ma11081320.

[39] X. Sánchez Sánchez and I. Date, "Application of ultrasonic micro injection molding for manufacturing of UHMWPE microparts," Thesis. School of Engineering and Sciences, Técnologico de Monterrey, 2017.

[40] T. Dorf, I. Ferrer, and J. Ciurana, "Characterizing Ultrasonic Micro-Molding Process of Polyetheretherketone (PEEK)," *Int. Polym. Process.*, vol. 33, no. 4, pp. 442–452, 2018, doi: 10.3139/217.3428.

[41] C. Espinoza-González, G. Martínez-Colunga, A. Maffezzoli, F. Lionetto, D. Bueno-Baqués, and C. Ávila-Orta, "An Overview of Progress and Current Challenges in Ultrasonic Treatment of Polymer Melts," *Adv. Polym. Technol.*, vol. 32, no. S1, pp. E582–E602, 2012, doi: 10.1002/adv.21303.

[42] B. Y. Jiang, J. L. Hu, W. Q. Wu, and S. Y. Pan, "Research on the Polymer Ultrasonic Plastification," *Adv. Mater. Res.*, vol. 87–88, pp. 542–549, 2009, doi: 10.4028/www.scientific.net/AMR.87-88.542.

[43] M. Janer, T. López, X. Plantà, and D. Riera, "Ultrasonic Nodal Point, a New Configuration for Ultrasonic Moulding Technology," *Ultrasonics*, vol. 114, no. June 2020, pp. 1–10, 2021, doi: 10.1016/j.ultras.2021.106418.

[44] M. Janer, D. Montes, J. A. Marfil, and F. J. Plantà, "Ultrasonic device For A polymer injector apparatus," European Patent, 17/264,028., 2021.
[45] M. Patrikios, R. S. Soloff, and S. Vargas, "System and method for mounting ultrasonic tools." U.S. Patent, 8,950,458, 2015.
[46] G. A. Pérez Llanos and others, "Application of ultrasound in twin-screw extrusion and microinjection molding: improvements of properties of processed materials and nanocomposites: polymers and biopolymers," PhD thesis, Universitat Politècnica de Catalunya, 2022.

# 11 Spark Plasma Sintering of Layered Structures and Powder Mixtures of Reactive Metals

*Vyacheslav I. Kvashnin and Dina V. Dudina*

## 11.1 INTRODUCTION

Spark plasma sintering (SPS) belongs to a group of field-assisted methods, which rely on the use of electromagnetic fields for the sintering enhancement [1,2]. SPS uses a pulsed direct current and uniaxial pressure applied simultaneously. Sintering is usually conducted under dynamic vacuum. The sample to be processed is placed in a graphite die between the graphite punches (Figure 11.1). Dies and punches made of other materials can also be used, depending on the goals of the experiment, required parameters, and the nature of the powder to be sintered. The punches are always made of an electrically conductive material, as they are to carry the electric current to the die and/or the sample.

The inherent feature of the processing of electrically conductive materials (powders, foils, plates) by SPS is a high heating rate enabled by the direct passage of an electric current through the sample. The possibility of the formation of local high-temperature regions due to high current densities at the inter-particle contacts is widely discussed in the literature. The theoretical analysis of model systems shows that, if the heat dissipation effects are taken into account, the temperature difference between the contact area and the particle volume is very small [3,4]. At the same time, experimental evidence of this effect exists, as specific microstructures attributable to local overheating have been reported in the SPS-processed materials [5–10]. In metals processed by electric-discharge sintering, which is similar to SPS, melting of the inter-particle contacts led to a local increase in hardness in the sintered material due to rapid cooling of the melts [11]. Furthermore, lower pressure at the pre-pressing stage yielded compacts of better quality (lower porosity and higher electrical conductivity) [12].

The formation of melt in the inter-particle regions during SPS allows conducting local liquid phase sintering of composite mixtures such that, in the sintered materials, a network of re-solidified material is formed [10]. In a metal-ceramic composite, this network is metal-rich; it surrounds the composite particles, the

 DOI: 10.1201/9781003364948-11

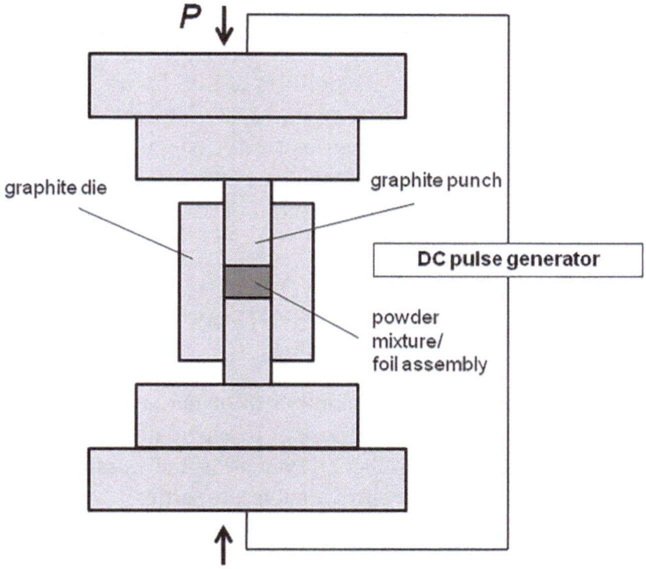


**FIGURE 11.1** Schematic of a spark plasma sintering (SPS) set-up.

volume of which remained in the solid state during SPS. It should be kept in mind that more intense heating of the inter-particle contacts relative to the particle center is possible but does not always occur, depending on the combination of several factors (particle size [4,13,14], particle morphology [8,9]). In electrically conductive materials, the phenomena of electromigration [15] and electroplasticity [16] can play a role in densification.

In order to prevent the chemical interaction between the sample and the graphite tooling during SPS, the inner surface of the die is lined with graphite foil. The foil is also placed between the flat ends of the punches and the sample. The contact with graphite can help reduce oxides present on the surface of the particles [16]. In some materials, the diffusion of carbon into the specimen's interior leads to the formation of the carbide phases [17,18].

The current-assisted heating has served the purposes of consolidation [19,20], synthesis [21,22], additive manufacturing [23], and materials joining [24]. Materials of different chemical nature have been successfully consolidated by SPS: metals [25–27], intermetallic [28], ceramics [29], and composites [20,30]. SPS has also been selected as a processing route for a variety of porous materials [31]. From the manufacturing standpoint, SPS is a versatile method as the specimen thicknesses can vary from below $10^{-3}$ m to $10^{-2}$ m. As an alternative to commercially available SPS facilities, micro field-assisted sintering devices have been developed [32]. Another sintering method closely related to SPS is current-activated tip-based sintering, in which an electric current is imposed locally to consolidate selected areas of a powder bed via the application of a tip electrode [33,34]. The behavior of materials during reactive SPS of metals may bear similarities to that in other electric current-assisted processes, including processes of micromanufacturing.

In this chapter, we focus on the possibilities of SPS for the synthesis of alloys, intermetallics, and composites from mixtures of powders of metals and processing of specimens of planar geometries (foils). Metallic foils are selected as starting materials for the fundamental studies of the effect of electric current on the reactivity of metals and the fabrication of laminated metal/intermetallic composites.

## 11.2 PROCESSING OF LAYERED SYSTEMS OF REACTIVE METALS BY SPARK PLASMA SINTERING

The interaction between metals in assemblies of planar geometry during SPS has been studied with the following purposes: 1) elucidating the electric current effect on the reactivity of materials; 2) evaluating the possibilities of the formation of laminated composites with valuable properties.

### 11.2.1 Elucidation of the Electric Current Effect on the Growth of Reaction Products: Model Studies and Methodological Issues

The formation of defects in metals under an applied current was studied in ref. [35]. Using in-situ depth-resolved positron annihilation spectroscopy, it was shown that the number of vacancies in Al-0.5 wt% Cu micrometer-sized lines increases when a direct current with a density of $8{\cdot}10^4$ A $cm^{-2}$ is applied. The increase in the vacancy concentration was substantially greater than that caused by the thermal generation alone.

The effect of electric current (exerted in addition to Joule heating) on the growth of reaction products (or formation of solid solutions) at the interface between metals was addressed in refs. [36–41].

Direct currents were used by Bertolino et al. [36,37] and Garay et al. [38] (experiments conducted in custom-made set-ups). Bertolino et al. [37] observed the influence of electric current on the rate of growth of the product layer between Al and Au foils and the incubation time. These effects were attributed to an increased diffusion flux due to electromigration and a faster nucleation process of the intermetallic phases in the presence of current. In their work, Garay et al. [38] suggested that electric current increases the mobility of defects of the crystalline lattice or increases their concentration over the thermal equilibrium value. Pulsed direct currents were used by Rudinsky et al. [39], Li et al. [40], and Abedi et al. [41] (experiments conducted in the SPS facilities). A schematic representation of the experimental configurations used in ref. [41] is shown in Figure 11.2a,b. The field-assisted and field-insulated tooling arrangements were designed. The product thickness can be seen in the cross-sectional images the Ni–Al diffusion pairs annealed in the presence of current and without current in a

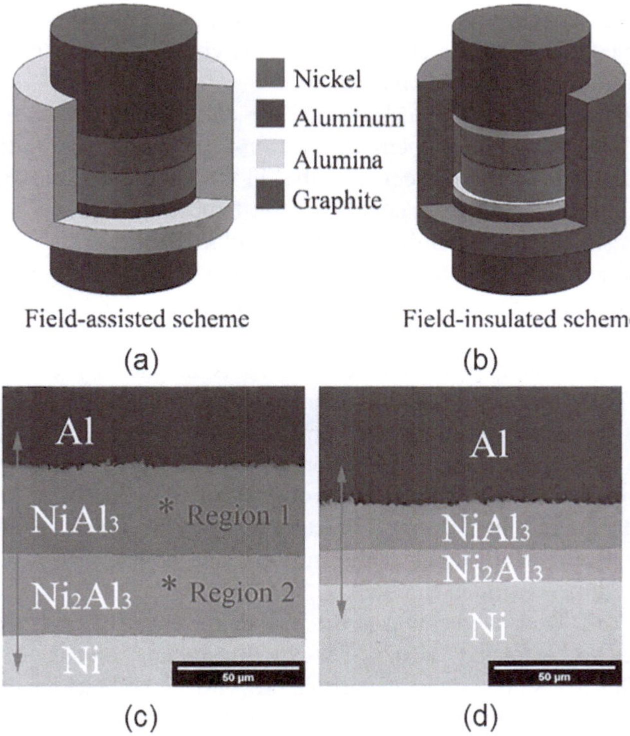


**FIGURE 11.2** Schematic representation of the experimental configurations used to study the effect of electric current on the product growth: (a) field-assisted scheme, (b) field-insulated scheme; cross-sectional images (scanning electron microscopy) of the Ni–Al diffusion pairs annealed for 18 ks at 863 K: (c) field-assisted scheme; (d) field-insulated scheme. (Reprinted from [41], Copyright (2022), with permission from Elsevier).

SPS device (Figure 11.2c,d): It can be concluded that the applied current accelerated the growth of the Ni-Al intermetallics.

For the growth of the layer of intermetallic products between Au and Al foils, the activation energy was found to be close to that measured in the experiments without electric current [37]. A decrease in the activation energies of the growth of intermetallics with the application of electric current was detected for several phases ($NiAl_3$ and $Ni_2Al_3$ [41]; NiTi and $NiTi_2$ [38]). The activation energy comparison made by Abedi et al. [41] is especially valuable, as the values with and without current were determined in the same series of experiments (using the same facility and the same starting materials). Rudinsky et al. [39] found that, under SPS conditions, the activation energy for diffusion in Cu-Ni diffusion couples was lower than that measured without current in samples annealed in a furnace. The activation energy measured for the case of electric current passing from Ni to Cu was lower than that measured in experiments with the opposite direction of current. The possible effect of thermal gradients between the pair of metals (formed due to a difference in their resistivities) on the activation energy of diffusion under SPS was proposed.

In the experiments aimed to determine the effect of electric current on the processes of consolidation or chemical reactions, the temperature in the current-assisted experiment should be equal to that in the experiment conduced in the absence of current. The direct control of the temperature of the specimen is, therefore, desirable for obtaining reliable results. However, it is difficult to measure the temperature directly in the growing interfacial layers. It is also important that the comparative studies of sintering of powder mixtures or layered structures by different methods be carefully designed (Table 11.1). The choice of the sample type and experimental conditions is dictated by the information sought. For example, when results of consolidation of a powder by SPS and hot pressing (HP) are compared (if the temperature is measured in the die, the temperature of the hot pressing experiments should be higher to make up for the sample temperature difference in the processes of SPS and HP), the possible consolidation enhancement during SPS can be due to the simultaneous action of two effects: Local overheating at the inter-particle contacts and the effect of electric current per se ("non-thermal").

The effect of local overheating at the inter-particle due to high local current densities can be "extracted" if the SPS experiments are conducted at different pressures. This is assuming that, at a certain pressure, an intimate contact between the particles is established, thus eliminating the local effects. For the same purposes, the use of pre-consolidated compacts made of powder mixtures is possible (provided pre-consolidation was conducted at a pressure high enough to establish an intimate contact between the particles). In these experiments, the measured parameter will be the thickness of the product layer or the concentration of the newly formed phase in the alloy or composite (in the case of powder mixtures).

Abedi et al. [42] conclude that the influence of the externally applied load during SPS is similar to that observed in the HP processes: The application of higher pressures leads to higher relative densities of the compacts. The influence of pressure on the growth of reaction product layers between particles of reacting metals is not so obvious: It can reduce the diffusion distances, accelerating growth, or reduce the local overheating effect, decelerating growth. When stacks of reactive metals of planar geometry are tested under a passing electric current and under conventional heating mode, the pressure at which the experiments are conducted should be specified and, if possible, varied. A higher applied pressure can be instrumental in eliminating the influence of "local" effects related to the surface roughness or the presence of the surface oxide films. The direct comparison of the thickness of reaction product layers formed between the deposited metals or foil stacks with the thickness of the reaction product layer formed between the powder particles in the compacts cannot be fully justified, as the concentration of oxides and other contaminants at the interfaces as well as the grain structure and defect states can be different in materials of the same elemental composition but prepared through different routes.

**TABLE 11.1**
**Methodological Considerations for the Comparative Experiments on Sintering of Reactive Powder Mixtures and Reactive Metals by Different Methods. Consolidation by Spark Plasma Sintering (SPS), Hot Pressing (HP), and Conventional (Pressure Less) Sintering (CS). Note that SPS Requires a Certain Level of Applied Pressure to Establish and Maintain the Electric Contact in the Assembly. The Thickness of the Product Layer or the Concentration of the Newly Formed Phase is to be Measured in the Samples**

| Samples Compared | | | Information Sought | Notes |
|---|---|---|---|---|
| powder mixture processed by SPS at $p_1$ | and | powder mixture processed by SPS at $p_2$ | the effect of local overheating at the inter-particle due to high local current densities | $p_1 < p_2$<br>if $p_2$ helps establish an intimate contact |
| powder mixture processed by SPS | and | pre-consolidated powder mixture processed by SPS | the effect of local overheating at the inter-particle due to high local current densities | if pre-consolidation helps establish an intimate contact |
| pre-consolidated powder mixture processed by SPS | and | pre-consolidated powder mixture processed by HP | the effect of electric current per se | – |
| stacks/deposited layers processed by SPS | and | stacks/ deposited layers processed by HP | the effect of electric current per se | – |
| pre-consolidated powder mixture processed by SPS | and | pre-consolidated powder mixture processed by CS | the effect of electric current per se | CS can be used if the effect of pressure on the sintering outcome can be neglected |
| stacks/deposited layers processed by SPS | and | stacks/deposited layers processed by CS | the effect of electric current per se | CS can be used if the effect of pressure on the sintering outcome can be neglected |

### 11.2.2 Formation of Laminated Composites by Treatment of Stacks of Foils

Laminated object manufacturing is an additive manufacturing technique [43]. Joining of the layers has been realized via gluing or adhesive bonding, thermal bonding, clamping, or ultrasonic welding. The processing of stacks of metallic foils or thin plates by SPS presents a promising alternative, as demonstrated in refs. [44–49]. The parameters of the process are adjusted to control the interfacial reactions and reaction layer thickness in the laminated composites. The foils can be modified by depositing a metallic layer on their surface prior to stacking and sintering [49].

Lazurenko et al. [46] reported the formation of Ti-$Al_3Ti$ composites by reactive SPS. Circles of titanium and aluminum foils were placed in a titanium container with an inner diameter equal to the diameter of the circles. The titanium container was covered with a cap. The assembly was placed in a graphite die with an inner diameter equal to that of the outer diameter of the cap. The electric current flowed through the sample and through the graphite die. Thin layers of reaction products formed at the interfaces between the metals during SPS. The titanium container prevented the aluminum leakage from the reaction zone. This design of the assembly allowed conducting the synthesis at high temperatures in a single stage. Aluminum was completely consumed, which resulted in the formation of $Al_3Ti$. Also, additional intermetallic interlayers ($Al_2Ti$, AlTi, and $AlTi_3$) and a Ti(Al) solid solution were formed during SPS as Ti reacted with $Al_3Ti$. Ceramic particle-reinforced intermetallic matrix composites with a layered structure were produced by SPS of stacks of metallic foils [47]. The ceramic reinforcements were introduced as powders and placed between the foils. The effect of pressure applied during SPS on the structure of laminated composites was studied in ref. [48]. Stacks of Ni–Al foils were sintered at different pressures. The liquid phase appears in the sample when aluminum melts at 660°C. The applied load can force the melt to move from the interlayer spaces. So, the necessary pressure was applied only at the final stage of the process. As seen in Figure 11.3, the number of defects within the layers of the reaction products decreased significantly with increasing pressure. Note that the Kirkendall porosity was not fully eliminated even at the highest pressure tested

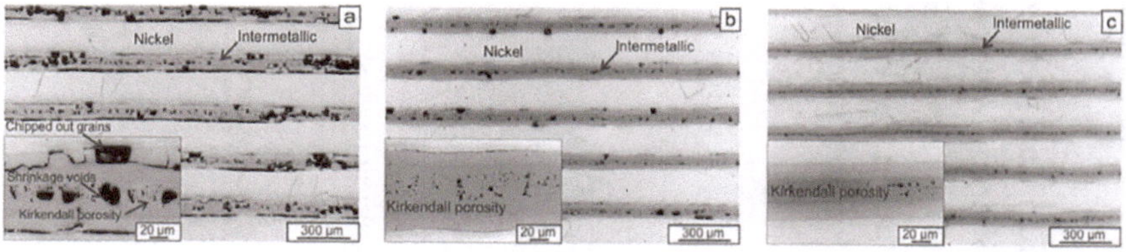


**FIGURE 11.3** Cross-sectional images of composites fabricated by SPS of stacks of Ni and Al foils at 1100°C and different pressures: 10 MPa (a), 20 MPa (b), 30 MPa (c). Insets show the structure of the reaction product layers. (Reprinted from [48], Copyright (2022), with permission from Elsevier).

(30 MPa). The composites consisted of the NiAl phase of the cubic modification (in the center of the intermetallic layers) and the tetragonal modification (NiAl-martensite) near the boundaries with nickel. The average thickness of the intermetallic layers formed at 10 MPa, 20 MPa, and 30 MPa was 87 μm, 78 μm, and 60 μm, respectively. The thickness and composition of the NiAl-martensite layer did not change significantly. The cubic NiAl phase was saturated with nickel and its thickness decreased as the pressure was increased. Also, layers of Ni(Al) solid solution appeared between the intermetallic and nickel layers. As the pressure increased, the solid solution layers became thicker and contained more aluminum.

Materials with a layered structure can be produced by sintering of layers formed directly from powders in a single step, as shown in refs. [50,51]. A layered structure is also formed when the compact obtained by SPS is further joined with a foil. The latter becomes a surface-modifying layer (coating). In the process of joining the foil to the compact, the possibility of surface carbidization of the compact during the preceding SPS consolidation should be taken into account. Metallic alloys containing carbide forming elements are prone to surface carbidization during SPS, as carbon of the graphite foil diffuses into the specimen during sintering [17,18]. Interestingly, the surface carbidization can be useful when the carbide-forming element in the alloy needs to be bound into a carbide phase. In the case of Ni-based alloys containing tungsten, surface carbidization helps prevent the formation of brittle Al-W intermetallics, when Al is applied to the surface. The interaction of aluminum with the SPS-consolidated Ni–W material was studied in ref. [17]. During SPS of a Ni(W) solid solution obtained by mechanical alloying, carbon diffused into the layer of the compact adjacent to the surface, which resulted in the formation of the WC phase. After SPS, the WC-containing layer was removed from one side of the compact and preserved at the other. Then a stack consisting of the sintered disk and circles of Al foil was placed in a SPS die and processed at 500°C for 20 min to join the foil to the Ni-W compact. During further annealing at 600°C (in a furnace), interfaces with dramatically different structures formed at the two flat ends of the compact (Figure 11.4). At the side, from which the WC-containing layer was removed, alternating layers rich in W and rich in Ni formed. The formation of W-rich and Ni-rich layers implies the diffusion mobility of tungsten in the metallic phase. Such mobility was obviously not characteristic of the WC-containing layer. Aluminum diffused into the WC-Ni composite, which resulted in the formation of Ni-Al intermetallics with WC particles preserved in the reaction layer.

A combination of a powder layer and a foil is useful in the SPS practice. Metallic foils inert to carbon can eliminate the influence of carbon of the tooling on the composition of the sintered material (help avoid the formation of surface carbides). When an oxidized metal powder is used and the oxide can be reduced by carbon under conditions of sintering, a direct contact of the compact with graphite (foil) lowers the concentration of oxygen in the material, as demonstrated for SPS of an oxidized Ni powder [16]. Note that the surface oxide films

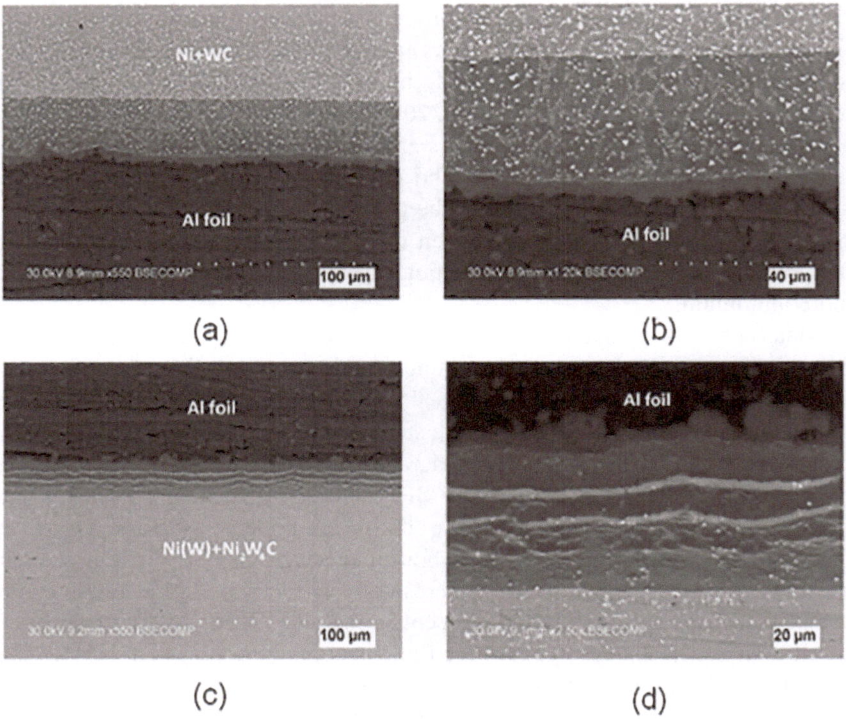


**FIGURE 11.4** Cross-sectional views of the interdiffusion zone formed between an Al foil and the flat ends of the spark plasma sintered Ni–W-based compact after SPS treatment at 500°C for 20 min and vacuum annealing at 600°C for 2 h: (a) and (b) the carbide layer was preserved on the compact after SPS with the graphite foil, different magnifications, (c) and (d) the carbide layer was removed, different magnifications. (Reprinted from [17], Copyright (2015)).

on the Ni powder were not reduced when it was sintered in contact with Cu foil placed between the sample and the punch.

## 11.3 SYNTHESIS OF MATERIALS BY SPARK PLASMA SINTERING OF MIXTURES OF REACTIVE METAL POWDERS

For the purpose of materials synthesis, powders are used more often than foils or plates. In this section, the possibilities of synthesis of materials by SPS of powder mixtures of reactive metals are shown. The selection of the SPS method for synthesis and consolidation over other sintering methods is dictated by the technological advantages of the facilities: Simultaneous heating and pressing combined with the possibilities of high heating rates (and small grain sizes of the final products). The enhanced reaction kinetics also serves as an attractive feature. However, as the model studies were carried out for a limited number of systems, the kinetic data may be simply unavailable. The number of materials obtained successfully (in terms of properties achieved) is much greater than the number of materials for which the diffusion studies have been conducted. Table 11.2 provides examples of alloys, intermetallics, and composites obtained by SPS accompanied by chemical reactions (at the laboratory scale).

The powder precursors used were in both blended and pre-milled states. More examples of the successful development of composites via SPS accompanied by reactions between metals can be found in ref. [74].

The characteristics of the microstructure of the synthesized materials attributable to the specific conditions realized during SPS (fast heating, local melting/over-heating) are discussed below. Specific features of the microstructure of composites with in-situ formed core-shell particles were identified by Hu et al. [65].

**TABLE 11.2**
**Selected Examples of Alloys, Intermetallics, and Composites Processed via SPS Accompanied by Alloying (Formation of Solid Solutions) or Synthesis Reactions (Formation of Compounds) between the Metals**

| Materials Formed by Reactive SPS of Metallic Powders | Reference |
|---|---|
| *Alloys* | |
| Al-Cu | [52,53] |
| Ti-Cu | [54] |
| Ni-Al | [55] |
| Fe-Al | [56] |
| Ti-Al | [57,58] |
| Fe-Al (porous) | [59,60] |
| Fe-Ti | [61] |
| Nb-Al | [62] |
| AlCoCrFeNi high-entropy alloy | [63] |
| $CoNi_{0.3}Fe_{0.3}Cr_{0.15}Al$ high-entropy intermetallic | [64] |
| *Composites* | |
| Al matrix composite reinforced with in-situ formed Ti/Ti-Al intermetallics/ fine-grained Al core-shell particles | [65] |
| Al matrix composite reinforced with in-situ formed Ti/Ti-Al intermetallics core-shell particles | [66] |
| Al reinforced with in-situ formed $TiAl_3$ | [67] |
| Al reinforced with in- situ formed Fe-based metallic glass/Fe-Al intermetallics core-shell particles | [68,69] |
| Al reinforced with core-shell particles formed from added Al@Cu | [70] |
| Al reinforced with W and in-situ formed $Al_4W$ and $Al_{12}W$ | [71] |
| Al reinforced with ω-Al-Cu-Fe and in situ- formed $Al_2Cu$ | [72] |
| Ni-Al intermetallic matrix composite reinforced with $TiB_2$ particles | [73] |

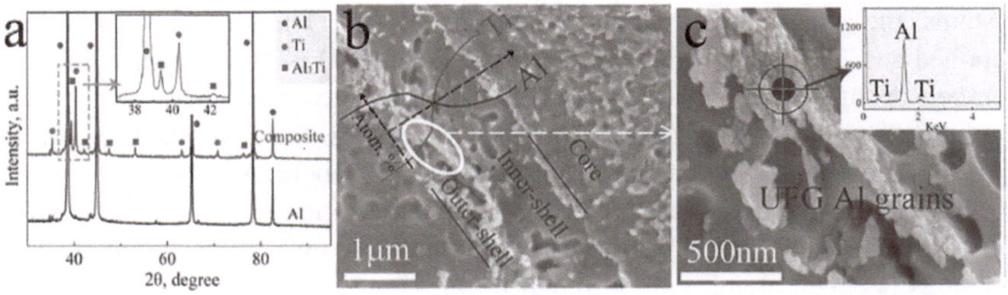


**FIGURE 11.5** (a) X-ray diffraction (XRD) patterns of Al and the in-situ composite formed by SPS of a Ti+Al mixture at 580°C and 30 MPa (b) the matrix/reinforcement interface, (c) enlarged view of the zone encircled by a yellow oval in (b). (Reprinted from [65], Copyright (2018), with permission from Elsevier).

Well-bonded ultrafine-grains of aluminum were observed in the outer shell of the core-shell reinforcing particles in the composite obtained by SPS of a Ti+Al mixture at 580°C (Figure 11.5). It was suggested, based on the microstructural evidence, that local-melting occurred at the particle edges and corners of large particles. Rapid solidification of the Al melt caused the formation of ultra-fine grains in the outer shell.

Usually, pre-alloyed powders of high-entropy alloys are processed by SPS to form bulk materials. Another possibility is a single-stage formation of the alloy by SPS of a powder blend. It was shown that, starting from a powder blend, it is possible to produce a high-entropy alloy [63]. A blend containing Al, Co, Cr, Fe, and Ni was alloyed and sintered by SPS at 1200°C for 20 min. The authors suggested that fast alloying was due to Joule heating and the effect of electric field. The formation of aluminum melt accelerated the alloy formation. The X-ray diffraction patterns of the powder and the sintered alloy are shown in Figure 11.6a. The microstructure of the sintered material together with the elemental maps (Figure 11.6b) indicates the formation of an alloy with a uniform

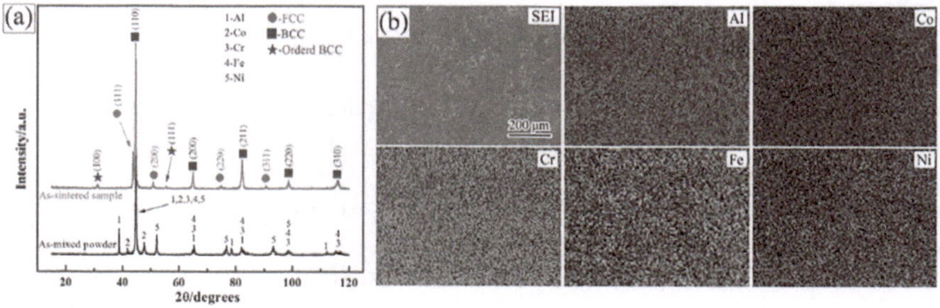


**FIGURE 11.6** (a) XRD patterns of the as-mixed powders and sintered AlCoCrFeNi alloy, (b) secondary electron image and elemental maps of the alloy. (Reprinted from [63], Copyright (2016), with permission from Elsevier).

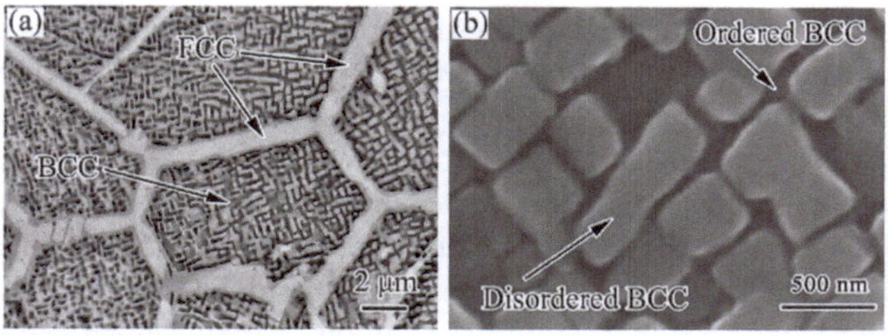


**FIGURE 11.7** Microstructure of the AlCoCrFeNi alloy: (a) low-magnification back-scattered electron image, (b) high-magnification secondary electron image of the body-centered cubic phase. (Reprinted from [63], Copyright (2016), with permission from Elsevier).

distribution of elements. At a higher magnification, it is seen that the alloy is composed of three phases (Figure 11.7).

The specific morphology of particles in reaction mixtures and interfaces in composites sintered by SPS should not be immediately related to the effect of electric current and accompanying phenomena [75,76]. In ref. [76], needle and platelet structures formed at the interface between the Al matrix and the shells of the core-shell particles formed as a result of the interaction of a Fe-based metallic glass with aluminum (Figure 11.8). Comparative experiments (SPS and conventional annealing) showed that those structures were not characteristic of the sintering method. They formed via crystallization from the melt. The formation of the latter was due to the evolution of heat in the course of exothermic reactions of the formation of Fe-Al intermetallic. In composites processed by SPS, the formation of melt was also facilitated by the passage of electric current. The dissipation of heat into the volume of the glassy particles occurs much slower than into the particles of the unalloyed crystalline metals, making a local

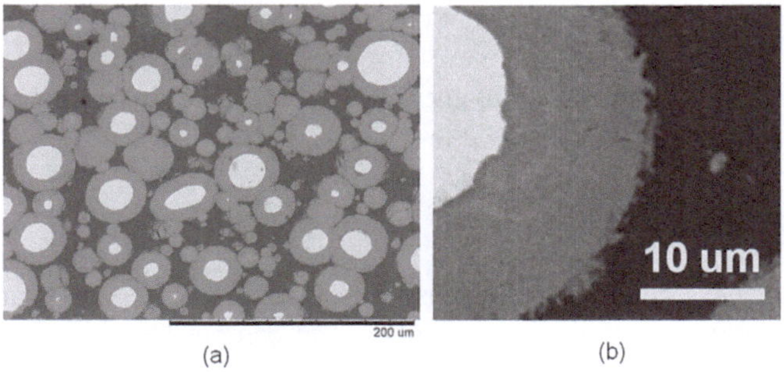


**FIGURE 11.8** Microstructure of the composite obtained by SPS of Al–20 vol.% $Fe_{66}Cr_{10}Nb_5B_{19}$ (a), the shell of a particle at a higher magnification (b). SPS conditions: Temperature—570 °C, holding time—3 min, applied pressure—40 MPa. Reprinted from [76]. (This article is an open access article distributed under the terms and conditions of the Creative Commons Attribution (CC BY) license (https://creativecommons.org/licenses/by/4.0/)).

temperature rise in the metallic glass-Al matrix inter-particle areas possible. In the case of local overheating of the material at the contacts between the reacting particles, the formation of core-shell particles during SPS is accelerated as compared with conventional heating.

Owing to the progress made in recent years in the area of the SPS processing of metallic materials, this sintering method is ready to enter the industrial sector. The formation of alloys by sintering may appear costly because of the high prices of the starting powders. At least partial solution of the problem can be the utilization of metallic scrap [77]. The production of metallic materials by the SPS-based powder metallurgy route is suitable for niche applications, in which the gain due to property enhancement is higher than the manufacturing costs.

## 11.4 FUTURE RESEARCH DIRECTIONS

In future research, it would be interesting to compare the effect of electric current on systems containing powders with oxide films of different thicknesses. Powders for such experiments can be obtained by partially oxidizing commercial metallic powders (by annealing in air at different temperatures and for different durations). Transmission electron microscopy has been proven to be a powerful method to study the evolution of metal-oxide systems subjected to an applied voltage [78]. In-situ investigations of the electric current-induced changes at the interface between the reacting metals at the atomic level should follow.

The electric current-related changes in the layered structures of metals formed by thermal spray present another promising research direction with implications for hybrid additive manufacturing processes. The treatment of the layers formed by the thermal spray methods (including cold spray) in SPS devices can improve the interface quality and induce chemical reactions [79].

An approach to the fabrication of bulk materials with harmonic structure via non-equilibrium consolidation suggested in ref. [80] can be extended to reactive systems, the reaction product layers presenting areas differing in the grain size and composition from the volume of the reactants (not fully consumed). The network-featuring composites obtained in ref. [10] are an example of harmonic structures formed in-situ during SPS.

## 11.5 SUMMARY

Recent years have witnessed an extensive accumulation of the experimental data on the processing of alloys, intermetallics, and composites via reactive SPS of metals at the laboratory scale. A variety of materials have been obtained via SPS of powders of reactive metals. In some cases, materials produced by SPS possessed structural features directly related to the specific conditions realized during the passage of an electric current through a powder compact. Metallic foils were used to synthesize and design metal-intermetallic composites with a laminated structure. Based on the accumulated data, a further development of SPS-based production technologies for the industrial sector is expected.

Progress has been made in the understanding of the physical and chemical processes occurring during SPS of powder mixtures and diffusion couples of metals. However, studies of the alloy formation between metals subjected to the action of an electric current are far from completion. The effect of electric current of the growth of the reaction products needs to be examined from different angles (kinetic parameters, grain size, and defect structure of the products). From the fundamental perspective, the influence of electric current on the reactivity of metals should be further elaborated.

## REFERENCES

[1] E. A. Olevsky, D. V. Dudina, *Field-Assisted sintering: science and applications*, Cham, Switzerland: Springer International Publishing, (2018); 425p. 10.1007/978-3-319-76032-2

[2] J. Trapp, B. Kieback, Fundamental principles of spark plasma sintering of metals: part I—Joule heating controlled by the evolution of powder resistivity and local current densities, *Powder Metall.* 62 (2019) 297–306. 10.1080/00325899.2019.1653532

[3] J. Trapp, A. Semenov, O. Eberhardt, M. Nöthe, T. Wallmersperger, B. Kieback, Fundamental principles of spark plasma sintering of metals: part II – about the existence or non-existence of the 'spark plasma effect', *Powder Metall.* 63 (2020) 312–328. 10.1080/00325899.2020.1829349

[4] C. Collard, Z. Trzaska, L. Durand, J. M. Chaix, J. P. Monchoux, Theoretical and experimental investigations of local overheating at particle contacts in spark plasma sintering, *Powder Technol.* 321 (2017) 458–470. 10.1016/j.powtec.2017.08.033

[5] X. Song, X. Liu, J. Zhang, Neck formation and self-adjusting mechanism of neck growth of conducting powders in spark plasma sintering, *J. Am. Ceram. Soc.* 89 (2006) 494–500. 10.1111/j.1551-2916.2005.00777.x

[6] T. T. Sasaki, T. Mukai, K. Hono, A high-strength bulk nanocrystalline Al–Fe alloy processed by mechanical alloying and spark plasma sintering, *Scr. Mater.* 57 (2007) 189–192. 10.1016/j.scriptamat.2007.04.010

[7] M. Kubota, B. P. Wynne, Electron backscattering diffraction analysis of mechanically milled and spark plasma sintered pure aluminium, *Scr. Mater.* 57 (2007) 719–722. 10.1016/j.scriptamat.2007.06.036

[8] T. M. Vidyuk, D. V. Dudina, M. A. Korchagin, A. I. Gavrilov, T. S. Skripkina, A. V. Ukhina, A. G. Anisimov, B. B. Bokhonov, Melting at the inter-particle contacts during Spark Plasma Sintering: direct microstructural evidence and relation to particle morphology, *Vacuum.* 181 (2020) 109566. 10.1016/j.vacuum.2020.109566

[9] T. M. Vidyuk, D. V. Dudina, M. A. Korchagin, A. I. Gavrilov, A. V. Ukhina, U. E. Bulanova, M. A. Legan, A. N. Novoselov, M. A. Esikov, A. G. Anisimov, Manufacturing of TiC-Cu composites by mechanical milling and spark plasma sintering using different carbon sources, *Surf. Interf.* 27 (2021) 101445. 10.1016/j.surfin.2021.101445.

[10] D. V. Dudina, T. F. Grigoreva, E. T. Devyatkina, S. V. Vosmerikov, A. V. Ukhina, V. V. Markushin, N. Z. Lyakhov, Structural features of tantalum carbide-copper composites obtained by liquid phase-assisted spark plasma sintering, *Ceram. Int.* 48 (2022) 32556–32560. 10.1016/j.ceramint.2022.07.322

[11] M. Z. Kol'chinskii, A. I. Raichenko, A model investigation of the sintering of metal powders with intense energy release at inter-particle contacts, *Soviet Powder Metall. Metal Ceram.* 16 (1977) 585–588. 10.1007/BF00790729

[12] M. Z. Kol'chinskii, N. F. Medvedenko, Yu. M. Solonin, F. P. Gnatush, Electric-discharge sintering of mixtures of aluminum and copper powders, *Soviet Powder Metall. Metal Ceram.* 16 (1977) 504–507. 10.1007/BF00790971

[13] Y. Ye, X. Li, K. Hu, Y. Lai, Y. Li, The influence of premolding load on the electrical behavior in the initial stage of electric current activated sintering of carbonyl iron powders, *J. Appl. Phys.* 113 (2013) 214902. 10.1063/1.4808339

[14] E. Olevsky, L. Froyen, Constitutive modeling of spark-plasma sintering of conductive materials, *Scr. Mater.* 55 (2006) 1175–1178. 10.1016/j.scriptamat.2006.07.009

[15] G. Lee, C. Manière, J. McKittrick, E. A. Olevsky, Electric current effects in spark plasma sintering: from the evidence of physical phenomenon to constitutive equation formulation, *Scr Mater.* 170 (2019) 90–94. 10.1016/j.scriptamat.2019.05.040

[16] D. V. Dudina, B. B. Bokhonov, Elimination of oxide films during Spark Plasma Sintering of metallic powders: A case study using partially oxidized nickel, *Adv. Powder Technol.* 28 (2017) 641–647. 10.1016/j.apt.2016.12.001

[17] B. B. Bokhonov, A. V. Ukhina, D. V. Dudina, A. G. Anisimov, V. I. Mali, I. S. Batraev, Carbon uptake during Spark Plasma Sintering: investigation through the analysis of the carbide "footprint" in a Ni–W alloy, *RSC Adv.* 5 (2015) 80228–80237. 10.1039/C5RA15439A

[18] D. V. Dudina, B. B. Bokhonov, A. V. Ukhina, A. G. Anisimov, V. I. Mali, M. A. Esikov, I. S. Batraev, O. O. Kuznechik, L. P. Pilinevich, Reactivity of materials towards carbon of graphite foil during spark plasma sintering: A case study using Ni–W powders, *Mater. Lett.* 168 (2016) 62–67. 10.1016/j.matlet.2016.01.018

[19] T. M. Vidyuk, M. A. Korchagin, D. V. Dudina, B. B. Bokhonov, Synthesis of ceramic and composite materials using a combination of self-propagating high-temperature synthesis and spark plasma sintering (Review). *Combust. Explos. Shock Waves* 57 (2021) 385–397. 10.1134/S0010508221040018

[20] D. V. Dudina, K. Georgarakis, E. A. Olevsky, Progress in aluminium and magnesium matrix composites obtained by spark plasma, microwave and induction sintering. *Int. Mater. Rev.* 2022, 10.1080/09506608.2022.2077029

[21] D. V. Dudina, A. K. Mukherjee, Reactive spark plasma sintering: successes and challenges of nanomaterial synthesis, *J. Nanomater.* 2013 (2013) 625218. 10.1155/2013/625218

[22] D. V. Dudina, T. M. Vidyuk, M. A. Korchagin, Synthesis of ceramic reinforcements in metallic matrices during spark plasma sintering: Consideration of reactant/matrix mutual chemistry, *Ceramics* 4 (2021) 592–599. 10.3390/ceramics4040042

[23] A. Nisar, C. Zhang, B. Boesl, A. Agarwal, Unconventional materials processing using spark plasma sintering, *Ceramics* 4 (2021) 20–39. 10.3390/ceramics4010003

[24] T. M. Vidyuk, D. V. Dudina, Electric current-assisted joining of similar/dissimilar materials // In: Joining Processes for Dissimilar and Advanced Materials, Woodhead Publishing Reviews: Mechanical Engineering Series, (2022), pp.151–176. 10.1016/B978-0-323-85399-6.00017-5

[25] N. S. Weston, B. Thomas, M. Jackson, Processing metal powders via field assisted sintering technology (FAST): A critical review, *Mater. Sci. Technol.* 35 (2019) 1306–1328. 10.1080/02670836.2019.1620538.

[26] J. P. Monchoux, A. Couret, L. Durand, T. Voisin, Z. Trzaska, M. Thomas, Elaboration of metallic materials by SPS: processing, microstructures, properties, and shaping, *Metals* 11 (2021) 322. 10.3390/met11020322
[27] M. Abedi, A. Asadi, S. Vorotilo, A. S. Mukasyan, A critical review on spark plasma sintering of copper and its alloys, *J. Mater. Sci.* 56 (2021) 19739–19766. 10.1007/s10853-021-06556-z
[28] N. F. Mogale, W. R. Matizamhuka, Spark plasma sintering of titanium aluminides: a progress review on processing, structure-property relations, alloy development and challenges, *Metals* 10 (2020) 1080. 10.3390/met10081080
[29] M. Stuer, P. Bowen, Z. Zhao, Spark plasma sintering of ceramics: from modeling to practice, *Ceramics* 3 (2020) 476–493. 10.3390/ceramics3040039
[30] K. Georgarakis, D. V. Dudina, V. I. Kvashnin, Metallic glass-reinforced metal matrix composites: design, interfaces and properties, *Materials* 15 (2022) 8278. 10.3390/ma15238278
[31] D. V. Dudina, B. B. Bokhonov, E. A. Olevsky, Fabrication of porous materials by spark plasma sintering: A review, *Materials* 12 (2019) 541. 10.3390/ma12030541
[32] K. Huang, Y. Yang, Y. Qin, G. Yang, D. Yin, A new densification mechanism of copper powder sintered under an electrical field, *Scr. Mater.* 99 (2015) 85–88. 10.1016/j.scriptamat.2014.12.002
[33] A. Numula, S. Kassegne, K. S. Moon, A. El-Desouky, K. Morsi, Reactive current-activated tip-based sintering of Ni–Al intermetallics, *Metallogr. Microstr. Anal.* 2 (2013) 148–155. 10.1007/s13632-013-0071-y
[34] M. Patel, K. S. Moon, S. K. Kassegne, K. Morsi, Effects of current intensity and cumulative exposure time on the localized current activated sintering of titanium nickelides, *J. Mater. Sci.* 46 (2011) 6690–6699. 10.1007/s10853-011-5622-5
[35] P. Asoka-Kumar, K. O'Brien, K. G. Lynn, P. J. Simpson, K. P. Rodbell, Detection of current-induced vacancies in thin aluminum-copper lines using positrons, *Appl. Phys. Lett.* 68 (1996) 406–408. 10.1063/1.116700
[36] N. Bertolino, J. Garay, U. Anselmi-Tamburini, Z. A. Munir, Electromigration effects in Al-Au multilayers, *Scr. Mater.* 44 (2001) 737–742. 10.1016/S1359-6462(00)00669-2
[37] N. Bertolino, J. Garay, U. Anselmi-Tamburini, Z. A. Munir, High-flux current effects in interfacial reactions in Au–Al multilayers, *Phil. Mag. B* 82 (2002) 969–985. 10.1080/13642810208218356
[38] J. E. Garay, U. Anselmi-Tamburini, Z. A. Munir, Enhanced growth of intermetallic phases in the Ni–Ti system by current effects, *Acta Mater.* 51 (2003) 4487–4495. 10.1016/S1359-6454(03)00284-2
[39] S. Rudinsky, R. Gauvin, M. Brochu, The effects of applied current on one-dimensional interdiffusion between copper and nickel in spark plasma sintering, *J. Appl. Phys.* 116 (2014) 154901. 10.1063/1.4898158
[40] R. Li, T. Yuan, X. Liu, K. Zhou, Enhanced atomic diffusion of Fe–Al diffusion couple during spark plasma sintering, *Scr. Mater.* 110 (2016) 105–108. 10.1016/j.scriptamat.2015.08.012
[41] M. Abedi, A. Asadi, S. Sovizi, D. Moskovskikh, S. Vorotilo, A. Mukasyan, Influence of pulsed direct current on the growth rate of intermetallic phases in the Ni–Al system during reactive spark plasma sintering, *Scr. Mater.* 216 (2022) 114759. 10.1016/j.scriptamat.2022.114759
[42] M. Abedi, S. Sovizi, A. Azarniya, D. Giuntini, M. E. Seraji, H. R. M. Hosseini, C. Amutha, S. Ramakrishna, A. Mukasyan, An analytical review on Spark Plasma Sintering of metals and alloys: from processing window, phase

transformation, and property perspective, *Critical Rev. Solid State Mater. Sci.*, (2022). 10.1080/10408436.2022.2049441

[43] I. Gibson, D. Rosen, B. Stucker, *Additive manufacturing technologies: 3D Printing, Rapid Prototyping, and Direct Digital Manufacturing – 2nd edition*, New York, London: Springer, (2015), 498 p.

[44] Y. Sun, S. K. Vajpai, K. Ameyama, C. Ma, Fabrication of multilayered Ti–Al intermetallics by spark plasma sintering, *J. Alloys Compd.* 585 (2014) 734–740. 10.1016/j.jallcom.2013.09.215

[45] K. Mizuuchi, K. Inoue, M. Sugioka, M. Itami, J.-H. Lee, M. Kawahara, Properties of Ni-aluminides-reinforced Ni-matrix laminates synthesized by pulsed-current hot pressing (PCHP), *Mater. Sci. Eng. A* 428 (2006) 169–174. 10.1016/j.msea.2006.04.113

[46] D. V. Lazurenko, V. I. Mali, I. A. Bataev, A. Thoemmes, A. A. Bataev, A. I. Popelukh, A. G. Anisimov, N. S. Belousova, Metal-intermetallic laminate Ti-$Al_3Ti$ composites produced by spark plasma sintering of titanium and aluminum foils enclosed in titanium shells, *Metall. Mater. Trans. A* 46, 4326–4334 (2015). 10.1007/s11661-015-3002-5

[47] D. V. Lazurenko, A. Stark, M. A. Esikov, J. Paul, I. A. Bataev, A. A. Kashimbetova, V. I. Mali, U. Lorenz, F. Pyczak, Ceramic-reinforced γ-TiAl-based composites: synthesis, structure, and properties. *Materials* 12 (2019) 629. 10.3390/ma12040629

[48] T. S. Ogneva, I. A. Bataev, V. I. Mali, A. G. Anisimov, D. V. Lazurenko, A. I. Popelyukh, Yu. Yu. Emurlaeva, A. A. Bataev, S. Tanaka, K. D. Yegoshin, Effect of sintering pressure and temperature on structure and properties of Ni-Al metal-intermetallic composites produced by SPS, *Mater. Charact.* 180 (2021) 111415. 10.1016/j.matchar.2021.111415

[49] S. P. Harimkar, T. Borkar, A. Singh, Spark plasma sintering of amorphous-crystalline laminated composites, *Mater. Sci. Eng. A* 528 (2011) 1901–1905. 10.1016/j.msea.2010.10.107

[50] W. Liu, M. Naka, In situ joining of dissimilar nanocrystalline materials by spark plasma sintering, *Scr. Mater.* 48 (2003) 1225–1230. 10.1016/S1359-6462(03)00074-5

[51] Y. Liu, Y. Ma, W. Liu, Y. Huang, L. Wu, T. Wang, C. Liu, L. Yang, The mechanical properties and formation mechanism of Al/Mg composite interface prepared by spark plasma sintering under different sintering pressures, *Vacuum* 176 (2020) 109300. 10.1016/j.vacuum.2020.109300

[52] K. Kim, D. Kim, K. Park, M. Cho, S. Cho, H. Kwon, Effect of intermetallic compounds on the thermal and mechanical properties of Al–Cu composite materials fabricated by spark plasma sintering, *Materials* 12 (2019) 1546. doi:10.3390/ma12091546

[53] D. Kim, K. Kim, H. Kwon, Investigation of formation behaviour of Al–Cu intermetallic compounds in Al–50vol%Cu composites prepared by spark plasma sintering under high pressure, *Materials* 14 (2021) 266. 10.3390/ma14020266

[54] O. O. Shichalin, V. N. Sakhnevich, I. Yu. Buravlev, A. O. Lembikov, A. A. Buravleva, S. A. Azon, S. B. Yarusova, S. N. Danilova, A. N. Fedorets, A. A. Belov, E. K. Papynov, Synthesis of Ti-Cu multiphase alloy by spark plasma sintering: mechanical and corrosion properties, *Metals* 12 (2022) 1089. 10.3390/met12071089

[55] M. Abedi, K. Kuskov, D. Moskovskikh, E. V. Zakharova, D. Belov, A. Mukasyan, Reactive spark plasma sintering of NiAl intermetallics: A comparative study, *Intermetallics* 152 (2023) 107750. 10.1016/j.intermet.2022.107750

[56] S. Paris, E. Gaffet, F. Bernard, Z. A. Munir, Spark plasma synthesis from mechanically activated powders: a versatile route for producing dense nanostructured iron aluminides, *Scripta Mater.* 50 (2004) 691–696. 10.1016/j.scriptamat.2003.11.019

[57] Y. Sun, K. Kulkarni, A. K. Sachdev, E. J. Lavernia, Synthesis of γ-TiAl by reactive spark plasma sintering of cryomilled Ti and Al powder blend, part I: Influence of processing and microstructural evolution, *Met. Mater. Trans. A* 45 (2014) 2750–2758. 10.1007/s11661-014-2215-3

[58] Y. Sun, K. Kulkarni, A. K. Sachdev, E. J. Lavernia, Synthesis of γ-TiAl by reactive spark plasma sintering of cryomilled Ti and Al powder blend, part II: Effects of electric field and microstructure on sintering kinetics, *Met. Mater. Trans. A* 45 (2014) 2759–2767. 10.1007/s11661-014-2216-2

[59] D. V. Dudina, M. A. Legan, N. V. Fedorova, A. N. Novoselov, A. G. Anisimov, M. A. Esikov, Structural and mechanical characterization of porous iron aluminide FeAl obtained by pressureless spark plasma sintering, *Mater. Sci. Eng. A* 695 (2017) 309–314. 10.1016/j.msea.2017.04.051

[60] D. V. Dudina, B. B. Bokhonov, M. A. Legan, A. N. Novoselov, I. N. Skovorodin, N. V. Bulina, M. A. Esikov, V. I. Mali, Analysis of the formation of FeAl with a high open porosity during electric current-assisted sintering of loosely packed Fe-Al powder mixtures, *Vacuum* 146 (2017) 74–78. 10.1016/j.vacuum.2017.09.031

[61] T. Nobuki, T. Moriya, M. Hatate, J.-C. Crivello, F. Cuevas, J.-M. Joubert, Synthesis of TiFe hydrogen absorbing alloys prepared by mechanical alloying and SPS treatment, *Metals* 8 (2018) 264. 10.3390/met8040264

[62] T. Murakami, A. Kitahara, Y. Koga, M. Kawahara, H. Inui, M. Yamaguchi, Microstructure of Nb–Al powders consolidated by spark plasma sintering process. *Mater. Sci. Eng. A* 239–240 (1997) 672–679. 10.1016/S0921-5093(97)00651-5

[63] A. Zhang, J. Han, J. Meng, B. Su, P. Li, Rapid preparation of AlCoCrFeNi high entropy alloy by spark plasma sintering from elemental powder mixture, *Mater. Lett.* 181 (2016) 82–85 10.1016/j.matlet.2016.06.014

[64] K. V. Kuskov, A. A. Nepapushev, S. Aydinyan, D. G. Shaysultanov, N. D. Stepanov, K. Nazaretyan, S. Kharatyan, E. V. Zakharova, D. S. Belov, D. O. Moskovskikh, Combustion synthesis and reactive spark plasma sintering of non-equiatomic CoAl-based high entropy intermetallics, *Materials* 16 (2023) 1490. 10.3390/ma16041490

[65] Z. Y. Hu, Z.-H. Zhang, H. Wang, S.-L. Li, S. Yin, Q. Song, X. W. Cheng, A rapid route for synthesizing Ti-($Al_xTi_y$/UFG Al) core-multishell structured particles reinforced Al matrix composite with promising mechanical properties, *Mater. Sci. Eng. A* 721 (2018) 61–64. 10.1016/j.msea.2018.02.065

[66] B. Guo, M. Song, X. Zhang, X. Cen, W. Li, B. Chen, Q. Wang, Achieving high combination of strength and ductility of Al matrix composite via in-situ formed Ti-$Al_3Ti$ core-shell particle, *Mater. Charact.* 170 (2020) 110666. 10.1016/j.matchar.2020.110666

[67] S. Vorotilo, A. A. Nepapushev, D. O. Moskovskikh, V. S. Buinevich, G. V. Trusov, D. Yu. Kovalev, A. O. Semenyuk, N. D. Stepanov, K. Vorotilo, A. Y. Nalivaiko, A. A. Gromov, Engineering of strong and hard in-situ Al-$Al_3Ti$ nanocomposite via high-energy ball milling and spark plasma sintering, *J. Alloys Compd.* 895 (2022) 162676. 10.1016/j.jallcom.2021.162676

[68] D. V. Dudina, B. B. Bokhonov, I. S. Batraev, Y. N. Amirastanov, A. V. Ukhina, I. D. Kuchumova, M. A. Legan, A. N. Novoselov, K. B. Gerasimov, I. A. Bataev, K. Georgarakis, G. Y. Koga, Y. Guo, W. J. Botta, A. Moreira Jorge Jr.

Interaction between $Fe_{66}Cr_{10}Nb_5B_{19}$ metallic glass and aluminum during spark plasma sintering, *Mater. Sci. Eng. A.* 799 (2021) 140165. 10.1016/j.msea.2020.140165.

[69] D. V. Dudina, B. B. Bokhonov, I. S. Batraev, V. I. Kvashnin, M. A. Legan, A. N. Novoselov, A. G. Anisimov, M. A. Esikov, A. V. Ukhina, A. A. Matvienko, K. Georgarakis, G. Y. Koga, A. Moreira Jorge Jr. Microstructure and mechanical properties of composites obtained by spark plasma sintering of Al–$Fe_{66}Cr_{10}Nb_5B_{19}$ metallic glass powder mixtures, *Metals* 11 (2021) 1457. 10.3390/met11091457

[70] R. Ali, F. Ali, A. Zahoor, R. Nawaz Shahid, N. ul Haq Tariq, T. He, M. Shahzad, Z. Asghar, A. Shah, A. Mahmood, H. Bin Awais, Effect of sintering path on the microstructural and mechanical behavior of aluminum matrix composite reinforced with pre-synthesized Al/Cu core-shell particles, *J. Alloys Compd.* 889 (2021) 161531. 10.1016/j.jallcom.2021.161531

[71] H. Zhang, P. Feng, F. Akhtar, Aluminium matrix tungsten aluminide and tungsten reinforced composites by solid-state diffusion mechanism, *Sci. Rep.* 7 (2017) 12391. 10.1038/s41598-017-12302-w

[72] A. Joseph, V. Gauthier-Brunet, A. Joulain, J. Bonneville, S. Dubois, J.-P. Monchoux, F. Pailloux, Mechanical properties of Al/ω-Al-Cu-Fe composites synthesized by the SPS technique, *Mater. Charact.* 145 (2018) 644–652. 10.1016/j.matchar.2018.09.025

[73] P. Hyjek, I. Sulima, P. Malczewski, K. Bryła, L. Jaworska, Effect of reactive SPS on the microstructure and properties of a dual-phase Ni-Al intermetallic compound and Ni-Al-$TiB_2$ composite, *Materials* 13 (2020) 5668. 10.3390/ma13245668

[74] D. V. Dudina, K. Georgarakis, Core-shell particle reinforcements—A new trend in the design and development of metal matrix composites, *Materials* 15 (2022) 2629. 10.3390/ma15072629

[75] D. V. Dudina, B. B. Bokhonov, A. K. Mukherjee, Formation of aluminum particles with shell morphology during pressureless Spark Plasma Sintering of Fe-Al mixtures: current-related or Kirkendall effect? *Materials* 9 (2016) 375. 10.3390./ma9050375

[76] D. V. Dudina, V. I. Kvashnin, A. A. Matvienko, A. A. Sidelnikov, A. I. Gavrilov, A. V. Ukhina, A. Moreira Jorge, Jr., K. Georgarakis, Towards a better understanding of the interaction of $Fe_{66}Cr_{10}Nb_5B_{19}$ metallic glass with aluminum: growth of intermetallics and formation of Kirkendall porosity during sintering, *Chemistry* 5 (2023) 138–150. 10.3390/chemistry5010011

[77] D. Paraskevas, K. Vanmeensel, J. Vleugels, W. Dewulf, Y. Deng, J. R. Duflou, Spark plasma sintering as a solid-state recycling technique: the case of aluminum alloy scrap consolidation, *Materials* 7 (2014) 5664–5687. 10.3390/ma7085664

[78] C. S. Bonifacio, T. B. Holland, K. van Benthem, Evidence of surface cleaning during electric field assisted sintering, *Scr. Mater.* 69 (2013) 769–772. 10.1016/j.scriptamat.2013.08.018

[79] T. M. Vidyuk, D. V. Dudina, M. A. Korchagin, A. I. Gavrilov, B. B. Bokhonov, A. V. Ukhina, M. A. Esikov, V. S. Shikalov, V. F. Kosarev, Spark plasma sintering treatment of cold sprayed materials for synthesis and structural modification: A case study using TiC-Cu composites, *Mater. Lett. X* 14 (2022) 100140. 10.1016/j.mlblux.2022.100140

[80] D. Orlov, H. Fujiwara, K. Ameyama, Obtaining copper with harmonic structure for the optimal balance of structure-performance relationship. *Mater. Trans.* 54 (2013) 1549–1553. 10.2320/matertrans.MH201320

# 12 Laser Micro-spot Welding of Magnesium Alloys for Bone Tissue Engineering Scaffolds

*Luis D. Cedeño-Viveros, Ciro A. Rodríguez, Elisa Vázquez-Lepe, Luis H. Olivas-Alanis, and Erika García-López*

## 12.1 INTRODUCTION: BACKGROUND

Bone Tissue Engineering Scaffolds (BTES) are 3D structures constructed with a biocompatible material and a designed internal porosity that allows bone growth. These scaffolds temporarily support the osteogenic cells and the new bone formation and stimulate cell proliferation and, consequently, bone tissue ingrowth [1–3]. Bone Tissue Engineering Scaffolds (BTE) combine principles of engineering materials, physicochemical factors, and the use of cells to develop biological substitutes to repair bone defects that are difficult to treat by conventional methods. The scaffold requirements for bone tissue engineering applications are classified into three categories biological, physical, and mechanical. Biological requirements include biocompatibility, osteoconductivity, osteoinductivity, and osteointegration. Biocompatibility supports cellular activity without evoking local or systemic adverse effects to the host, such as cytotoxicity, allergic, and cancerogenic [2,4]. Osteoconductivity induces new bone formation through biomolecular signaling and recruiting progenitor cells allowing differentiation in a controlled phenotype [1,5]. Osteoinductivity constructs and mineralizes the extracellular matrix for the new bone tissue with osteoblasts and mesenchymal cells. Osteointegration forms strong bonds with the surrounding bone allowing proper load transfer and continuity [1].

The physical requirements of a scaffold are related to the bone tissue structure, meaning that the network must be porous and hierarchical [2,6]. The mechanical requirements are related to the match between scaffold stiffness and the host bone stiffness; otherwise, stress shielding can be caused. Figure 12.1 presents the scaffold requirements for BTE classified as biological, physical, and mechanical.

Laser micro-welding (LMW) is a capable manufacturing method to produce welds with excellent quality and hermeticity in the medical field. LMW has

DOI: 10.1201/9781003364948-12

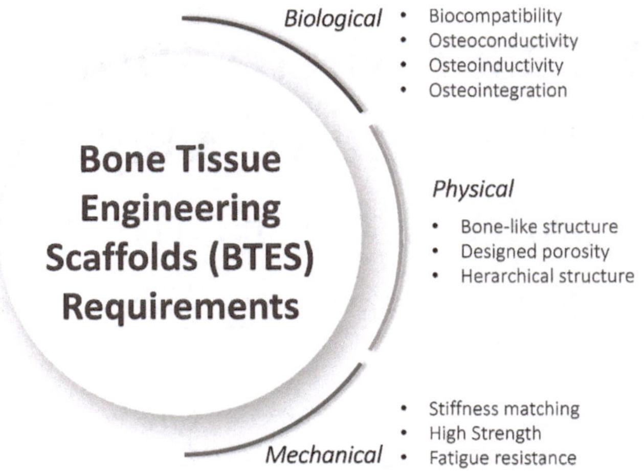


**FIGURE 12.1** Main biological, physical, and mechanical requirements for Bone Tissue Engineering Scaffolds (BTES).

several advantages related to the non-contact workpiece, which prevents contamination using a low heat input. Compared with Resistance Micro Welding (RMW) and Micro-TIG welding processes (MTW), this method generates a low mechanical distortion and excellent quality [7,8]. LMW has presented higher efficiency, higher depth-to-width penetration ratio, lower residual stresses, smaller heat affected zone (HAZ), and low-temperature gradients compared to other technologies like micro-plasma welding, Electron beam (EB) welding, and friction welding, among others [9–13]. Additionally, LMW can be easily automated compared with other manufacturing methods which impact precision [12,14–16]. Some medical applications where the LMW process is used include pacemakers [12], defibrillators, and cochlear devices [16]. LMW can be performed in three regimes (i.e., conduction, open keyhole, and close keyhole) [17,18]. The conduction regime allows the substrate to fuse it by heat conduction, and keyhole regimes form a plasma plume inside the weld pool, which is fully penetrated (open keyhole) or partially penetrated (closed keyhole). The energy absorption in the keyhole regime is nearly 100% higher than the partial absorption obtained in the conduction regime; this is due to the multiple reflections inside the plasma plume, which leads to a more significant amount of energy absorbed [12]. Keyhole welding has an aspect ratio (i.e., weld depth to weld width ratio) larger than conduction; however, conduction welding offers the advantages of a larger process window, faster optimization, and fewer welding defects [11]. For medical devices, a material conduction regime is recommended due to the small thickness of the raw material [12]. Figure 12.2 presents a) the laser welding process, b) the process parameters, and c, d, e) the three regimens for LMW. Additionally, according to the material position (Figure 12.3), welds can be classified as a) bead on plate, b) butt weld, c) overlap, and d) edge weld.

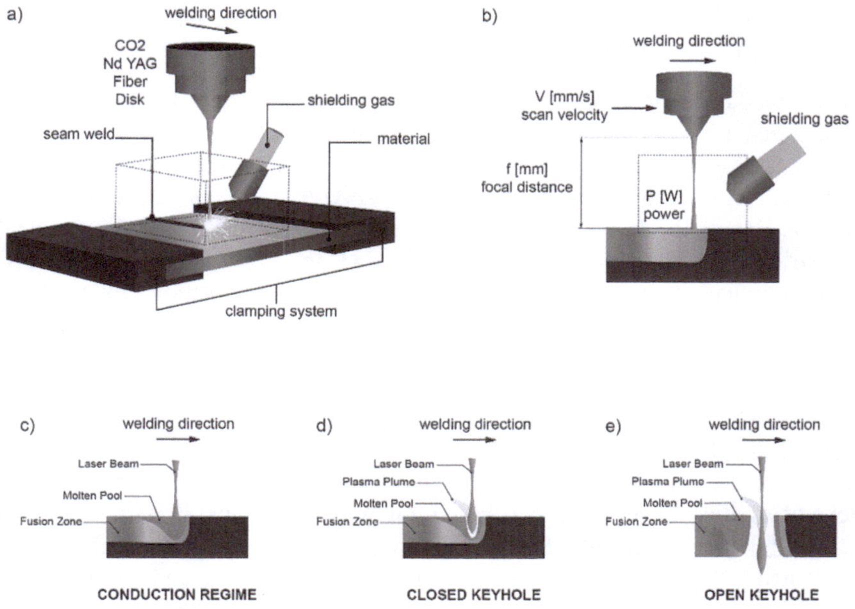


**FIGURE 12.2** a) Laser welding process, b) laser welding parameters, and laser welding regimes c) conduction regime, d) closed keyhole regime, and e) open keyhole regime.

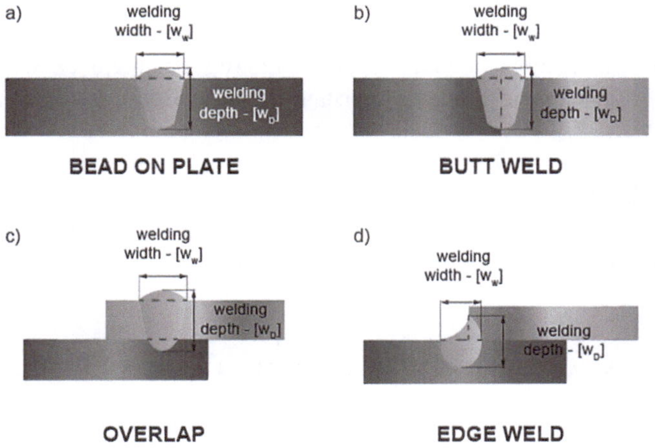


**FIGURE 12.3** Welding types: a) bead on plate, b) butt weld, c) overlap, and d) edge weld.

## 12.2 OVERVIEW OF MAGNESIUM ALLOYS FOR BTES APPLICATIONS

In the medical field, Mg alloys have shown great potential because of their biodegradability and capacity to induce no inflammatory or systemic response. Besides, these alloys are osteoconductive and present suitable cell attachment [2,19–21]. Among the Mg alloys, the series AM, AZ, and ZK (alloys based on Al, Zn, Mn, and Zr elements, respectively) are the most used by medical industries [13]. The main applications of AZ31 and AZ91 alloys are in fabricating coronary stents [22,23], bone scaffolds [24], and fixation devices [25]. However, their main disadvantage is poor corrosion resistance and high degradation rates [2]. Recent investigations have tested coatings in AZ31 alloy to improve corrosion resistance reducing the degradation rates [20,26]. The degradation rate varies according to the medical application.

For example, the degradation of spinal fusion implants is ~9 months, while for cranium maxillofacial implants, degradation is related to 3 – 6 months [1,2,4]. Several studies have been performed on animal models in terms of biocompatibility requirements. For example, an AZ91D magnesium scaffold was implanted in the condyle of a rabbit's femur in the left knee, and a bone defect was used as a control in the right knee. [24,27–29]. Their results reveal that the scaffold was rapidly degraded with similar mechanical properties to cancellous bone, allowing mechanical support during the first 12 weeks of cartilage repair. Even though of the magnesium scaffold's fast degradation, no damage or disorder in the surrounding cartilage was observed regarding tissue morphology or collagen matrix composition [24]. Besides, new bone formation at the interface scaffold bone indicated that AZ91D magnesium scaffolds are osteoconductive, and the remaining aluminum is absorbed through phagocytosis and macrophages. Yu et al. demonstrated that applying an MgF coating over AZ31 scaffold helped to improve their corrosion resistance and deal with the fast rate degradation problem.

Moreover, the coated scaffolds showed enhanced osteogenic activity compared to AZ31 [30]. The mechanical behavior of scaffolds has been studied for several works with mechanical properties near cancellous bone properties. Table 12.1 shows the magnesium scaffolds' mechanical properties found in the literature.

## 12.3 LASER WELDING OF MAGNESIUM ALLOYS

Laser Welding (LW) of magnesium alloys has been widely explored in the last decade due to their physical properties, such as high strength-to-weight ratio, high thermal conductivity, good castability, and excellent recyclability [9,11,13,42]. Several laser sources have been used to manufacture magnesium alloys in the last two decades. Specifically, AZ31 alloy is the most common material used for LW [43]. Figure 12.4 and Figure 12.5 present studies focused on LW of magnesium aluminum zinc sheets with thicknesses between 0.3 mm to

**TABLE 12.1**
**Summary of the Mechanical Properties of Several Mg Scaffolds in the Literature**

| Reference | Material | Porosity [%] | Relative Density | Apparent Comp Strength [MPa] | Apparent Young Modulus [GPa] |
|---|---|---|---|---|---|
| [31] | Mg | 0.0% | 1.00 | 100.0 | 44.0 |
| | Mg | 55% | 0.45 | 15.0 | 3.6 |
| [32] | Mg | 43% | 0.57 | 12.0 | 0.6 |
| | Mg | 50% | 0.50 | 9.0 | 0.5 |
| | Mg | 51% | 0.49 | 8.4 | 0.4 |
| [33] | W4 alloy | 61% | 0.40 | 15.0 | 0.8 |
| | W4 alloy | 61% | 0.40 | 22.0 | 2.9 |
| [34] | Mg | 6.% | 0.94 | 79.0 | 3.1 |
| | Mg | 8% | 0.92 | 55.0 | 2.5 |
| [35] | Mg 4% Zn | 15% | 0.85 | 50.0 | 5.0 |
| | Mg 4% Zn | 25% | 0.75 | 35.0 | 4.0 |
| | Mg 4% Zn | 35% | 0.65 | 25.0 | 4.0 |
| | Mg 4% Zn | 45% | 0.55 | 15.0 | 2.5 |
| | Mg 6% Zn | 15% | 0.85 | 50.0 | 6.0 |
| | Mg 6% Zn | 25% | 0.75 | 35.0 | 5.0 |
| | Mg 6% Zn | 35% | 0.65 | 25.0 | 4.0 |
| | Mg 6% Zn | 45% | 0.55 | 17.5 | 3.0 |
| [36] | MgP | 50% | 0.50 | 4.8 | 0.03 |
| [37] | Mg | 33% | 0.67 | 12.0 | 0.7 |
| | Mg | 48% | 0.52 | 11.0 | 0.6 |
| | Mg | 54% | 0.46 | 10.0 | 0.2 |
| | Mg | 33% | 0.67 | 30.0 | 0.4 |
| | Mg | 48% | 0.52 | 15.0 | 0.2 |
| | Mg | 54% | 0.46 | 10.0 | 0.05 |
| [38] | Mg | 54% | 0.45 | 4.3 | 0.5 |
| | Mg | 51% | 0.49 | 4.6 | 0.6 |
| | Mg | 43% | 0.57 | 6.2 | 1.0 |
| [5] | Mg | 30% | 0.70 | 103.0 | 2.2 |
| | Mg | 41% | 0.59 | 63.0 | 2.1 |
| | Mg | 55% | 0.45 | 63.0 | 1.5 |
| [39] | WE43 alloy | 86% | 0.13 | 9.4 | 0.4 |
| | | | | 2.1 | 0.1 |
| [40] | Compact bone | – | – | 88–164 | 3.–18.7 |
| [41] | Cancellous bone | – | – | 0.2–8 | 0.01–2 |

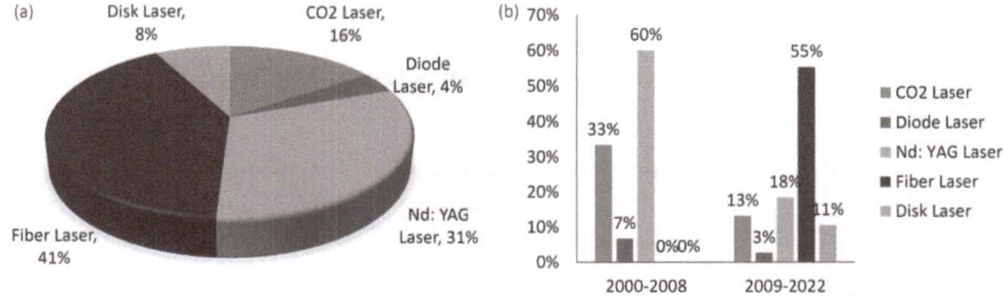


**FIGURE 12.4** a) Percentage of use of different laser types for LW of AZ31 and AZ91, b) Percentage of use of laser types from 2000 to 2017.

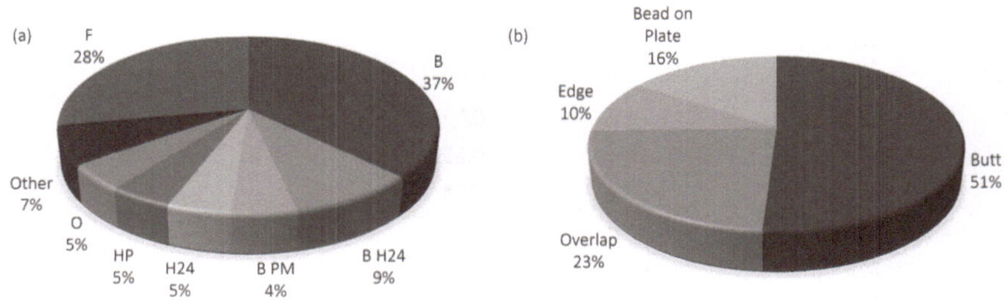


**FIGURE 12.5** a) Common hardening and heat treatment process of raw materials; materials and; b) Common weld joints found in the literature of AZ31 LW (* B - wrought fabrication, HP - strain hardening, H24 - half strain hardened and partially annealed, O - annealed, F - as fabricated, Other preparation procedure before welding).

10 mm. This material is preferably used in welding automotive and aeronautical parts due to aluminum toxicity. Other alloys, such as WE43 [39] and JDBM (Mg-Nd-Zn) alloy [44], have emerged for medical devices. However, laser welding studies are limited.

Figure 12.4.a. illustrates the types of sources used by the researchers from 2000 until the present for the LW of AZ31 and AZ91 alloys. There is a predominant use of fiber lasers (41 %) followed by NdYAG Lasers (31%), and only two studies use diode lasers [11,45]. Figure 12.4.b presents the different types of lasers in two time periods. The presence of Nd: YAG lasers is predominant in the first period (2000–2008) with 60%, while in the second period (2009–2022), the use of fiber lasers is popularized with a total of 55%. For the manufacture of medical devices in magnesium alloys, the most common laser type is the Nd YAG laser. Nd YAG lasers present higher energy absorption of the laser beam in solids and metals than CO2 lasers [12]. While, fiber laser has higher efficiency than other laser types and a compact design, simplifying the installation and smaller beam focus diameters [15]. For the manufacture of magnesium alloys, inert gases such as argon and helium are mainly used to control the process. Argon gas is preferred due to its high purity. However, helium gas has the advantage of good thermal conductivity and a high ionization potential (24.5 eV) [42].

Additionally, combinations of argon and helium gases have been performed to cover the weld's top and bottom, respectively [46]. Some studies have evaluated the influence of the shielding gas flow rate applied on weld quality through the inclination of the shielding gas pipe [11]. Zhu et al. compared two gas delivery methods using a box [17 L/min] and an inclined shielding gas pipe [24 L/min]. The shielding box resulted in better quality because the concentrated flux of the pipe caused side blows in the welds [11]. Figure 12.5.a. presents a study of the heat treatments in magnesium alloys before the welding process based on the ASTM B296-03 [47], and Figure 12.5.b. the type of weld joint used in magnesium alloys of studies reported from 2000 to 2022. From our results, around 37% of the materials used in research came from a wrought fabrication. In comparison, 9% are H24 which means a half-strain hardening process combined with a partial annealed treatment as the fabrication process of the material. The second major group is the AZ31 material as fabricated – F (28%) with no hardening or heat treatment specifications. Besides, only two studies have used AZ31 with a thickness below 1 mm (i.e., 0.3 mm) [48,49]. According to our results in Figure 12.5.b, the most common is the butt joints (51%), followed by the overlap (23%), the bead on plate (16%), and the edge joint (10%) in weld joints.

LW technology can improve seam wealds such as excellent visual quality, higher scan speed, finer grain microstructure, better corrosion resistance, and mechanical properties equal to or slightly higher than the base materials. In LW, the temperature gradients or cooling rates strongly determine the weld's grain morphology, microstructure, and mechanical properties. An increase in the laser intensity, which means a decrease of the beam radius rB [mm] or an increase in the laser power P [W], results in low gradient temperatures [13]. Controversially, magnesium at high temperatures liberates hydrogen gases, which escape to melting pool-producing pores [12]. ISO 13919–2 [50] is a good guide for the visual quality evaluation of the laser beam welds of sheets 1 mm thick or above. Regarding LMW, there is no standard for this topic; however, the same standard can be used as a reference. Figure 12.6 presents the defects found in LMW [11] and the corresponding best practice to reduce the defects inside the weld.

The shielding gas in LMW covers the melting pool, protects the weld from contaminants, and insulates it from oxygen in the environment. For AZ31 alloy, oxygen promotes Mg oxidation and causes loss of Mg content, causing defects such as shrinkages or holes in the laser weld [11,13]. Oxide layers are common in the AZ31B plates and sheet; these layers could be a problem, especially in overlapped welds, because of the gases formed in the interphase. It is produced when the scan velocity – V[mm/s] is fast, and there is not enough time for gases to escape the melting pool through the keyhole [51]. To deal with these layers, there are some methods to remove; sandblasting, which is time-consuming and costly for industrial applications [52,53] and a preheating technique using a dual beam laser [54], plasma arc preheating [51], and a re-melting using laser on the

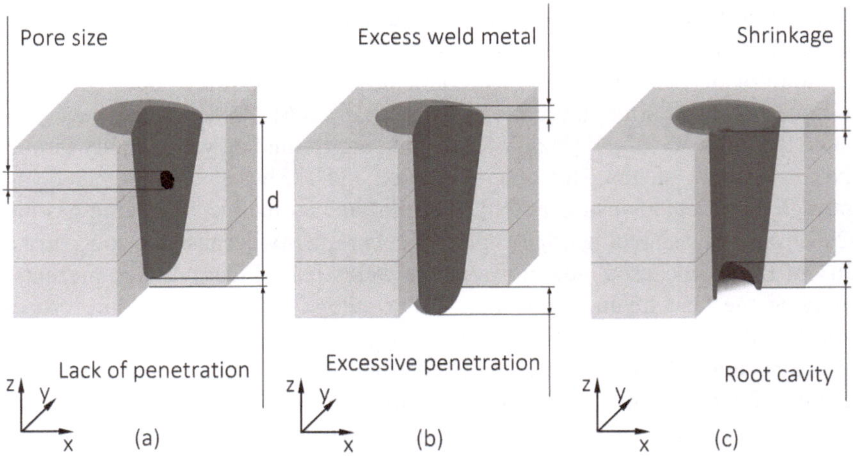


**FIGURE 12.6** Common laser welding defects according to ISO 13919: a) pore size, lack of penetration, b) excess weld metal and excessive penetration, c) incomplete filled groove and root cavity.

surface [55]. Cracks in LMW can be avoided using a strong clamping system as part of the experimental set [11] and sinusoidally modulated wave operating mode to improve the keyhole stability, resulting in no pore formation inside the melting pool [56].

In LMW, three zones are identified with different grain morphology: Fusion Zone (FZ), Heat Affected Zone (HAZ), and Base Material (BM). A narrow and columnar transition region between the FZ and BM is called the Partial Melting Zone (PMZ) [57,58]. The average grain sizes reported in the literature for the BM are 50 µm [57], 58 µm [9], and 10 µm [59]. The FZ presents refined equiaxed grains compared to the BM [59]. In the LW of AZ31, the finer grain in FZ, the higher strength of the weld, besides if the precipitates are intra-granular scattered hardening, is achieved [58]. Srinivasan et al. investigated AZ61 wire as filler material and reported a recovery in the laser weld's ductility, increasing the UTS's base material [60]. These results coincided with other tests performed with AZ90 alloy with a sheet of AZ91D with 1.3 mm in thickness; the filler material improved the mechanical properties of the weld and reduced the porosity levels in the FZ [61]. In the PMZ, dendritic grain growth is formed due to the high cooling rates; in this region, the grains have no preferred orientation [58,59]. The reported grain sizes for the FZ are 6 µm [62], 7 µm [9], and 12 µm [57,60]. For the PMZ, the reported value of region width is 50 µm [57, obtained with a heat input of 85.7 [J/mm]. Besides, the formation of the precipitate for AZ31 depends also on the cooling rates. The distribution of precipitates goes from intra-granular scattered to inter-granular packed as the cooling rates decrease. The precipitates formed during rapid cooling are nanosized with dimensions of 50 to 150 nm for Al8Mn5, 500 to 1500 nm for Mg17Al12, and

500 to 1500 nm for Mg17(Al Zn)12. Additionally, Mg17(Al Zn)12 is coarser in the FZ than in the BM [62]. The precipitates grow around Al8Mn5 particles rich in Mn [59]. Also, they are brittle, and their excessive concentrations in the weld could cause the fragilization of the joint [58]. Heat input H [J/mm] plays an essential role in the final microstructure of the weld. An increase in the heat input results in lower cooling rates, coarser grains in the FZ, and the distribution of precipitates are intergranular packed. In contrast, if scan velocity decreases and power remains constant, PMZ columnar bandwidth decreases [58]. When the cooling rate is low, the precipitates are accumulated at the grain boundary, leading to a microstructure with intergranular brittle packets [58]. Cooling rates have a strong influence on the final microstructure and mechanical properties. For this reason, it is essential to have a way to measure the cooling rates or control them. It is critical and valuable for the industry to have a monitoring system to detect defects in the laser welds in real time and nondestructively [52]. Another method of dynamic monitoring with a CCD camera and a spectrometer allows us to see the pore formation in overlap AZ31 LW in real-time. It also helps control the thermal stability along the welding direction [51,52]. The use of monitoring systems helped improve the thermal stability of LW and the quality of laser weld cost-effectively and reliably [52].

## 12.4 JOINING STACKING BASED LASER MICRO-SPOT WELDING (JS-LMSW) OF MAGNESIUM ALLOYS

Joining Staking – Laser Micro-Spot Welding (JS-LMSW) uses micro laser spot welding to stack layers of metal tubes to form a three-dimensional porous structure. Mainly it takes advantage of the high processing speed, high mechanical resistance, and reduced heat-affected zones of the laser micro welding process. Also, compared to laser powder bed fusion processes, the volume of material to be melted is reduced from entire layers to discrete spots. This reduction represents approximately 80% less volume, translated into less probability of defects in the final parts, such as wrapping and non-homogeneous solidified areas. The raw materials used for this process are tubes or sheets with a thickness below 1 mm, which impacts raw material handling. Figure 12.7 displays examples of micro-spot welds obtained in WE43 sheets with 0.25 mm thickness. It is observed that several pores were formed in the interlayer surface between the sheets. This pore formation has been reported by Harooni et al. in laser seam welds of AZ31 sheets in an overlap configuration. According to the results, this pore indicates the formation of magnesium hydroxide inside the micro-weld bead during the fusion process. The oxide layer mainly causes this pore formation at the faying surfaces of the sheets [51]. This type of defect can be avoided by preheating the material. Also, incomplete filled grooves are observed [54,55]. Shrinkage defects were identified in micro-spot welds produced with 210 W and 220 W of laser power. This defect indicates keyhole instabilities due to the lack of gap space between sheets during the fusion. According to Hong et al., a potential solution is to increase this gap space to improve the keyhole stability [63].

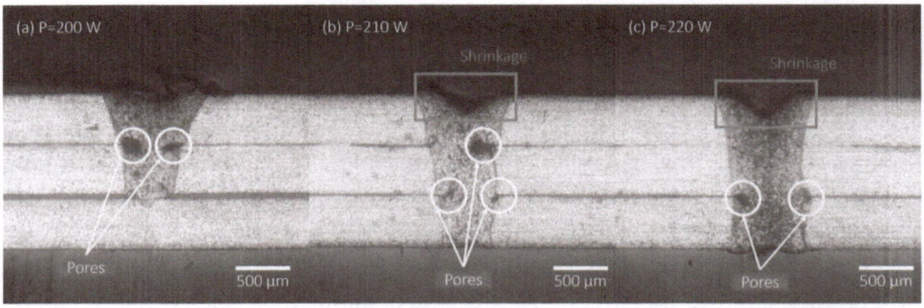


**FIGURE 12.7** Laser micro-spot welds obtained on WE43 sheets with 2 ms of exposure time and a) 200 W, b) 210 W, c) and 220 W of laser power.

Scaffolds with a fully interconnected porosity of 85% and dimensions 8 × 8 × 6 mm were manufactured with magnesium minitubes. The prototypes agreed with the design and demonstrated the feasibility of fabricating porous structures in a reliable and repeatable approach. After the mechanical tests, the obtained compressive properties were close to the human cancellous bone. Furthermore, the mechanical strength of the fabricated scaffolds indicated low risks of implant failure due to a lack of resistance or stress shielding [39]. The JS-LMSW is a pioneering method in the manufacture of porous magnesium scaffolds. Compared with the laser powder bed fusion process (LPBF) of magnesium alloys, the investigations are preliminary and oriented to study the influence of the process parameters. At the same time, its application in porous structures is still unexplored.

## REFERENCES

[1] Velasco MA, Narvaez-Tovar CA, Garzon-Alvarado DA (2015) Design, materials, and mechanobiology of biodegradable scaffolds for bone tissue engineering. *Biomed Res Int* 2015:729076. 10.1155/2015/729076

[2] Alvarez K, Nakajima H (2009) Metallic scaffolds for bone regeneration. *Materials* 2:790–832. 10.3390/ma2030790

[3] Salgado AJ, Coutinho OP, Reis RL (2004) Bone tissue engineering: State of the art and future trends. *Macromol Biosci* 4:743–765. 10.1002/mabi.200400026

[4] Bose S, Roy M, Bandyopadhyay A (2012) Recent advances in bone tissue engineering scaffolds. *Trends Biotechnol* 30:546–554. 10.1016/j.tibtech.2012.07.005

[5] Md Saad AP, Prakoso AT, Sulong MA, et al (2019) Impacts of dynamic degradation on the morphological and mechanical characterisation of porous magnesium scaffold. *Biomech Model Mechanobiol* 18:797–811. 10.1007/s10237-018-01115-z

[6] Wu S, Liu X, Yeung KWK, et al (2014) Biomimetic porous scaffolds for bone tissue engineering. *Materials Science & Engineering R* 80:1–36. 10.1016/j.mser.2014.04.001

[7] Cao X, Jahazi M, Immarigeon JP, Wallace W (2006) A review of laser welding techniques for magnesium alloys. *J Mater Process Technol* 171:188–204. 10.1016/j.jmatprotec.2005.06.068

[8] Demir A, Akman E, Canel T, et al (2007) Optimization of Nd: YAG laser welding of magnesium. *Journal of Laser Micro Nanoengineering* 2:108–113. 10.2961/jlmn.2007.01.0020
[9] Padmanaban G, Balasubramanian V (2010) Metallurgical characterization of pulsed current gas tungsten arc, friction stir and laser beam welded AZ31B magnesium alloy joints. *Mater Chem Phys* 125:686–697. 10.1016/j.matchemphys.2010.09.072
[10] Dhahri M, Masse JE, Mathieu JF, et al (2000) CO2 Laser Welding of Magnesium alloys. In: Proceedings of SPIE. Proceedings of SPIE, Osaka, pp 725–732.
[11] Zhu J, Li L, Liu Z (2005) CO2 and diode laser welding of AZ31 magnesium alloy. *Appl Surf Sci* 247:300–306. 10.1016/j.apsusc.2005.01.162
[12] Xie J (2013) Laser hermetic welding of implantable medical devices. In: *Joining and Assembly of Medical Materials and Devices*. Woodhead Publishing Limited, pp 211–235.
[13] Quan Y, Chen Z, Gong X, Yu Z (2008) CO2 laser beam welding of dissimilar magnesium-based alloys. *Materials Science and Engineering A* 496:45–51. 10.1016/j.msea.2008.04.065
[14] Ely KJ, Hall P, Zhou Y (2013) *Microwelding methods in medical components and devices*. Woodhead Publishing Limited.
[15] Quintino L, Costa A, Miranda R, et al. (2007) Welding with high power fiber lasers - A preliminary study. 28:1231–1237. 10.1016/j.matdes.2006.01.009
[16] Pequegnat A, Zhou Y (2013) Joining of platinum (Pt) alloy wires to stainless steel wires for electronic medical devices. In: *Joining and Assembly of Medical Materials and Devices*. Woodhead Publishing Limited, pp 154–177.
[17] Krasnoperov MY, Pieters RRGM, Richardson IM (2004) Weld pool geometry during keyhole laser welding of thin steel sheets. *Science and Technology of Welding and Joining* 9:501–506. 10.1179/136217104225021733
[18] Cedeño-Viveros LD, Vázquez-Lepe E, Rodríguez CA, García-López E (2021) Influence of process parameters for sheet lamination based on laser micro-spot welding of austenitic stainless steel sheets for bone tissue applications. *International Journal of Advanced Manufacturing Technology* 115:247–262. 10.1007/s00170-021-07113-3
[19] Li X, Liu X, Wu S, et al (2016) Design of magnesium alloys with controllable degradation for biomedical implants: From bulk to surface. *Acta Biomater* 46:2–30. 10.1016/j.actbio.2016.09.005
[20] Yazdimamaghani M, Razavi M, Vashaee D, et al (2017) Porous magnesium-based scaffolds for tissue engineering. *Material Science and Engineering C* 71:1253–1266. 10.1016/j.msec.2016.11.027
[21] Witte F (2015) Reprint of: The history of biodegradable magnesium implants: A review. *Acta Biomater* 23:S28–S40. 10.1016/j.actbio.2015.07.017
[22] García-López E, Ibarra-Medina JR, Siller HR, et al (2018) Surface finish and back-wall dross behavior during the fiber laser cutting of AZ31 magnesium alloy. *Micromachines (Basel)* 9:1–15. 10.3390/mi9100485
[23] Nuñez-Nava M, Vazquez E, Ortega-Lara W, et al (2021) An assessment of magnesium AZ31 coronary stents manufacture. *Mater Res Express* 8:. 10.1088/2053-1591/ac16f2
[24] Witte F, Reifenrath J, Müller PP, et al (2006) Cartilage repair on magnesium scaffolds used as a subchondral bone replacement. *Materwiss Werksttech* 37:504–508. 10.1002/mawe.200600027
[25] Sekar PSN, Desai V (2021) Recent progress in in vivo studies and clinical applications of magnesium based biodegradable implants – A review. *Journal of Magnesium and Alloys* 9:1147–1163.

[26] Hanas T, Sampath Kumar TS, Perumal G, et al (2017) Electrospun PCL/HA coated friction stir processed AZ31/HA composites for degradable implant applications. *Journal of Materials Processing Tech* 252:398–406. 10.1016/j.jmatprotec.2017.10.009
[27] Witte F, Ulrich H, Rudert M, Willbold E (2006) Biodegradable magnesium scaffolds: Part I: Appropriate inflammatory response. *Journal of Biomedical Materials Reasearch Part A* 33:97–103. 10.1002/jbm.a
[28] Witte F, Ulrich H, Palm C, Willbold E (2007) Biodegradable magnesium scaffolds: Part II: Peri-implant bone remodeling. *Journal of Biomedical Materials Reasearch Part A* 81:757–765.
[29] Reifenrath J, Palm C, Müller P, et al (2005) Subchondral plate reconstruction by fast degrading magnesium scaffolds influence cartilage repair in osteochondral defects. In: 51st Annual Meeting of the Orthopedic Research Society. p 1347.
[30] Yu W, Zhao H, Ding Z, et al (2016) In vitro and in vivo evaluation of MgF 2 coated AZ31 magnesium alloy porous scaffolds for bone regeneration. *Colloids Surf B Biointerfaces* 149:330–340. 10.1016/j.colsurfb.2016.10.037
[31] Zhuang H, Han Y, Feng A (2008) Preparation, mechanical properties and in vitro biodegradation of porous magnesium scaffolds. *Materials Science and Engineering C* 28:1462–1466. 10.1016/j.msec.2008.04.001
[32] Geng F, Tan L, Zhang B, et al (2009) Study on beta-TCP Coated Porous Mg as a Bone Tissue Engineering Scaffold Material. *J Mater Sci Technol* 25:123–129.
[33] Bobe K, Willbold E, Morgenthal I, et al (2013) In vitro and in vivo evaluation of biodegradable, open-porous scaffolds made of sintered magnesium W4 short fibres. *Acta Biomater* 9:8611–8623. 10.1016/j.actbio.2013.03.035
[34] Zhang X, Li X, Li J, Sun X (2013) Processing, microstructure and mechanical properties of biomedical magnesium with a specific two-layer structure. *Progress in Natural Science: Materials International* 23:183–189. 10.1016/j.pnsc.2013.03.006
[35] Seyedraoufi ZS, Mirdamadi S (2013) Synthesis, microstructure and mechanical properties of porous Mg-Zn scaffolds. *J Mech Behav Biomed Mater* 21:1–8. 10.1016/j.jmbbm.2013.01.023
[36] Lee J, Farag MM, Park EK, et al (2014) A simultaneous process of 3D magnesium phosphate scaffold fabrication and bioactive substance loading for hard tissue regeneration. *Materials Science & Engineering C* 36:252–260. 10.1016/j.msec.2013.12.007
[37] Zhang X, Li X-W, Li J-G, Sun X-D (2014) Preparation and mechanical property of a novel 3D porous magnesium scaffold for bone tissue engineering. *Materials Science & Engineering C* 42:362–367. 10.1016/j.msec.2014.05.044
[38] Jiang G, He G (2014) A new approach to the fabrication of porous magnesium with well-controlled 3D pore structure for orthopedic applications. *Materials Science & Engineering C* 43:317–320. 10.1016/j.msec.2014.07.033
[39] Cedeño-Viveros LD, Olivas-Alanis LH, Lopez-Botello O, et al (2022) A novel method for the fabrication of tubular WE43 magnesium scaffold based on laser micro-spot welding. *Engineering Science and Technology, an International Journal* 34:. 10.1016/J.JESTCH.2022.101096
[40] Bayraktar HH, Morgan EF, Niebur GL, et al (2004) Comparison of the elastic and yield properties of human femoral trabecular and cortical bone tissue. *J Biomech* 37:27–35. 10.1016/S0021-9290(03)00257-4
[41] Gibson Lorna J (1985) The mechanical behaviour of cancellous bone. *J Biomech* 18:317–328. 10.1016/0021-9290(85)90287-8

[42] Dhahri M, Masse JE, Mathieu JF, et al (2001) Laser welding of AZ91 and WE43 magnesium alloys for automotive and aerospace industries. *Adv Eng Mater* 3:504–507. 10.1002/1527-2648(200107)3:7<504::AID-ADEM504>3.0.CO;2-3

[43] Hong K-M, Shin YC (2017) Prospects of laser welding technology in the automotive industry: A review. *Journal of Materials Processing Tech* 245:46–69. 10.1016/j.jmatprotec.2017.02.008

[44] Ding W (2016) Opportunities and challenges for the biodegradable magnesium alloys as next-generation biomaterials. *Regen Biomater* 3:79–86.

[45] Chowdhury SM, Chen DL, Bhole SD, et al (2011) Microstructure and mechanical properties of fiber-laser-welded and diode-laser-welded AZ31 magnesium alloy. *Metallurgical and Materials Transactions A* 42:1974–1989. 10.1007/s11661-010-0574-y

[46] Scintilla LD, Tricarico L, Brandizzi M, Satriano AA (2010) Nd: YAG laser weldability and mechanical properties of AZ31 magnesium alloy butt joints. *Materials Processing Technology* 210:2206–2214. 10.1016/j.jmatprotec.2010.08.005

[47] ASTM (2014) ASTM B296–03 Temper Designations of Magnesium Alloys, Cast and Wrought.

[48] Ishak M, Yamasaki K, Maekawa K (2010) Microstructure and corrossion behavior of laser welded Magnesium Alloys with Silver Nanoparticles. *Mechanical and Mechatronics Engineering* 7:853–858.

[49] Ishak M, Maekawa K, Yamasaki K (2011) The characteristics of laser welded magnesium alloy using silver nanoparticles as insert material. *Materials Science and Engineering A* 536:143–151. 10.1016/j.msea.2011.12.092

[50] ISO (2019) ISO-13919 - 2: Electron and laser-beam welded joints. Requirements and recommendations on quality levels for imperfections. *Aluminium, magnesium and their alloys and pure copper.*

[51] Harooni M, Carlson B, Strohmeier BR, Kovacevic R (2014) Pore formation mechanism and its mitigation in laser welding of AZ31B magnesium alloy in lap joint configuration. *Mater Des* 58:265–276. 10.1016/j.matdes.2014.01.050

[52] Harooni M, Carlson B, Kovacevic R (2014) Detection of defects in laser welding of AZ31B magnesium alloy in zero-gap lap joint configuration by a real-time spectroscopic analysis. *Opt Lasers Eng* 56:54–66. 10.1016/j.optlaseng.2013.11.015

[53] Sahul M, Sahul M, Lokaj J (2016) Effect of surface layer on the properties of AZ31 magnesium alloy welded joints. *Proceedings of Materials Today* 3:1150–1155. 10.1016/j.matpr.2016.03.015

[54] Harooni M, Carlson B, Kovacevic R (2013) Dual-beam laser welding of AZ31B magnesium alloy in zero-gap lap joint configuration. *Opt Laser Technol* 56:247–255. 10.1016/j.optlastec.2013.08.018

[55] Harooni M, Ma J, Carlson B, Kovacevic R (2015) Two-pass laser welding of AZ31B magnesium alloy. *Materials Processing Technology* 216:114–122. 10.1016/j.jmatprotec.2014.08.028

[56] Zhang L, Zhang X, Ning J (2014) Modulated fiber laser welding of high reflective AZ31. *Advanced Manufacturing Technology.* 10.1007/s00170-014-6303-8

[57] Wang Z, Gao M, Tang H, Zeng X (2011) Characterization of AZ31B wrought magnesium alloy joints welded by high power fiber laser. *Mater Charact.* 10.1016/j.matchar.2011.07.002

[58] Quan Y, Chen Z, Gong X, Yu Z (2008) Effects of heat input on microstructure and tensile properties of laser welded magnesium alloy AZ31. *Mater Charact* 59:1491–1497. 10.1016/j.matchar.2008.01.010

[59] Commin L, Dumont M, Rotinat R, et al (2010) Texture evolution in Nd: YAG-laser welds of AZ31 magnesium alloy hot rolled sheets and its influence on mechanical properties. *Materials Science and Engineering A* 528:2049–2055. 10.1016/j.msea.2010.11.061

[60] Srinivasan PB, Riekehr S, Blawert C, et al (2011) Mechanical properties and stress corrosion cracking behaviour of AZ31 magnesium alloy laser weldments. *Transactions of Nonferrous Metals Society of China* 21:1–8. 10.1016/S1003-6326(11)60670-5

[61] Wahba M, Mizutani M, Kawahito Y, Katayama S (2012) Laser welding of die-cast AZ91D magnesium alloy. *Mater Des* 33:569–576. 10.1016/j.matdes.2011.05.016

[62] Coelho RS, Kostka A, Pinto H, et al (2008) Microstructure and mechanical properties of magnesium alloy AZ31B laser beam welds. *Materials Science and Engineering A* 485:20–30. 10.1016/j.msea.2007.07.073

[63] Hong KM, Shin YC (2017) The effects of interface gap on weld strength during overlapping fiber laser welding of AISI 304 stainless steel and AZ31 magnesium alloys. *International Journal of Advanced Manufacturing Technology* 90:3685–3696. 10.1007/s00170-016-9681-2

# 13 Mixing Enhancement in Low Amplitude Wavy Channel Micromixers (LAWC-M) with Different Obstacles

*Ranjitsinha R. Gidde, Amarjit P. Kene, and Rajni Patel*

## 13.1 INTRODUCTION

The microfluidic plays a significant role in numerous applications, including Lab-On-a-Chip (LOC), Micro-Total-Analysis-System (μTAS), Point of Care Testing (POCT), etc. In these applications, micromixers are used for mixing applications as mixing is an essential phenomenon for analysis in microfluidic devices. The efficient and rapid mixing of fluids in microfluidic devices is important. Hence, the design of micromixers with high mixing efficiency is essential, which helps in rapid analysis. There are two main categories of micromixers: Passive and active. In the passive micromixer, the geometrical modifications are used to enhance the mixing, and in the active micromixers, external perturbations are used to improve the mixing. Due to the complex design of active micromixers, passive micromixers are preferred as a better option for the mixing process due to their simplicity in integrating with microfluidic devices. Mixing at the microscale is achieved mainly by molecular diffusion.

Many researchers have proved that the microchannels' shape can help enhance interfacial contact and mixing efficiency [1]. Qualitative and quantitative flow and mixing analysis using computational fluid dynamics is reliable, and it also helps to visualize flow structures, species concentration, and mixing performance [2]. The mixing performance of the micromixer with serpentine microchannel and semicircular obstacle was investigated using computational fluid dynamics. The simulation and experimental results revealed that the obstacles help in enhancing the mixing [3]. The mixing enhancement in convergent and divergent sinusoidal microchannels with different ratios of amplitude and wavelength was carried out through computational and

DOI: 10.1201/9781003364948-13

experimental approaches. The study results showed that the amplitude ratio to the wavelength significantly impacts the mixing performance [4]. Trapezoidal-Zigzag Micromixer (TZM) helps to create multiple mixing mechanisms, namely, fluid twisting, transversal flow, vortices, and chaotic advection [5]. The micromixer having a fractal-like tree network micromixer based on Murray's law was investigated for the effect of hierarchical structure on mixing efficiency. The results showed that increasing the hierarchical levels increases the mixing efficiency [6]. Computational fluid dynamics and a response surface approach were proposed to optimize the grooved micromixer with obstructions. The micromixer designs with circular and diamond-shaped obstructions showed better performance [7]. A novel 3-D T-Junction passive micromixer design with helicoidal flows was studied. The increased mixing index and performance were obtained using a novel design [8].

Nimafar et al. and Viktorov et at. [9,10] performed a number of experiments and simulations to analyze the performance characteristics of passive micromixers and found that the H-micromixer's efficiency is very high due to the SAR process. Sadegh et al. [11] studied planar micromixers to identify an optimum micromixer having short mixing length. The study results revealed that the chamber geometry manifests the mixing efficiency and the pressure drop at low Reynolds number. However, the obstacle geometry is significant with respect to the mixing efficiency and the pressure drop at a high Reynolds number. The micromixer with triangular baffles was proposed. The designed micromixer showed enhanced mixing due to the split and recombination of streams and vortex generation [12]. The passive H-C micromixer, which uses folding, rotation, expansion, and construction to improve mixing, was investigated for a wide range of Reynolds number. The study results showed that the H-C micromixer exhibited a mixing efficiency greater than the tear-drop and chain micromixers [13]. Wong et al. and Cortes-Quiroz et al. [14,15] conducted experiments to study the mixing characteristics in T-Junction Straight Microchannel. They observed that rapid mixing is achieved by generating secondary flow, swirling flow, and vortices in the micromixer.

Numerous works have been carried out using numerical and experimental approaches, demonstrating the influences of the modification in channel geometry on the mixing characteristics of passive micromixers. Mixing at a small scale depends mainly on molecular diffusion; a relatively long channel is necessary to achieve the desired mixing. A simple and innovative design of a T-junction channel with a wavy structure has been proposed, which triggers secondary flow for enhanced chaotic mixing. Accordingly, a comparative study for T-junction straight channel, wavy channel, and wavy channels with different obstacles has been carried out. CFD has been used to simulate the flow and mixing phenomenon inside the microchannel. The effect of channel length, obstacle geometry, and channel shape on the mixing efficiency and the pressure drop has been studied both qualitatively and quantitatively.

## 13.2 MICROCHANNEL GEOMETRY

The schematics of T-junction straight channel, T-junction wavy channels, and T-junction wavy channels with different obstacles are shown in Figure 13.1. The three types of obstacles namely rectangular, semicircular, and triangular, with an aspect ratio of 2, are used to produce secondary flow and enhance the mixing process. The dimensions of the channel and obstruction are depicted in Table 13.1, and distances of planes at different distances from the confluence points are shown in Table 13.2.

## 13.3 NUMERICAL ANALYSIS AND EVALUATION SCHEME

The numerical analysis was carried out using the Computational Fluid Dynamics (CFD) software COMSOL Multiphysics.

### 13.3.1 Governing Equations

**Fluid flow:** The governing equations, the Navier-Stokes and the Continuity equations, Eqs. (13.1) & (13.2) were used for modelling the fluid flow phenomenon [16,17].

$$\nabla u = 0 \tag{13.1}$$

$$\rho\left[\frac{\partial u}{\partial t} + (u.\ \nabla)u\right] = -\nabla p + \mu\nabla^2 u \tag{13.2}$$

Where u is the fluid velocity in m/s, p is the fluid pressure in Pa, ρ is the fluid density in $kg/m^3$, and μ is the fluid viscosity in Pa.s.

**Mass transport:** For modelling the mass transport of fluids with constant viscosity and density, the convection–diffusion equation, Eq. (13.3), was used [18].

$$\frac{\partial c}{\partial t} = D\nabla^2 c - u.\ \nabla c \tag{13.3}$$

Where D is the diffusivity in $m^2/s$ and c is the molar concentration $mol/m^3$.

### 13.3.2 Boundary Conditions

At inlet 1, a solution in the form of pure water with a concentration of 0 $mol/m^3$ was employed. In inlet 2, a solute in the form of ethanol with a concentration of 1 $mol/m^3$ was used. The density and the dynamic viscosity of the water are $9.998 \times 10^2$ $kg/m^3$ and $0.9 \times 10^{-3}$ Pa.s, respectively, employed for the simulation. The ethanol's density and dynamic viscosity are $7.890 \times 10^2$ $kg/m^3$ and

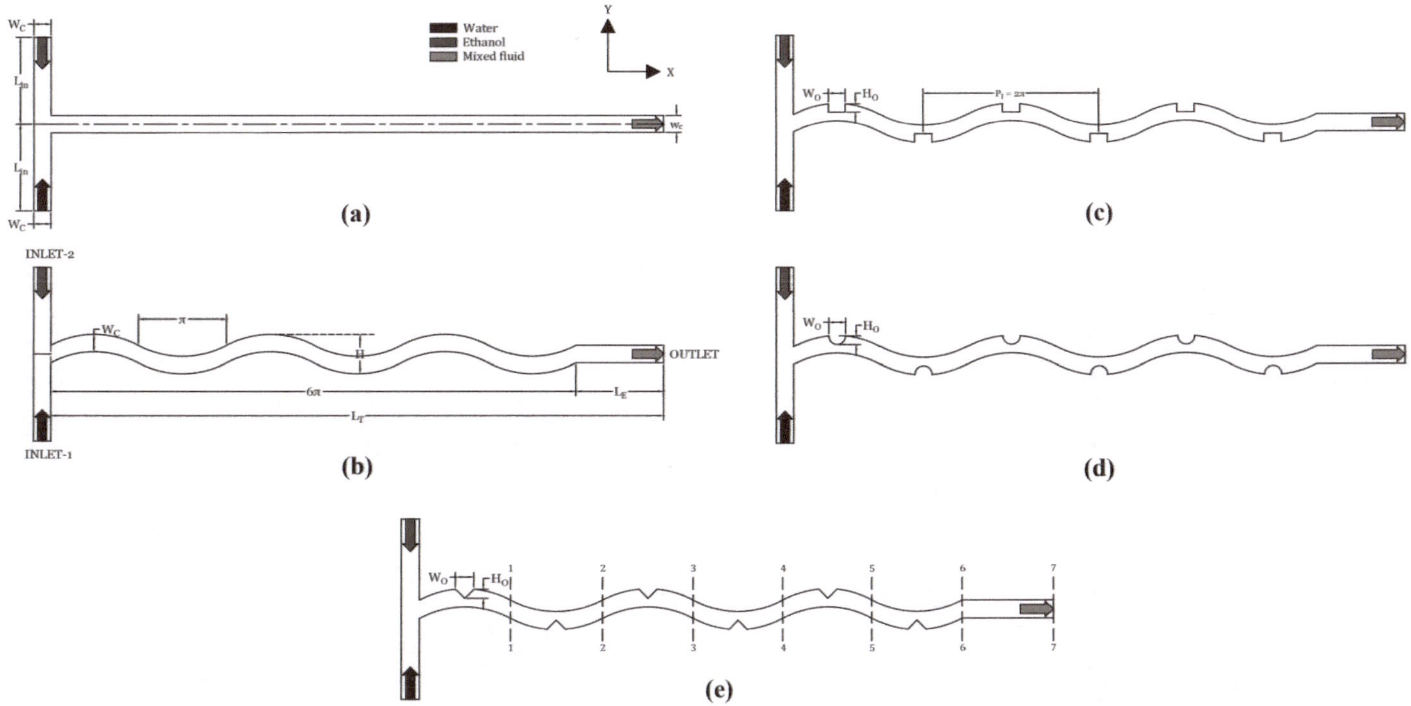


**FIGURE 13.1** Schematics of T-junction straight channel and Low Amplitude Wavy Channel micromixers (LAWC-M) (a) T-Junction straight channel, (b) Wavy channel, (c) Wavy channel with rectangular obstacles, (d) Wavy channel with semi-circular obstacles and (e) Wavy channel with triangular obstacles.

**TABLE 13.1**
**Dimensions of the T-junction Straight and Wavy Channels**

| Dimension | $L_{in}$ | $W_C$ | $L_E$ | $L_T$ | $f_{wavy}$ | $P_l$ | $W_O$ | $H_O$ |
|---|---|---|---|---|---|---|---|---|
| Size | 2 | 400 | 2 | 14–18 | 6π–8π | 2π | 400 | 200 |
| Unit | mm | μm | mm | mm | – | – | μm | μm |

**TABLE 13.2**
**Distances of y-z Planes from the Confluence Point**

| Designation of plane | 1–1 | 2–2 | 3–3 | 4–4 | 5–5 | 6–6 | 7–7 | 8–8 | 9–9 | 10–10 |
|---|---|---|---|---|---|---|---|---|---|---|
| Distance in mm | 0 | 2 | 4 | 6 | 8 | 10 | 12 | 14 | 16 | 18 |

*Note:* 1–1 to 8–8 for wavy channels with $f_{wavy} = 6\pi$; 1–1 to 10–10 for wavy channels with $f_{wavy} = 8\pi$.

$1.2 \times 10^{-3}$ Pa.s, respectively. The diffusivity of the ethanol in water is $1.2 \times 10^{-9}$ $m^2/s$. The boundary conditions employed at the inlets and the outlet were velocity and pressure, respectively. The no-slip boundary condition was applied to the channel walls [19,20].

### 13.3.3 Evaluation Scheme

The micromixer designs were evaluated using the two major performance characteristics, i.e., the mixing efficiency and the pumping power. The mixing quality in the form of mixing efficiency at the desired plane (which is perpendicular to the direction of fluid flow) was calculated using standard deviation and maximum standard deviation. Eq. (13.4) and Eq. (13.5) were used to calculate the mixing efficiency [21].

$$\sigma = \sqrt{\frac{1}{N_S}\sum_{i=1}^{N_S}(C_i - C_{opt})^2} \tag{13.4}$$

$$\eta_m = \left(1 - \frac{\sigma}{\sigma_{max}}\right) \times 100 \tag{13.5}$$

Where $\sigma$, $N_S$, $C_i$, $C_{opt}$, $\sigma_{max}$ and $\eta_m$ denote the standard deviation, the number of sampling points, the mass fraction of $i$th sampling point, the optimal mass fraction, the maximum standard deviation and the mixing efficiency, respectively. Note that the optimal mass fraction, $C_{opt,}$ for the completely mixed fluids is 50 % of the concentration of the solute, and it is 0.5 mol/m$^3$ (in our case), and $\sigma_{max}$ is the maximum standard deviation is the initial unmixed state in the micromixer (0.5 in our case). Note that 0% and 100% mixing efficiencies indicate perfect segregation and perfect mixing, respectively.

Another vital performance characteristic is the pressure drop. This is the difference between the area-weighted average of the total pressures at the inlet cross-sectional plane ($P_{in}$) in Pa and the outlet cross-sectional plane ($P_{out}$) in Pa, respectively. It is calculated using Eq. (13.6) [22].

$$\Delta P = P_{In} - P_{Out} \tag{13.6}$$

## 13.4 RESULTS AND DISCUSSION

The COMSOL Multiphysics was used for simulating the flow and mixing fields for a wide range of Reynolds numbers from 0.1 to 75. The grid independency study was carried out for the representative Channel geometry.

### 13.4.1 Qualitative Analysis: Mixing Field/Concentration Plots

#### 13.4.1.1 T-junction Straight Channel and Wavy Channel: Channel Length = 14 mm; Wavy Frequency = 6π

Mixing field analysis was carried out using concentration plots on y-z planes along the flow direction for T-junction channel and wavy channel with wavy frequency of 6π. The concentration plots for T-junction straight and wavy channels were studied to assess the mixing quality. The ethanol mass concentration plots on eight y-z planes for T-junction straight channel and wavy channel without obstacles are shown in Figure 13.2. From Figure 13.2(a), it can be seen the ethanol mass concentration plot for plane 8–8 is uniform at Re of 0.1. This is due to diffusion dominated mixing. However, at Re > 0.1, interface between the water and ethanol can clearly identified. This is due to the less resident time with increase in the Reynolds number. From Figure 13.2(b), due to wavy channel, the nature of the interface between the water and ethanol is influenced as a result of secondary flow at Re > 10. Here, the chaotic advection dominated mixing is observed and it increases with increase in the Reynolds number. In the case of wavy channels with obstacles, a geometrical change influences mixing mechanism beyond Re = 10.

#### 13.4.1.2 T-Junction Wavy Channels with Different Obstacles: Wavy Frequency = 6π

Figure 13.3 represents ethanol mass fraction plots for wavy channels of wavy frequency of 6π and different obstacles on its wall. It can be seen

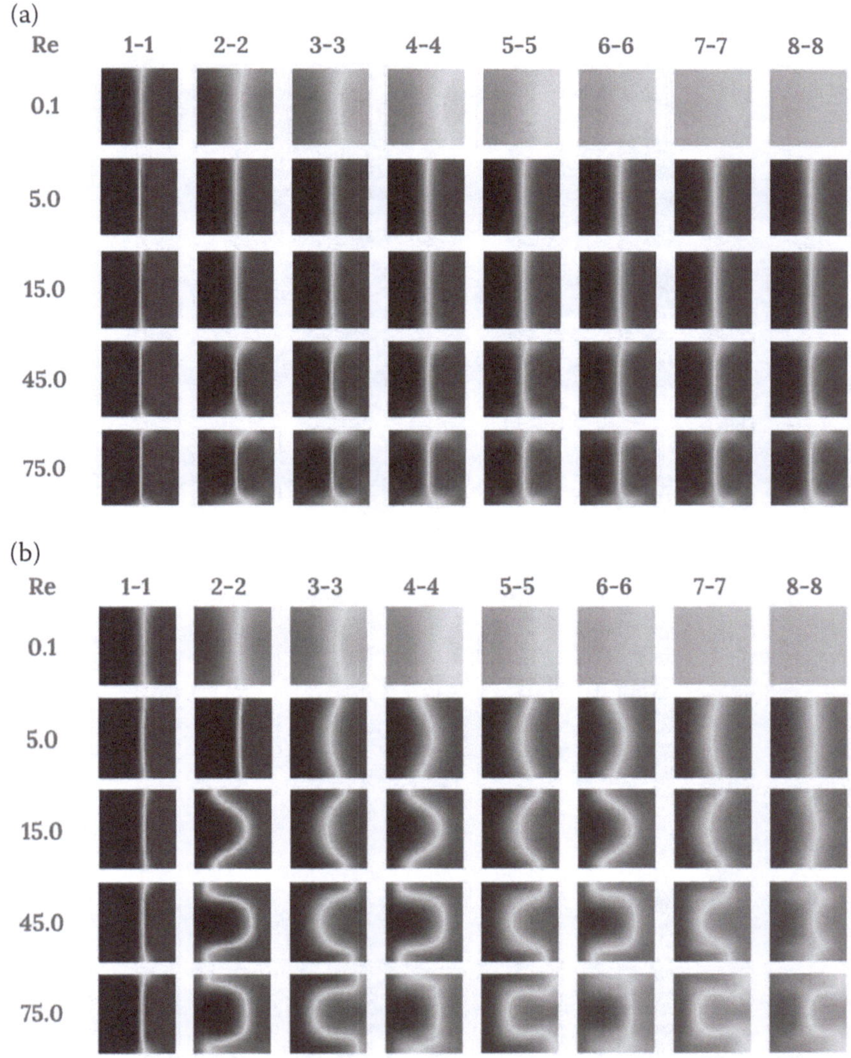


**FIGURE 13.2** Ethanol mass fraction plots on y-z planes along the direction of the flow for (a) T-Junction Straight Channel, and (b) T-Junction Wavy Channel at Reynolds numbers 0.1, 5.0, 15.0, 45.0, and 75.0 for channel length 14 mm.

from these plots, at Re = 0.1, the diffusion dominated mixing observed for all the wavy channel with obstacles. However, at Re ≥ 5, the mixing is chaotic advection dominated for all wavy channel with obstacles. In case of wavy channels with obstacles, a geometrical change influences mixing mechanism beyond Re = 5.

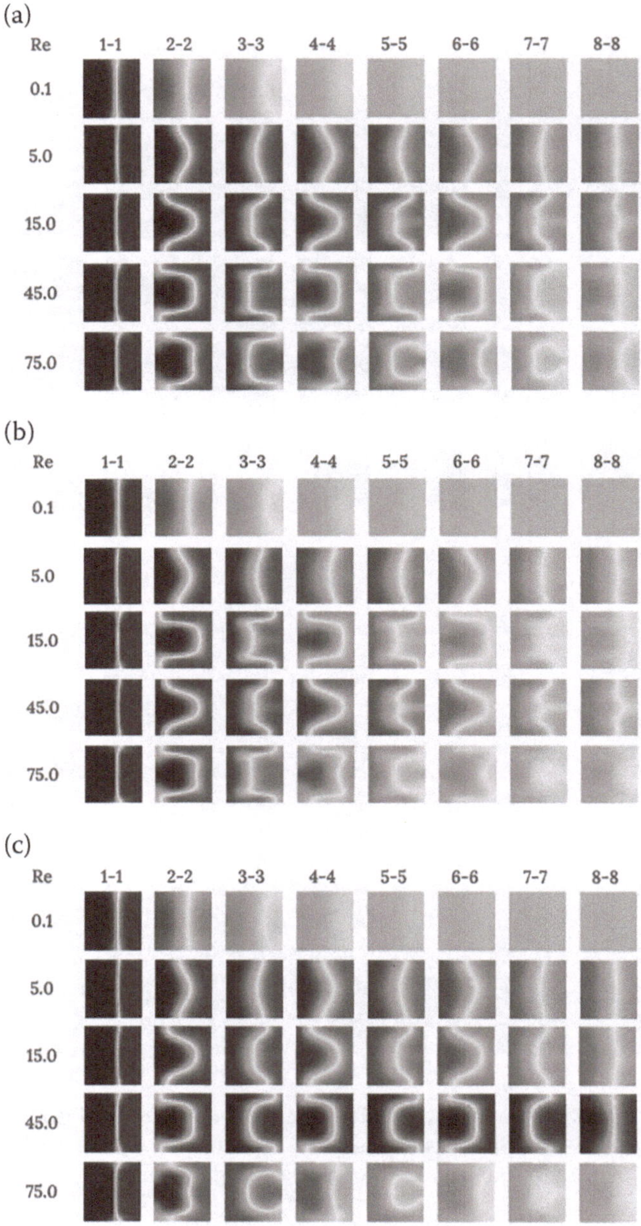


**FIGURE 13.3** Ethanol mass fraction plots on y-z planes along the direction of the flow for wavy channel with obstacles at Reynolds numbers 0.1, 5.0, 15.0, 45.0 and 75.0 for channel length of 14 mm (a) Rectangular obstacles, (b) Semi-circular obstacles, and (c) Triangular obstacles.

#### 13.4.1.3 T-Junction Straight Channel and Wavy Channel: Channel Length = 18 mm; Wavy Frequency = 8π

The mixing field analysis was carried out using concentration plots on y-z planes along the flow direction for T-junction channel and wavy channel with wavy frequency of $8\pi$. The ethanol mass concentration plots on ten y-z planes for T-junction straight channel and wavy channel without obstacles are shown in Figure 13.4. From Figure 13.4(a), it can be seen the ethanol mass concentration plot for planes 8–8 to 10–10 is uniform and same in nature at Re of 0.1. However, at Re > 0.1, interface between the water and ethanol can clearly

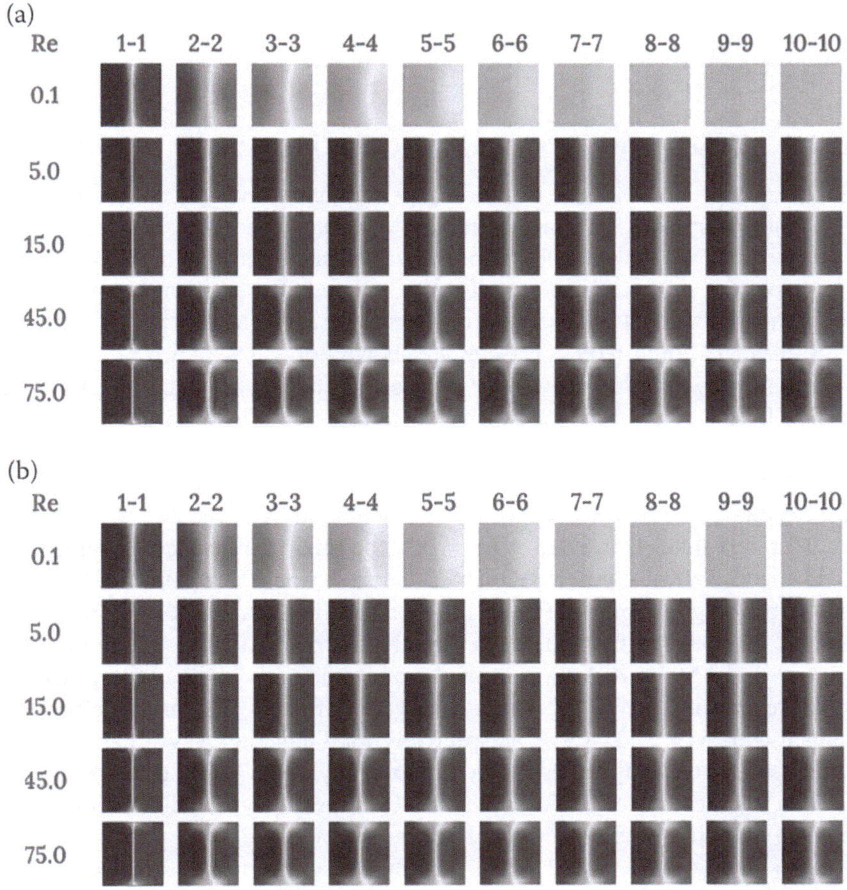


**FIGURE 13.4** Ethanol mass fraction plots on y-z planes along the direction of the flow for (a) Simple T-shaped microchannel, and (b) T-shaped wavy microchannel at Reynolds numbers 0.1, 5.0, 15.0, 45.0 and 75.0 for channel length 18 mm.

identified. This is due to the lower resident time with increase in the Reynolds number. From Figure 13.4(b), due to wavy channel, the nature of the interface between the water and ethanol is influenced as a result of secondary flow at Re > 10. At Re > 10, the mixing process mainly depends on the chaotic advection inside channel.

#### 13.4.1.4 T-Junction Wavy Channels with Different Obstacles: Wavy Frequency = 8π

Figure 13.5 represents ethanol mass fraction plots for wavy channels with wavy frequency of 8π and different obstacles on its wall. At Re = 0.1, it can be seen from ethanol mass fraction plots that the diffusion dominated mixing observed for all the wavy channel with obstacles. However, at Re ≥ 5, the mixing mechanism changes from diffusion to the chaotic advection for all wavy channel with obstacles.

### 13.4.2 Quantitative Analysis: Mixing Efficiency and Pressure Drop

The quantitative analysis for the mixing performance was carried out. The two significant characteristics, namely mixing efficiency and pressure drop, were evaluated as a function of the length of the wavy channel and the Reynolds number and shape of the obstacle. The results of the mixing efficiency and pressure drop for T-junction straight channel micromixers and wavy channel micromixer and wavy channel with different obstacles are shown in Figures 13.6, 13.7.

From the Figure 13.6(a), it can be seen that Re = 15 is critical Reynold number for the T-junction straight channel. At Re = 15, the mixing efficiency of 12.65% and 14.30% can be attained for T-junction straight channel with 14 mm and 16 mm lengths, respectively. At Re = 75, mixing efficiency attained is 22.41% and 23.35%, respectively for T-junction straight channel with 14 mm and 16 mm length. However, at Re = 0.1, the mixing efficiencies attained are 95.43% and 98.03%, respectively for T-junction straight channel with 14 mm and 16 mm length. From the Figure 13.6(b), it can be seen that Re = 10 is critical Reynold number for the T-junction straight channel. At Re = 10, the mixing efficiency of 33.03% and 39.15% can be attained for T-junction wavy channel wavy frequencies of 6π and 8π, respectively. At Re = 75, mixing efficiency attained is 62.53% and 71.91%, respectively for T-junction wavy channel wavy frequencies of 6π and 8π, respectively. However, at Re = 0.1, the mixing efficiencies attained are 96.82% and 98.80%, respectively for T-junction wavy channel wavy frequencies of 6π and 8π, respectively.

From the Figure 13.6(c), it can be seen that Re = 5 is critical Reynold number for the T-junction straight channel. At Re = 5, the mixing efficiency of 57.80% and 66.60% can be attained for T-junction wavy channels and rectangular obstacles for channel with wavy frequency of 6π and 8π, respectively. At Re = 75, mixing efficiency attained is 74.61% and 82.76%, respectively for T-junction wavy channels and rectangular obstacles. However, At Re = 0.1, the mixing efficiencies attained are 99.03% and 99.74%, respectively for T-junction wavy

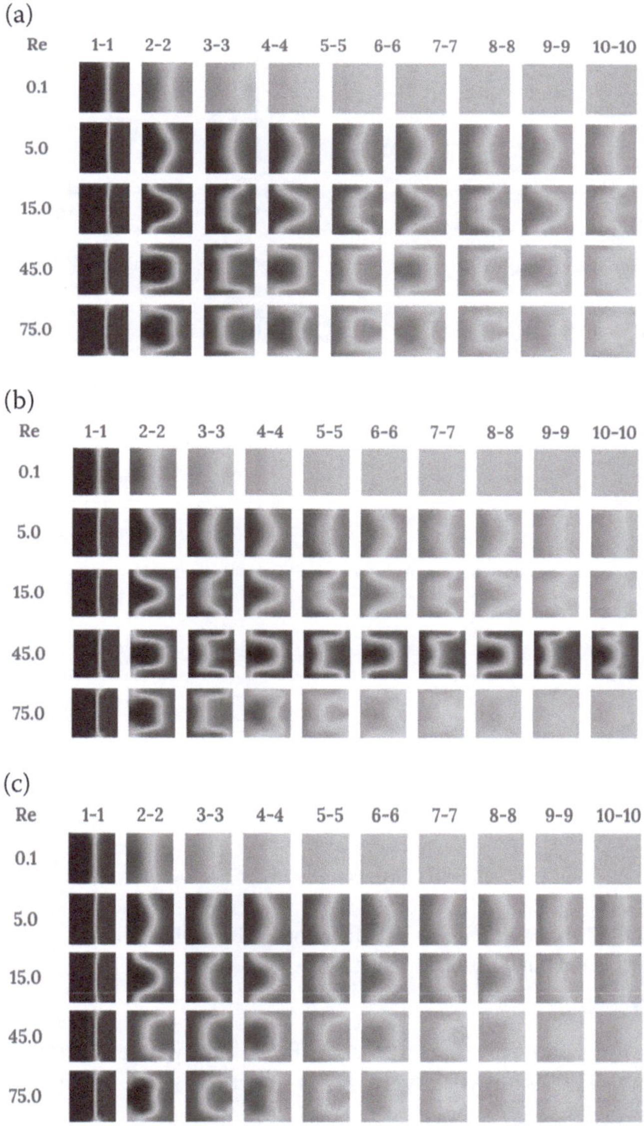


**FIGURE 13.5** Ethanol mass fraction plots on y-z planes along the direction of the flow for wavy Channel with obstacles at Reynolds numbers 0.1, 5.0, 15.0, 45.0 and 75.0 for channel length of 18 mm (a) Rectangular obstacles, (b) Semi-circular obstacles, and (c) Triangular obstacles.

channel wavy frequencies of 6π and 8π, respectively. From Figure 13.6(d), it can be seen that Re = 5 is critical Reynold number for the T-junction straight channel. At Re = 5, the mixing efficiency of 63.98% and 72.65% can be attained for T-junction wavy channels and semi-circular obstacles for channel with wavy

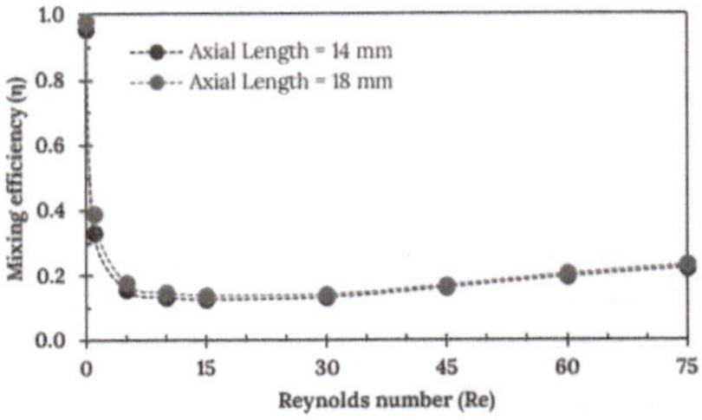


(a) T-junction straight channel

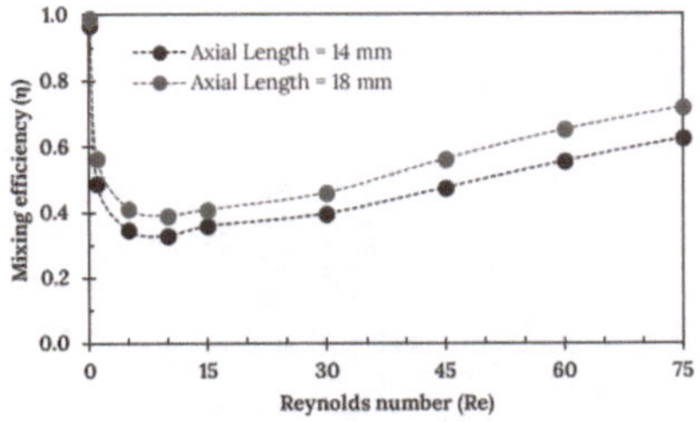


(b) T-junction wavy channel

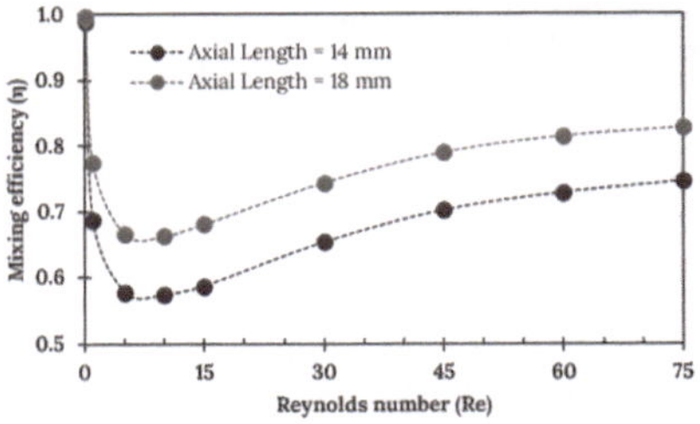


(c) T-junction wavy channel with rectangular obstacles

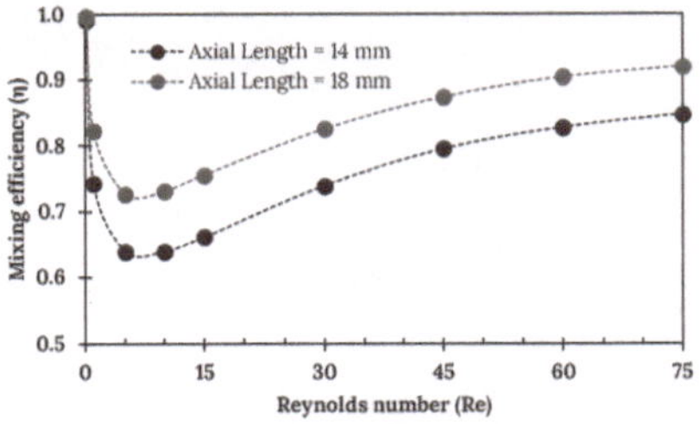


(d) T-junction wavy channel with semi-circular obstacles

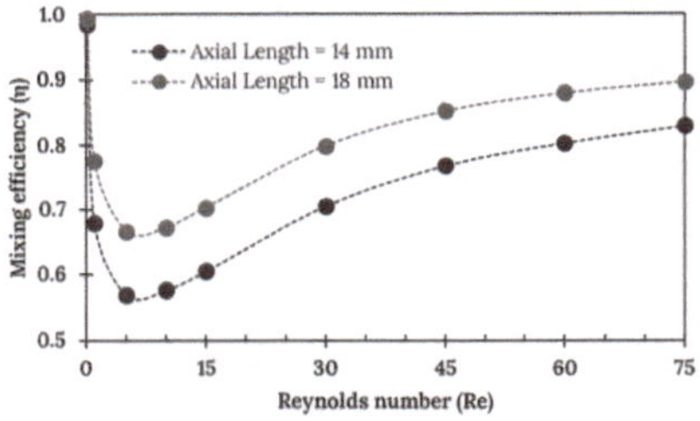


(e) T-junction wavy channel with triangular obstacles

**FIGURE 13.6** Mixing efficiency as a function of Reynolds number, length, and wavy frequency of channel and obstacle shape.

frequency of 6π and 8π, respectively. At Re = 75, mixing efficiency attained is 84.76% and 91.99%, respectively for T-junction wavy channels and semi-circular obstacles. However, At Re = 0.1, the mixing efficiencies attained are 99.11% and 99.76%, respectively for T-junction wavy channel wavy frequencies of 6π and 8π, respectively. From Figure 13.6(e), it can be seen that Re = 5 is critical Reynold number for the T-junction straight channel. At Re = 5, the mixing efficiency of 56.96% and 66.69% can be attained for T-junction wavy channels and triangular obstacles for channel with wavy frequency of 6π and 8π, respectively. At Re = 75, mixing efficiency attained is 82.98% and 89.67%, respectively for T-junction wavy channels and triangular obstacles. However, at Re = 0.1, the mixing efficiencies attained are 98.61% and 99.79%, respectively for T-junction wavy channel wavy frequencies of 6π and 8π, respectively.

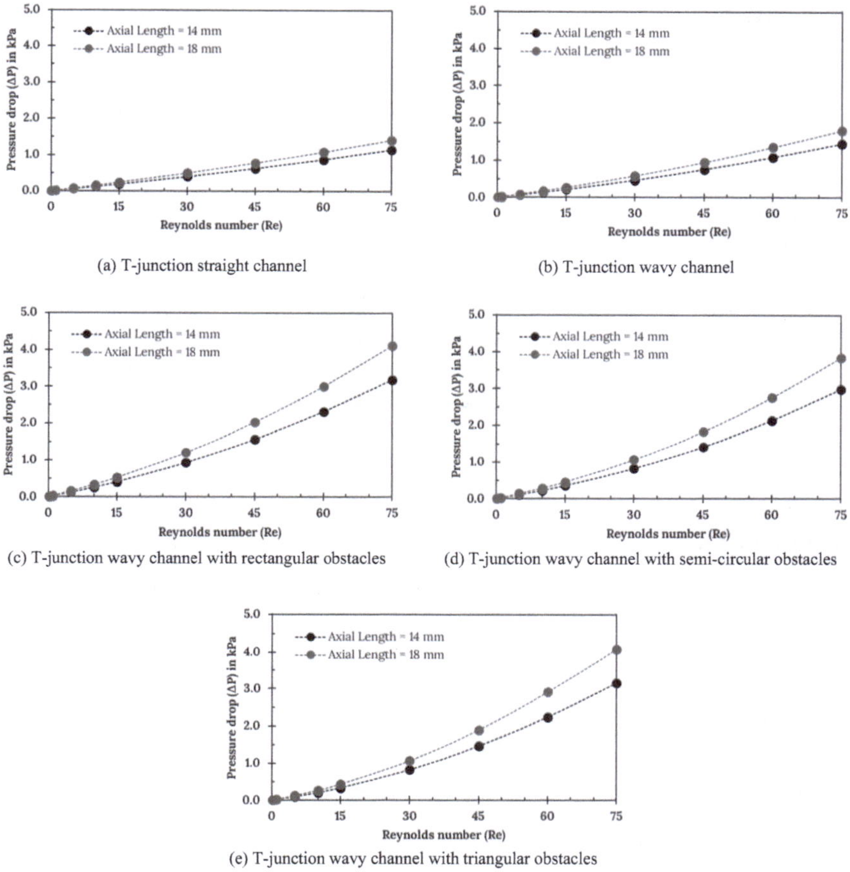


**FIGURE 13.7** Pressure drop as a function of Reynolds number, length, and wavy frequency of channel and obstacle shape.

## 13.5 CONCLUSIONS

A comparative analysis of the T-Junction straight channel, wavy channel and wavy channel with different obstacles has been carried out for a wide range of Reynolds numbers from 0.1 to 75. The mixing efficiency and pressure drop have been analyzed. The conclusions drawn from the proposed study are as follows:

1. Compared with the T-Junction straight micromixer of channel lengths 14 mm and 18 mm, the mixing efficiency of T-Junction wavy micromixers at Re = 75 is increased by 64.14% and 68.83% for wavy channels of $6\pi$ and $8\pi$, respectively. Compared with the T-Junction wavy micromixer, the mixing efficiency of T-Junction wavy micromixers with semi-circular

obstacles at Re = 75 is increased by 26.22% and 21.22% for wavy frequencies of 6π and 8π, respectively.

2. Compared with the T-Junction straight micromixers of channel lengths 14 mm and 18 mm, the pressure drop of T-Junction wavy micromixers increased by 20.55% and 21.75% for wavy frequencies of 6π and 8π, respectively. Compared with the T-Junction wavy micromixer, the pressure drop of T-Junction wavy micromixers with semi-circular obstacles increased by 51.45% and 53.03% for wavy frequencies of 6π and 8π, respectively.
3. The wavy micromixer with semi-circular obstacles can induce better mixing performance. The values of mixing efficiency achieved at 0.1 are 99.11 and 99.76 for wavy channel of wavy frequency 6π and 8π, respectively. However, the values of mixing efficiency achieved at Re = 75 are 84.76% and 91.99% for wavy frequencies of 6π and 8π, respectively.

## REFERENCES

[1] Chen, X., Li, T., & Hu, Z. (2017). A novel research on serpentine microchannels of passive micromixers. *Microsystem Technologies*, 23(7), 2649–2656. 10.1007/s00542-016-3060-7

[2] Chen, X., Zhang, Z., Yi, D., & Hu, Z. (2017). Numerical studies on different two-dimensional micromixers basing on a fractal-like tree network. *Microsystem Technologies*, 23(3), 755–763. 10.1007/s00542-015-2742-x

[3] Wangikar, S. S., Patowari, P. K., & Misra, R. D. (2018). Numerical and experimental investigations on the performance of a serpentine microchannel with semicircular obstacles. *Microsystem Technologies*, 24(8), 3307–3320. 10.1007/s00542-018-3799-0

[4] Khosravi Parsa, M., Hormozi, F., & Jafari, D. (2014). Mixing enhancement in a passive micromixer with convergent-divergent sinusoidal microchannels and different ratio of amplitude to wave length. *Computers and Fluids*, 105, 82–90. 10.1016/j.compfluid.2014.09.024

[5] The, H. Le, Ta, B. Q., Thanh, H. Le, Dong, T., Thoi, T. N., & Karlsen, F. (2015). Geometric effects on mixing performance in a novel passive micromixer with trapezoidal-zigzag channels. *Journal of Micromechanics and Microengineering*, 25(9), 94004. 10.1088/0960-1317/25/9/094004

[6] Xu, J., & Chen, X. (2019). Mixing performance of a fractal-like tree network micromixer based on Murray's law. *International Journal of Heat and Mass Transfer*, 141, 346–352. 10.1016/j.ijheatmasstransfer.2019.06.070

[7] Rahmannezhad, J., & Mirbozorgi, S. A. (2019). CFD analysis and RSM-based design optimization of novel grooved micromixers with obstructions. *International Journal of Heat and Mass Transfer*, 140, 483–497. 10.1016/j.ijheatmasstransfer.2019.05.107

[8] Okuducu, M. B., & Aral, M. M. (2019). Novel 3-D T-Junction passive micromixer design with Helicoidal Flows. *Processes*, 7(9). 10.3390/pr7090637

[9] Nimafar, M., Viktorov, V., & Martinelli, M. (2012). Experimental comparative mixing performance of passive micromixers with H-shaped sub-channels. *Chemical Engineering Science*, 76, 37–44. 10.1016/j.ces.2012.03.036

[10] Viktorov, V., & Nimafar, M. (2013). A novel generation of 3D SAR-based passive micromixer: Efficient mixing and low pressure drop at a low Reynolds number. *Journal of Micromechanics and Microengineering*, 23(5). 10.1088/0960-1317/23/5/055023

[11] Sadegh Cheri, M., Latifi, H., Salehi Moghaddam, M., & Shahraki, H. (2013). Simulation and experimental investigation of planar micromixers with short-mixing-length. *Chemical Engineering Journal*, 234, 247–255. 10.1016/j.cej.2013.08.067

[12] Santana, H. S., Silva, J. L., & Taranto, O. P. (2019). Optimization of micromixer with triangular baffles for chemical process in millidevices. *Sensors and Actuators, B: Chemical*, 281(April 2018), 191–203. 10.1016/j.snb.2018.10.089

[13] Viktorov, V., Mahmud, M. R., & Visconte, C. (2016). Numerical study of fluid mixing at different inlet flow-rate ratios in Tear-drop and Chain micromixers compared to a new H-C passive micromixer. *Engineering Applications of Computational Fluid Mechanics*, 10(1), 183–193. 10.1080/19942060.2016.1140075

[14] Wong, S. H., Ward, M. C. L., & Wharton, C. W. (2004). Micro T-mixer as a rapid mixing micromixer. *Sensors and Actuators, B: Chemical*, 100(3), 359–379. 10.1016/j.snb.2004.02.008

[15] Cortes-Quiroz, C. A., Azarbadegan, A., & Zangeneh, M. (2017). Effect of channel aspect ratio of 3-D T-mixer on flow patterns and convective mixing for a wide range of Reynolds number. *Sensors and Actuators, B: Chemical*, 239, 1153–1176. 10.1016/j.snb.2016.08.116

[16] Gidde, R. R. (2019). On the computational analysis of short mixing length planar split and recombine micromixers for microfluidic applications. *International Journal of Environmental Analytical Chemistry*, 00(00), 1–16. 10.1080/03067319.2019.1660875

[17] Gidde, R. R. (2020a). Design optimization of micromixer with circular mixing chambers (M-CMC) using Taguchi-based grey relational analysis. *International Journal of Chemical Reactor Engineering*, 18(9). 10.1515/ijcre-2020-0057

[18] Gidde, R. R. (2021). Concave wall-based mixing chambers and convex wall-based constriction channel micromixers. *International Journal of Environmental Analytical Chemistry*, 101(4), 561–583. 10.1080/03067319.2019.1669585

[19] Gidde, R. (2022). On the study of teardrop shaped split and collision (TS-SAC) micromixers with balanced and unbalanced split of subchannels. *International Journal of Modelling and Simulation*, 42(1), 168–177. 10.1080/02286203.2020.1858239

[20] Gidde, R. R. (2020b). On effects of shape, aspect ratio and position of obstacle on the mixing enhancement in micromixer with hexagonal-shaped chambers. *International Journal of Chemical Reactor Engineering*, 18(10–11). 10.1515/ijcre-2020-0054

[21] Gidde, R. R., Pawar, P. M., Gavali, S. R., & Salunkhe, S. Y. (2019). Flow feature analysis of an eye shaped split and collision (ES-SAC) element based micromixer for lab-on-a-chip application. *Microsystem Technologies*, 25(8), 2963–2973. 10.1007/s00542-018-4271-x

[22] Gidde, R. R., & Pawar, P. M. (2020). Flow feature and mixing performance analysis of RB-TSAR and EB-TSAR micromixers. *Microsystem Technologies*, 26(2), 517–530. 10.1007/s00542-019-04498-w

# 14 Prediction of Machined Slot Areas Using Artificial Neural Network and Adaptive Neural Fuzzy Interference System by WEDM for Different Slant Angles

*I.V. Manoj and S. Narendranath*

## 14.1 INTRODUCTION: BACKGROUND AND DRIVING FORCES

Surface integrity, cutting speed, surface morphology, and material removal are not only the process related parameters. The input parameters also influence the accuracy and profile of the machined components. Especially in cutting precise components and features on hard metals where wire electric discharge machining is used the input parameters play an important role [1–3]. Natarajan et al. [4] analyzed that increasing wire tension dramatically improves cutting speed and accuracy during wire electric discharge of stainless steel. Sawada et al. [5] showed that vibration and deflection of the wire electrode exert a large impact on machining accuracy. To achieve higher accuracy the WEDM simulation was performed to estimate the actions of the wire electrode during machining. It was also observed that the kerf width and wire feed speed also contribute to machining errors. Chou et al. [6] reported that wire rupture not only increases the processing time but may deteriorate surface integrity. The authors used artificial neural network (ANN) for the prediction of wire rupture for different input parameters. The precision of Wire Electrical Discharge Machining (WEDM) was enhanced with the use of a novel technique that enabled real-time prediction of

 DOI: 10.1201/9781003364948-14

wire breakage. Lun li et al. [7] used the gap discharge state monitoring system for better machining performance. There was an improvement in accuracy and material removal rate as servo feed rate control system received radio frequency from signal monitoring system as a feedback signal. Li Z et al. [8] adapted finite element method (FEM) during the single pulse discharge machining (WEDM) process to simulate the flow field and distribution of temperature. It was observed that the minimum gap between the width between the electrode wire and the initial surface of the workpiece determined the discharge position. It was observed that there was an average error of 8.26% in surface roughness compared to the simulation and experimental results. Ľuboslav Straka and Ivan Čorný [9] found that the circular profiles are very difficult to machine due the errors in circular profiles caused by the vibration of the wire electrode. They observed that by eliminating the critical frequency of the produced during discharges there can be a decrease in vibration amplitude of the tool wire electrode. They increased the accuracy, time, and productivity of the wire electric discharge machining by controlling the discharge energy and acoustic emission during WEDM process. Ghasempour-Mouziraji et al. [10] reported that the geometrical deviation of machined parts with many input characteristics is a major concern in the industry. So they employed artificial neural networks (ANN) and non-dominated sorting genetic algorithm (NSGA) machine learning techniques for minimizing the geometric errors. The authors found that the best optimal solution was derived by the combination of two methods. Yang et al. [11] monitored the wire electrical discharge machining using acoustic emission signals. The authors used a deep learning method like batch relevance temporal convolution neural network for batch relevance, encoder, noise extractor, and decoder for prediction discharge status. By understanding the relation between acoustic emission and pulse time series and predicting the discharge the smoothness of processing can be ensured with machining quality. Manoj et al. [12] investigated the prediction of cutting velocity and surface roughness using response surface methodology and artificial neural network. It was observed that the artificial neural network was more accurate in all the aspects of prediction compared to response surface methodology and the error percentage lesser than 6%. Phate et al. [13] utilized the multi-response optimization during machining of aluminum-based metal matrix composites. The TOPSIS-ANFIS predicted closeness of coefficient which increased from 0.656793 to 0.772138. Hemalatha et al. [14] used ANFIS model in WEDM for stir casted aluminum metal matrix composites with Boron nitride reinforcements. The performance features of dimensional deviation and material removal rate were examined in this study. ANFIS models were used to map and assess these qualities rate.

From the literature review we can see different output parameters like cutting speed, surface roughness, diameter, recast layer thickness, etc. Different soft computing methods like ANN, RSM ANFIS, Fuzzy logic, etc. are used for prediction of output parameters. These prediction techniques can be used for prediction of output parameters by input parameters. This eliminates the trial and error experimentation which reduces the experimental cost and time. Both ANN and ANFIS can be used for prediction of the machined area.

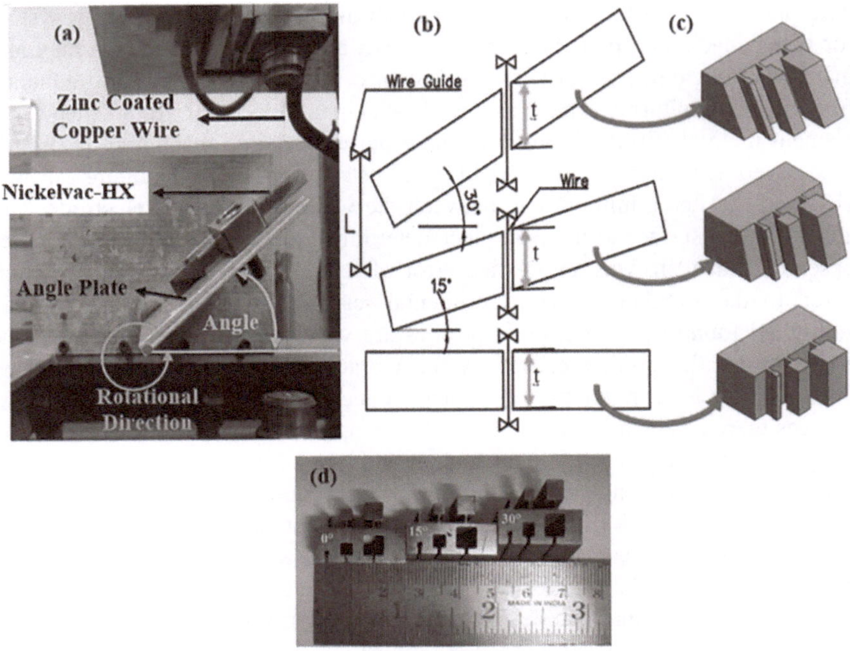


FIGURE 14.1 (a) Machining procedure followed using slant type fixture, (b) & (c) Dimensional view of machined slots (d) Machined slots with different dimensions.

## 14.2 EXPERIMENTAL PROCEDURE

Nickelvac HX is a nickel based super alloy used in many high temperature applications. The WEDM was used to machine the profile on Nickelvac HX alloy with the help of a fixture. The slant type fixture was used to obtain angular machining slots. Nickelvac HX was fixed to the slant type taper fixture. The zinc coated copper wire was used to machine the square slots as shown in Figure 14.1(a) and (b). The fixture position and angular machining on the WEDM was as followed in Manoj et al. [15,16]. The heat treatment was solution annealing which was heated to 2150°F (1177°C) and rapid cooled. Figure 14.1c and d shows the different slots obtained after WEDM for different angles.

## 14.3 CHARACTERIZATION PARTICULARS

After the machining using WEDM the machined slots were dried using a blower and preserved in airtight containers. The slots were characterized using CMM as shown in Figure 14.2. The sensor on the CMM was made to traverse on the boundaries of the machined slots as shown in Figure 14.3. The sensors detected the sides and corners of the co-ordinates of the slot that was machined. The sides of the machined slots were calculated based on the co-ordinates. Based on the sides from the CMM co-ordinates the area was calculated.

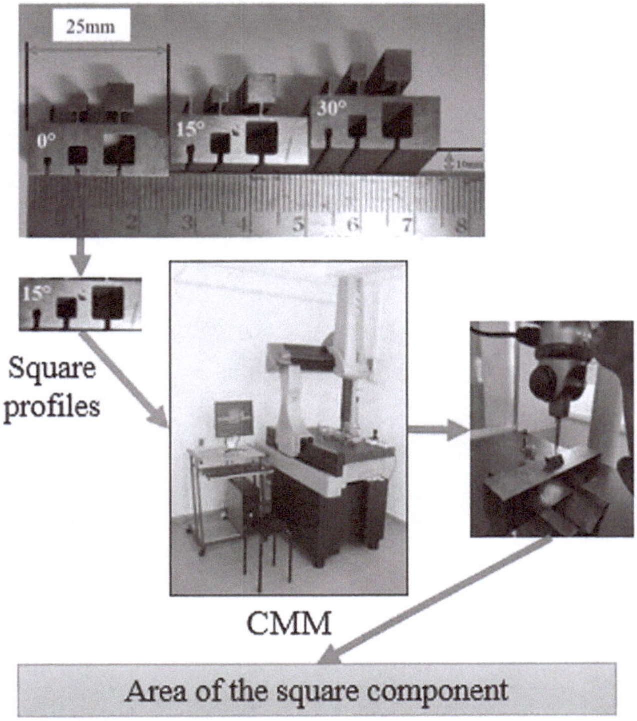


**FIGURE 14.2** Characterization steps using CMM.

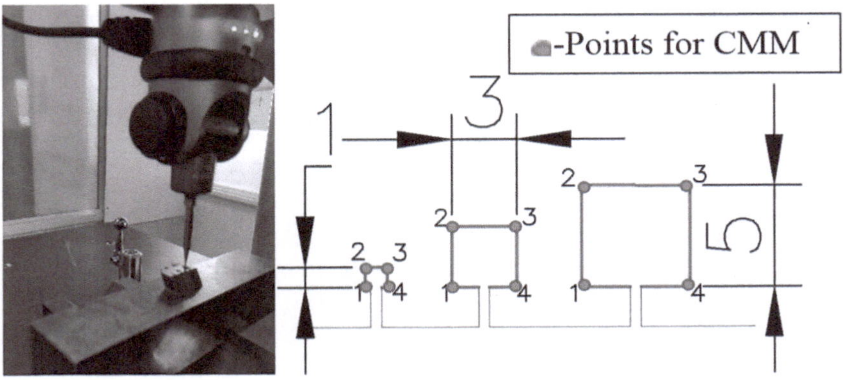


**FIGURE 14.3** CMM sensor movement along the machined slot.

## 14.4 RESULTS AND DISCUSSION

Different parameter combinations were used to cut using slant fixture. The areas after CMM measurement are shown in Table 14.1. The table also shows in that

**TABLE 14.1**
**Areas Obtained from Different Parameters by CMM**

| Sl. No. | σ (mm) | λ (s) | Ω (μm) | d (%) | Area of holes in $mm^2$ | |
|---|---|---|---|---|---|---|
| | | | | | 3mm | 5mm |
| | | | 0° taper angle | | | |
| 1 | 40 | 0 | 0 | 31 | 10.86 | 28.386 |
| 2 | 40 | 33 | 40 | 54 | 10.822 | 28.269 |
| 3 | 40 | 66 | 80 | 77 | 11.465 | 28.412 |
| 4 | 40 | 99 | 120 | 100 | 11.997 | 29.431 |
| 5 | 50 | 0 | 40 | 77 | 10.592 | 27.623 |
| 6 | 50 | 33 | 0 | 100 | 10.394 | 27.214 |
| 7 | 50 | 66 | 120 | 31 | 11.889 | 29.584 |
| 8 | 50 | 99 | 80 | 54 | 11.449 | 28.732 |
| 9 | 60 | 0 | 80 | 100 | 10.872 | 28.332 |
| 10 | 60 | 33 | 120 | 77 | 11.338 | 28.933 |
| 11 | 60 | 66 | 0 | 54 | 10.189 | 27.126 |
| 12 | 60 | 99 | 40 | 31 | 10.876 | 27.725 |
| 13 | 70 | 0 | 120 | 54 | 11.406 | 28.692 |
| 14 | 70 | 33 | 80 | 31 | 11.156 | 28.066 |
| 15 | 70 | 66 | 40 | 100 | 10.077 | 27.43 |
| 16 | 70 | 99 | 0 | 77 | 9.987 | 27.054 |
| | | | 15° taper angle | | | |
| 1 | 75 | 0 | 0 | 31 | 12.219 | 30.786 |
| 2 | 75 | 33 | 40 | 54 | 12.072 | 30.669 |
| 3 | 75 | 66 | 80 | 77 | 12.461 | 30.912 |
| 4 | 75 | 99 | 120 | 100 | 12.654 | 31.731 |
| 5 | 85 | 0 | 40 | 77 | 11.755 | 30.023 |
| 6 | 85 | 33 | 0 | 100 | 11.493 | 29.513 |
| 7 | 85 | 66 | 120 | 31 | 12.803 | 31.884 |
| 8 | 85 | 99 | 80 | 54 | 12.418 | 31.132 |
| 9 | 95 | 0 | 80 | 100 | 12.105 | 30.632 |
| 10 | 95 | 33 | 120 | 77 | 12.462 | 31.333 |
| 11 | 95 | 66 | 0 | 54 | 11.414 | 29.426 |
| 12 | 95 | 99 | 40 | 31 | 11.684 | 30.125 |
| 13 | 105 | 0 | 120 | 54 | 12.294 | 31.092 |
| 14 | 105 | 33 | 80 | 31 | 12.072 | 30.466 |
| 15 | 105 | 66 | 40 | 100 | 11.611 | 29.630 |
| 16 | 105 | 99 | 0 | 77 | 11.212 | 29.454 |
| | | | 30° taper angle | | | |
| 1 | 100 | 0 | 0 | 31 | 12.360 | 31.246 |
| 2 | 100 | 33 | 40 | 54 | 12.392 | 31.505 |
| 3 | 100 | 66 | 80 | 77 | 12.965 | 32.060 |

**TABLE 14.1 (Continued)**
**Areas Obtained from Different Parameters by CMM**

| Sl. No. | σ (mm) | λ (s) | Ω (μm) | d (%) | Area of holes in mm² | |
|---|---|---|---|---|---|---|
| | | | | | 3mm | 5mm |
| 4 | 100 | 99 | 120 | 100 | 13.397 | 33.169 |
| 5 | 110 | 0 | 40 | 77 | 12.062 | 30.882 |
| 6 | 110 | 33 | 0 | 100 | 11.794 | 30.145 |
| 7 | 110 | 66 | 120 | 31 | 13.289 | 32.956 |
| 8 | 110 | 99 | 80 | 54 | 12.749 | 31.684 |
| 9 | 120 | 0 | 80 | 100 | 12.372 | 31.509 |
| 10 | 120 | 33 | 120 | 77 | 12.838 | 32.163 |
| 11 | 120 | 66 | 0 | 54 | 11.689 | 30.028 |
| 12 | 120 | 99 | 40 | 31 | 12.346 | 30.863 |
| 13 | 130 | 0 | 120 | 54 | 12.906 | 31.733 |
| 14 | 130 | 33 | 80 | 31 | 12.656 | 31.681 |
| 15 | 130 | 66 | 40 | 100 | 11.647 | 30.215 |
| 16 | 130 | 99 | 0 | 77 | 11.387 | 29.495 |

there were four input parameters with 48 iterations in the analysis. A comprehensive investigation was conducted by Manoj et al. [16] to examine the distinct behaviors and impacts of each parameter on the result.

The prediction of the slot areas by artificial neural network and adaptive neuro fuzzy interference system used the 48 different iterations at various combinations of input parameters. The input parameters were machining parameters and angle and the output parameter is 3 mm and 5 mm areas that were generated at different parameters. The artificial neural network employs 70% of the iteration for training, 15% for testing, and, 15% for validation. Figure 14.4 shows different

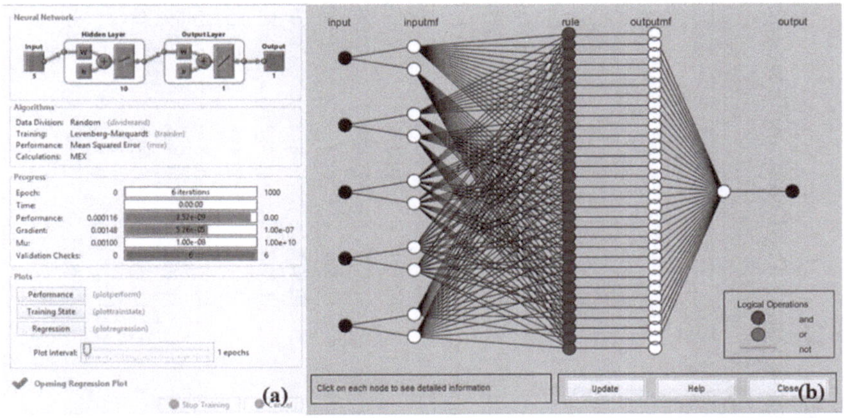


**FIGURE 14.4** (a) ANN optimal model (b) ANFIS model used for prediction of areas.

**TABLE 14.2**
**Experimental, Predicted and Percentage Error in Areas**

| Sl. No. | Area of Holes in $mm^2$ | | | | | | | | | |
|---|---|---|---|---|---|---|---|---|---|---|
| | 3mm | | | | | 5mm | | | | |
| | E | ANN-P | % | ANFIS-P | % | E | ANN-P | % | ANFIS-P | % |
| | 0° taper angle | | | | | | | | | |
| 1 | 10.86 | 10.86 | 0.036 | 10.861 | 0.009 | 28.386 | 28.436 | 0.177 | 28.376 | 0.035 |
| 2 | 10.822 | 10.822 | 0.149 | 10.832 | 0.092 | 28.269 | 28.201 | 0.241 | 28.279 | 0.035 |
| 3 | 11.465 | 11.465 | 0.059 | 11.467 | 0.017 | 28.412 | 28.453 | 0.144 | 28.421 | 0.032 |
| 4 | 11.997 | 11.997 | 0.176 | 11.997 | 0.000 | 29.431 | 29.399 | 0.108 | 29.531 | 0.339 |
| 5 | 10.592 | 10.592 | 0.289 | 10.591 | 0.009 | 27.623 | 27.614 | 0.032 | 27.625 | 0.007 |
| 6 | 10.394 | 10.394 | 0.124 | 10.393 | 0.010 | 27.214 | 27.186 | 0.105 | 27.224 | 0.037 |
| 7 | 11.889 | 11.889 | 0.507 | 11.889 | 0.000 | 29.584 | 29.589 | 0.015 | 29.597 | 0.044 |
| 8 | 11.449 | 11.449 | 0.001 | 11.45 | 0.009 | 28.732 | 28.751 | 0.067 | 28.712 | 0.070 |
| 9 | 10.872 | 10.872 | 0.103 | 10.882 | 0.092 | 28.332 | 28.260 | 6.372 | 28.001 | 1.182 |
| 10 | 11.338 | 11.338 | 0.229 | 11.338 | 0.000 | 28.933 | 28.893 | 0.137 | 28.943 | 0.035 |
| 11 | 10.189 | 10.189 | 0.172 | 10.129 | 0.592 | 27.126 | 27.234 | 0.396 | 27.726 | **2.164** |
| 12 | 10.876 | 10.876 | 2.154 | 10.856 | 0.184 | 27.725 | 28.046 | 1.146 | 27.925 | 0.716 |
| 13 | 11.406 | 11.406 | 0.568 | 11.407 | 0.009 | 28.692 | 28.841 | 0.517 | 28.699 | 0.024 |
| 14 | 11.156 | 11.156 | 1.204 | 11.156 | 0.000 | 28.066 | 28.025 | 0.146 | 28.266 | 0.708 |
| 15 | 10.077 | 10.077 | 0.219 | 10.077 | 0.000 | 27.43 | 27.488 | 0.211 | 27.431 | 0.004 |
| 16 | 9.987 | 9.987 | 0.105 | 9.988 | 0.010 | 27.054 | 26.926 | 0.474 | 27.014 | 0.148 |

| | | | | | | | | | | |
|---|---|---|---|---|---|---|---|---|---|---|
| | | | | | **15° taper angle** | | | | | |
| 1 | 12.219 | 12.219 | 0.008 | 12.229 | 0.082 | 30.786 | 30.541 | **8.215** | 30.186 | 1.988 |
| 2 | 12.072 | 12.072 | 0.058 | 12.075 | 0.025 | 30.669 | 30.523 | 0.478 | 30.669 | 0.000 |
| 3 | 12.461 | 12.461 | 0.051 | 12.462 | 0.008 | 30.912 | 30.894 | 0.058 | 30.212 | **2.317** |
| 4 | 12.654 | 12.654 | 0.220 | 12.644 | 0.079 | 31.731 | 31.805 | 0.233 | 31.731 | 0.000 |
| 5 | 11.755 | 11.755 | 0.199 | 11.756 | 0.009 | 30.023 | 30.180 | 6.703 | 30.013 | 0.033 |
| 6 | 11.493 | 11.493 | 4.232 | 11.491 | 0.017 | 29.513 | 29.446 | 0.227 | 29.513 | 0.000 |
| 7 | 12.803 | 12.803 | 0.203 | 12.807 | 0.031 | 31.884 | 31.904 | 0.063 | 31.884 | 0.000 |
| 8 | 12.418 | 12.418 | 0.174 | 12.438 | 0.161 | 31.132 | 31.005 | 0.409 | 31.232 | 0.320 |
| 9 | 12.105 | 12.105 | **8.825** | 12.155 | 0.411 | 30.632 | 30.608 | 0.080 | 30.732 | 0.325 |
| 10 | 12.462 | 12.462 | 0.096 | 12.442 | 0.161 | 31.333 | 31.301 | 0.101 | 31.353 | 0.064 |
| 11 | 11.414 | 11.414 | 0.001 | 11.424 | 0.088 | 29.426 | 29.613 | 0.631 | 29.436 | 0.034 |
| 12 | 11.684 | 11.684 | 0.037 | 11.654 | 0.257 | 30.125 | 30.146 | 0.069 | 30.226 | 0.334 |
| 13 | 12.294 | 12.294 | 7.862 | 12.394 | 0.807 | 31.092 | 31.103 | 0.035 | 31.122 | 0.096 |
| 14 | 12.072 | 12.072 | 0.278 | 12.072 | 0.000 | 30.466 | 30.541 | 0.245 | 30.566 | 0.327 |
| 15 | 11.611 | 11.611 | 0.111 | 11.621 | 0.086 | 29.630 | 29.600 | 0.100 | 29.632 | 0.007 |
| 16 | 11.212 | 11.212 | 0.101 | 11.202 | 0.089 | 29.454 | 29.437 | 4.485 | 29.254 | 0.684 |
| | | | | | **30° taper angle** | | | | | |
| 1 | 12.360 | 12.36 | 0.142 | 12.36 | 0.000 | 31.246 | 31.101 | 0.466 | 31.546 | 0.951 |
| 2 | 12.392 | 12.392 | 0.137 | 12.392 | 0.000 | 31.505 | 31.549 | 0.140 | 31.515 | 0.032 |
| 3 | 12.965 | 12.965 | 7.079 | 12.965 | 0.000 | 32.060 | 32.034 | 0.082 | 32.062 | 0.006 |
| 4 | 13.397 | 13.397 | 0.062 | 13.398 | 0.007 | 33.169 | 33.160 | 0.027 | 33.129 | 0.121 |
| 5 | 12.062 | 12.062 | 0.115 | 12.362 | **2.427** | 30.882 | 30.947 | 0.210 | 30.812 | 0.227 |
| 6 | 11.794 | 11.794 | 0.171 | 11.784 | 0.085 | 30.145 | 30.312 | 0.550 | 30.155 | 0.033 |
| 7 | 13.289 | 13.289 | 0.085 | 13.279 | 0.075 | 32.956 | 33.004 | 0.145 | 32.986 | 0.091 |

*(Continued)*

**TABLE 14.2 (Continued)**
**Experimental, Predicted and Percentage Error in Areas**

| Sl. No. | Area of Holes in mm² | | | | | | | | | |
|---|---|---|---|---|---|---|---|---|---|---|
| | 3mm | | | | | 5mm | | | | |
| | E | ANN-P | % | ANFIS-P | % | E | ANN-P | % | ANFIS-P | % |
| 8 | 12.749 | 12.749 | 0.058 | 12.759 | 0.078 | 31.684 | 31.741 | 0.179 | 31.784 | 0.315 |
| 9 | 12.372 | 12.372 | 0.091 | 12.572 | 1.591 | 31.509 | 31.480 | 0.091 | 31.529 | 0.063 |
| 10 | 12.838 | 12.838 | 0.093 | 12.878 | 0.311 | 32.163 | 32.304 | 0.436 | 32.263 | 0.310 |
| 11 | 11.689 | 11.689 | 0.087 | 11.699 | 0.085 | 30.028 | 30.009 | 0.064 | 30.128 | 0.332 |
| 12 | 12.346 | 12.346 | 4.631 | 12.386 | 0.323 | 30.863 | 30.801 | 0.200 | 30.8623 | 0.002 |
| 13 | 12.906 | 12.906 | 0.042 | 12.916 | 0.077 | 31.733 | 31.784 | 0.162 | 31.1733 | 1.795 |
| 14 | 12.656 | 12.656 | 0.032 | 12.676 | 0.158 | 31.681 | 31.609 | 0.229 | 31.611 | 0.221 |
| 15 | 11.647 | 11.647 | 0.055 | 11.644 | 0.026 | 30.215 | 30.129 | 5.956 | 30.1215 | 0.310 |
| 16 | 11.387 | 11.387 | 2.528 | 11.317 | 0.619 | 29.495 | 29.458 | 0.126 | 29.1495 | 1.185 |

E-Experimental, P- Predicted, %-Error %

**TABLE 14.3**
**Evaluation Matrix for the Regression Model**

| Sl. No. | Measurement Index | 3mm | | 5mm | |
|---|---|---|---|---|---|
| | | ANN | ANFIS | ANN | ANFIS |
| 1 | RMSE | 0.297 | 0.057 | 0.675 | 0.208 |
| 2 | MSE | 0.088 | 0.003 | 0.456 | 0.043 |
| 3 | MAE | 0.117 | 0.023 | 0.272 | 0.112 |

architectures that were generated from MATLAB for optimal combination. The 5, 10, 1, 1 architecture was chosen as the optimal architecture for prediction of output parameters. The mean square error was used as performance evaluation index. The Levenberg-Marquardt training method with feed forward back-propagation method were employed. The adaptive neuro fuzzy interference system has 32 rules with five input layers and one output layer. The membership function type of trapmf and linear variation was used with two hidden layers (22222). An epoch of 1000 was given to the model for training. The model uses 38 training and ten testing data for the optimal model. Table 14.2 shows that the error in ANN model varies from 0–8% and ANFIS varies from 0–3%. Table 14.3 shows the root mean square, mean square error, and absolute mean square for both ANN and ANFIS model in 1 mm and 5 mm. From the above tables we can conclude that in ANFIS the errors are 2 times lower compared to ANN models. Similar functions are used in Manoj et al. [17–19].

## 14.5 CONCLUSIONS

The Nickelvac-HX alloys were machined in WEDM with different dimensions of 1 mm, 3 mm, and 5 mm. These machined slots were measured using a co-ordinate measuring machine where the area is calculated for different parameters. ANN and ANFIS models were used to predict the areas of the machined slots.

1. The optimal models in ANN of 5, 10, 1, 1 architecture and ANFIS with 32 rules with five input layers and one output layer architecture can be used for prediction of areas for different areas of the machined slots.
2. The errors for ANN and ANFIS models ranged from approximately 0–9% and 0–3% respectively and ANFIS models are more accurate than ANN models.
3. The RMSE, MSE, and MAE shows ANFIS models are more efficient in the prediction of machined profile areas than ANN models.

## REFERENCES

[1] Mandal A, Dixit AR, Chattopadhyaya S, Paramanik A, Hloch S, Królczyk G Improvement of surface integrity of Nimonic C 263 super alloy produced by WEDM through various post-processing techniques. *Int J Adv Manuf Technol.* 2017;93(1–4):433–443. doi:10.1007/s00170-017-9993-x

[2] Manoj IV Slant type taper profiling and prediction of profiling speed for a circular profile during in wire electric discharge machining using Hastelloy-X. 2021. doi:10.1177/0954406221992398.

[3] Usman, M, Ishfaq, K, Rehan, M, Raza, A, Mumtaz, J An in-depth evaluation of surface characteristics and key machining responses in WEDM of aerospace alloy under varying electric discharge environments. *Int J Adv Manuf Technol.* 2022. 10.1007/s00170-022-10608-2.

[4] Natarajan K, Ramakrishnan H, Gacem A, Vijayan V, Karthiga K, Elhosiny Ali H, Prakash B, and Mekonnen A Study on optimization of WEDM process parameters on stainless steel. *Journal of Nanomaterials, Hindawi.* 2022. 10.1155/2022/6765721.

[5] Sawada T, Kunieda M, Matsuura K, Hanawa T Development of WEDM simulation with parameters adaptively changed using wire electrode displacement sensor. *Procedia CIRP.* 2022;113:137–142. 10.1016/j.procir.2022.09.121.

[6] Chou PH, Hwang YR, Yan BH The study of machine learning for wire rupture prediction in WEDM. *Int J Adv Manuf Technol.* 2022;119: 1301–1311. 10.1007/s00170-021-08323-5

[7] Li ZL, Xi XC, Chu H-Y, Xu L-Y, Gao Q, Zhao W-S Improving RT-WEDM performance with a radio frequency signal monitoring system. *Int J Adv Manuf Technol.* 2022;118: 391–404. 10.1007/s00170-021-07983-7.

[8] Li Z, Liu Y, Cao B, Li W Modeling of material removal morphology and prediction of surface roughness based on WEDM successive discharges. *Int J Adv Manuf Technol.* 2022;120: 2015–2029. 10.1007/s00170-022-08870-5.

[9] Straka Ľ, Čorný I Identification of geometric errors of circular profiles at WEDM caused by the wire tool electrode vibrations and their reduction with support of acoustic emission method. *Engineering Failure Analysis.* 2022;134. 10.1016/j.engfailanal.2022.106040.

[10] Ghasempour-Mouziraji M, Hosseinzadeh M, Hajimiri H, Najafizadeh M, Shirkharkolaei EM Machine learning-based optimization of geometrical accuracy in wire cut drilling. *Int J Adv Manuf Technol.* 2022;123: 4265–4276. 10.1007/s00170-022-10351-8.

[11] Yang X, Liu C, Peng L, Peng S, Zhang Y, Xie N, Zhong RY A new BRTCN model for predicting discharge status of WEDM based on acoustic emission. *Journal of Manufacturing Systems.* 2022;64: 409–423, 10.1016/j.jmsy.2022.07.003.

[12] Manoj IV, Soni H, Narendranath, Mashinini M, Kara M Examination of Machining Parameters and Prediction of Cutting Velocity and Surface Roughness Using RSM and ANN Using WEDM of Altemp HX. *Advances in Materials Science and Engineering, Hindawi,* 10.1155/2022/5192981.

[13] Phate M, Toney S, Phate V, Tatwawadi V Multi-response optimization of Al/GrCp10 MMC performance in WEDM using integrated TOPSIS-ANFIS approach. *J. Inst. Eng. India Ser. D.* 2022;103: 249–261. 10.1007/s40033-021-00302-0.

[14] Hemalatha, A, Reddy, VD, Prasad, K Evolution of regression and ANFIS models for wire spark erosion machining of aluminium metal matrix composites for aerospace applications. *Int J Interact Des Manuf.* 2022. 10.1007/s12008-022-01012-x

[15] Manoj IV, Joy R, Narendranath S Investigation on the effect of variation in cutting speeds and angle of cut during S=slant type taper cutting in WEDM of Hastelloy X. *Arab J Sci Eng*. 2020;45(2):641–651. doi:10.1007/s13369-019-04111-2

[16] Manoj IV Slant type taper profiling and prediction of profiling speed for a circular profile during in wire electric discharge machining using Hastelloy-X. 2021. doi:10.1177/0954406221992398.

[17] Manoj IV, Narendranath S Wire electric discharge machining at different slant angles during slant type taper profiling of microfer 4722 superalloy. *J Mater Eng Perform*. 2022;31(1):697–708. doi:10.1007/s11665-021-06168-3

[18] Manoj IV, Narendranath S Evaluation of WEDM performance characteristics and prediction of machining speed during taper square profiling on Hastelloy-X. *Aust J Mech Eng*. 2021;00(00):1–16. doi:10.1080/14484846.2021.1960670.

[19] Maher I, Ling LH, Sarhan AAD, Hamdi M Improve wire EDM performance at different machining parameters - ANFIS modeling. *IFAC-PapersOnLine*. 2015;48(1): 105–110, ISSN 2405-8963, 10.1016/j.ifacol.2015.05.109.

# 15 Nanocomposite Surface Modification

*Haifeng Xie, Gaoying Hong, Qiyue Zhou, Yumin Wu, Ying Chen, Ran You, Shijia Tang, and Fei Han*

## 15.1 OVERVIEW

Nanocomposites are a kind of material compounded by nanofillers and other materials [1]. Nanocomposites doped with nanofilled phases have lower density than conventional fillers, and the addition of nanofilled phases enables the composites superior mechanical, thermal, and electrochemical properties, thus nanocomposites are widely used in industry, energy storage, electronic communication, transportation, aerospace, biomedicine, tissue engineering, etc. [2,3].

The types and properties of nanocomposites vary depending on the matrix used for nanocomposites, and are usually classified as metal-based, ceramic-based, and polymer-based nanocomposites [4]. When nanoparticles enter into the matrix as a filler phase, the nanocomposites have optical, electrical, magnetic, acoustic, thermal, and chemically active functions due to effects such as small nano size, and give the material good overall properties. However, due to the small particle size, large surface energy and easy agglomeration of nanoparticles, it is difficult to fill the nanofilled phase into the matrix as nano-scale particles with poor dispersion by conventional methods [5]. Therefore, in order to reduce the surface energy of nanoparticles, eliminate/weaken their surface charge and surface polarity, and improve the dispersibility of nanoparticles and their affinity to organic phases, many scholars have tried to perform surface modification of nanocomposites through advanced nanotechnology and fabrication processes to improve the material properties [6]. The surface-modified nanofilled phase also does not change its role in improving the fracture elongation of the nanocomposites since it does not change its nanoscale range. The surface-modified nanoparticles can act as cross-linking points of polymer molecular chains with energy transfer effect and substantially improve the tensile strength of the nanocomposites. Moreover, the surface-modified nanoparticles have a balancing effect of stress concentration and stress radiation, so that the nanocomposite matrix has no obvious stress concentration, thus achieving a mechanically balanced state [7].

After surface modification, chemical changes occur at the interface of the nanofilled phase to form a mono- or multi-molecular barrier film on its surface


DOI: 10.1201/9781003364948-15

where the nanoparticles cannot agglomerate, which is the main principle of surface modification of nanofilled phases. The surface modification of nanofilled phases can be as simple as physical adsorption, chemical or biological affinity immobilization, which is mainly done by employing various coupling agents, surfactants or grafting functional groups [8]. The surface-modified substances can be inorganic, organic, small molecule compounds, polymeric compounds, polar compounds, non-polar compounds, ionic compounds or neutral molecular compounds. The molecular structure of the modifier must have characteristic groups that are easy to interact with the nanofilled phase. Surface modifiers with specific structures have strong encapsulation ability, and are easy to form a certain space barrier on the surface of the nano-filled phase to prevent the aggregation and flocculation of nanoparticles; moreover, comb-like, blocked polymer modifiers have stronger modifying effects [9]. Silane coupling agent is the classical representative, which has been maturely applied to the surface modification of nanocomposites such as nano bioactive glass, nano resin ceramics, nano complex ceramics and nano hydroxyapatite [10].

In addition to surface modification of the nanofilled phase of nanocomposites, the material itself can also be modified: Grafting with functional groups that promote cell adhesion and growth on the surface of nanocomposites by grafting, including coupling grafting, chemical initiation, ozone initiation, radiation initiation, and light radiation initiation, effectively enhancing the biocompatibility of nanocomposites [11]. Photochemical immobilization is the coupling of molecules to the surface of nanocomposites using chemically linked components with thermally active and photoactive groups under the action of UV or visible light. This modification technique has fast reaction rate, simple operation, universal applicability and significantly reduces the disordered cross-linking generated by the destination coating reagent on the surface of the nanocomposite [12]. Small molecular functional groups can also be generated on the polymer surface by using the reaction of groups already present in the polymer matrix, or by the reaction of certain reactive groups or atoms on the side groups of the main chain. The attraction of these reactive groups can improve the biocompatibility of the nanocomposite by appropriately increasing its hydrophilicity [13]. Plasma technology has been applied to surface modification by virtue of its low impact on the intrinsic properties of the nanocomposite and the ease of changing and controlling the modification conditions. This technique enables the grafting of monomers on the nanocomposite surface by generating reactive groups such as peroxy and amine groups, etc., further initiating the polymerization of monomers on the material surface [14]. The surface modification of titanium implants with micro-nano structure by non-thermal plasma (NTP) technology can activate high level of reactive oxygen species/nitrogen species (ROS/RNS), high density of -OH and $-NH_2$ groups on titanium surface, reduce carbon contamination and increase surface hydrophilicity. This not only facilitates cell adhesion, proliferation and differentiation, but also reduces the adhesion of hydrophobic bacteria [15]. However, the surface of nanocomposites modified by plasma technology gradually degrades over time, which proves to be the drawback that cannot be overcome by plasma modification technology itself [16]. In recent years, researchers have tried to slow down or eliminate the aging of the plasma technology by exploring the effects of aging factors such as nanocomposite type, placement temperature and placement atmosphere [17]. Bioactive molecule immobilization introduces bioactive molecules on the surface of nanocomposites to promote cell adhesion and growth, and usually includes physical adsorption (i.e., bioactive molecules containing multiple negative charges can be immobilized on positively charged parts of the nanocomposite by electrostatic adsorption, which is one of the easiest ways to introduce active molecules on the nanocomposite surface) and chemical immobilization (certain groups in the bioactive molecule are chemically bonded to reactive groups on the substrate surface). The latter allows the bioactive molecule to be firmly immobilized on the surface of the nanocomposite, thus obtaining long-term histocompatibility, which cannot be achieved through physical adsorption [18]. In recent years, self-assembled surface modification has attracted increasing attention [19]. Polymer molecules are spontaneously constructed into assemblies with specific structures and shapes by molecular self-assembly techniques, driven by interaction forces such as hydrogen bonding, electrostatic interactions, hydrophobic-lipophilic interactions, and van der Waals forces [20]. This is an interdisciplinary discipline integrating chemistry, physics, life sciences and materials science, opening up a new field of materials chemistry with a wide application potential in the field of surface modification of nanocomposites [21].

Notably, for metal-based nanocomposites, nano-reinforced phases, such as carbon nanotubes (CNTs) combine low density, extraordinary mechanical strength, nano-size, high aspect ratio, and excellent electrical and thermal conductivity, making them ideal reinforcements for lightweight, high-strength multifunctional composites [22]. To achieve a homogeneous composite between the nano-reinforced phase and the metal matrix, researchers have performed surface oxidation (high concentrations of strong oxidants such as $HNO_3$, $H_2SO_4$, $HNO_3$, $KMnO_4$ and $O_2/O_3$) [23] and coating treatments (electroplating, thermal spraying, sol-gel method, chemical vapor deposition, physical vapor deposition, atomic layer deposition, etc.) [24]. The modified transition layer should be able to infiltrate well with both the metal matrix and the surface of the nano-reinforced phase [25]. However, the plated layer obtained by physical vapor deposition does not have any chemical reaction with the surface of the nano-reinforced phase, therefore, the bonding state between them is poor, which limits its application in practical application. The chemical vapor deposition can control the composition of the coated film, but the thermal damage caused by the high reaction temperature may damage the lattice of the nano-reinforced phase, thus affecting the mechanical properties of the nanocomposite, and bringing higher coating costs. Moreover, most of the coatings are distributed in a continuous film on the surface of the nano-reinforced phase, which improves the interfacial bonding to some extent, but the thickness of the coating is larger, compared with the diameter of the nano-reinforced phase, resulting in the introduction of too much coating. More importantly, the introduction of continuous coating completely blocks the contact between the nano-reinforced phase and the metal matrix, which is not conducive to the material densification process and interfacial bonding, nor is it conducive to the performance of the load-bearing or functional conduction of the nano-enhanced phase, making it difficult to achieve effective enhancement. It has been shown that the interfacial reaction between the nano-reinforced phase and the metal matrix occurs preferentially from locations such as defects and ends [26]. From this perspective, local modification by introducing stable nanoparticles on the surface of the nano-reinforced phase can be attempted to passivate the reactivity of the defective points at the interface between the nano-reinforced phase and the metal matrix. The amount of nanoparticles introduced is much smaller than that of continuous coating, which can simultaneously regulate the interfacial reaction, densification process and strengthening benefits, thus avoiding the interfacial reaction without affecting the enhancement effect of the nano-reinforced phase. Meanwhile, a certain amount of alloying elements, such as Si Cr, etc., can be added to the metal matrix through metal matrix alloying. Through the interfacial reaction between the alloying elements and the nano-reinforced phase, a certain thickness of interfacial reaction layer can be formed, which not only improves the interfacial bonding strength but also isolates the interfacial reaction between the nano-reinforced phase and the metal matrix [27].

This chapter focuses on the surface modification technology of nanocomposites and its application progress and future prospects. Taking nanocomposite biomaterials for hard tissue defect repair as an example, the representative polymer-based, ceramic-based, and metal-based nanocomposite biomaterials are selected to illustrate the innovative development of surface modification technology in its performance improvement and application expansion, and summarize the challenges and new opportunities for surface modification of nanocomposites.

## 15.2 APPLICATION OF DOPAMINE IN SURFACE MODIFICATION OF NANOCOMPOSITE SCAFFOLDS FOR BONE TISSUE ENGINEERING

Biomaterials are a general term for a class of materials that interact with biological systems to diagnose, treat, repair or replace tissues or organs in the body or to enhance their functions without produce adverse reactions to human tissues and blood [28]. With the continuous development of tissue engineering technology in recent years, it provides us with a new means of repairing and reconstructing organs. Tissue engineering technology is used to form the desired tissue (organ) by implanting seed cells into a three-dimensional biological scaffold, which is then implanted in the human body or in an environment that simulates body fluids, with the gradual degradation of the material and the adhesion of the cells by secreting extracellular matrix. A scaffold required for tissue engineering is a scaffold in which materials can be combined with tissues, cells, and proteins to culture normal tissues *in vitro*. Along with the advancement of tissue engineering research, new tissue engineering scaffolds have emerged, driving the continuous development of tissue engineering scaffolds [29].

The selection and preparation of the tissue engineering scaffold material, which is a central part of bone tissue engineering for the repair of bone defects and artificial bone reconstruction, is of particular importance. The ideal bone tissue engineering scaffold material must be biocompatible and facilitate the implantation and growth of seed cells. In view of this, the search for better biocompatible materials and the necessary surface modification of existing materials to improve the biocompatibility or surface properties of the scaffold material to achieve the desired cell adhesion and growth is also a hot topic of research in bone tissue engineering [30]. Surface modification is the process of improving the biocompatibility or surface properties of a material by immobilizing some cytokines on the material surface or changing the local structural features of the material surface. The selection of suitable surface modification materials, and the targeted surface modification and surface modification of bone tissue engineering scaffolds to improve cell adhesion and promote cell growth and osteogenic differentiation on the surface of scaffold materials are important for bone tissue engineering scaffold materials [31,32].

Nanoscience and nanotechnology today have raised the level of human knowledge and research scales in the field of tissue engineering and have provided new ideas for the design of tissue engineering scaffolds [33], namely that tissue engineering should be carried out simultaneously at the macro (super cell >100 μM and cell >l0 μM), micro (subcellular 0.1–10 μM) and nano (<0.1 μM) scales. While there is a more comprehensive understanding at the macroscopic and microscopic levels, there is limited understanding at the nanoscopic level, where it is actually more important. This is because the natural extracellular matrix (ECM) *in vivo* is a three-dimensional nanofibrous structure with nanoscale pores, fibers, and bulges in the ECM. In addition, cell surface receptors are nanostructured and the functional structural domains on the cell surface are also at the nanoscale. Many interactions between biofunctional molecules, ECM components and cells occur at the nanoscale, i.e., at the molecular level, and these interactions directly affect the behavior and function of cells. From a biological point of view, almost all human tissues and organs have certain form and structure of nanofibers, such as bone, collagen, cartilage, and skin. Scaffolds with a three-dimensional nanofiber structure can therefore mimic the structure of the natural extracellular matrix to the maximum possible extent, and thus be functionally integrated with the muscle tissue [34]. As a result, increasing applications of nanomaterials in the design and construction of tissue engineering scaffolds for bone tissue have led to the formation of nanocomposite scaffolds, providing them with more complex and diverse functions.

Nanocomposite scaffolds for bone tissue engineering often require a combination of different matrix materials and nanomaterials depending on the characteristics of the defect site, different functional requirements, etc. Moreover, their function as artificial extracellular matrix and the characteristics of the implanted host determines the extensive complexity of the cell-biomaterial surface interactions, which also reinforces the importance of surface modification of such composite scaffold materials. From the materials science perspective, surface modification of bone tissue engineering scaffolds imparts new properties to the material, for example, for brittle scaffold materials, surface modification can increase their strength and reduce their brittleness, thus making them more suitable for use as bone graft materials. In terms of clinical applications in tissue engineering, surface of the scaffolds acts as an artificial extracellular matrix for the interface and transition between the adherent tissue and the implanted biomaterial, which is a key to clinical applications [35].

Therefore, targeting surface modification of the scaffolds to provide good scaffold-cell and scaffold-tissue interaction interfaces is an important part of bone tissue engineering. Dopamine (DA) and its polymer polydopamine (PDA) contain a large number of active functional groups such as catechol groups and amino groups, which provide them with good adhesion, biocompatibility, reactivity, and reducibility, and are widely used for the modification of the surfaces of medical device materials, tissue engineering materials, marine materials, sensor device materials, drug transport and other materials, among which the research on the modification of the surfaces of tissue engineering materials is particularly promising. Dopamine has excellent adhesion and chemical reactivity due to its active functional groups, and can combine bone repair materials and growth factors by covalent or non-covalent bonding, significantly increasing the binding force and thus the regenerative capacity of the bone tissue [36]. Dopamine can also act as a bridge between the engineered scaffold and the cells, effectively promoting the adhesion and proliferation of cells on the surface of the scaffold [37].

Although it has been shown that the adhesion behavior of DA to the surface of the substrate originates from the catechol and amino functional groups of it [38,39], the exact adhesion mechanism is still unclear. The current studies show that the adhesion mechanism between dopamine and its polymers and various substrate materials is firmly related to the surface chemical composition of the substrate, and that the adhesion is mainly divided into two types of adhesion: Covalent and non-covalent binding. The covalent binding mainly refers to the Michael addition or Schiff base reaction between some surfaces containing amino or sulfhydryl functional groups and dopamine under alkaline conditions, while the non-covalent binding refers to the adhesion process through metal coordination or chelation, hydrogen bonding, $\pi$-$\pi$ stacking and van der Waals forces [40,41]. Metal ions can chelate with the phenolic group of PDA, and the metal or metal oxide surface is usually hydroxylated, so the adhesion of DA on metal or metal oxide surfaces is generally due to ligand and chelate bonding. The mechanism of DA adhesion to organic surfaces is not clear. Lee et al. demonstrated with the aid of atomic force microscopy that the adhesion of dopamine to organic surfaces is achieved by the oxidation of dopamine to quinone under alkaline conditions and the formation of covalent bonds by Michael addition reactions of quinone with organic surfaces [42]. This section will summarize the application modalities of dopamine in the surface modification of tissue-engineered nanocomposite scaffolds.

### 15.2.1 Direct Application as a Coating

Dopamine (DA) readily undergoes a self-polymerization reaction in the presence of oxygen in alkaline aqueous solutions, cross-linking to form polydopamine (PDA) on the surface of the material in which it is immersed [43] (Figure 15.1). PDA is rich in -OH and -NH2 groups and possesses excellent hydrophilicity, bioactivity, and stability [44], and the modification of substrates with PDA can significantly improve cell adhesion, which is also conducive to the immobilization of osteogenic factors, resulting in improved cell proliferation, differentiation, and osteogenic gene expression functions [45,46]. Wang et al. performed surface modification of keratin/poly (lactic acid) nanofiber membranes by UV-initiated oxidative self-polymerization of DA to generate PDA in order to improve their biocompatibility and mechanical properties. Dopamine can chemically react with the side chain groups of keratins, while the self-polymerization, surface modification, and adhesion of DA can not only compensate the shortcomings of non-biologically active polymers such as PLA and improve the affinity of material cells and tissue adhesion [47–49], but also improve the mechanical properties of the composites [50,51]. The results showed that PDA successfully adhered to the surface of the fibers, the average diameter of the surface-modified fibers increased, the elastic modulus and elongation at break increased, and the cellular activity increased significantly after 24 h of inoculation. Bioactive glass, as one of the important bone regeneration materials, is widely used in bone tissue engineering due to its high osteogenic bioactivity,

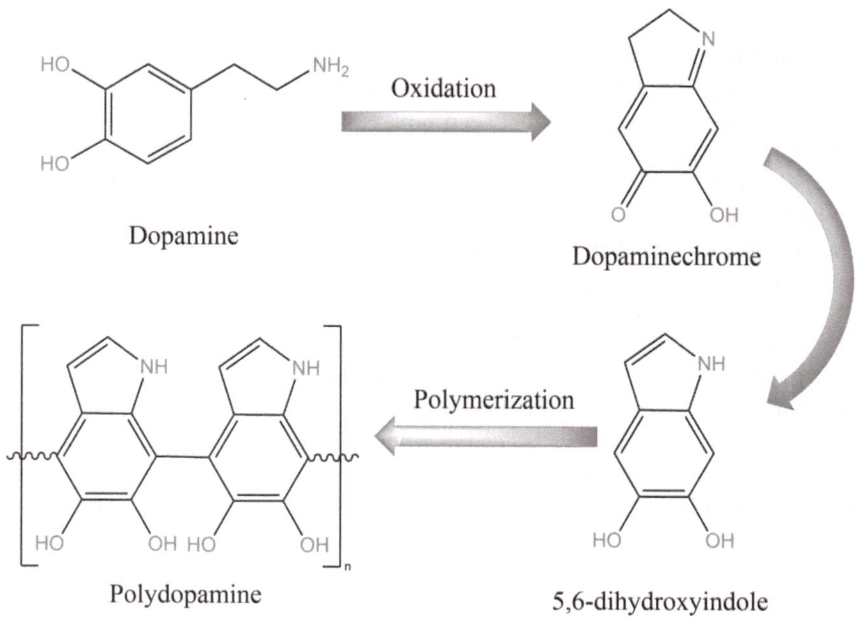


**FIGURE 15.1** Schematic diagram of polydopamine synthesis.

gene activation ability, and manageable biodegradation. In particular, the high surface area and pore volume are more favorable for bioactive nanoparticles as carriers to deliver guest molecules (drugs, proteins, and genes), which are expected to promote bone regeneration. However, the poor physiological stability of bioactive glasses greatly reduces the intracellular delivery efficiency. Improvements such as molecular grafting and surface coating are currently available to functionalize and modify inorganic nanoparticles to enhance their physiological stability [52]. Zheng et al. investigated the properties of Cu-MBG@PDA nanocomposites by changing the DA concentration and pH, which indicated that the optimal pH of the DA solution was 8.5, and the dosage of PDA attached to the surface of bioglass increased accordingly when the DA solution concentration was increased from 1 g/L to 2 g/L, but when the DA solution concentration was increased to 6 g/L, there was no significant increase in the amount of PDA attached to the surface of the bioglass. In vitro bioactivity studies showed that the modification of PDA improved the ability of copper-doped bioglass to induce hydroxyapatite production, thus improving its bioactivity [53].

### 15.2.2 Application as an Adhesive

Polydopamine is rich in functional groups and can graft polymers containing sulfhydryl and amino groups as well as some bioactive factors through Michael addition reaction and Schiff base reaction [54,55], leading to the formation of a new material layer on the surface of polydopamine coating to achieve modification of the material surface. Wang et al [56] fabricated surface-modified porous titanium metal composite scaffold materials (Ti/Tacrolimus@HPDA NPs) by depositing Tacrolimus@HPDA nanoparticles loaded with tacrolimus on the surface of 3D-printed porous titanium scaffolds using the viscosity of polydopamine. The modification of the porous titanium scaffold surface by Tacrolimus@HPDA NPs significantly enhanced the hydrophilicity of the scaffold surface. The enhanced hydrophilicity and the nanoscale rough structure of the scaffold surface both facilitate the adhesion and proliferation of rabbit bone marrow mesenchymal stem cells, which effectively contribute to the repair of bone defects, and the composite scaffold material is capable of promoting bone formation in vivo and has a better biosafety. Zhao et al [57] used electrostatic spinning technology to prepare polycaprolactone (PCL) nanofiber membranes, using the adhesion effect of polydopamine (PDA) to uniformly attach a polydopamine coating on the surface of PCL fiber membranes to form polydopamine-modified PC nanofiber membranes (PCL/PDA), and using the positive charge on the surface of PDA to adsorb and fix 45 nm gold nanoparticles on the PCL/PDA composite nanofiber membranes through electrostatic interactions. The PDA treatment not only enhanced the physicochemical properties of the fibrous membranes, but also enabled the gold nanoparticles to be successfully and firmly attached to the fibrous membranes, further improving the mechanical strength of the nanofibrous membranes and enhancing their hydrophilic properties, thus forming a composite scaffold material with good biocompatibility and showing good potential for potential to promote bone tissue regeneration.

### 15.2.3 Electroless Metallization

The catechol group of dopamine has a good binding ability to metals and the polydopamine composite layer exhibits a strong reduction ability to metal ions. The modified material with a polydopamine layer deposited on the surface will reduce the positive metal ions from the solution and deposit them on the surface of the material when immersed in a metal salt solution, thus achieving surface metallization of the material without electroplating [58]. Tan et al [59] used titanium as the substrate material and successfully prepared titanium sheets with silver surface loading by immersing the PDA surface modified titanium sheets in silver nitrate solution. The results showed that the antibacterial and cyto-compatible properties of the PDA-modified titanium flakes were favorable. The PDA coating was used to reduce the adsorbed $[Ag(NH_3)_2]^+$ to Ag nanoparticles in situ, and Ag nanoparticle coatings were successfully prepared on the surface of polyether ether ketone (PEEK) materials, and demonstrated the superior antibacterial properties of PEEK-PDA-Ag materials [60]. Zhang et al. successfully prepared polydopamine-nanosilver modified coatings on porous and smooth titanium surfaces, concluding that the constructed modified coatings solved the infection prone problem of porous surfaces and that DA-mediated mineralization effectively improved surface osteocyte compatibility [61]. Fu Chuan et al. used the reducing property of PDA itself to reduce gold tetrachloride ions to gold nanoparticles (AuNPs) and deposited them on the surface of L-lysine functionalized graphene oxide and PLGA co-blended composite scaffolds to prepare surface gold-loaded scaffold materials (AuNPs-PDA@PLGA/Lys-g-GO) [62]. The modified composite scaffolds possessed good hydrophilicity, mechanical strength, and antibacterial properties, which could significantly enhance cellular behaviors such as osteoblast adhesion, proliferation, osteogenic differentiation, and calcium deposition *in vitro* and promote new bone and collagen matrix production *in vivo*.

## 15.3 SURFACE MODIFICATION STRATEGY OF NANO-HYDROXYAPATITE FOR TISSUE ENGINEERING

Large-scale bone defects caused by periodontitis, trauma, tumor resection, genetic diseases, and systemic diseases have been challenging clinical problems. According to statistics, millions of bone transplantations are used every year to treat bone defects in orthopedics, neurosurgery, and oral and maxillofacial surgery worldwide [63]. The materials used for repairing bone defects mainly include autologous bone, allogeneic bone, and bone substitute materials [64]. Autologous bone transplantation is the "golden standard" of bone transplantation at present, but it has some disadvantages such as limited amount of bone, need to open a second operation area, and unpredictable bone absorption in the long term. The immune rejection and virus infection of allogeneic bone often lead to transplantation failure [65,66]. Recently, with the rise of tissue engineering and biomaterials, artificial bone substitutes have attracted more and more attention.

Hydroxyapatite (HAP) is a synthetic biomaterial, whose chemical composition is comparable to the mineral composition of natural bone, and has been widely valued and applied in advanced hard tissue engineering [67,68]. In fact, 70% of the natural skeleton is composed of 20–80 nm long and 2–5 nm wide nano HAP (nHAP) crystals, so nHAP is more popular than micron HAP. Previous studies have found that nHAP has high interface activity and exhibits higher biological activities, including promoting protein adhesion, cell adhesion proliferation, and bone regeneration [69,70]. The chemical composition of HAP is $Ca_{10}(PO4)_6(OH)_2$, and the atomic molar ratio of calcium and phosphorus is 1.67. HAP is mainly composed of hydroxy and apatite, in which hydroxy can be replaced by fluoride, chloride, and carbonate ions to form fluoroapatite or chloroapatite; The calcium ion can be replaced by various metal ions through ion exchange reaction to form apatite corresponding to different metal ions. The nHAP is mainly prepared by biomimetic mineralization, hydrothermal synthesis, co deposition, sol gel method, and other technologies [71,72].

HAP has good biocompatibility, biological activity and bone conductivity. Its surface hydroxyl structure enables it to form chemical bonds to host tissue directly. However, pure HAP has shortcomings such as high brittleness, low bending strength and slow degradation *in vivo*, which limits its application in bone tissue engineering [73,74]. Therefore, researchers have been trying to prepare bone regeneration materials by combining hydroxyapatite with polymers, fully incorporating the excellent properties of the two materials [75,76]. Common polymers are divided into natural polymers and synthetic polymers. Natural polymers mainly include polysaccharides (such as chitosan, sodium alginate, gelatin) and proteins (such as fibrin, collagen), while synthetic polymers include polyglycolic acid (PGA), polylactic acid (PLA), polylactic acid-glycolic acid copolymer (PLGA) [77,78]. However, the physical and chemical properties of hydroxyapatite and polymer materials are quite different, resulting in poor interfacial compatibility, and the biological activities and mechanical properties of the composite materials fail to achieve the desired effect [79]. In addition, nHAP will gather and form clusters due to the van der Waals force and hydrogen bond between them, which will further affect its combination with polymers.

Therefore, in order to enhance the compatibility between the two phases and reduce particle aggregation, researchers have made significant progress in improving the dispersion and colloidal stability of particles and enhancing the interaction with organic matrix through surface modification of nHAP. Surface modification can also endow nHAP with new biological activities, such as antibacterial activity and promoting angiogenesis, which broaden the application range of nHAP(such as drug carriers and ion adsorption carriers) [80,81]. In addition, surface modification can also affect the physical properties of nHAP by adjusting the length scale, surface roughness, morphology, and crystallization order of nHAP [82]. At present, many chemicals have been manipulated to

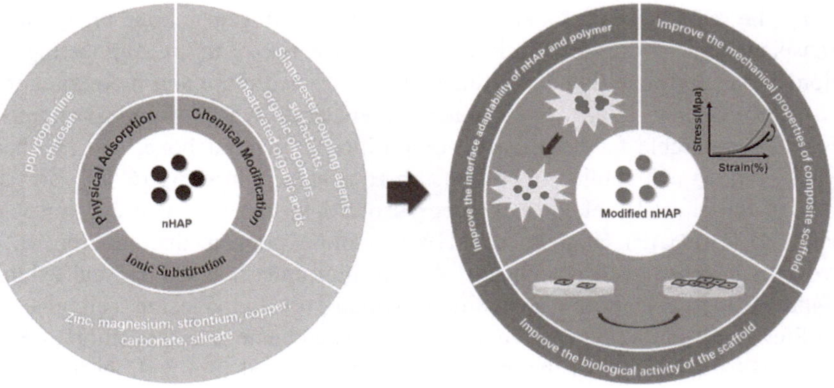


**FIGURE 15.2** Surface modification methods and functions of nHAP in tissue engineering composite scaffold.

change the surface properties of nHAP, including acids, alcohols, bases, polymers, proteins, silane coupling agents, natural polysaccharides, and peptides [83–85]. Here, we introduce and summarize the research on surface modification of nHAP in tissue engineering composite scaffold (Figure 15.2).

### 15.3.1 Physical Adsorption

Physical adsorption depends on the van der Waals force, hydrogen bond, or electrostatic attraction between molecules. Common adhesion molecules include polydopamine (PDA), chitosan (CS), which have good biocompatibilities and film-forming properties [86,87]. In order to increase the interaction between HAP and matrix in the composite scaffold, quaternized chitosan (qCS) was used as a template to realize the biomimetic nucleation and growth process of HAP via electrostatic interaction between tetrahedral structure of qCS and $PO_4^{3-}$. Then the porous gradient alginate-dopamine/qCS-HAP scaffold was prepared by "iterative layering" freeze-drying technology. The scaffold has a seamless integrated layer structure with high porosity (77.5%) and high compression modulus (1.7 MPa) [87]. Hu et al. coated the surface of HAP with PDA to obtain functional nHAP@PDA nanoparticles. TEM results showed that the average diameter of nanoparticles increased from 21.54 nm (nHAP) to 30.42 nm (nHAP@PDA). Subsequently, the nanocomposite fiber scaffold polylactic acid (PLA)/ nHAP@PDA was prepared by electrospinning. Compared with PLA and PLA/nHAP, the thermal stability, and static and dynamic mechanical properties of PLA/nHAP@PDA were significantly improved, due to the fact that the PDA coating promoted the interfacial adhesion between HA nanoparticles and PLA matrix. The high Young's modulus (112.76 MPa) of the composite scaffold can meet the mechanical requirements of human cancellous bone. In addition, the attachment and viability of mouse embryonic osteoblasts (MC3T3-E1) cultured on surface of PLA/nHAP@PDA were significantly improved than those of PLA and PLA/nHAP [88]. Tian et al. added tannic acid (TA) modified nHAP into polyurethane (PU) to prepare a composite scaffold. *In vivo* rat skull defects experiment revealed that the PU/HAP scaffold significantly enhanced bone mineral density and angiogenesis compared with the control group and the PU/HAP group [89].

In addition to being able to adhere to almost all types of materials, the above coatings can also be used as anchors to graft secondary functional biopolymers via Michael addition or Schiff base reaction via mercaptan and amine. Sun et al. prefabricated a PDA coating on the surface of nHAP, and then chemically grafted the bone morphogenetic peptide derived from BMP-7 with catechol. Compared with nHAP, peptide modified nHAP can significantly promote the adhesion, proliferation, and differentiation of MG-63 cells, especially at the concentration of 100 g/mL [24]. The bioactive molecules in the coating are usually released explosively at the initial stage, due to physical adsorption with weak force. In order to improve drug release, Zou et al. developed a coating that can effectively slow the initial release of drugs. Using vanillin as a non-toxic crosslinking agent, CS was cross-linked with antibiotic berberine to prepare coating on the surface of nHAP/polyamide66 scaffold [90].

### 15.3.2 Chemical Modification

Chemical modification is the most commonly used surface modification method, which changes the physical and chemical properties and biological activity of nHAP surface through the chemical combination of the active functional group in the surface modifier molecule with the hydroxyl or $Ca^{2+}$ on the surface of hydroxyapatite under certain conditions. The common surface chemical modifiers used for nHAP mainly include silane coupling agents (and ester coupling agents), surfactants, organic oligomers, and unsaturated organic acids [91–93].

#### 15.3.2.1 Silane Coupling Agent/Ester Coupling Agent

Silane coupling agents have been widely used in the surface modification of nHAP without increasing the cytotoxicity of the compound. Silane coupling agents are special compounds that can connect inorganic and organic substances. Their general formula can be expressed as $Y\text{-}R\text{-}Si\text{-}X_3$, where Y represents the active group that can have affinity with organic polymer molecules, such as amino, vinyl; X represents groups that can be hydrolyzed, such as methoxy, ethoxy, chloro. These groups can be hydrolyzed to form silanols, which combine with hydroxyl groups on the surface of hydroxyapatite to form siloxanes. At the same time, the films with network structure formed by mutual polymerization of silanols cover the surface of nHAP.

Silane coupling agents change the physical and chemical properties of nHAP surface, thus enhancing the interface adaptability between nHAP and organic polymers, improving the dispersion of nHAP in the matrix, and promoting the structural stability and mechanical properties of nanocomposite scaffolds. Wang et al. firstly obtained γ- MPS-modified HAP (MNHAP) by hydrothermal method, then prepared MNHAP/polylactic acid (PLLA) nanocomposite scaffold via thermal induced phase separation. The scaffold had nanofiber network (fiber size 100–800 nm), interconnected microporous structure (1–8 μm), and high porosity (>90%). When the content of nHAP in PLLA-based composite scaffold was less than 20wt%, MNHAP was evenly distributed in the scaffold, while NHAP tended to aggregate. When the content of MNHAP was increased to 30%, MNHAP particles appeared obvious agglomeration, which reduced the uniformity of composite scaffold structure [91]. Ma et al. used silane coupling agent γ-(2,3-epoxypropyl) propyltrimethoxysilane to modify HAP and prepared HAP/PEEK composites by hot-pressing. The tensile strength of HAP/PEEK composite containing 5 wt% modified HAP reached the maximum value, which was 23% higher than that of pure PEEK specimen [94]. Wang et al. modified nHAP with tetraethyl orthosilicate (TEOS) and (3-Aminopropyl) triethoxysilane (APTES) at the same time to further improve the dispersion of nanoparticles in gelatin (Gel)-oxidized gellan gum, and the composite scaffold had ultra-high compressive stress [95]. Jaramillo et al. compounded APTES-modified nHAP with epoxy resin. The APTES-HAP was covalently anchored by the amino group and R-epoxy resin, which improved the mechanical properties and thermal stability of the scaffold. However, the mechanical performance of the modified nHAP scaffold added with 10 wt% was not as good as that of the 5wt% scaffold, which was suspected that the high content of nHAP led to aggregation in the polymer [96]. In order to enhance the drug loading capacity, degradation control, and mechanical properties of bone regeneration tissue engineering scaffolds, He et al. synthesized polymethylene carbonate (PTMC)/APTES-modified HAP nanoparticles (APTES-HAP)/vancomycin hydrochloride (VH) porous microsphere novel scaffold. APTES-HAP particles showed enhanced compression modulus, improved structural stability, and significantly reduced swelling capacity. Due to the interaction between APTES-HAP and PTMC, the porous microsphere scaffold showed controllable biodegradability and mechanical properties. After 4, 8 and 12 weeks of treatment on rat femoral defect, PTMC/APTES-HAP/VH scaffold can promote bone regeneration by reducing peripheral inflammatory reaction [97]. In addition to promoting the chemical binding between nHAP and polymer matrix, the special terminal group of silanes can also improve the biological behavior of cells on the scaffold surface. Atak et al. prepared chitosan (CS), nHAP/CS composite scaffold (CS+nHAP), and amine-modified nHAP/CS composite scaffold (CS+nHAP-$NH_2$) by freeze-drying method. The modified nHAP was evenly distributed in the scaffold. Compared with CS and CS+nHAP scaffold, CS+nHAP - $NH_2$ scaffold had an effective surface for improving the adhesion, survival, and osteogenic differentiation of hBM-MSCs cells [98].

Through coupling agent modification, different active functional groups can be added to the HAP band, and these functional groups can be further grafted with different polymers. APTES was used to couple nHAP with citric acid, and then the modified nHAP was crosslinked with poly (propylene fumarate) and polyethylene glycol to obtain a composite scaffold with better interface binding performance. The scaffold has suitable compression modulus (100–300 MPa), biodegradability, and drug release capacity [99]. Taking APTES-nHAP as the core, Huysal et al. synthesized different generations of PAMAM dendritic modified HAP nanoparticles through repeated Michael addition and amination reactions to promote the combination of nHAP and other polymers [100]. Dai et al. coupled APTES on the surface of nHAP, and then modified the surface of nHAP by ring-opening polymerization (ROP) of L-phenylalanine N-carboxylic anhydride (Pha-NCA), effectively improving its colloidal stability in organic solvents, and preventing the aggregation of nHAP [101]. Liao et al. obtained the poly (γ-benzyl-L-glutamate) grafted HAP (PBLG-g-HAP)/PLLA scaffold by ring-opening polymerization of γ-benzyl-L-glutamate N-carboxyanhydride triggered by amino functionalized HAP. PLLA scaffold containing PBLG-g-HAP was more conducive to promoting the proliferation and differentiation of BMSC (increased mRNA expression of RUNX2, ALP, and OCN) [102]. Lee et al. prepared coating containing nHAP pretreated with β-cyclodextrin (β-CD) on the surface of Poly (*l*-Lactic Acid)/Gelatin (PLLA/PG) scaffold, where β-CD and the scaffold surface form a host-guest interaction via APTES. Compared with pure nHAP, the coating containing β-CD modified nHAP showed higher drug load [103].

#### 15.3.2.2 Surface Active Agent

Surfactants with hydrophobic group (nonpolar) and hydrophilic group (polar) located at each end of molecular structure respectively produce steric and electrostatic steric hindrance by adsorbing groups on the surface of particles, thus preventing the agglomeration of particles. Surfactants mainly change the surface characteristics of nanoparticles, and have a significant impact on the size and shape of nanoparticles. The common surfactants used on nHAP include non-ionic surfactants, such as polyethylene glycol (PEG), dodecanol, 12-hydroxystearic acid, and ionic surfactants sodium dodecyl sulfate (SDS), hexadecyl trimethylammonium bromide (CTAB) [92,104–107]. The chemical binding mechanism between different surfactants and HAP is different.

Non-ionic surfactants form complexing reaction with $Ca^{2+}$ in HAP through multi-hydroxyl structure. Fu et al. coated PEG on the surface of nHAP to enhance its dispersion in polycaprolactone-polyglycol-polycaprolactone (PCL-PEG-PCL, PCEC) matrix. SEM observation showed that the uniformly distributed circular pores on the surface of PCEC film with 20% PEG-nHAP decreased. With the increase of nHAP content to 30% and 40%, the surface morphology of the composite film changed significantly, and the pore size changed irregularly. In addition, some drop pits appeared on the surface of the composite film, and HAP agglomerates with size of 2 to 10 μm could be observed at the bottom of the drop pits. It shows that the content of nHAP is too high, which results in the formation of agglomeration and disrupts the continuous crosslinked PCEC matrix. The test of mechanical properties showed that PEG-HAP could significantly enhance the strength of the composite film, and the tensile modulus increased from 352.49 ± 59.57 MPa of pure PCEC film to 693.70 ± 30.82 MPa [92]. Ding et al. prepared 12-hydroxystearic acid surface-modified nHAP particles by co-precipitation method and blending method respectively. The long-hydrocarbon non-polar chain of 12-hydroxystearic acid was used to reduce the surface energy of HAP and increase its hydrophobicity to improve dispersion in PLGA. Thermogravimetric analysis showed that more 12-hydroxystearic acid was grafted onto the surface of HAP prepared by solution blending. The morphology of nHAP obtained by blending method (BI-HAP) did not change significantly, and it was needle-like, 110–130 nm long, and 10–20 nm wide. However, the morphology and size of nHAP modified by co-precipitation method (Co-HAP) changed greatly, with a length of about 60–80 nm and a width of 6–8 nm. The mechanical properties of the composite with modified nHAP were significantly improved, especially the tensile strength of the composite scaffold with 5 wt% nHAP which was 23.81% higher than that of pure PLGA scaffold [104]. nHAP surface modified with dodecyl alcohol via esterification reaction could promote the dispersion of nHAP in the PLA matrix and improve interfacial interactions between PLA and nHAP, which benefits from the hydrophobic surface of dodecyl alcohol. The tensile strength of 1.0 wt% modified HAP/PLA scaffold was the highest, 9.7% higher than that of pure PLA. Adding nHAP nanoparticles into PLA could effectively improve the hydrophilicity, cell adhesion, diffusion, and proliferation of the scaffold [105].

Gheysari et al. treated HAP with ionic surfactant SDS and CTAB, and TEM results showed that single-phase HAP nanorods with aspect ratio > 10 and high bioactivity could be synthesized in the presence of SDS. The prepared high-porous gelatin/carboxymethyl cellulose/HAP nanocomposites could improve the activity and adhesion of cells [106]. In addition, Yahia et al. also used different concentrations of CTAB and SDS surfactants to fabricate HAP with controlled morphology and small size, and found that the modified HAP has good binding ability with protein [107].

### 15.3.2.3 Organic Oligomers/Unsaturated Organic Acids

The modification by organic polymers can significantly reduce the aggregation degree of hydroxyapatite particles, increase the dispersion stability, and endow HAP with more new characteristics. Polymers such as polylactic acid (PLLA), acrylic acid, vinyl phosphonic acid β-CD has a wide range of applications in HAP/polymer composites [93,108–110].

Most organic polymers can chemically bond to hydroxyl groups on the surface of nHAP. In order to improve the binding between nHAP particles and PLLA, and promote the mechanical properties of PLLA/HAP composite as a potential bone substitute, HAP nanoparticles were pre-grafted with PLLA (PLLA-HAP). PLLA molecules were grafted on the surface of HAP as bridging molecules, which played an important role in improving the bonding strength of PLLA/HAP scaffold. When the content of PLLA-HAP was low (< 4wt%), the PLLA/PLLA-HAP nano-composites exhibited higher bending strength and impact energy than the pure PLLA, which indicated that the PLLA-HAP particles had both reinforcing and toughening effects in the composites [111]. Zhang et al. prepared PLLA-modified HAP by ring-opening polymerization of LLA on the surface of HAP particles through the action of stannous octoate (Sn (Oct)) catalyst, and then fabricated PLLA-modified HAP/PLGA three-dimensional porous scaffold by solvent casting/particle leaching method. After surface modification, the distribution of PLLA-modified HAP particles in PLGA was more uniform, but the calcium exposure was reduced. The intramuscular implantation results indicated that the PLLA-modified HAP/PLGA scaffold was more stable and had more suitable biodegradability than the PLGA scaffold [112]. Foroughi et al. improved the mechanical properties of nHAP/ poly-3-hydroxybutyrate (P3HB) composite scaffolds by coating 2, 4, and 6 wt% of P3HB on the surface of nHAP respectively. The compressive strength and modulus of the scaffold containing 50 wt% HAP coated with 6 wt% P3HB were 1.51 MPa and 22.73 MPa respectively, which matched to those of bone tissue [85]. Kumar et al. grafted hydroxyethylenediphosphonic acid onto the surface of nHAP, and the morphology of nanoparticles changed from irregular shape to flaky shape. With the concentration of modified nHAP increasing from 0 wt% to 30 wt%, the compressive strength of the polyurethane/nHAP nanocomposites increased from 0.094 MPa to 22.4 MPa. The compressive strength (22.4 MPa) of the nanocomposites with 30 wt% filler concentration was very close to human trabecular bone [109].

In addition to meeting the appropriate biodegradability and mechanical properties, some amino acids have been used to modify HAP to improve the biocompatibility and bioactivity of composite scaffolds for tissue engineering. Jiang et al. designed an effective surface grafting of nHAP assisted by L-lysine, and incorporated the modified nHAP into PLGA nanocomposites. The grafting of L-lysine improved the dispersion of nHAP and made the scaffold better mechanical properties and biocompatibility [113]. Timpu et al. used arginine (Arg) or polyethyleneimine (branched-bPEI, or linear-LPEI) as cationic modifiers and dispersants to obtain functional nHAP through wet chemical technology. Among them, LPEI functionalized nHAP has good similarity with biological apatite and the best DNA binding ability [114]. Haider et al. first grafted L-glutamic acid on the surface of nHAP, then dipped it into water-soluble carbon dioxide to activate the carboxyl group on the surface of nHAP, and chemically combined with bone morphogenetic protein (BMP-2) at last. BMP-2 grafted nHAP/PLGA nanofiber scaffold greatly stimulated the growth of osteoblasts, compared with the control nHAP/PLGA and pure PLGA scaffold [84].

Limited active hydroxyl groups and weak reactivity affect the grafting of organic polymer onto nHAP surface. In order to improve the grafting ratio of PLLA to HAP surface, carboxyl end modified PLLA was used to treat nHAP as a particle emulsifier in 3D printing high internal phase lotion [115]. Nichols et al. prepared a 1–2 nm thick degradable acrylic film on the surface of nHAP by radio frequency plasma polymerization technology. The activated carboxyl groups ensure the covalent binding between nHAP and polymer matrix PLGA [108]. Lee et al. grafted caprolactone with three different hydroxyl groups on the surface of nHAP (the hydroxyl group of unmodified nHAP, the hydroxyl group of nHAP modified by lactic acid, and the hydroxyl group of nHAP modified by ethylene glycol). Compared with unmodified hap/PCL and pure PCL, the grafted HAP/PCL nanocomposites exhibit stronger tensile strength and toughness [116].

The organic polymer can not only directly combine with hydroxyl groups on the surface of hydroxyapatite, but also chelate with $Ca^{2+}$ on the surface of nHAP. Yi et al. chelated hydroxyapatite with citric acid, followed by cross-linking between the hydroxyl groups of β-CD, and further in situ ring-opening polymerization of PLA over the hydroxyl group. HAP was successfully modified with PLA/β-CD/citrate acid network. The modified HAP showed uniform distribution, relatively high hydrophobicity, and surficial positive charges. The synthesized PLA/β-CD/HAP showed better adhesion and higher activity of mesenchymal stem cells (MSCs), and greater osteoinductivity [110]. Carboxymethyl (CM)-β-CD has amphiphilic structure, whose outer surface is hydrophilic and central cavity is hydrophobic. Tang et al treated HAP with CM-β-CD to improve the dispersion of HAP in PLGA matrix and form a good interface binding with PLGA [117]. Kumar et al. obtained HAP modified by triethanolamine (TEA) through chemical coprecipitation. After TEA modification, the morphology of nano-hydroxyapatite changed from the initial crystal to irregular sheet/plate (30–40 nm). Compared with pure PU, PU/TEA-nHAP nanocomposites had improved cell compatibility and cell viability [118].

### 15.3.3 Ionic Substitution Modification

In recent years, surface modification by doping exogenous ions in nHAP through chemical coprecipitation, hydrothermal method, sol-gel method has attracted extensive attention [119,120]. The special structure of nHAP makes it easy to be replaced by ions at different crystal points (such as phosphate and calcium sites). Zinc ($Zn^{2+}$), magnesium ($Mg^{2+}$), strontium ($Sr^{2+}$), copper ($Cu^{2+}$), carbonate ($CO_3^{2+}$), and silicate ($SiO_4^{4+}$) have been doped into nHAP [121–129]. The doping of these ions does not destroy the structure of nHAP skeleton, but changes the crystallinity, morphology and solubility of nHAP. Some studies have shown that loading the above elements in bioceramics based on calcium phosphate (CaP) will greatly improve the biological activity, biocompatibility, and conductivity of bioceramics [130].

Heidari et al. prepared pure HAP scaffold and ZnO modified HAP scaffold by cold isostatic pressing. Compared with pure HA scaffold, HAP/ZnO scaffold has higher compressive strength, fracture toughness, and density, but lower hardness. After the scaffold was immersed in SBF solution, the surface roughness of HAP/ZnO scaffold increased from 773.9 nm to 1503 nm, indicating that ZnO can increase the apatite deposition on the surface of HAP scaffold. Although in vitro biological analysis shows that HA/ZnO scaffold has quiet good ALP activity and biocompatibility, the proliferative activity of cell cultured on the surface of HA/ZnO has decreased, because the early excessive release of Zn ions brings adverse effects on cells [121]. Elsayed et al. developed a cellulose acetate electrospun nanofiber scaffold containing HAP doped with different content of Cu ions to promote tissue healing via antibacterial properties of Cu ions. The addition of Cu ion promoted the adhesion, diffusion, and proliferation of fibroblasts, but did not significantly improve the mechanical properties of the scaffold [122]. Similarly, Sr has been used to modify HAP (Sr-HAP) as a common material in bone regeneration due to its good bone conductivity and high alkaline phosphatase activity. In a 3D in vitro bone model system, the physical and chemical characteristics, cell compatibility, and osteogenic induction ability of different nano bioceramic materials (such as HAP, BCP, Sr-HAP, and Si-HAP) were compared, revealing the superior performance of Sr-HAP [123].

Although some physical-chemical properties or biological functions of HAP replaced by single ions have been improved to some extent, its practical application is still limited. Therefore, researchers began to try to modify HAP through multi-ion substitution. It is worth noting that the scaffolds for bone defects usually require high compression modulus. Although Sr HAP nanoparticles have good biological activity, the composite scaffolds prepared by them do not have ideal mechanical properties. Therefore, Hassan et al. synthesized strontium and zinc doped nano hydroxyapatite (Sr/Zn HAP) using the hydrosol gel technology to obtain excellent mechanical properties and osteogenic activity [124]. El-Hamshary et al. blended different concentrations of Ag ion with vanadium (V) salt ion and HAP to modify the prepared Ag/V-HAP@PCL. The fracture strength of the support increased from $2.51\pm0.35$ MPa to $4.23\pm0.64$ MPa. Moreover, the scaffold showed significant antibacterial activity, promoted cell adhesion, and the ability to adsorb protein [125]. Tomoaia et al. synthesized nHAP-Si, nHAP-Si-Mg, and nHAP-Si-Mg-Zn by chemical precipitation, respectively. TEM and SEM observation showed that the morphology of the synthesized nHAP-Si changed from needle to spherical particles. nHAP-Si-Mg and nHAP-Si-Mg-Zn presented a micro-domain (aggregate) formed by cauliflower-like nanoparticles, in which there was a porous system for identifying interconnection. In addition, in vitro cytological experiments confirmed that the incorporation of a small amount of Si, Mg, and Zn into the nHAP lattice could improve the biological activity of human osteoblasts on the HAP/COL/CS scaffold [126]. Tithito et al. prepared CS grafted polymethylmethacrylate (CS-g-PMMA)/HAP loaded with mineral ions composite scaffold, which has good drug release and promote the inward growth of bone cells. HAP loaded with mineral ions was obtained by placing HAP in simulated body fluid for hydrothermal treatment. The element analysis of the content of the synthesized HAP sample revealed the presence of Ca, P, K, Mg, Na, and Cl in the sample. XRD results showed that in addition to apatite crystal phase, there was tricalcium phosphate crystal phase on the surface of composite nHAP, and the existence of two phases might increase the biodegradability of biomaterials [127].

In addition to doping inorganic ions in the framework of nHAP, some scholars have blended polysaccharide polymer with HAP to improve the biodegradability and biological activity of nHAP. Chen et al. prepared gelatin-modified nHAP by chemical coprecipitation method. The gelatin-modified nHAP was observed under TEM as nanoparticles with a length of 30 to 50 nm and a width of 5 to 15 nm. According to the FTIR spectrum, the gelatin-modified nHAP spectrum showed a $CO_3^{2-}$ peak, which meant that the prepared nHAP was hydroxy-carbonate apatite, similar to the HAP in natural bone. It is speculated that during the synthesis of gelatin-modified nHAP, the alkaline reaction solution environment will introduce carbon dioxide in the air into the reaction system, replacing $PO_4^{3-}$. This substitution will reduce the crystallinity of HAP and enhance the degradability of HAP. This is beneficial to the scaffold because HAP with good crystallinity is too stable and the degradation cycle is too long [128]. Shebi et al. extracted pectin from natural balsam pear, and functional sites such as carboxyl and hydroxyl groups in pectin combined with $Ca^{2+}$ in HAP to form carboxylate ions, thus initiating crystal nucleation and growth. The use of natural balsam pear containing pectin can effectively stabilize or wrap the nanoparticles after the nucleation process, thus reducing the particle size [129].

### 15.3.4 Conclusion

HAP nanocomposite scaffold plays an important role in tissue engineering, especially in bone tissue engineering, due to its excellent biological properties. However, hydroxyapatite is easy to aggregate, resulting in poor interfacial compatibility and stability of the composite, which limits the further development and application of this kind of composite materials. Therefore, it is recommended to pre-modify the surface of nHAP when it is combined with polymer to prepare composite scaffold. The modification methods are mainly divided into: (1) physical adsorption modification; (2) Chemical surface coating modification; (3) Ion substitution modification. The binding strength of physical adsorption is low, and there are drawbacks of initial explosive release. Chemical surface binding is more effective for a long time, allowing continuous regulation of cell behavior. However, the method is usually multi-step and complicated, such as surface activation or functionalization steps requiring chemical treatment. In addition, the denaturation of bioactive molecules by toxic chemicals and multiple reaction steps may be another problem that needs to be solved, which limits its clinical application. In addition, the amount of ion substitution is a problem that needs to be emphasized. Although previous studies have shown that metal ions, such as Zn, Sr, can promote the proliferation and differentiation of osteoblasts, the excessive substitution rate in nHAP will lead to the explosive release of ions at the initial stage, producing certain biological toxicity to cells. To sum up, first of all, the concentration of modified substances and the amount of grafting should be strictly and accurately controlled; Secondly, when the effect of a single modification method is limited, the combination of two or more modification methods and modifiers may bring synergistic effect.

## 15.4 SURFACE MODIFICATION OF TITANIUM/TITANIUM IMPLANTS WITH MICRO-NANO STRUCTURE

### 15.4.1 Introduction

Since the discovery of osseointegration, titanium dental implants are widely used for restoring missing tooth for patients. Titanium implants have several favorable properties such as mechanical strength, chemical stability and biocompatibility, but still have a failure rate of 4% - 6% in clinical application [131–133]. The most common causes of implant failure are aseptic loosening caused by poor osseointegration and peri-implant inflammation caused by bacterial infection [132,133]. Therefore, the research of implant materials mainly focuses on two important issues: How to improve the early osseointegration performance and how to avoid peri-implant inflammation.

A large number of studies *in vivo* have showed that the early osseointegration was largely influenced by the implant surface roughness, and moderately rough (Ra=1–2 μm) surfaces showed the strongest bone responses [134]. As a result, sand-blasting and acid-etching (SLA) implants were developed and put on the market. The common procedures include blasting with large size alumina particles (0.25–0.50 μm), and then etching with $HCL/H_2SO_4$ [5]. Micro-sized pits are formed through sand-blasting, which makes the titanium surface reach the preset optimum roughness. Acid-etching helps remove the residual contamination on the surface, smooth sharp edges, and activate the surface to promote protein adsorption and cell adhesion [135]. SLA implants have showed successful clinical application for decades, yet it is generally believed that the osseointegration performance and anti-infection ability of SLA implants should be further improved, especially in recent years when materials science has been developed rapidly.

SLA surfaces are hydrophobic (with contact angles of 90°–150°), which is not conducive to early bone contact [136]. It is widely accepted that hydrophilic surfaces are more beneficial to the adhesion, proliferation, and mineralization of osteoblasts than hydrophobic surfaces [137]. In order to achieve superior osseointegration performance, a variety of chemical methods have been applied on SLA implants to increase surface hydrophilicity and have made progress [138]. Furthermore, some researchers found that titanium implants will gradually lose their hydrophilicity due to continuous pollution of hydrocarbons in air or water after production, which is called biological aging [139]. Therefore, the storage of hydrophilic SLA implants and the chair-side activation of biological aging surfaces are also worth investigation.

Another problem with SLA implants is the high risk of peri-implant inflammation after implantation. Researches showed that hydrophobic titanium implants with rough surface (such as SLA surface) would increase bacterial attachment and colonization, thus increasing the risk of peri-implant inflammation when expose to external environment such as oral cavity [140]. When it comes to the long-term prognosis of implants restoration, especially for patients such as periodontitis (CP) and coronary heart disease (CHD), peri-implant inflammation is the main factor that threatens the stability of implants [141]. Dental implants generally face the risk of bacterial infection during surgery or blood-borne bacterial transmission *in vivo*. The surface characteristics of the implant greatly affect the initial adhesion of bacteria. Therefore, surface modification is the key strategy for preventing peri-implant inflammation [142].

SLA technology cannot improve the corrosion or friction resistance of titanium surface [143]. In the peri-implant microenvironment, inevitable interaction between titanium surface and tissue fluid (composed of $Na^+$, organic acid, protein, etc.) and/or saliva (composed of $H_2O$, $Cl^-$, etc.) may cause metal corrosion through electrochemical oxidation reaction [144]. It has been proved that local release of free metal particles/ions mainly induced by corrosion into the surrounding bone tissue will cause inflammatory reaction and allergic reaction, and eventually lead to the aseptic loosening of implants. The bacterial colonization of peri-implant inflammation will undergo electrical erosion, which will further decrease the corrosion resistance of titanium surface [145,146]. Therefore, improving the physicochemical properties of SLA titanium surface is another important direction to promote its long-term effect.

In conclusion, our requirements for titanium implants are constantly changing from mechanical properties matching with bone, to biocompatibility, then to osteoinduction and antibacterial properties, according to technological developments. With recent advancement in molecular biology, nanomedicine, nanotechnology, and other subjects, researchers are now developing multifunctional implants. Considering the successful clinical application of SLA titanium implants for many years, it is more reasonable to upgrade the SLA surface using new technology than to introduce a new material or a new primary modification method to replace it. The aim of this review is to present recent progress on advancing performances of titanium SLA implants using new surface engineering approaches, and provides a prospective outlook for the future modification of SLA implants (Figure 15.3).

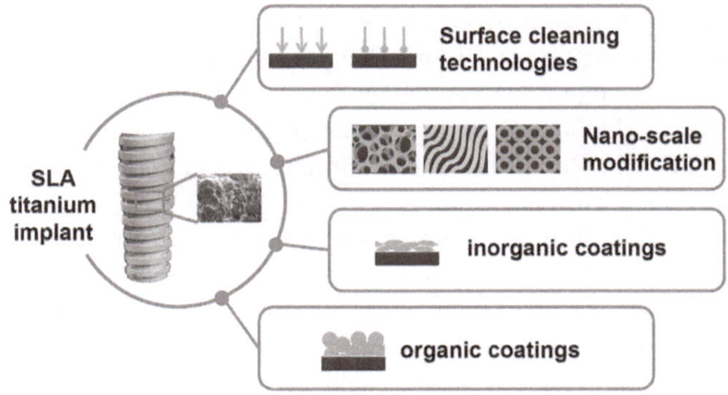


**FIGURE 15.3** Surface modification methods of SLA titanium implants.

### 15.4.2 Surface Cleaning Technologies

The oxide layer on the surface of SLA implants will constantly absorb organic hydrocarbon pollutants in storage, thus leading to lower surface properties [147]. However, surface cleaning before clinical use can remove pollutants, endow titanium surface with super hydrophilicity without changing its morphology, and also have considerable sterilization function. At present, the most common methods of SLA surface cleaning are ultraviolet (UV) functionalization and non-thermal plasma (NTP) treatment.

#### 15.4.2.1 UV Functionalization

Photofunctionalization (PhF) refers to the modification of titanium surface after UV treatment, including the change of physicochemical properties and the enhancement of biological capabilities [148]. UV radiation is categorized into UVA (wavelength $\lambda$ = 320–400 nm), UVB ($\lambda$ = 280–320 nm), and UVC ($\lambda$ = 200–280 nm). UVA and UVC can both effectively improve the hydrophilicity of titanium surface and eliminate hydrocarbon pollution [149].

Mercury lamps and bacteria lamps were used in many experiments before to induce PhF, and usually took a long time for treatment. Shen JW et al. employed 15 W bacteria lamp to irradiate SLA implant for 48h, and detected improvement of bone-to-implant contact and osseointegration *in vivo* [150]. After 48 hours of exposure with mercury vapor lamp (UVA/UVC, 500 J/cm$^2$), the reduction of surface carbon pollution and the enhancement of hydrophilicity were also observed [151].

TheraBeam ultraviolet ray treatment meter is currently used in many experiments, which can provide UV light as a mixture of UVA ($\lambda$ = 360 nm, intensity 0.05 mW/cm$^2$) and UVC ($\lambda$ = 250 nm, intensity 2 mW/cm$^2$) [152]. It can significantly improve the early-phase stability of SLA implants and promote osseointegration after 12–15 min treatment [153–155].

#### 15.4.2.2 NTP Treatment

Plasma, known as the fourth state of matter other than liquid, gas, and solid, refers to gas containing plasma particles (such as free electrons, ions, and various other active atoms or molecular radicals). Among them, plasma at room temperature is called non-thermal plasma (NTP) [156]. NTP is usually produced by electric discharges of different nature. Plasma jet (only hit small target areas) and dielectric barrier discharge (DBD) (for larger areas) are the most common NTP sources for medical applications.

Duske et al. applied Plasma jet (INP Greifswald, Greifswald, Germany) to modify titanium surface with argon (5 slm, standard liter per minute) and oxygen (1.0%) at a distance of 5 mm for 2 min [157]. SLA surface was completely hydrophilic after plasma modification, with better cell adhesion and spreading compared with untreated samples. *In vivo* experiments also showed enhanced early osseointegration. Compressed air (oxygen 16%, 1% hydrogen, 78% nitrogen) was employed instead of argon for surface treatment in the follow-up study [158]. Although the surface properties of SLA titanium were improved, the effect was not as significant as that of argon.

Wang et al. used the atmospheric pressure glow discharge plasma system AS400+PFW10 (Plasma Treatment GmbH, Stein Hagen, Germany) to modify the titanium plate at a distance of 2 cm for 90 s with argon (300 L/h) and oxygen (5%) [159]. Removal of surface carbon pollution and increasement of hydrophilicity were observed, and the adhesion, proliferation, and mineralization of osteoblasts were significantly improved. Lee et al. made a Plasma jet device to generate a plasma stream with nitrogen-containing ammonia (1000 sccm), and treated the titanium plate at 3 mm for 10 s [160]. The modified surface showed excellent antibacterial properties.

The small nozzle device described above produces plasma flow in a small area, with an effective area of 15 mm, and the NTP effect decreases as distance increases from the central area [157–160]. Therefore, it is necessary to move the specimen manually in order to obtain uniform NTP functionalization. Especially for screw implants for clinic use, relatively complex and uncontrollable treatment process will lead to heterogeneity of treatment effect. To solve this problem, Dong et al. adopted DBD technology to directly use titanium implants as coaxial internal electrodes, and inserted dielectric materials between the internal and external electrodes to generate cold plasma [161]. This method can make uniform plasma flow submerge the entire implant, facilitating rapid hydrophilic treatment. The device uses argon (3000 sccm, standard cubic centimeter per minute) and oxygen (0.3%) as working gases to activate a 20w–50 w plasma source. After treatment of 30 s, the hydrophilicity of SLA implant was increased, and the bone integration was enhanced *in vivo*.

The high level of reactive oxygen/nitrogen species (ROS/RNS), high-density level of -OH and -NH2 groups produced by NTP treatment can remove carbon contamination on titanium surface and increase hydrophilicity [162]. The enhancement of hydrophilicity not only facilitates cell adhesion, proliferation, and differentiation, but also reduces the adhesion of hydrophobic bacteria. ROS/RNS are bactericidal active particles processing abilities to cause physical damage to the microbial cell membrane and trigger programmed cell death.

One of the clinical application advantages of NTP treatment is that it can achieve rapid chair-side activation of implant surface. It has been proven that 12 min of UV treatment or 1 min of NTP treatment could equally improve surface hydrophilicity and increase osteoblasts adhesion [152]. This shortening of the processing time makes NTP treatment more practical for clinical routines.

### 15.4.3 Nano-scale Modification

Nanostructures (<100 nm) are believed to be capable of regulating cell differentiation, gene expression, and tissue regeneration, owing to similar size with collagen, protein, and cell membrane receptors in extracellular matrix (ECM) [163]. They have unique influence on molecules and cell activities and can ultimately improve the osteointegration of implants through integrin and protein adsorption signal pathway.

Nanomorphology can also solve the problem of increased bacterial adhesion on SLA surface. It is well known that surface hydrophobicity/hydrophilicity and effective contact area are two key factors leading to different adhesion behaviors of bacteria [140]. Nanostructures can remarkably reduce bacterial adhesion by constructing super hydrophilic structure and reducing effective contact area. In addition, fine and dense nanowires, nanoneedles, and nanopillars show bactericidal activity conducted by physically piercing the bacterial cell membrane [164,165]. These surface physical modification overcoming the drawbacks of short action duration and drug resistance of chemical antibacterial agents, are expected to become new strategies for anti-infection of biomaterials.

In conclusion, nano-scale modification on SLA surface is a vital strategy to improve its surface properties. This micro/nano-scale topographies modification can reduce bacterial adhesion, add drug delivery functions, and enhancing bone adhesion at the same time. Up to now, a variety of technologies have been applied to nano-scale modification of SLA surface, such as alkali heat treatment, laser modification, electrochemical anodizing, etc., which can produce different forms of structures, such as pits, holes, tubes, columns, lines, etc., and exhibit a variety of biological functions.

#### 15.4.3.1 Alkali-Heat Treatment

Alkali-heat treatment includes soaking titanium materials with strong alkali solution (usually NaOH), and heating them at high temperature for a certain time to obtain titanium nano-nets (TNNs) [166]. Altering the parameters of alkali-heat treatment, such as temperature, time, concentration, etc., can produce surfaces with different morphologies.

The concentration of NaOH affect the morphology of treated surface. As for 5 hours treatment at 110 °C, 1 M NaOH aqueous solution treatment would generate nanowire structure surface, while 5 M and 10 M treatment would generate nanonet structure with pore diameter of 264 nm and 188 nm respectively [167]. Whereas the change of processing temperature would affect the pore size of the nanonet: The pore size was about 50~100 nm after treatment in 10 M NaOH aqueous solution for 24 h at 30 °C; As the temperature rise above 40 °C, the pore size would increase to 100~200 nm [168]. Nowadays, alkali-heat treatment has been applied by a large number of researchers as an elementary surface treatment technology before other coating strategies, but no uniform treatment parameter has been recognized. 5 M and 10 M NaOH aqueous solution, 70–80 °C treatment temperature and 12–24 h processing time are most commonly used [166–171].

Su et al. found that 600 °C is the best temperature for heat treatment on titanium surface [168]. Excessive temperature would destroy the structure formed by alkali corrosion. However, many studies omitted this step and conduct subsequent experiments directly after alkali treatment, which also showed superior biological activity.

A large number of experiments have confirmed that alkali heat treatment on SLA surface will not affect its original micro-scale morphology, but add nano-scale structure on SLA surface [170,171]. The micro/nano-scale structure surface obtained by this strategy can further increase wettability of SLA implants, improve cell proliferation and protein production, and show better osteointegration *in vivo*.

#### 15.4.3.2 Laser Modification

Laser modification is a new surface topography micromachining technology, which can melt or evaporate selected parts of titanium surface to obtain specific surface morphology through thermal effect and photon effect [172]. Compared with chemical and mechanical methods, laser technology presents unique characteristics of flexibility, simplicity, controllability, and reproducibility, making it a very promising technique to achieve specific patterns or texture on titanium implants for promoting biological functions [173].

Among various laser sources, long-pulse fiber lasers, including millisecond lasers and microsecond lasers, are suitable for complex micro-scale surface modification and have been widely used in industry [174]. In recent years, femtosecond lasers have become an advanced technic for metal surface processing, as it produces controlled features with fewer thermal effects and less collateral damage [175]. An important feature of femtosecond laser is that it can create self-organized quasi-periodic patterns, commonly known as laser induced periodic surface structure (LIPSS), which can produce anisotropic surface properties and perform various biological functions.

Cunha et al. used Yb: KYW chirped-pulse regenerative amplification laser system (Amplitude Syst è mes s-Pulse HP; Bordeaux, France) with a central wavelength of 1030 nm and a pulse duration of 500 fs to produce surface texture. Periodic ripples of about 700 nm were formed on titanium surface with fluence of 0.5 J/cm$^2$. While micro-columnar texture was formed with fluence between 0.5 and 2 J/cm$^2$ [176]. In the follow-up study, uniform LIPSS surface of parallel nano-ripples and nano-pillar texture surface can be produced by changing the fluence and moving samples by means of a computer-controlled XYZ stage (PI miCos; Eschbach-Germany). *In vitro* experiments showed that both textures improved cell adhesion and enhanced matrix mineralization [177].

Dumas et al. employed Ti: Sa laser chain (Bright Thales) to provide 120 fs, 800 nm pulses at a repetition rate of 5 kHz. By different parameter designs, including mean power (P), beam diameter, fluence (F), and speed, three topographies are produced: 1) pits and nano bottom, 2) pits and nano top, 3) nanostructures (ripples). All morphologies showed enhanced cell adhesion and osteogenesis, and the first two morphologies showed the ability to induce cell directional movement as well [178].

In general, all laser parameters (fluence, number of pulses, beam diameter, wavelength, scanning speed, etc.) can be easily altered. Therefore, expected geometric pattern can be created by gradually modifying titanium on submicron scale, which is helpful to control cell reaction more accurately.

Erbium-Doped Yttrium Aluminum Garnet Laser (Er: YAG) Irradiation has been applied to the modification of SLA surface. Laser irradiation enhanced the deposition of $TiO_2$ layer on SLA surface and reduced the surface roughness, resulting in reduced bacterial adhesion with osteoinductive activity retained [179]. Unfortunately, laser technologies are mainly applied on polished titanium surface at present. Applying more advanced laser technologies to SLA surface modification is a potential way to further enhance its properties.

#### 15.4.3.3 Electrochemical Anodizing

Anodic oxidation is an electrochemical method which can result in a porous and chemically modified titanium surface [180]. Anodization uses titanium specimen as anode and platinum/titanium as cathode, both of which are immersed in electrolyte. Under preset current and voltage conditions, the oxidation reaction occurs on the titanium surface, and finally generates titanium nanotubes (TNTs) [180–182]. These nano-scale tubes endow anodized surface with capability of loading and delivery various drugs, including antibiotics, anti-inflammatory drugs, proteins, and growth factors [182].

Yang et al. treated the SLA surface with electrochemical anodization treatment in 5 M NaOH solution under anodic currents of I1 (<0.1 A) or I2 (<0.2 A) in less than 30 min. The obtained structures contain nanotubes of about 100 nm on SLA surface [183]. Compared with traditional SLA topography, this micro-nano structure proved highly effective in enhancing osteogenic responses *in vitro* and *in vivo*.

Huang et al. performed anodic oxidation procedure in aqueous electrolyte containing 0.15 M $NH_4F$ and 0.5 M $(NH_4)_2SO_4$ at 20 V for 30 min to obtain a similar surface morphology [184]. The incorporation of nanotubes on SLA surface increased the hydrophilicity, enhanced cell differentiation, and increased osteogenesis. However, nano-scale TNT demonstrated better biological activities than TNT/SLA surface. Therefore, whether the combination of SLA and TNT can achieve the best effect remains to be studied.

The future research may concentrate on exploring the optimal parameters of TNT manufacturing on SLA surface, and to load active molecules on the surface to achieve controlled peri-implant drug release *in vivo*.

### 15.4.4 Inorganic Coating

Loading inorganic coating is one of the classic strategies to improve the osteogenic activity of titanium surface. Hydroxyapatite (HA), which composition is similar to natural bone, can not only improve the bioactivity of titanium implant surface, but also inhibit electrical corrosion [185]. Other popular coating materials include α-Tricalcium phosphate (α-TCP), β-Tricalcium phosphate (β-TCP), zirconium hydrogen phosphate, etc. [186]. Different grain size, pore size, wettability, and other characteristics of HA coating can improve the adhesion and osteogenic activity of osteoblasts through regulating the interaction with proteins [187].

Obtaining controllable deposition of minerals on titanium surface is the main focus of the researches on inorganic coatings. Common deposition techniques include plasma spraying, sol-gel coating, micro arc oxidation (MAO), chemical vapor deposition (CVD), electrophoretic deposition, and physical vapor deposition magnetron sputtering (PVDS) [188]. Each of these methods has its advantages and disadvantages. Currently, electrophoretic deposition has been used to construct HA coating on SLA surface and has made progress.

Chi et al. prepared HA coating by electrophoretic deposition (25 V, 60 s) on SLA surface, which had a positive effect on the adhesion, proliferation, and differentiation of osteoblasts [189]. Zhao et al. obtained HA coating by electrochemical deposition method (3.0 V, 85 ° C, 30 min), which showed potential benefits for improving bone integration with implant surfaces after 4 and 8 weeks after surgery [190]. The electrodeposited HA coating presented as thin layers; thus, they would not influence the micro-scale morphology of SLA surface, but further improve its osteogenic activity.

HA coatings can be reinforced by adding other inorganic particles. For example, reinforcing with Cu, Ag, Zn, and other metallic elements provides antibacterial and anti-inflammatory properties to HA-coated implants [191]. These inorganic antibacterial agents usually have dose-dependent antibacterial activity and cytotoxicity at the same time. Therefore, special attention should be paid to achieve a balance between sufficient antibacterial effect and biological compatibility in application [192]. Doping these ions or particles in the HA coatings can ensure the optimal concentration of metallic elements in peri-implant environment. On the other hand, Bone and other hard tissues are composites of HA and traces of other minerals like Zn, Sr, Mg, and others [193]. Therefore, the current coating is more inclined to replace HA coating with ions with similar properties as bone mineral composition, which is conducive to biomedical applications and is worth further exploration.

### 15.4.5 Organic Coatings

Organics coated on titanium surface include a variety of molecules, from polypeptides, and active small molecule substance to proteins, polymers, and other substances with various functions of adhesion promoting, osteoinduction properties, antibacterial activity, and so on [194].

As a kind of bionic strategy, proteins of specific functions in bone extracellular matrix (ECM), such as fibronectin and collagen, as well as specific peptide sequences, are used for SLA titanium surface modification [195]. Cell adhesion motif (such as RGD sequence) is the function basis of most ECM proteins [196]. It can recognize and bind to integrin receptors expressed by eukaryotic cells, thus promoting cell adhesion and proliferation. In addition, regulatory molecules in osteogenesis related pathways, such as transforming growth factor (TGF-β1), bone morphogenetic protein (BMP), platelet-derived growth factor (PDGF), insulin-like growth factor (IGF), fibroblast growth factor (FGF), vascular endothelial growth factor (VEGF), etc. [197,198] have also been applied to SLA surface modification. Such active molecules can enhance the activity of peri-implant osteoblasts, fibroblasts, and vascular endothelial cells through different signal pathways, and promote osseointegration. Although proteins of large molecular weight have good biological effects, their low chemical stability and high-cost lead to a long distance from the clinical application. Meanwhile, small molecular polypeptides have a greater application prospect because of their strong stability and convenience for artificial synthesis [199].

Organic coatings based on cell adhesion components seem to provide a strategy that should not increase bacterial colonization. However, studies have shown that bacteria have similar adhesion mechanisms with eukaryotic cells and may bind to ECM proteins, such as fibronectin or collagen, as well [200]. In order to resist bacterial colonization, improving antibacterial properties is a vital issue to be solved for this kind of coatings.

Antibiotics are commonly used to prevent and treat peri-implant infections. Antibiotics can not only be administered systemically but also be loaded on biomaterials (such as bone cement, temporary bone cement spacer, etc.) which are already on the market for several years [201]. The main advantage of antibiotics (such as vancomycin, gentamycin, etc.) loaded biomaterials is that such local administration can reduce potential toxicity and maximize drug activity [202]. However, the use of antibiotics increases the risk of bacterial drug resistance, and it is difficult to ensure sustained and appropriate drug release after implantation. In many existing studies, loaded antibiotics were initially released in large quantities, and then continuously released below the minimum inhibitory concentration (MIC), which not only reduces the bacteriostatic activity, but also increases the risk of drug resistance. This drug release problem may also have a potential adverse impact on bone integration [203].

Nowadays, there are two strategies to achieve better antibacterial properties of titanium surfaces instead of antibiotics. Passive coatings are usually based on anti-adhesion polymers that prevent adhesion of proteins and cells (such as bacteria). Polymers that can inhibit bacterial adhesion include polyethylene glycol (PEG), poly (methacrylic acid) (PMAA), dextran, or hyaluronic acid [204]. The adhesion-resist property of these polymers is related to their flexible hydrophilic chains, which form a wide range of structures that hinder protein adsorption and cell adhesion, thus are very effective in repelling bacterial adhesion. However, these materials can also inhibit the adhesion of eukaryotic cells at the same time [205]. Therefore, cell adhesion sequences usually need to be added when these polymers are applied, to maintain the function of biomaterials [206]. Another strategy is applying substances with direct bactericidal effects, such as chitosan, fungicides (such as chlorhexidine), and antimicrobial peptides (AMP) [203]. The barrier to the application of this strategy also lies in how to maintain the sustained release of local drugs. Besides, there are some multifunctional substances, such as polyphenols extracted from plants, chitosan, hyaluronic acid, etc., which also have multiple biological functions, and can be applied to SLA surface modification [207].

The bonding strategies of organic coatings are mainly divided into two types: Physical adsorption and covalent bonding. The covalent binding strategies, including maleimide mercaptan chemistry, carbodiimide chemistry, and silane chemistry [208], are relatively stable on the titanium surface. While physical adsorption strategies, mainly by wrapping active molecules in polymer hydrogels and loading them on the implant surface by gel-sol method, are often used for continuous drug release [209]. Such polymers include dopamine, chitosan, gelatin methacrylate (GelMA), etc. The new intelligent hydrogel also has temperature or pH responsiveness, which can control drug release more precisely [210].

### 15.4.6 Discussion and Conclusion

This review summarizes several modification technologies that have been applied or have application potential on the SLA surface. Surface cleaning technologies, nano-scale modification, inorganic coatings, and organic coatings have their own advantages and disadvantages. Further investigation should focus on achieving a multifunctional implant surface, which will be able to promote osteogenesis, resist infection, as well as modulate inflammation.

## 15.5 SURFACE MODIFICATION OF NANOCOMPOSITE CERAMICS

Dental ceramics have become an important material for the restoration of dental defects due to their good aesthetics, biocompatibility, and wear resistance. The increasing clinical demand, especially the development of computer aided design/ computer aided manufacturing (CAD/CAM) technology, has greatly improved the mechanical properties of ceramics. However, the high fragility of ceramic materials has always been a major problem for clinicians, which leads to the fracture of prosthetics and failure of restoration under the effect of oral occlusal force. In the 1990s, Niihara et al. first reported the high toughness and low-temperature superplastic behavior of nanocomposite ceramics, suggesting researchers to introduce nanoparticles to overcome the fragility and workability of ceramic materials [211–213].

Generally, nanocomposite ceramics refer to the composite materials obtained by effective dispersion and recombination so that the heterogeneous phase (second phase) nanoparticles are uniformly dispersed and retained in the ceramic matrix. Different from single-phase nano-ceramics, nanocomposite ceramics can inhibit the growth of crystalline grains due to the interleaving and nailing between different grains during sintering, as well as the difference in the growth rate of nano-reinforced particles and matrix grains, thus keeping the particle size of one or even multiphase at the nanometer level. A large number of studies have shown that the multiphase composite of ceramic materials is the most effective way to strengthen and toughen mechanical properties, and also the best way to realize the superposition of properties and complementary advantages through the combination of microstructure [214–216].

However, when the particle size reaches submicron or nano states, its size is between atom, molecule, and block/particle like, with surface molecular arrangement, electron distribution, and crystal structure changed [217,218]. Single ultrafine particles are often in an unstable state due to increased surface energy and high surface activity, and the particles are easily attracted to each other to form particle aggregation. After sintering, the particle aggregation will become the main source of crack generation, which seriously reduces the fracture strength and toughness of ceramics. Therefore, a series of measures need to be taken to make the particles in a sufficient dispersion state. The surface modification could improve the physical and chemical properties of the particle surface, and increase the repulsive force and dispersion between the particles, so as to effectively prevent the tendency of aggregation, with new characteristics on the material meanwhile. Surface modification usually includes physical modification, chemical modification, and particle coating, covering physical, chemical, mechanical, and other deep processing of particle surface. In this section, the surface modification of nanocomposite ceramics in recent years is arranged. It should be noted in advance that the particle coating technology is summarized as a separate part because of the varied technologies, including physical and chemical treatment.

## 15.5.1 Surface Physical Modification

The modification agent is adsorbed on the surface of the particle by the intermolecular force (including van der Waals force, hydrogen bond, etc.) with different processing methods, which reduce the surface tension and change the surface polarity of the particles, thus reducing the aggregation of ultrafine particles.

### 15.5.1.1 Ball Grinding Mixing Method

Different nanoparticles are mixed by mechanical mixing method, including ball grinding, vibrating ball grinding, agitating grinding, colloid grinding, and nano airflow grinding, etc. This method is simple and convenient, without excessive equipment requirements. During the mechanical ball grinding process, the powders are milling through the friction force between the ball and powder, and mixed as evenly as possible. However, ball grinding cannot completely destroy the aggregation of nanoparticles, or achieve sufficient and uniform dispersion of multiple components. And the re-aggregation may occur due to the higher surface energy of the refined powder in the long time of ball grinding. Therefore, it is necessary to use high-power ultrasonic or introduce external magnetic field on the basis of mechanical mixing to destroy aggregation. Alternatively, the dispersion of the final particles can be improved by adjusting the pH value of the system, controlling the atmosphere and using appropriate dispersant [219]. In addition, the ball grinding process can also cooperate with the introduction of other particles.

For example, Garegin et al. modified zirconia powder with aluminum particles by using planetary ball grinding. Due to the continuous oxidation of aluminum particles during the grinding process, the stable $Al_2O_3$-$ZrO_2$ tetragonal solid solution is synthesized under the continuous mechanical reaction with nano-zirconia powder, which can avoid the transformation of zirconia crystal from tetragonal to monoclinic phase under the effect of low temperature aging [220]. Zhigachevy et al. modified the surface of nano-zirconia particle with different amounts of calcium ions by high-energy ball grinding method [221]. The results show that the sintered nanocomposite zirconia ceramics approach the theoretical density when the content of calcium ions is between 4 and 10 mol%, and the Vickers hardness increases 76.2% with the increase of calcium content. When the content of calcium ion is 6.6 mol%, the fracture toughness reaches the maximum.

#### 15.5.1.2 Mixed Precipitation Method

The modified nanoparticle was obtained by physical precipitation reaction, of which the nanoparticle suspension and modified agent were mixed directly with or without gelling agent. Before precipitation, the nanoparticles need to be evenly dispersed through ball grinding mixing or other methods. If the modified agent has opposite charge to the nanoparticle in the matrix suspension, there is no need to add gelling agent for sedimentation, and the deposition of materials is realized automatically by electrostatic attraction. For example, when carbon nanotubes (CNT) are combined with $Al_2O_3$ ceramic matrix, the negatively charged CNT are deposited on the surface of positively charged $Al_2O_3$ due to electrostatic attraction. Subsequently, due to the electrostatic attraction, CNT and $Al_2O_3$ particles would not separate during drying and barely aggregation occurred between CNTs [222].

### 15.5.2 Surface Chemical Modification

Some new functional groups are introduced on the surface of nanoparticle, through esterification, amination, halogenation, coupling reaction, or other chemical reactions between modified agent and nanoparticles, created with the changed surface structure state, improved dispersion of nanoparticles and some new physical or chemical properties of the whole ceramic.

#### 15.5.2.1 Coupling Agent

Coupling agent is one of the most commonly applied surface modified agent, whose molecular structure usually contains two groups with different chemical properties, which can react with inorganic nanoparticles. There are mainly two types most commonly used for particle surface modification in coupling agents: Silane coupling agent and titanate coupling agent.

The structure of silane coupling agent is generally Y-R-Si-$X_3$, which is usually used to modify the surface of inorganic particle with hydroxyl group or strong polarity. For example, when preparing carbon nanotubes-silicon carbide nanocomposite ceramics (CNTs/SiC), the CNTs were treated with mixed acid and grafted with negatively charged functional groups (mainly carboxyl groups) on their surfaces and ends, so as to achieve the purification of carboxyl functional groups and prevent the aggregation of CNTs. And SiC particles were modified with silane coupling agent to obtain positive charge on the surface, which facilitated the combination with negative charged CNTs after acidification.

In addition, silane coupling agents can also be used to treat metal oxides. Joanna Kujawa et al. applied perfluoroalkyl silanes (PFAS) to modify the surface of zirconia powders to make them hydrophobic [223]. When in contact with water molecules, the grafted PFAS chains are twisted, and when in contact with butanol molecules, the chains are straight. Generally, silane coupling agent need to be prepared into aqueous solution before applying, and its pH value is commonly controlled between 3–5, as too high pH could lead to self-condensation and failure of silane. Bangi et al. reported using n-propanol zirconium as the precursor to prepare nano-zirconia particle by sol-gel method, in which the silane coupling agent with pH=4–4.5 was used to modify the surface of the prepared zirconia gel [224]. And the results showed that the surface chemical modification was helpful for the prepared nanoparticle to maintain its structural characteristics after the final drying. In addition, the specific surface area of the modified particle increases obviously.

Titanate coupling agent is also widely used in inorganic particle for surface modification. The isopropoxy group in the molecule can react with the hydroxyl or other groups on the surface of inorganic particle. For example, the general formula of reaction of titanate with zirconia particle is:

$$R' - O - Ti - (OR)_3 + ZrO_2 - OH \rightarrow ZrO_2 - O - Ti - (OR)_3 + R'OH$$

Liu et al. modified the surface of zirconia powder by adding 1 wt% titanate coupling agent in the water-soluble ceramic injection molding process [225]. The results proved that the modified particle would not aggregate obviously in the organic agent, with densification temperature of 1450 °C, which is about 100 °C lower than the original powder. However, it is worth noting that most titanate coupling agents will have transesterification reactions with ester plasticizers to varying degrees. Contact with ester plasticizers should be avoided in the process of use to prevent side reactions and affect the modification effect.

#### 15.5.2.2 Esterification, Amination, Halogenation, and other Surface Modification

In addition to coupling agents, surface modification can also be achieved through esterification, amination, halogenation and other reactions with the surface of particle. The most common is hydrophobization of nanoparticles through the above-mentioned reactions.

Generally, hydrophobic treatment selects organic substances with hydrophobic groups (such as long chain alkyl, alkyl chain, and naphthenic alkyl, etc.) to replace the hydroxyl on the surface of particle, so that the alkyl could be firmly combined on the surface, thus presenting a strong hydrophobicity. The modified particles showed improved dispersibility, decreasing aggregation, and good dewatering effect. For example, polyvinyl alcohol (PVA) is added for surface modification during the preparation of nano-metal oxides. Since PVA contains a large number of free hydroxyl groups with strong polarity, these groups produce chelating bonds with metal ions in aqueous solution, and are tightly coated around metal ions to form a finite structure with limited shape of PVA chain, so that the size of the synthesized nanoparticles is limited.

In addition, stearic acid (SA) is also commonly used as a hydrophobic treatment agent for metal oxide particles with a coupling agent-like effect, which could improve the affinity and dispersion of inorganic particles. Besides, due to the lubricating effect of SA, this modification could also reduce the friction force within the composite system and improve the flowing performance overall. Liu et al. reported that after surface modification of nano-zirconia particle with SA, esterification reaction occurred between the carboxyl group of SA and the hydroxyl group on the surface of zirconia, thus forming SA layer around the zirconia [226]. This SA coating changed the particle surface from hydrophilic to hydrophobic, thus limiting the aggregation of zirconia particle.

Wen et al. used oleic acid to modify the surface of zirconia powder [227]. Firstly, oleic acid was dissolved in ethanol with a mass ratio of 1:8, and the modified zirconia powder was gradually added to the oleic acid/alcohol mixture solution, which was wet ground for 16 h with a planetary ball grinding. The results showed that the zirconia powder changes from hydrophilic to hydrophobic when the oleic acid is added at 2 wt%, and the modified ceramic showed an increase in its relative density and bending strength after sintering.

The surface modification of inorganic particles by fluoride normally requires heating and pressure process. Sergey Lermontova et al. modified the surface of zirconia particle aerogel by supercritical drying method using hydrophobic fluorinated groups in fluoro-containing isopropyl alcohol [228]. The sample was put into a stainless-steel autoclave, heated to 210–215 °C, and the pressure in the autoclave was set at 4.5–5 MPa. After modification, nuclear magnetic resonance detection (NMR) was conducted on the surface of the sample. It was found that the modified powder contained hexafluoroisopropoxy group. Implantation of hydrophobic fluoride groups on the surface of the oxide particle was carried out to prevent the aggregation of nanoparticles.

### 15.5.3 Particle Coating Modification

Certain substances are coated on the surface of the nanoparticles to form a core-shell like complex through the physical or chemical reaction between the modified components with solution or matrix powder, which could reduce or shield the aggregation of nanoparticles. At the same time, the steric hindrance repulsion generated by the inclusion, prevents the agglomeration and abnormal growth of the particles in the sintering process.

#### 15.5.3.1 Layer-by-Layer Self-assembly Technology

Layer-by-layer self-assembly technology can alternately adsorb polyelectrolyte on the surface of charged material by adjusting the pH of solution. It uses electrostatic attraction between dissimilar charges to generate driving force to form a double or multi-layer polyelectrolyte on the surface of the material, so as to change the charge properties of the particle surface, increase the Zeta potential value, and achieve improved dispersity through the electrostatic mutual repulsion between particles and steric hindrance stability. In recent years, a large number of studies have shown that the layer-by-layer self-assembly method is promising for the applications in electromagnetic information transformation, bioengineering, surface engineering, and other aspects.

In the research of Lu et al., two layers of polyelectrolyte were adsorbed on the surface of diatomite particles respectively through layer-by-layer assembly technology, so as to improve the dispersion of diatomite particles, prevent aggregation and obtain stronger adsorption capacity [229]. The modified diatomite particles were then used to prepare nano-composite ceramic powders with nano-zirconia, forming a tight internal adsorption structure. The modified diatomite showed reduced particle size, narrowed size distribution range, sufficient dispersity without obvious aggregation.

The operation of the layer-by-layer self-assembly technology is relatively simple. The multi-layer polyelectrolyte could be successfully assembled by immersing the material alternately in the polyelectrolyte solution with opposite charge and ultrasonic vibrated for a period of time, which does not require overly complex equipment, and there are no restrictions on the size and shape of the material.

#### 15.5.3.2 In Situ Chemical Encapsulation

In situ, chemical encapsulation refers to the method of generating precursor on the outer surface of micron or sub-micron particle, and the precursor is decomposed under high temperature with laser or plasma as the heat source, followed with the enhanced nano-phase is grown at the original location of the raw material [230]. For example, Vivekanandhan et al. prepared nano-$ZrO_2$ powders by acrylamide-assisted citric acid polymer combustion using polyacrylic acid and ethylene glycol as precursors [231]. An ultra-thin layer of $MoO_3$ was coated on the surface of nano-$ZrO_2$ powders by polymeric resin method at 500°C, which enhanced the corrosion resistance of zirconia and reduced the agglomeration among the particles. Niihara et al. mixed $Si_3N_4$ powder, boric acid, urea, and ethanol through conventional wet ball grinding method, dried, followed with hydrogen reduction reaction, and then nitriding to produce BN-coated $Si_3N_4$/BN(n) nano-composite powder [213].

#### 15.5.3.3 Sol-gel Method

The compounds containing high chemically active components go through solution, sol-gel, and curing states, and then are processed with heat treatment to form oxides or other compounds as solids. The process of using sol-gel method to prepare core-shell composite powder is as follows: First of all, select the appropriate matrix ceramic powder, and remove impurities through pretreatment before the reaction. And then immerse the powders into the sol, forming a layer of functional membrane structure on the surface of the ceramic powder with chemical bonding or physical adsorption. The modified composite powder coated with multiple layers can be prepared by repeating the above process [232]. For example, Strauss et al. synthesized mesoporous $TiO_2$-$SiO_2$ nano-composite powders with a diameter of about 20 nm with good photocatalytic activity by this method [233].

#### 15.5.3.4 Surfactant

According to the properties of the surface charge of nanoparticles, surfactants with cations or anionic ions are added to form a coating layer with carbon and oxygen chains extending outwards on the surface. Surfactant is a kind of organic compound with hydrophilic and oleophilic structures, which can reduce the surface tension and surface energy, as well as emulsify, moisten, and form film. The characteristics of hydrophilic group of surfactant including solid adsorption, chemical reaction activity, and reducing the surface tension can control the hydrophilic, lipophilic, and surface activity of nanoparticles, with the surface modification: Firstly, the hydrophilic groups combine with the surface groups of the particle to form a new structure, giving the surface of the nano-particle new functionality; Secondly, lowering the surface energy of nanoparticles can keep them in a stable state. Thirdly, the lipophilic group of the surfactant forms steric hindrance on the surface of the particle to prevent the aggregation of the nano-particles [234]. Thereby, applying surfactants could improve the dispersion of nanoparticles in different media, increase surface reaction activity and ameliorate surface structure of nanoparticles.

For example, a positively charged organic polymer dispersible, such as Polyetherimide (PEI), which is added to a ceramic suspension formed by nano-metal oxide powder and an organic solvent, develops a positive charge on the surface of the modified particles so that they repel each other and form stable and dispersed particles, which greatly prevent the agglomeration of ceramic particles [235]. In previous studies, triethanolamine, sodium polyacrylate, sodium dodecyl benzene sulfonate, and polyethylene glycol were used to modify the surface of nano-$TiO_2$, which improved the dispersion state of nano-$TiO_2$ and effectively prevented its agglomeration [236].

#### 15.5.3.5 Inorganic Coating Modification

The combination between the inorganic modification agents and the nano-particles depends on physical methods or van der Waals force, without chemical reaction. Generally, inorganic compounds are used to precipitate on the surface of nanoparticles, forming precipitates and coating nanoparticles to form core-shell complexes. After a series of treatments, the coating is fixed on the particle surface, which can change the dispersion and stability of the particle in different media, reduce the surface activity and prevent their aggregation. For example, the metal oxides such as $SiO_2$ and $Al_2O_3$ are often used for surface modification of inorganic nanoparticles. Carbonate, oxalate, hydroxide, etc. are mostly used as precipitators since the $OH^-$ in the solution of these substances will affect the pH value directly or indirectly, which in turn affect the precipitation process [237]. Therefore, pH is a quite important parameter in the precipitation process.

For example, nano-SiC particles are added to an Al-containing solution for mixing by the ball grinding or ultrasonic dispersion, after which a precipitant (aqueous ammonia solution) is added to form the SiC-$Al(OH)_3$ composite precipitate [238]. Followed with drying, calcination, and ball grinding, the SiC particles coated with $Al(OH)_3$ can be obtained. This method is also applicable to the surface modification of other metallic oxide particles. Ohmori et al. used ethyl tetrasilicate as the raw material to uniformly coat the surface of $Fe_2O_3$ particles with a layer of $SiO_2$ by optimizing hydrolysis conditions, making the particles easy to disperse in non-aqueous media, effectively preventing aggregation and hydrolysis [239].

### 15.5.4 Summary

Surface modification of nanoparticles is the key to reduce aggregation and improve comprehensive performance of ceramics. The surface modification methods commonly used in the current researches mainly include physical modification, chemical modification, and surface coating, which improved the surface properties of nanoparticles and avoid the adverse effects caused by agglomeration, increased particle size, etc. Inevitably, each method has its own limitations and shortcomings, and most of them only have a good effect in a certain aspect of application. With the expansion of the application range of nano-composite ceramics, the surface modification mechanism, modification methods and equipment, and modification effect need to be further improved.

## 15.6 SURFACE MODIFICATION OF RESIN-BASED CERAMICS

Resin-based ceramics generally are a mixture of resin matrix and inorganic ceramic materials. This kind of composite materials are industrially polymerized under high temperature and pressure, which are usually processed using computer-aided design and fabrication. Therefore, resin-based ceramics are not considered ceramic materials in the narrow definition of ceramics [240,241]. In 2013, the American Dental Association (ADA) redefined dental ceramic materials as "materials made from inorganic refractory components by means of hot die-casting, sintering or cutting" [242]. Based on this definition, resin-based ceramics are often used internationally as a class of all-ceramic materials in recent years because their main component is inorganic fillers, usually with a mass fraction of more than 50%, and their mechanical and aesthetic properties are similar to those of ceramic materials [243].

Resin-based ceramics can be divided into two main categories according to the internal microstructure and composition of the materials. One type is a ceramic network with added resin polymer, forming a polymer infiltrated ceramic network (PICN), e.g., Vita Enamic [244]. Another type is resin nanoceramic (RNC), also known as nanoparticle-filled resins, which are nanoceramic particles embedded in a highly cross-linked resin matrix, e.g., Lava Ultimate, Cerasmart, Hyramic[243,245]. However, as resin nanoceramics are formed by a special hot-pressing process in which resin matrix and a high content of inorganic fillers with a progressive fine structure are compounded together and polymerized under high temperature and pressure, with a high monomer conversion, so their physical properties and color stability are higher than those of conventionally polymerized resins [241]. The nanoceramic particles, on the other hand, provide a significant increase in translucency, color stability, and superior aesthetic properties. The presence of the organic resin matrix gives the material some excellent properties of a composite resin at the same time, with a modulus of elasticity of approximately 12 GPa-28 GPa and a flexural strength of approximately 200 MPa, overcoming the shortcomings of traditional ceramic materials which are brittle and prone to fracture [246–248]. All these features have greatly improved the clinical performance of resin nanoceramic restorations and have led to an increasing interest in the clinical application of this material.

Nevertheless, for most dental ceramic materials, the relatively smooth surface of the material without any surface treatment has a small contact area with the resin cement and does not allow for an effective bond to be formed. The resin nanoceramic composites are polymerized with fewer residual monomers that can participate in the bonding process and cannot form sufficient C=C required for bonding to meet the demands of clinical applications [249–251]. Consequently, surface modification of resin nanoceramic materials is required to improve the bonding properties in order to obtain better bonding results. Different surface treatments can significantly affect the surface physicochemical and adhesive properties of the restoration and can help to guide the clinicians to choose a best treatment method for different types of composite materials to obtain better bonding results. Common surface modification methods include sandblasting, hydrofluoric acid etching, silane coupling agents, and laser.

### 15.6.1 Sandblasting

Sandblasting is one of the most common and easiest mechanical surface modification methods, commonly used on zirconia and composite resin material surfaces. It has been shown that sandblasting can increase the surface roughness, surface area, and surface energy of resin nanoceramics.

A certain intensity of sandblasting can remove the tarnished layer produced by the cutting process of the material and provide a clean surface. On the other hand, physical modification is achieved by changing the surface morphology (e.g., making grooves, patterns, etc.) and improving the surface roughness [252–254]. A certain degree of surface roughening could increase the micromechanical locking force between the material and the resin cement, as well as increasing the surface energy and wettability, and creating a residual compressive stress layer on the surface of the material, thus improving the bond strength [255]. At the same time, sandblasting exposes the inorganic fillers in the resin matrix, thus allowing the silane coupling agent to interact better with the inorganic fillers [256]. Currently, for resin nanoceramics, sandblasting is one of the common clinical surface treatments, but the exact blasting conditions are inconclusive and varied considerably from study to study. In the existing studies, the blasting pressure mostly ranged from 0.1 MPa to 0.4 MPa; the size of the blasting particle was uneven, with some studies using blasting particle as low as 25 μm and others using alumina blasting up to 120 μm; the blasting time ranged from 5 s to 40 s; and other factors such as the distance between the blasting apparatus nozzle and the material (5 mm–10 mm), the angle between the nozzle and the material (45°–90°) also varied. It was found that sandblasting on resin nanoceramic specimens with 110 μm $Al_2O_3$ particles at a pressure of 0.2 MPa was effective in improving the bond strength [257]. However, several studies have also reported that sandblasting causes microcracks of 1 μm–10 μm on the surface and subsurface of the ceramics, which usually occur within the resin matrix and the interface between the resin matrix and the inorganic ceramic [258,259].

The surface morphology of the resin nanoceramics after sandblasting was observed by field emission scanning electron microscopy, which revealed that after sandblasting, the surface of the resin nanoceramics disappeared from scratches and became uneven [260]. Furthermore, when the sandblasting pressure was high (0.2 MPa) and the sandblasting particles were large (110 μm), a large number of micro-cracks appeared on the surface of the resin nanoceramics, accompanied by sandblasting particles embedded, which may have caused some damage to the structure and strength of the material [258]. Meanwhile, sandblasting significantly increases the surface roughness of resin nanoceramics. The larger the sandblast particles and the higher the pressure, the greater the surface roughness and the greater the subsequent micromechanical retention provided with the bonding material [261].

The wettability of a material surface can be demonstrated by measuring the surface water contact angle and surface energy. The smaller the water contact angle, the better the surface wettability and the more favorable for bonding. Strasser et al. found that resin nanoceramics blasted with 50 μm alumina at 0.1 MPa significantly improved roughness and surface energy [255,262], and Reymus et al. used the same blasting conditions on resin nanoceramics for surface modification, resulting in a significant increase in the tensile bond strength of the resin nanoceramics [257].

Existing research advances in the blasting treatment of silicified alumina coatings is the use of silicified alumina particles blasting, with the CoJet and

SilJet systems used most commonly clinically [263]. This sandblasting medium is the alumina particle coated with a layer of silica, which not only increases the surface roughness of the material, but also forms a special coating on the surface of the ceramic substrate. In addition to roughening the ceramic surface, it also has the equivalent of a silanization treatment, which enhances the chemical bonding. The use of silicon oxide coated sandblasted composite materials can significantly improve the bond strength, comparable to that of sandblasting with aluminium oxide particles [264–267].

In summary, it is recommended to use small (≤50 μm) alumina particles for sandblasting on resin nanoceramic surfaces at low pressure (0.1 MPa–0.2 MPa), and a combination of silane coupling agent treatment is commonly suggested.

### 15.6.2 Hydrofluoric Acid Etching

Hydrofluoric acid is often used as an acid etchant for all-ceramic materials in dentistry as a highly reactive compound [268,269]. The inorganic filler of resin nanoceramics contains silica in its ceramic network structure. Hydrofluoric acid dissolves the silica in the following reaction: $4HF+SiO_2=SiF_4\uparrow+2H_2O$, and the resulting silicon tetrafluoride reacts with other hydrofluoric acid molecules to form a soluble complex ion, hexafluorosilicate. Subsequently, hydrogen ions react with the hexafluorosilicate complex to form the water-soluble substance fluorosilicic acid, which is carried away with the fluorosilicic acid by rinsing, thus forming a porous morphology on the surface and increasing the roughness [270–272].

Lise et al. conducted scanning electron microscopy on the surface of resin nanoceramics after acid etching with hydrofluoric acid, and the inorganic filler particles on the surface were dissolved and dislodged, forming a honeycomb porous structure [273]. The uneven microstructure formed on the acid-etched surface of the composite not only increases the bonding area, but also forms micro-mechanical retention with the bonding resin cement, increasing the bond strength. For resin nanoceramics, it has been found that after acid etching with hydrofluoric acid, nanoceramic particles embedded in the intact resin matrix can be observed on the material surface [274].

Nevertheless, it has been argued that resin nanoceramics are whether suitable for hydrofluoric acid etching, and both Park et al. and Sagsoz et al. showed that the bond strength of hydrofluoric acid-etched resin nanoceramics was lower than that of sandblasted ones [266,275]. The lower number of micropores on the surface of resin nanoceramics after hydrofluoric acid etching observed by Park et al. under scanning electron microscopy, which could be attributed to the presence of zirconium oxide nanoparticles in the resin nanoceramics [266]. Zirconium oxide is resistant to the acid etching action of hydrofluoric acid and is virtually uncorroded [276–278]. Duzyol et al. showed that the bond strength of resin nanoceramics obtained after hydrofluoric acid etching was significantly lower than that obtained after sandblasting [279], and Cekic-Nagas et al. concluded that the inorganic fillers on the surface of resin nanoceramics were completely dissolved and removed after hydrofluoric acid etching, and the resin matrix structure lost the support of inorganic components, which made the surface structure unstable and susceptible to collapse. This is not conducive to the formation of a good and solid bond [280].

Although there are many studies proving that the application of hydrofluoric acid can improve the bonding properties of resin-based ceramics, it is true that the acid-etching treatment does not show significant advantages compared with other treatments in the available research results. At the same time, the problems on toxicity and environmental pollution of hydrofluoric acid also limit its wide application [281]. In addition, the adverse effects of excessive acid etching with hydrofluoric acid on the bonding of composite materials need to be taken seriously. Attention should be paid to the control of acid etchant concentration and action time in clinical application. It is generally recommended that the commonly used hydrofluoric acid concentration is 2.5%–10% and the time is 0.5–3 min to obtain the best bonding effect.

### 15.6.3 Silane Coupling Agent

Broadly speaking, silane refers to a saturated compound with four substituents in a collection of silicon elements, in which one or more silicon atoms are linked to other elements. Silanes contain synthetic silicon-carbon bonds, which are quite stable chemically [282]. Silane coupling agents used in dentistry contain trialkoxysilanes, such as 3-methacryloxypropyltrimethoxysilane (MPS) as the key reactive component [283].

The silane coupling agent can generate two reactive groups under specific conditions. The methoxy at one end forms a silanol group (Si-OH) under hydrolysis conditions, which can condense with the hydroxyl group on the surface of the silica contained in the inorganic filler of the resin nanoceramic to form a siloxane bridge (Si-O-Si) and connect with the surface of the restoration; and the organic group at the other end can react with the resin monomer in a copolymerization reaction and form a chemical connection with the surface of the resin cement, thus bonding the restoration closely to the resin cement. Silane coupling agents can also produce several monolayers which cross-link and condense with each other to form silicone oligomers [284]. Silane can play a good role in chemical modification of the surface of the material, and improve the wettability of the material surface, which facilitates the spreading of the resin cement and increases the bond strength [285].

However, many academics believe that the chemical bonding formed by silane coupling agents is significantly less than mechanical bonding, and the silane coupling agent treatment should be used in conjunction with other mechanical treatment methods, such as sandblasting or hydrofluoric acid etching on the surface of resin-ceramic composites followed by the application of silane coupling agents [286,287]. A number of studies have shown that the bond strengths of resin nanoceramic obtained by sandblasting followed with applying silane coupling agents are considerably stronger than those obtained by sandblasting alone. Lise et al. reported the inorganic fillers are visible on the surface of resin nanoceramics after sandblasting [273]. And Park et al. investigated the elemental analysis of the surface which showed that the surface silicon content of the untreated group was 22.16 wt%, while the surface silicon content remained as high as 19.14wt % after sandblasting [266]. These results indicate that there are silicate particles on the surface of resin nanoceramics before and after sandblasting, which can react with the silane coupling agent to form a chemical bond for further improving the bond strength.

However, it was found that the bond strength of the resin nanoceramics was not obviously improved, or even reduced, after hydrofluoric acid etching in combination with silane coupling agent treatment compared to hydrofluoric acid etching alone [261,288]. Peumans et al. showed no significant difference in bond strength between hydrofluoric acid etching and silane coupling agent treatment of resin nanoceramics compared to hydrofluoric acid etching only; and a similar conclusion was obtained in the study by Elsaka et al [260,264]. This may be due to the fact that after the acid etching treatment with hydrofluoric acid, the silica component on the surface of the resin nanoceramics dissolves and falls off, forming a large number of tiny pores. On the one hand, these tiny pores may be filled with silane coupling agent, which prevents the resin cement from infiltrating and penetrating into them, affecting the micro-mechanical locking force. On the other hand, the silane coupling agent loses the object of chemical reaction and cannot form an effective chemical bond with the resin nanoceramics. This was confirmed by the study of Park et al [289–291]. The untreated group had a surface elemental silica content of 22.16 wt%, which was reduced to 8.89 wt% after acid etching with hydrofluoric acid [266].

In overview, it is currently controversial whether the use of silane coupling agents alone has a significant facilitating effect on the bonding of resin nanoceramic materials. In order to improve the bonding performance of non-silicone-based materials, a specific pretreatment of the material surface must be selected prior to the use of silane coupling agents. Most scholars believe that silane coupling agents in combination with other mechanical treatments can further enhance chemical bonding, except for acid etching.

### 15.6.4 Laser

Laser etching can produce a unique morphology on the surface of the ceramic materials, and obtain surface pits with accurate size, depth, and spacing, which are easy to handle without environmental pollution. The types of lasers commonly used in the current researches include Er: YAG lasers, Er, Cr: YSGG lasers, Nd: YAG lasers, and $CO_2$ lasers, etc. [291]. Among these, the Er: YAG laser is the most widely used in clinical for periodontal surgery, decay of dental tissue, cavity preparation, surface modification of dental materials, etc. Studies have also been conducted to explore the application of laser etching for surface pre-treatment of restorations [292].

The mechanism of action of the laser is to convert light energy into heat energy. During etching, the instantaneous high temperature or pressure action destroys the inorganic particles on the surface of the material as well as the resin matrix, while the laser beam is absorbed by the water provided by the laser system itself, creating scattered pits on the surface of the composite material and increasing the roughness for increased mechanical retention [293]. For resin nanoceramics, lasers can effectively increase the surface roughness. It has been shown that the bond strength of resin nanoceramics modified by Er: YAG laser and Nd: YAG laser is significantly higher than that of hydrofluoric acid treatment, similar to or even higher than that of sandblasting, without damaging the microstructure of the resin nanoceramics or causing microcracking. Compared to chemical surface modification methods such as hydrofluoric acid, the mechanical surface modification methods such as sandblasting and laser are more effective in improving the bond strength of resin nanoceramics by increasing the roughness [259,294]. But Maawadh et al. reported that the bond strength of Er, Cr: YSGG laser-treated resin nanoceramics was comparable to that of hydrofluoric acid etching and sandblasting. Scanning electron microscopy showed irregular microscopic cracks on the laser-treated resin nanoceramic surface and localized sintered lumps formed after the resin matrix had been heated and melted [287].

There are still relatively few studies related to the application of lasers to new resin-ceramic composites, and the effects of different laser types, power densities and pulse patterns on the surface treatment of resin nanoceramic restorative materials still need further research.

### 15.6.5 Cold Atmospheric Plasma Treatment

Cold atmospheric plasma (CAP) is a non-thermodynamic equilibrium plasma that is generated and acts at atmospheric pressure and controlled low temperatures, with high treatment efficiency and simple operation. The application in dentistry is also one of the important branches of plasma medicine, with promising applications in traditional industrial material manufacturing modifications as well as in biomedicine [295–298]. There have been many studies on the application of plasma in dentistry, such as sterilization in dental and periodontal treatment, as well as surface modification of various dental ceramic materials and dental hard tissues [299].

Cold atmospheric plasma jets contain a large number of reactive groups and are chemically active and prone to chemical reactions. Helium (He) and argon (Ar) gases are common discharge gases for generating plasma [295,296]. Both of two are inert gases, belonging to non-reactive plasmas, which can transfer energy to surface molecules and become sub-stable elements. After release, the secondary chemical reactions occur with air to introduce reactive groups to the surface of the material [300].

Both helium (He) and argon (Ar) cold atmospheric plasma treatment of resin nanoceramics did not change the surface micromorphology or surface roughness, but significantly reduced the surface water contact angle and improved the surface wettability [301]. And the surface carbon content of the material is remarkably lower and the oxygen content is significantly higher, which explained why plasma treatment of resin nanoceramics enhances the wettability of the material surface and has an effective chemical modification effect on the resin nanoceramic and is conducive to improving the bonding properties [300–302]. After the treatment, hydroxyl radical groups (-OH) are detected as predominant on the surface of the resin nanoceramics and the element Si is not only present in the form of pure $SiO_2$, but may also form partially reactive groups such as Si-N. This suggests that reactive groups such as hydroxyl groups in cold atmospheric plasma jets can be present and act on the surface of resin nanoceramics [303]. The content of hydroxyl radical groups on the surface of the material treated by Ar gas plasma is more than that treated by He plasma, which may be more conducive to the formation of chemical bonds with resin cement. Moreover, cold atmospheric plasma can remove residual organic contaminants from the surface of resin nanoceramics, and the contents of C-OH, C=O, and COOH groups are reduced to varying degrees after treatment. Therefore, it can be assumed that He- and Ar-gas cold atmospheric plasma have some cleaning effect on the surface of material [301]. In contrast, excessive radiation may form damage layers and microcracks on the surface, which in turn would reduce the bond strength.

Cold atmospheric plasma modification treatment can improve the bonding properties of resin nanoceramic materials to a certain extent, and the Ar plasma combined with silane coupling agent treatment obtains better bonding results, which can be applied as a new surface modification method of resin-based ceramic material.

### 15.6.6 Preheating

Problems commonly associated with the clinical use of resin nanoceramics include marginal breakage, secondary caries, and postoperative sensitivity, which inevitably lead to restoration failure. The fundamental cause of these problems lies in the polymerization shrinkage of the materials. Among the different techniques used to reduce polymerization shrinkage, preheating is the most promising treatment at present [304,305].

Preheating of resin nanoceramics to 54°C was suggested by Choudhary et al. to increase their adaptability and reduce the total surface area [304]. Another study reported that preheating can improve the mechanical properties of resin nanoceramic restorations. Using infrared spectroscopy, Daronch et al. found that preheating the composite to 60°C increased the monomer conversion at the surface and the depth of 2 mm below, and the conversion rate increased with increasing temperature, but only within a certain temperature range (no higher than 90°C) [306,307]. Resin nanoceramics with a low monomer conversion rate are more prone to color change. The staining resistance of preheated resin nanoceramics was also found to be improved by Sousa et al [308]. In another study, Taufin et al. found less surface staining and higher color stability in preheated resin nanoceramics, which can be attributed to the higher monomer conversion rate [309]. Dionysopoulos et al. also observed an increase in the surface microhardness values of resin nanoceramics preheated to 55°C [310]. Therefore, based on existing research it is known that preheating has a positive effect on the surface microhardness and color stability of resin nanoceramics, while its effect on bonding properties has not been investigated and needs to be further explored.

### 15.6.7 Summary

The surface pretreatment of resin-based ceramics is important to improve the bonding properties and maintain the durability of the restoration. The common surface modification methods such as hydrofluoric acid etching, sandblasting, and silanization can improve the bond strength of resin nanoceramic composites to varying degrees. Laser treatment, cold atmospheric plasma treatment, and preheating are newer surface modification methods available and their effect on the bonding performance needs further investigation. It is important for clinicians to select appropriate surface modification according to the structural component properties of the composite materials. With the rapid development of dental materials and the emergence of new restorative materials, the in-depth exploration of new surface modification methods is of great significance for clinical applications.

## 15.7 THE STRATEGY AND DEVELOPMENT PROSPECT OF SURFACE FUNCTIONALIZATION OF NANO-BIOACTIVE GLASS

Bioactive glass (BAG), with its excellent biocompatibility and bioactivity, has been widely used in the fields of soft and hard tissue engineering applications, to promote tissue healing, accelerate bone defect repair and cancer therapy, etc. [311–315]. This kind of biomaterial generally refers to silicate glass composed of $SiO_2$, $Na_2O$, CaO, $P_2O_5$, and other inorganic components. Once in contact with human body fluids (such as plasma), it can locally release specific metal ions at therapeutic concentrations to promote the process of tissue healing, and can also mineralize on the surface to form a low crystallinity hydroxycarbonate apatite (HCA) layer similar to inorganic minerals in human bone [316–319]. Because its chemical composition is similar to that of biological bone, it is easy to form a strong chemical bond and stable osseointegration with the surrounding bones. So, it has excellent bone inductivity, bone conductivity, and biological activity [320–322].

The earliest bioactive glass was discovered by Professor Hench in Florida, USA, in 1969. At that time, 45S5 glass (mass fraction of 24.5% $Na_2O$-24.5% CaO-45% $SiO_2$-6% $P_2O_5$) obtained by the high-temperature melting method was implanted into the human body, and surprisingly, its surface did not form a fiber coating, but formed a close combination with human bone, and the mechanical strength of the bonding surface was even higher than that of bone and material itself [323–325].

However, this preparation process requires to be carried out at extremely high temperatures (1300–1500 °C). For a long time, it has been found that the final reaction product particles agglomerate seriously and have a small specific surface area, and the particle surface is dense and non-porous, which leads to unreasonable biodegradation efficiency and ion release. At the same time, in the preparation process, due to the low purity of raw materials and the characteristics of the melting method itself, a large number of impurities will inevitably be introduced [326–328]. Due to many defects of the melting method, a new technology for preparing bioglass by the sol-gel method has been developed. This method has the characteristics of low energy consumption, mild reaction conditions, and easy process control, and the product has a high specific surface area and ordered mesoporous channels, becoming the key preparation technology of bioglass at present [329–331]. However, there are also some problems with this technology, including the difficulty of meeting the particle size requirements for a wide range of biological applications and the relatively homogeneous nature of the material [332].

In recent years, with the support of relevant research, bioactive glass materials have been continuously improved, and modified bioactive glass such as nanoporous biomaterials and organic/inorganic composite materials have emerged. Compared with the traditional bioactive glass mentioned above, the nano-bioactive glass (nBAG) material has a regular spherical nanostructure and can prepare uniform pores on the surface. It has a larger specific surface area and can provide more surface-active sites, so it has more ideal biological activity, degradation performance, drug loading, and antibacterial properties [333–335]. At present, the main methods for preparing nano bioactive glass materials include sol-gel method, template synthesis method, microemulsion method, flame spray method, 3D printing, etc. In order to further improve the mechanical and biological properties of nano-bioactive glass material and expand its application range, magnesium, zinc, strontium, copper, silver and other elements can also be added to the material [336]. Through surface functionalization or surface modification [337], its performance expression can be further enhanced, and more ideal particle size characteristics and functional expansion can be obtained, to meet the needs of various biological applications, especially human body applications.

Generally, the surface of nBAGs can be modified by physical (e.g., change in surface topography) and chemical/biochemical (e.g., adsorption of molecules by atomic layer deposition, covalent grafting of biomolecules/drugs) approaches. On the one hand, the chemical structure of nBAGs provides the possibility of incorporating various metallic and non-metallic elements into its basic components, thus generating composite materials with extended biological effects, for example, silver (Ag) modified bioactive glass can provide antibacterial properties [338], and cerium (Ce) doped bioactive glass can product antioxidant materials with strong catalase activity [339].

On the other hand, the surface of nBAGs is also considered to be a suitable site for grafting bioactive macromolecules (such as polyphenols). By combining bioactive glass particles with gelatin, chitosan, polylactic acid, polylactide and other polymer materials, organic/inorganic composite materials with both bioactivity and excellent mechanical properties can be prepared [340]. In addition, bioactive bone cement materials can be prepared by combining bioactive glass with calcium sulfate and calcium phosphate [341].

In short, surface functionalization can not only guarantee the typical properties of nBAGs (e.g., ion release, bioactivity), but also artificially combine them with new customized functional features (such as antibacterial, antioxidation, and anti-cancer activities), to optimize the biological activity and biological reactivity of materials, or improve the compatibility of nBAGs particles with other phases to obtain composite biomaterials with stable structure [342].

### 15.7.1 Mechanisms of Surface Modification of nBAGs

The surface of nanomaterials plays a key role in many physical and chemical processes, and the surface ligands or molecules combined with the surface are also one of the important components of nanomaterials, affecting the interaction between the material itself and other materials or biological systems. Figure 15.4 shows the main ways of surface modification of bioactive glass. The physical properties of the material surface, including morphology, particle size, porosity, pore size, and charge, can determine the interaction between biological components (such as protein) and biomaterials [343–345]. Its reaction mechanism involves adjusting the surface roughness of micron and nanometer to trigger favorable biological reactions, such as improving cell adhesion or tissue growth [346]. For nBAGs, surface nano-texturing seems to be less important, and the inherent reactivity of bioactive glasses upon contact with biological fluids will not be different due to the presence or absence of surface pores [347]. Strategies

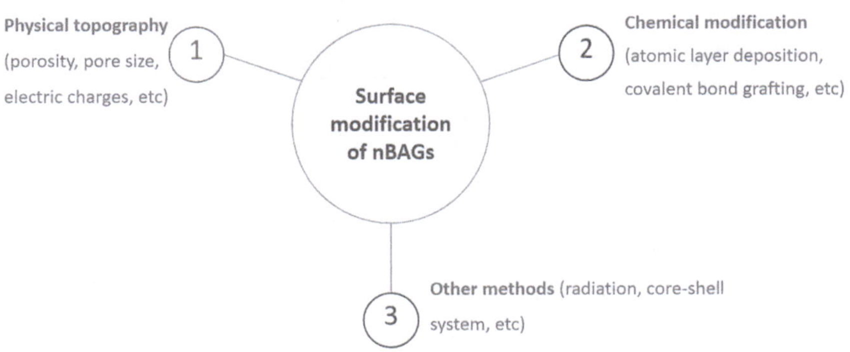


**FIGURE 15.4** Main ways of surface modification of bioactive glass.

of chemical surface modification using coupling agents or pH changes are more suitable and effective. Other approaches to surface modification involve the deposition of coatings, the use of radiations, and the development of core-shell systems.

### 15.7.2 Surface Modification by Chemical Agents

Among various methods used for surface modification of biomaterials, the use of surfactants is considered to be the most powerful with satisfactory outcomes [348] (Figure 15.5). Moreover, the functionalizing process using surfactants is simple, straightforward, and without high costs. Several functional groups (e.g., $NH_2$ or COOH) have been introduced on the surface of BGs in order to act as active sites for specific properties or further molecular grafting. Common chemical reagents include functional organic silane such as 3-aminopropyltriethoxysilane (APTS), glutaraldehyde (GA), tetraethoxysilane (TEOS), etc.

As a common silane agent containing amino groups, APTS is an ideal choice for the surface functionalization of biomaterials. When inorganic phase nBAGs are compounded with biological macromolecules, their compatibility is poor and it is difficult to disperse uniformly, which affects the interfacial properties of the composite. Silane coupling agents can significantly improve the interfacial compatibility of inorganic powder and polymer materials. For example, aminated bioactive glass nanoparticles can be integrated into gelatin-based tissue adhesive in the form of covalent bonds to improve the mechanical properties, adhesion properties, biocompatibility, and osteogenic ability of the adhesive system [349]. In addition, it can promote the formation of spherical HCA agglomerates without decreasing the bioactivity of the glasses. In an aqueous

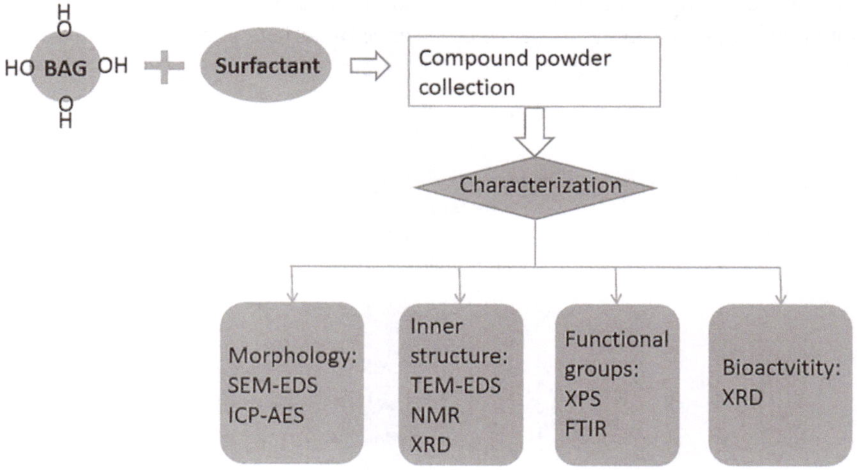


**FIGURE 15.5** Physical and chemical characterization strategies for modified products.

environment, the hydrolysis of APTS will also lead to the formation of silanol groups, which could react with hydroxyl groups and enrich to the surface [350]. Besides, APTS has also been proved to improve other characteristics of nBAGs, such as cell compatibility. In addition to silica based bioactive glass, APTS has also been successfully grafted onto phosphate and borate glass [351], thus opening up more opportunities for surface functionalization of these materials.

In terms of physical and chemical characterization, SEM and EDS analysis can be used to determine the changes in morphological (surface layer formation) and chemical composition changes (variation of Si, N surface content) attributable to the presence of APTS [352]. XPS analysis can also be used to detect the silane molecules on the glass surface and the typical functional groups of silanes [353]. The calculation of atomic ratios between the characteristic elements of APTS molecules (e.g., N) and the substrate ones can indicate the silane layer thickness. Similarly, the comparison between the chemical composition results obtained with EDS and XPS, which have different penetration depths (1 μm vs. a few nanometers, respectively) can give indications on the thickness of the organic layer. In addition, FTIR spectroscopy has been used to explore the APTS functional groups on the glass surface (mainly powder), and good results have been obtained [354]. Contact angle measurement can be used as a fast and effective method to determine the existence of silane molecules on the surface of glass [355]. Its principle is that nBAGs are highly hydrophilic, and there are many -OH groups on the surface of the material, while APTS is a hydrophobic organic molecule. Therefore, after the modification of APTS, the original wettability of the material will change significantly, and this change is also easier to monitor.

Glutaraldehyde (GA) is another reagent for surface modification of nBAGs, which can promote protein adhesion to the glass surface and improve protein binding capacity [356]. Some studies have shown that the surface functionalization of GA and APTS is conducive to the formation of hydroxyapatite when the glass is immersed in simulated body fluid. As a protein coupling agent, GA is used for surface modification of nBAGs in order to improve the stability of protein adhesion and induce the polymerization of protein molecules (such as hemoglobin). Generally, electron-paramagnetic resonance (EPR) spectroscopy combined with a fixed-point spin labeling program (SDSL) can be used to study the bioactive glass samples before and after protein adhesion, and analyze their conformation and dynamics changes. Fourier transform infrared spectroscopy (FTIR) and X-ray photoelectron spectroscopy (XPS) can be used to analyze the change of surface element composition before and after modification while scanning electron microscopy can be used to observe the change of surface structure, that is, the change of surface morphology after protein coating.

Tetraethoxysilane (TEOS), as the source of silicate, is also one of the most widely used compounds in the synthesis of nBAGs. Studies have proved that the surface functionalization of bioactive glass with tetraethoxysilane (TEOS) and APTS can improve its binding ability with biological molecules (such as drugs, proteins, and peptides). The functionalization of TEOS and APTS could promote the increase of -OH groups and the formation of $-NH_2$ groups on the glass surface, respectively. Infrared spectroscopy and elemental analysis are used to study the presence of -OH or $-NH_2$ groups. In addition, the bioactivity of these modified bioglass materials can be evaluated in vitro in simulated body fluid (SBF), and the results have shown that they were functionalized while maintaining their bioactivity.

### 15.7.3 Physico-Chemical Techniques

Methods belonging to this category include the use of plasma, coating procedures or ion exchange strategies to change the surface structure of nBAGs. The combination of the ordered deposition of polymer coating and plasma activation of nBAGs surface is considered to be an effective strategy for correctly customizing the surface characteristics of implants. Generally, plasma polymerization refers to the formation of polymer materials under the influence of partially ionized gas (plasma), which provides another option for traditional coating process. This method has been applied to the surface modification of hydroxyapatite particles with acrylic acid as a monomer, and successfully achieved the surface modification of particles, reducing the tendency of particles to form agglomerates, thus improving the mechanical properties of the composite and improving the thermal stability of the bioglass filled composite system.

The advantage of plasma polymerization technology is that it does not use organic solvents, reducing the consumption of reagents (monomers) and saving the time of the whole reaction process. Therefore, the plasma polymerization method is considered a fast way to modify nBAGs in the wet synthesis process [357]. For example, the surface wettability and adhesion of the bioactive glass coating in hyaluronic acid (HA), alginate (AL) or mixed solution could be improved by using air plasma treatment in the low-temperature plasma system. In terms of characterization, the surface of modified bioglass particles can be characterized by thermogravimetric analysis (TGA), X-ray photoelectron spectroscopy (XPS) and Fourier transform infrared spectroscopy (FTIR). Dynamic and isothermal thermogravimetric analysis (TGA) and gas permeation chromatography (GPC) were used to study the thermal degradation behavior of bioglass composites. In order to study the performance of the composite material system in the actual thermoplastic processing technology, the mechanical properties of the final product can be measured by tensile test after hot pressing.

The formation of a core-shell system refers to another important technology for developing glass and glass-ceramics with ideal surface characteristics [358]. Having a more active shell on the surface of nBAGs may further accelerate the formation of the HCA layer, thus accelerating the attachment of bone implants to tissues. In addition, some studies have shown that selectively changing the chemical composition of nBAGs may produce new and more reactive bioactive glass in the shell (core-shell system) surrounding the constant core. Through this core-shell system, the formation of silica gel layer is promoted, the formation of HCA layer is accelerated, and the biological activity is improved.

### 15.7.4 Radiation-Based Methods

A relatively underexplored physical strategy for changing the surface of nBAGs involves the use of radiation of various wavelengths. After irradiation, the solubility of nBAGs increases, resulting in the generation of non-bridging oxygen in the glass network, which has a significant impact on bioactivity and biocompatibility [359]. Research shows that γ-irradiation (25kGy) has an impact on the biological response of cells, resulting in increased proliferation of normal fibroblasts [360]. Applying an infrared laser beam on the surface of fused 45S5 bioactive glass to conduct micro and nano texture treatment has been confirmed to improve biological activity characteristics [361]. For example, nanosecond laser beam irradiation led to the formation of porous microstructure (pore size: 50 nm~2 μm) on the surface of 45S5 bioactive glass in the SBF immersion test, and it showed that the formation of an HCA layer on the sample surface was significantly improved, which could effectively promote bone integration and accelerate bone healing. On the other hand, the surface roughness and wettability of the glass after laser treatment increased, thus affecting the surface area of the samples. In the relevant research on antibacterial effect, it was found that the interaction between bacteria and nBAGs surface was mainly determined by the micron and submicron levels of surface roughness, so the samples with the highest surface roughness could exert the maximum inhibition effect on bacterial adhesion. At the same time, the cell compatibility of the samples did not decrease after laser treatment, further confirming the applicability of this method in endowing BAG-based biological implants and devices with the required characteristics [362].

### 15.7.5 Biological Improvements Carried by Surface Functionalization

The surface treatment of nBAGs is an effective approach to enhance the biological characteristics of nBAGs, which is conducive to cell proliferation, osteogenesis, anticancer and antibacterial activities. This functionalization process will not adversely affect the texture and properties of the bioactive glass itself, including layered pore structure (from meso to macro scale) and total porosity. Through surface functionalization, nBAGs combine some specific biological properties in addition to maintaining their excellent apatite forming ability as nano-grained glass. For example, SH-nBAGs and $NH_2$-nBAGs scaffolds have similar apatite mineralization ability and cell compatibility compared with untreated nBAGs scaffolds, and the modified SH-nBAGs and $NH_2$-nBAGs scaffolds have significantly improved the attachment, proliferation and differentiation of human bone marrow mesenchymal stem cells (hBMSCs), confirming the beneficial effect of this surface functionalization on bone tissue engineering applications [363].

Furthermore, anticancer applications using functionalized nBAGs include the combination of surface grafting of chemotherapeutic drugs with specific targeting molecules such as folate. Functionalization of bioactive glass using folic acid (FA), for example, can be used to prepare local site-specific anticancer delivery systems [364]. Ferrous magnetic bioactive microcrystalline glasses containing magnetite as the crystal phase have been functionalized in combination with chemotherapeutic drugs (doxorubicin and cisplatin) to facilitate the organic integration of hyperthermia with chemotherapy [365]. Natural polyphenols grafted onto the surface of bioactive glasses have also shown the ability to promote the growth of healthy osteoblasts and to selectively kill cancer cells [366].

To sum up, surface functionalization is a versatile tool for the modification of the outermost surface layer of biomaterials. In this way, the interaction of the material with the biological environment can be tailored, with this modification occurring mainly at the interface without changing the bulk properties of the material and allowing starting features of the specific material to be combined with those of the surface addition. As it stands, surface modification has been widely studied in the fields of polymers and metallic materials, but relatively little in the field of bioactive glasses. nBAGs are widely used because of their specific biological reactivity, ability to chemically bond to bone and soft tissue and their ability to release specific ions that modulate biological responses. Research efforts on surface functionalization of BGS have confirmed that the functionalization process can generally maintain their superior biological activity and, based on that, impart new specific properties leading to multifunctional materials with high added value. Moreover, the surface reactivity of BGS can be effectively exploited for the development of grafting procedures. With these factors in mind, surface functionalization of BGS holds some research promise to further tailor a certain characteristic of bioactive glasses for the development of novel composites for tailoring therapeutic applications according to the needs of clinical practical applications.

## REFERENCES

[1] Bharadwaz A, Jayasuriya AC. Recent trends in the application of widely used natural and synthetic polymer nanocomposites in bone tissue regeneration. *Mater Sci Eng C Mater Biol Appl.* 2020;110:110698. doi: 10.1016/j.msec.2020.110698.

[2] Shende P, Shah P. Carbohydrate-based magnetic nanocomposites for effective cancer treatment. *Int J Biol Macromol.* 2021;175:281–293. doi: 10.1016/j.ijbiomac.2021.02.044.

[3] Darwish MSA, Mostafa MH, Al-Harbi LM. Polymeric nanocomposites for environmental and industrial applications. *Int J Mol Sci.* 2022;23:1023. doi: 10.3390/ijms23031023.

[4] Cai Y, Chang SY, Gan SW, Ma S, Lu WF, Yen C-C. Nanocomposite bioinks for 3D bioprinting. *Acta Biomater.* 2022;151:45–69. doi: 10.1016/j.actbio.2022.08.014.

[5] Ansari F, Berglund LA. Toward semistructural cellulose nanocomposites: The need for scalable processing and interface tailoring. *Biomacromolecules.* 2018;19:2341–2350. doi: 10.1021/acs.biomac.8b00142.

[6] Dubey R, Dutta D, Sarkar A, Chattopadhyay P. Functionalized carbon nanotubes: synthesis, properties and applications in water purification, drug delivery, and material and biomedical sciences. *Nanoscale Adv*. 2021;3:5722–5744. doi: 10.1039/d1na00293g.
[7] Kim SW, Chung WS, Sohn K-S, Son C-Y, Lee S. Improvement of flexure strength and fracture toughness in alumina matrix composites reinforced with carbon nanotubes. *Materials Science and Engineering: A*. 2009;517:293–299. doi: 10.1016/j.msea.2009.04.035.
[8] Megiel E. Surface modification using TEMPO and its derivatives. *Adv Colloid Interface Sci*. 2017;250:158–184. doi: 10.1016/j.cis.2017.08.008.
[9] Novak S, Kovac J, Drazic G, Ferreira JMF, Quaresma S. Surface characterisation and modification of submicron and nanosized silicon carbide powders. *J Eur Ceram Soc*. 2007;27:3545–3550. doi: 10.1016/j.jeurceramsoc.2007.02.199.
[10] Ravarian R, Zhong X, Barbeck M, Ghanaati S, Kirkpatrick CJ, Murphy CM,, Schindeler A, Chrzanowski W, Dehghani F. Nanoscale chemical interaction enhances the physical properties of bioglass composites. *ACS Nano*. 2013;7:8469–8483. doi: 10.1021/nn402157n.
[11] Lee XJ, Show PL, Katsuda T, Chen W-H, Chang J-S. Surface grafting techniques on the improvement of membrane bioreactor: State-of-the-art advances. *Bioresour Technol*. 2018;269:489–502. doi: 10.1016/j.biortech.2018.08.090.
[12] Yan M. Photochemically initiated single polymer immobilization. *Chemistry*. 2007;13:4138–4144. doi: 10.1002/chem.200700317.
[13] Jariwala DH, Patel D, Wairkar S. Surface functionalization of nanodiamonds for biomedical applications. *Mater Sci Eng C Mater Biol Appl*. 2020;113:110996. doi: 10.1016/j.msec.2020.110996.
[14] Dou S, Tao L, Wang R, El Hankari S, Chen R, Wang S. Plasma-assisted synthesis and surface modification of electrode materials for renewable energy. *Adv Mater*. 2018;30:e1705850. doi: 10.1002/adma.201705850.
[15] Long L, Zhang M, Gan S, Zheng Z, He Y, Xu J, Fu R, Guo Q, Yu D, Chen W. Comparison of early osseointegration of non-thermal atmospheric plasma-functionalized/ SLActive titanium implant surfaces in beagle dogs. *Front Bioeng Biotechnol*. 2022;10:965248. doi: 10.3389/fbioe.2022.965248.
[16] Chen W, Zhou X, Zhang X, Bian J, Shi S, Nguyen T, Chen M, Wan J. Fast enhancement on hydrophobicity of poplar wood surface using low-pressure dielectric barrier discharges (DBD) plasma. *Appl Surf Sci*. 2017;407:412–417. doi: 10.1016/j.apsusc.2017.02.048.
[17] Wang C, Chen J-R. Studies on surface graft polymerization of acrylic acid onto PTFE film by remote argon plasma initiation. *Appl Surf Sci*. 2007;253:4599–4606. doi: 10.1016/j.apsusc.2006.10.014.
[18] Qian Y, Chen H, Xu Y, Yang J, Zhou X, Zhang F, Gu N. The preosteoblast response of electrospinning PLGA/PCL nanofibers: effects of biomimetic architecture and collagen I. *Int J Nanomedicine*. 2016;11:4157–4171. doi: 10.2147/IJN.S110577.
[19] Liu R, Zhao M, Zheng X, Wang Q, Huang X, Shen Y, Chen B. Reduced graphene oxide/TiO2(B) immobilized on nylon membrane with enhanced photocatalytic performance. *Sci Total Environ*. 2021;799:149370. doi: 10.1016/j.scitotenv.2021.149370.
[20] Shi Q, Qian Z, Liu D, Liu H. Surface modification of dental titanium implant by layer-by-layer electrostatic self-assembly. *Front Physiol*. 2017;8:574. doi: 10.3389/fphys.2017.00574.

[21] Hasani-Sadrabadi MM, Pouraghaei S, Zahedi E, Sarrion P, Ishijima M, Dashtimoghadam E, Jahedmanesh N, Ansari S, Ogawa T, Moshaverinia A. Antibacterial and osteoinductive implant surface using layer-by-layer assembly. *J Dent Res*. 2021;100:1161–1168. doi: 10.1177/00220345211029185.

[22] Price GJ, Nawaz M, Yasin T, Bibi S. Sonochemical modification of carbon nanotubes for enhanced nanocomposite performance. *Ultrason Sonochem*. 2018;40:123–130. doi: 10.1016/j.ultsonch.2017.02.021.

[23] Balasubramanian K, Burghard M. Chemically functionalized carbon nanotubes. *Small*. 2005;1:180–192. doi: 10.1002/smll.200400118.

[24] Rajukumar LP, Belmonte M, Slimak JE, Elias AL, Cruz-Silva E, Perea-Lopez N, Morelos-Gomez A, Terrones H, Endo M, Miranzo P, Terrones M. 3D Nanocomposites of covalently interconnected multiwalled carbon Nanotubes with SiC with enhanced thermal and electrical properties. *Adv Funct Mater*. 2015;25:4985–4993. doi: 10.1002/adfm.201501696.

[25] Ye J-S, Cui HF, Liu X, Lim TM, Zhang W-D, Sheu F-S. Preparation and characterization of aligned carbon nanotube-ruthenium oxide nanocomposites for supercapacitors. *Small*. 2005;1:560–565. doi: 10.1002/smll.200400137.

[26] Ci L, Ryu Z, Jin-Phillipp NY, Ruehle M. Investigation of the interfacial reaction between multi-walled carbon nanotubes and aluminum. *Acta Mater*. 2006;54:5367–5375. doi: 10.1016/j.actamat.2006.06.031.

[27] Laha T, Kuchibhatla S, Seal S, Li W, Agarwal A. Interfacial phenomena in thermally sprayed multiwalled carbon nanotube reinforced aluminum nanocomposite. *Acta Mater*. 2007;55:1059–1066. doi: 10.1016/j.actamat.2006.09.025.

[28] Lee J, Byun H, Madhurakkat Perikamana SK, Lee S, Shin H. Current Advances in immunomodulatory biomaterials for bone regeneration. *Adv Healthc Mater*. 2019;8(4):e1801106. doi: 10.1002/adhm.201801106.

[29] Holzapfel BM, Rudert M, Hutmacher DW. Scaffold-based bone tissue engineering. *Orthopade*. 2017;46(8):701–710. doi: 10.1007/s00132-017-3444-0.

[30] Kim BS, Mooney DJ. Development of biocompatible synthetic extracellular matrices for tissue engineering. *Trends Biotechnol*. 1998;16(5):224–230. doi: 10.1016/s0167-7799(98)01191-3.

[31] Wubneh A, Tsekoura EK, Ayranci C, Uludağ H. Current state of fabrication technologies and materials for bone tissue engineering. *Acta Biomater*. 2018;80:1–30. doi: 10.1016/j.actbio.2018.09.031.

[32] Hu C, Ashok D, Nisbet DR, Gautam V. Bioinspired surface modification of orthopedic implants for bone tissue engineering. *Biomaterials*. 2019;219:119366.

[33] Arora P, Sindhu A, Dilbaghi N, Chaudhury A, Rajakumar G, Rahuman AA. Nano-regenerative medicine towards clinical outcome of stem cell and tissue engineering in humans. *J Cell Mol Med*. 2012;16(9):1991–2000. doi: 10.1111/j.1582-4934.2012.01534.x.

[34] Jun Y, Kang E, Chae S, Lee SH. Microfluidic spinning of micro- and nano-scale fibers for tissue engineering. *Lab Chip*. 2014;14(13):2145–2160.

[35] Ahn TK, Lee DH, Kim TS, Jang GC, Choi S, Oh JB, Ye G, Lee S. Modification of titanium implant and titanium dioxide for bone tissue Engineering. *Adv Exp Med Biol*. 2018;1077:355–368. doi: 10.1007/978-981-13-0947-2_19.

[36] Yan H-H. *Preparation of Mussel-Inspired Conductive Oligoaniline Biomaterials for Tissue Engineering Applications*. Hefei: University of Science and Technology of China. 2020.

[37] Qin Z-Z. *Design And Preparation of Injectable Hydrogel with High Mechanical Strength*. Hefei: University of Science and Technology of China. 2019.

[38] Coyne KJ, Qin XX, Waite JH. Extensible collagen in mussel byssus: a natural block copolymer. *Science*. 1997;277(5333):1830–1832. doi: 10.1126/science.277.5333.1830.

[39] Tsai W-B, Chien C-Y, Thissen H, Lai J-Y. Dopamine-assisted immobilization of poly(ethylene imine) based polymers for control of cell-surface interactions. *Acta Biomater*. 2011;7(6):2518–2525. doi: 10.1016/j.actbio.2011.03.010.

[40] Ye Q, Zhou F, Liu W. Bioinspired catecholic chemistry for surface modification. *Chem Soc Rev*. 2011;40(7):4244–4258. doi: 10.1039/c1cs15026j.

[41] Anderson TH, Yu J, Estrada A, Hammer MU, Waite JH, Israelachvili JN. The contribution of DOPA to substrate-peptide adhesion and internal cohesion of mussei-inspired synthetic peptide films. *Adv Funct Mater*. 2010;20(23):4196–4205. doi: 10.1002/adfm.201000932.

[42] Lee H, Scherer NF, Messersmith PB. Single-molecule mechanics of mussel adhesion. *Proc Natl Acad Sci USA*. 2006;103(35):12999–13003. doi: 10.1073/pnas.0605552103.

[43] Lian C, Zheng A, Shi X. Research progress of dopamine for surface modification. *Synthetic Rubber Industry*. 2014;37(04):327–330.

[44] Yan J, Wu R, Liao S, Jiang M, Qian Y. Applications of polydopamine-modified scaffolds in the peripheral nerve tissue engineering. *Frontiers in Bioengineering and Biotechnology*. 2020;8:590998. doi: 10.3389/fbioe.2020.590998.

[45] Wu JP, Liu YT, Cao QD, Yu T, Zhang J, Liu QY, Yang XY. Growth factors enhanced angiogenesis and osteogenesis on polydopamine coated titanium surface for bone regeneration. *Materials & Design*. 2020;196:109162. doi. 10.1016/j.matdes.2020.109162.

[46] Pacelli S, Chakravarti AR, Modaresi S, Subham S, Burkey K, Kurlbaum C, Fang M, Neal CA, Mellott AJ, Chakraborty A, Paul A. Investigation of human adipose-derived stem-cell behavioruUsing a cell-instructive polydopamine-coated gelatin-alginate hydrogel. *Journal of Biomedical Materials Research Part A*. 2021;109(12):2597–2610. doi: 10.1002/jbm.a.37253.

[47] Zuppolini S, Cruz-Maya I, Guarino V, Borriello A. Optimization of polydopamine coatings onto poly-ε-caprolactone electrospun fibers for the fabrication of bio-electroconductive interfaces. *Journal of Functional Biomaterials*. 2020;11(1):19. doi: 10.3390/jfb11010019.

[48] Yang W, Zhang X, Wu K, Liu X, Jiao Y, Zhou C. Improving cytoactive of endothelial cell by introducing fibronectin to the surface of poly LLactic acid fiber mats via dopamine. *Materials Science and Engineering C*. 2016;69:373–379. doi: 10.1016/j.msec.2016.07.006.

[49] Zhao J, Han F, Zhang W, Yang Y, You D, Li L. Toward improved wound dressings: Effects of polydopamine-decorated poly(lactic-co-glycolic acid) electrospinning incorporating basic fibroblast growth factor and ponericin G1. *RSC Advances*. 2019;9(57): 33038–33051. doi: 10.1039/c9ra05030b.

[50] Kim HJ, Song JH. Improvement in the mechanical properties of carbon and aramid composites by fiber surface modification using polydopamine. *Composites Part B: Engineering*. 2019;160:31–36. doi: 10.1016/j.compositesb.2018.10.027.

[51] Lin CM, Wen YS, Huang XN, Yang ZJ, Gong FY, Zhao X, Hao SL, Pan LP, Ding L, Li J, Guo SY. Tuning the mechanical performance efficiently of various LLM-105 based PBXs via bioinspired interfacial reinforcement of polydopamine modification. *Composites B: Engineering*. 2020;186:107824. doi: 10.1016/j.compositesb.2020.107824.

[52] Xue Y, Niu W, Wang M, Chen M, Guo Y, Lei B. Engineering a biodegradable multifunctional antibacterial bioactive nanosystem for enhancing tumor photothermo-chemotherapy and bone regeneration. *ACS Nano*. 2020;14(1):442–453. doi: 10.1021/acsnano.9b06145.

[53] Zheng YH. *Preparation and in vitro bioactivity of copper-doped bioactive glass composite scaffold by polydopamine functionalization*. Zhejiang Sci-Tech University, 2021.

[54] LaVoie MJ, Ostaszewski BL, Weihofen A, Schlossmacher MG, Selkoe DJ. Dopamine covalently modifies and functionally inactivates parkin. *Nat Med*. 2005;11(11):1214–1221. doi: 10.1038/nm1314.

[55] Lee H, Rho J, Messersmith PB. Facile conjugation of biomolecules onto surfaces via mussel adhesive protein inspired coatings. *Adv Mater*. 2009;21(4):431–434. doi: 10.1002/adma.200801222.

[56] Lu W. *Tacrolimus-loaded hollow polydopamine nanoparticles modified Ti scaffolds for jaw bone defect repair through the FAK/ERK signaling pathway*. Jilin University. 2021.

[57] Zhao HL. *Preparation of gold nanoparticles modified composite polycaprolactone nanofiber membrane and research on its osteogenic properties*. Guangzhou Medical University. 2021.

[58] Lee H, Dellatore SM, Miller WM, Messersmith PB. Mussel-inspired surface chemistry for multifunctional coatings. *Science*. 2007;318(5849):426–430. doi: 10.1126/science.1147241.

[59] Ying T, Guoxin T, Chengyun N, Xicang R, Yu Z. Silver nanoparticles Modified with polydopamine on Titanium Surface and Their antibacterial and cytocompatibility. *Journal of Inorganic Materials*, 2014,29(12):1320–1326.

[60] Wang Y. *Preparation, antibacterial properties and biocompatibility of polyether ether ketone with silver coating on the surface of polydopamine assisted deposition*. Soochow University. 2016.

[61] Zhang WB. *Construction of Dopamine-mediated titanium surface nano silver coating and evaluation of its antibacterial and osteogenic activity*. Shandong University. 2019.

[62] Fu C. *Application of graphene oxide composites modified with gold nanoparticles and polydopamine coating in bone defect repair*. Jilin University. 2021.

[63] Confalonieri D, Schwab A, Walles H, Ehlicke F. Advanced therapy medicinal products: A guide for bone marrow-derived MSC application in bone and cartilage tissue engineering. *Tissue Eng Part B Rev*. 2018;24(2):155–169. doi: 10.1089/ten.TEB.2017.0305.

[64] Artas G, Gul M, Acikan I, Kirtay M, Bozoglan A, Simsek S, Yaman F, Dundar S. A comparison of different bone graft materials in peri-implant guided bone regeneration. *Braz Oral Res*. 2018;32:e59. doi: 10.1590/1807-3107bor-2018.vol32.0059.

[65] Hollý D, Klein M, Mazreku M, Zamborský R, Polák Š, Danišovič Ľ, Csöbönyeiová M. Stem cells and their derivatives-implications for alveolar bone regeneration: A comprehensive review. *Int J Mol Sci*. 2021;22(21):11746. doi: 10.3390/ijms222111746.

[66] Chanchareonsook N, Junker R, Jongpaiboonkit L, Jansen JA. Tissue-engineered mandibular bone reconstruction for continuity defects: a systematic approach to the literature. *Tissue Eng Part B Rev*. 2014;20(2):147–162. doi: 10.1089/ten.TEB.2013.0131.

[67] Abdul Halim NA, Hussein MZ, Kandar MK. Nanomaterials-upconverted hydroxyapatite for bone tissue engineering and a platform for drug delivery. *Int J Nanomedicine*. 2021;16:6477–6496. doi: 10.2147/IJN.S298936.

[68] Venkatesan J, Kim SK. Nano-hydroxyapatite composite biomaterials for bone tissue engineering--a review. *J Biomed Nanotechnol*. 2014;10(10):3124–3140. doi: 10.1166/jbn.2014.1893.

[69] Mohd Zaffarin AS, Ng SF, Ng MH, Hassan H, Alias E. Nano-hydroxyapatite as a delivery system for promoting bone regeneration in vivo: A systematic review. *Nanomaterials (Basel)*. 2021;11(10):2569. doi: 10.3390/nano11102569.

[70] Zhou H, Lee J. Nanoscale hydroxyapatite particles for bone tissue engineering. *Acta Biomater*. 2011;7(7):2769–2781. doi: 10.1016/j.actbio.2011.03.019.

[71] Molino G, Palmieri MC, Montalbano G, Fiorilli S, Vitale-Brovarone C. Biomimetic and mesoporous nano-hydroxyapatite for bone tissue application: a short review. *Biomed Mater*. 2020;15(2):022001. doi: 10.1088/1748-605X/ab5f1a.

[72] Fox K, Tran PA, Tran N. Recent advances in research applications of nanophase hydroxyapatite. *Chemphyschem*. 2012;13(10):2495–2506. doi: 10.1002/cphc.201200080.

[73] Shi J, Dai W, Gupta A, Zhang B, Wu Z, Zhang Y, Pan L, Wang L. Frontiers of hydroxyapatite composites in bionic bone tissue engineering. *Materials (Basel)*. 2022;15(23):8475. doi: 10.3390/ma15238475.

[74] Jagadeeshanayaka N, Awasthi S, Jambagi SC, Srivastava C. Bioactive surface modifications through thermally sprayed hydroxyapatite composite coatings: a review of selective reinforcements. *Biomater Sci*. 2022;10(10):2484–2523. doi: 10.1039/d2bm00039c.

[75] Bharadwaz A, Jayasuriya AC. Recent trends in the application of widely used natural and synthetic polymer nanocomposites in bone tissue regeneration. *Mater Sci Eng C Mater Biol Appl*. 2020;110:110698. doi: 10.1016/j.msec.2020.110698.

[76] Li J, Li K, Du Y, Tang X, Liu C, Cao S, Zhao B, Huang H, Zhao H, Kong W, Xu T, Shao C, Shao J, Zhang G, Lan H, Xi Y. Dual-nozzle 3D printed nano-hydroxyapatite scaffold loaded with vancomycin sustained-release microspheres for enhancing bone regeneration. *Int J Nanomedicine*. 2023;18:307–322. doi: 10.2147/IJN.S394366.

[77] Guo L, Liang Z, Yang L, Du W, Yu T, Tang H, Li C, Qiu H. The role of natural polymers in bone tissue engineering. *J Control Release*. 2021;338:571–582. doi: 10.1016/j.jconrel.2021.08.055.

[78] Coenen AMJ, Bernaerts KV, Harings JAW, Jockenhoevel S, Ghazanfari S. Elastic materials for tissue engineering applications: Natural, synthetic, and hybrid polymers. *Acta Biomater*. 2018;79:60–82. doi: 10.1016/j.actbio.2018.08.027.

[79] Roy P, Sailaja RR. Mechanical, thermal and bio-compatibility studies of PAEK-hydroxyapatite nanocomposites. *J Mech Behav Biomed Mater*. 2015;49:1–11. doi: 10.1016/j.jmbbm.2015.04.022.

[80] Son J, Kim J, Lee K, Hwang J, Choi Y, Seo Y, Jeon H, Kang HC, Woo HM, Kang BJ, Choi J. DNA aptamer immobilized hydroxyapatite for enhancing angiogenesis and bone regeneration. *Acta Biomater*. 2019;99:469–478. doi: 10.1016/j.actbio.2019.08.047.

[81] Qin L, Yao S, Meng W, Zhang J, Shi R, Zhou C, Wu J. Novel antibacterial dental resin containing silanized hydroxyapatite nanofibers with remineralization capability. *Dent Mater*. 2022;38(12):1989–2002. doi: 10.1016/j.dental.2022.11.014.

[82] Abe Y, Okazaki Y, Hiasa K, Yasuda K, Nogami K, Mizumachi W, Hirata I. Bioactive surface modification of hydroxyapatite. *Biomed Res Int.* 2013;2013:626452. doi: 10.1155/2013/626452.
[83] Wang J, Gao H, Hu Y, Zhang N, Zhou W, Wang C, Binks BP, Yang Z. 3D printing of Pickering emulsion inks to construct poly (D,L-lactide-co-trimethylene carbonate)-based porous bioactive scaffolds with shape memory effect. *J Mater Sci.* 2021;56:731–745. doi: 10.1007/s10853-020-05318-7.
[84] Haider A, Kim S, Huh MW, Kang IK. BMP-2 grafted nHA/PLGA hybrid nanofiber scaffold stimulates osteoblastic cells growth. *Biomed Res Int.* 2015;2015:281909. doi: 10.1155/2015/281909.
[85] Foroughi MR, Karbasi S, Ebrahimi-Kahrizsangi R. Mechanical evaluation of nHAp scaffold coated with poly-3-hydroxybutyrate for bone tissue engineering. *J Nanosci Nanotechnol.* 2013;13(2):1555–1562. doi: 10.1166/jnn.2013.6019.
[86] Sun Y, Deng Y, Ye Z, Liang S, Tang Z, Wei S. Peptide decorated nano-hydroxyapatite with enhanced bioactivity and osteogenic differentiation via polydopamine coating. *Colloids Surf B Biointerfaces.* 2013;111:107–116. doi: 10.1016/j.colsurfb.2013.05.037.
[87] Shi D, Shen J, Zhang Z, Shi C, Chen M, Gu Y, Liu Y. Preparation and properties of dopamine-modified alginate/chitosan-hydroxyapatite scaffolds with gradient structure for bone tissue engineering. *J Biomed Mater Res A.* 2019;107(8):1615–1627. doi: 10.1002/jbm.a.36678.
[88] Hu S, Wu J, Cui Z, Si J, Wang Q, Peng X. Study on the mechanical and thermal properties of polylactic acid/hydroxyapatite@polydopamine composite nanofibers for tissue engineering. *J Appl Polym SCI.* 2020;137(39):49077. doi: 10.1002/app.49077.
[89] Tian X, Yuan X, Feng D, Wu M, Yuan Y, Ma C, Xie D, Guo J, Liu C, Lu Z. In vivo study of polyurethane and tannin-modified hydroxyapatite composites for calvarial regeneration. *J Tissue Eng.* 2020;11:2041731420968030. doi: 10.1177/2041731420968030.
[90] Zou Q, Li J, Niu L, Zuo Y, Li J, Li Y. Modified n-HA/PA66 scaffolds with chitosan coating for bone tissue engineering: cell stimulation and drug release. *J Biomater Sci Polym Ed.* 2017;28(13):1271–1285. doi: 10.1080/09205063.2017.1318029.
[91] Wang X, Song G, Lou T. Fabrication and characterization of nano-composite scaffold of PLLA/silane modified hydroxyapatite. *Med Eng Phys.* 2010;32(4):391–397. doi: 10.1016/j.medengphy.2010.02.002.
[92] Fu S, Wang X, Guo G, Shi S, Fan M, Liang H, Luo F, Qian Z. Preparation and properties of nano-hydroxyapatite/PCL-PEG-PCL composite membranes for tissue engineering applications. *J Biomed Mater Res B Appl Biomater.* 2011;97(1):74–83. doi: 10.1002/jbm.b.31788.
[93] Hong Z, Qiu X, Sun J, Deng M, Chen X, Jing X. Grafting polymerization of l-lactide on the surface of hydroxyapatite nano-crystals.*Polymer.* 2004;45(19): 6699–6706. doi: 10.1016/j.polymer.2004.07.036.
[94] Ma R, Li Q, Wang L, Zhang X, Fang L, Luo Z, Xue B, Ma L. Mechanical properties and in vivo study of modified-hydroxyapatite/polyetheretherketone biocomposites. *Mater Sci Eng C Mater Biol Appl.* 2017;73:429–439. doi: 10.1016/j.msec.2016.12.076.
[95] Wang L, Li M, Li X, Liu J, Mao Y, Tang K. A biomimetic hybrid hydrogel based on the interactions between amino hydroxyapatite and gelatin/gellan gum. *Macromol Mater Eng.* 2020;305:2000188. doi: 10.1002/mame.202000188.

[96] Jaramillo AF, Medina C, Flores P, Canales C, Maldonado C, Rivera PC, D.Rojas, Meléndrez MF. Improvement of thermomechanical properties of composite based on hydroxyapatite functionalized with alkylsilanes in epoxy matrix. *Ceram Int.* 2020;46(6): 8368–8378. doi: 10.1016/j.ceramint.2019.12.069.
[97] He J, Hu X, Xing L, Chen D, Peng L, Liang G, Xiong C, Zhang X, Zhang L. Enhanced bone regeneration using poly (trimethylene carbonate)/vancomycin hydrochloride porous microsphere scaffolds in presence ofthe silane coupling agent modified hydroxyapatite nanoparticles. *J Ind Eng Chem.* 2021;99:134–144. doi: 10.1016/j.jiec.2021.04.021.
[98] Atak BH, Buyuk B, Huysal M, Isik S, Senel M, Metzger W, Cetin G. Preparation and characterization of amine functional nano-hydroxyapatite/ chitosan bionanocomposite for bone tissue engineering applications. *Carbohydr Polym.* 2017;164:200–213. doi: 10.1016/j.carbpol.2017.01.100.
[99] Victor SP, VM V, Komeri R, Selvam S, Muthu J. Covalently cross-linked hydroxyapatite–citric acid–based biomimetic polymeric composites for bone applications. *Journal of Bioactive and Compatible Polymers.* 2015;30(5):524–540. doi:10.1177/0883911515585181.
[100] Huysal M, Şenel M. Dendrimer functional hydroxyapatite nanoparticles generated by functionalization with siloxane-cored PAMAM dendrons. *J Colloid Interface Sci.* 2017;500:105–112. doi: 10.1016/j.jcis.2017.04.004.
[101] Dai Y, Xu M, Wei J, Zhang H, Chen Y. Surface modification of hydroxyapatite nanoparticles by poly(L-phenylalanine) via ROP of L-phenylalanine N-carboxyanhydride (Pha-NCA). 2012;258(7): 2850–2855. doi: 10.1016/ j.apsusc.2011.10.147.
[102] Liao L, Yang S, Richard JM, Wei J, Zhang Y, Zhang M. In vitro characterization of PBLG-g-HA/PLLA nanocomposite scaffolds. *Journal of Wuhan University of Technology-Mater. Sci. Ed.* Aug. 2014;29(4):841–847. doi: 10.1007/s11595-014-1006-4.
[103] Lee JB, Kim JE, Balikov DA, Bae MS, Heo DN, Lee D, Rim HJ, Lee DW, Sung HJ, Kwon IK. Poly(l-lactic acid)/gelatin fibrous scaffold loaded with simvastatin/beta-cyclodextrin-modified hydroxyapatite inclusion complex for bone Tissue Regeneration. *Macromol Biosci.* 2016;16(7):1027–1038. doi: 10. 1002/mabi.201500450.
[104] Ding H, Jiang L, Tang C, Tang S, Ma B, Zhang N, Wen Y, Zhang Y, Sheng L, Su S, Hu X. Study on the surface-modification of nano-hydroxyapatite with lignin and the corresponding nanocomposite with poly (lactide-co-glycolide). *Front Chem Sci. Eng.* 2021;15:630–642. doi: 10.1007/s11705-020-1970-5.
[105] Diao H, Si Y, Zhu A, Ji L, Shi H. Surface modified nano-hydroxyapatite/poly (lactide acid) composite and its osteocyte compatibility. *Mater Sci Eng C Mater Biol Appl.* 2012;32(7):1796–1801. doi: 10.1016/j.msec.2012.04.065.
[106] Gheysari H, Mohandes F, Mazaheri M, Dolatyar B, Askari M, Simchi A. Extraction of Hydroxyapatite Nanostructures from Marine Wastes for the Fabrication of Biopolymer-Based Porous Scaffolds. *Mar Drugs.* 2019;18(1):26. doi: 10.3390/md18010026.
[107] Yahia IS, Keshk S, Shkir M, Darwish R, Alshahrie A. An effect of cationic CTAB and Anionic SDS surfactants on the physicochemical properties of hydroxyapatite nanostructures for bone tissue engineering. *J Nanoelectron Optoe.* 2020;15:316–324. doi: 10.1166/jno.2020.2753.
[108] Nichols HL, Zhang N, Zhang J, Shi D, Bhaduri S, Wen X. Coating nanothickness degradable films on nanocrystalline hydroxyapatite particles to improve the bonding strength between nanohydroxyapatite and degradable

polymer matrix. *J Biomed Mater Res A*. 2007;82(2):373–382. doi: 10.1002/jbm.a.31066.
[109] Kumar L, Ahuja D. Preparation and characterization of aliphatic polyurethane and modified hydroxyapatite composites for bone tissue engineering. *Polym Bull*. 2020;77:6049–6062. doi: 10.1007/s00289-019-03067-5.
[110] Yi W, Li L, He H, Hao Z, Liu B, Zi-Sheng C, Yi S. Synthesis of poly(l-lactide)/β-cyclodextrin/citrate network modified hydroxyapatite and its biomedical properties. *New J Chem*. 2018;42:14729–14732. doi: 10.1039/C8NJ01194J.
[111] Hong Z, Zhang P, He C, Qiu X, Liu A, Chen L, Chen X, Jing X. Nanocomposite of poly(L-lactide) and surface grafted hydroxyapatite: mechanical properties and biocompatibility. *Biomaterials*. 2005;26(32):6296–6304. doi: 10.1016/j.biomaterials.2005.04.018.
[112] Zhang P, Hong Z, Yu T, Chen X, Jing X. In vivo mineralization and osteogenesis of nanocomposite scaffold of poly(lactide-co-glycolide) and hydroxyapatite surface-grafted with poly(L-lactide). *Biomaterials*. 2009;30(1):58–70. doi: 10.1016/j.biomaterials.2008.08.041.
[113] Liuyun J, Lixin J, Chengdong X, Lijuan X, Ye L. Effect of l-lysine-assisted surface grafting for nano-hydroxyapatite on mechanical properties and in vitro bioactivity of poly(lactic acid-co-glycolic acid). *J Biomater Appl*. 2016;30(6):750–758. doi: 10.1177/0885328215584491.
[114] Timpu D, Sacarescu L, Vasiliu T, Dinu MV, David G. Surface cationic functionalized nano-hydroxyapatite – Preparation, characterization, effect of coverage on properties and related applications. *Eur Polym J*. 2020;132:109759. doi: 10.1016/j.eurpolymj.2020.109759
[115] Hu Y, Wang J, Li X, Hu X, Zhou W, Dong X, Wang C, Yang Z, Binks BP. Facile preparation of bioactive nanoparticle/poly(ε-caprolactone) hierarchical porous scaffolds via 3D printing of high internal phase Pickering emulsions. *J Colloid Interface Sci*. 2019;545:104–115. doi: 10.1016/j.jcis.2019.03.024.
[116] Lee H, Choi H, Kim K, Lee S. Modification of hydroxyapatite nanosurfaces for enhanced colloidal stability and improved interfacial adhesion in nanocomposites. *Chem Mater*. 2006;18(21):5111–5118. doi: 10.1021/cm061139x.
[117] Chunyan T, Haojie D, Shuo T, Liuyun J, Bingli M, Yue W, Na Z, Liping S, Shengpei S. A combined-modification method of carboxymethyl β-cyclodextrin and lignin for nano-hydroxyapatite to reinforce poly(lactide-co-glycolide) for bone materials. *Int J Biol Macromol*. 2020;160:142–152. doi: 10.1016/j.ijbiomac.2020.05.142.
[118] Kumar L, Ahuja D. 3D porous polyurethane (PU)/ triethanolamine modified hydroxyapatite (TEA-HA) nano composite for enhanced bioactivity for biomedical applications. *J Polym Res*. 2022;29:17. doi: 0.1007/s10965-021-02861-y.
[119] Wu T, Li B, Wang W, Chen L, Li Z, Wang M, Zha Z, Lin Z, Xia H, Zhang T. Strontium-substituted hydroxyapatite grown on graphene oxide nanosheet-reinforced chitosan scaffold to promote bone regeneration. *Biomater Sci*. 2020;8(16):4603–4615. doi: 10.1039/d0bm00523a.
[120] Zima A, Czechowska J, Siek D, Olkowski R, Noga M, Lewandowska-Szumieł M, Ślósarczyk A. How calcite and modified hydroxyapatite influence physicochemical properties and cytocompatibility of alpha-TCP based bone cements. *J Mater Sci Mater Med*. 2017;28(8):117. doi: 10.1007/s10856-017-5934-3.
[121] Heidari F, Bazargan-Lari R, Razavi M, Fahimipour F, Vashaee D, Tayebi L. Nano-hydroxyapatite and nano-hydroxyapatite/zinc oxide scaffold for bone tissue engineering application. *Int J Appl Ceram Technol*. 2020;17:2752–2761. doi: 10.1111/ijac.13596.

[122] Elsayed MT, Hassan AA, Abdelaal SA, Taher MM, Ahmed MK, Shoueir KR. Morphological, antibacterial, and cell attachment of cellulose acetate nanofibers containing modified hydroxyapatite for wound healing utilizations. *J Mater Res Technol.* 2020;9(6):13927–13936. doi: 10.1016/j.jmrt.2020.09.094.
[123] Sundar R, Joseph J, Babu S, Varma H, John A, Abraham A. 3D-bulk to nanoforms of modified hydroxyapatite: Characterization and osteogenic potency in an in vitro 3D bone model system. *J Biomed Mater Res B Appl Biomater.* 2022;110(5):1151–1164. doi: 10.1002/jbm.b.34989.
[124] Hassan M, Sulaiman M, Yuvaraju PD, Galiwango E, Rehman IU, Al-Marzouqi AH, Khaleel A, Mohsin S. Biomimetic PLGA/strontium-zinc nano hydroxyapatite composite scaffolds for bone regeneration. *J Funct Biomater.* 2022;13(1):13. doi: 10.3390/jfb13010013.
[125] El-Hamshary H, El-Naggar ME, El-Faham A, Abu-Saied MA, Ahmed MK, Al-Sahly M. Preparation and characterization of nanofibrous scaffolds of Ag/vanadate hydroxyapatite encapsulated into polycaprolactone: morphology, mechanical, and in vitro cells adhesion. *Polymers (Basel).* 2021;13(8):1327. doi: 10.3390/polym13081327.
[126] Tomoaia G, Soritau O, Tomoaia-Cotisel M, Pop LB, Pop A, Mocanu A, Horovitz O, Bobos LD. Scaffolds made of nanostructured phosphates, collagen and chitosan forcell culture. *Powder Technol.* 2013;238:99–107. doi:10.1016/j.powtec.2012.05.023.
[127] Tithito T, Suntornsaratoon P, Charoenphandhu N, Thongbunchoo J, Krishnamra N, Tang IM, Pon-On W. Fabrication of biocomposite scaffolds made with modified hydroxyapatite inclusion of chitosan-grafted-poly(methyl methacrylate) for bone tissue engineering. *Biomed Mater.* 2019;14(2):025013. doi: 10.1088/1748-605X/ab025f.
[128] Chen X, Wu D, Xu J, Yan T, Chen Q. Gelatin/Gelatin-modified nano hydroxyapatite composite scaffolds with hollow channel arrays prepared by extrusion molding for bone tissue engineering. *Mater Res Express* 2021;8:015027. doi: 10.1088/2053-1591/abde1f.
[129] Shebi A, Lisa S. Pectin mediated synthesis of nano hydroxyapatite-decorated poly(lactic acid) honeycomb membranes for tissue engineering. *Carbohydr Polym.* 2018;201:39–47. doi: 10.1016/j.carbpol.2018.08.012.
[130] Gallo M, Le Gars Santoni B, Douillard T, Zhang F, Gremillard L, Dolder S, Hofstetter W, Meille S, Bohner M, Chevalier J, Tadier S. Effect of grain orientation and magnesium doping on β-tricalcium phosphate resorption behavior. *Acta Biomater.* 2019;89:391–402. doi: 10.1016/j.actbio.2019.02.045.
[131] Chrcanovic BR, Kisch J, Albrektsson T, Wennerberg A. A retrospective study on clinical and radiological outcomes of oral implants in patients followed up for a minimum of 20 years. *Clin. Implant Dent. Relat. Res.* 2018;20:199–207. doi: 10.1111/cid.12571.
[132] Antoun H, Karouni M, Abitbol J, Zouiten O, Jemt T. A retrospective study on 1592 consecutively performed operations in one private referral clinic. Part I: Early inflammation and early implant failures. *Clin. Implant Dent. Relat. Res.* 2017;19:404–412. doi: 10.1111/cid.12477.
[133] Chrcanovic BR, Kisch J, Albrektsson T, Wennerberg A. Factors influencing early dental implant failures. *J. Dent. Res.* 2016;95:995–1002. doi: 10.1177/0022034516646098.
[134] Bagno A, Di Bello C. Surface treatments and roughness properties of Ti-based biomaterials. *J Mater Sci Mater Med.* 2004;15(9):935–949. doi: 10.1023/B:JMSM.0000042679.28493.7f.

[135] Hao CP, Cao NJ, Zhu YH, Wang W. The osseointegration and stability of dental implants with different surface treatments in animal models: a network meta-analysis. *Sci Rep*. 2021;11(1):13849. doi: 10.1038/s41598-021-93307-4.
[136] Wang Q, Zhou P, Liu S, Attarilar S, Ma RL, Zhong Y, Wang L. Multi-scale surface treatments of titanium implants for rapid osseointegration: A review. *Nanomaterials (Basel)*. 2020;10(6):1244. doi: 10.3390/nano10061244.
[137] Spriano S, Sarath Chandra V, Cochis A, Uberti F, Rimondini L, Bertone E, Vitale A, Scolaro C, Ferrari M, Cirisano F, Gautier di Confiengo G, Ferraris S. How do wettability, zeta potential and hydroxylation degree affect the biological response of biomaterials? *Mater Sci Eng C Mater Biol Appl*. 2017;74:542–555. doi: 10.1016/j.msec.2016.12.107.
[138] Rupp F, Liang L, Geis-Gerstorfer J, Scheideler L, Hüttig F. Surface characteristics of dental implants: A review. *Dent Mater*. 2018;34(1):40–57. doi: 10.1016/j.dental.2017.09.007.
[139] Iwasa F, Tsukimura N, Sugita Y, Kanuru RK, Kubo K, Hasnain H, Att W, Ogawa T. TiO2 micro-nano-hybrid surface to alleviate biological aging of UV-photofunctionalized titanium. *Int J Nanomedicine*. 2011;6:1327–1341. doi: 10.2147/IJN.S22099.
[140] Mas-Moruno C, Su B, Dalby MJ. Multifunctional coatings and nanotopographies: Toward cell instructive and antibacterial implants. *Adv Healthc Mater*. 2019;8(1):e1801103. doi: 10.1002/adhm.201801103.
[141] Wang Q, Zhou P, Liu S, Attarilar S, Ma RL, Zhong Y, Wang L. Multi-scale surface treatments of titanium implants for rapid osseointegration: A review. *Nanomaterials (Basel)*. 2020;10(6):1244. doi: 10.3390/nano10061244.
[142] Losic D. Advancing of titanium medical implants by surface engineering: recent progress and challenges. *Expert Opin Drug Deliv*. 2021;18(10):1355–1378. doi: 10.1080/17425247.2021.1928071.
[143] Wennerberg A, Ide-Ektessabi A, Hatkamata S, Sawase T, Johansson C, Albrektsson T, Martinelli A, Södervall U, Odelius H. Titanium release from implants prepared with different surface roughness. *Clin Oral Implants Res*. 2004;15(5):505–512. doi: 10.1111/j.1600-0501.2004.01053.x.
[144] Asri RIM, Harun WSW, Samykano M, Lah NAC, Ghani SAC, Tarlochan F, Raza MR. Corrosion and surface modification on biocompatible metals: A review. *Mater Sci Eng C Mater Biol Appl*. 2017;77:1261–1274. doi: 10.1016/j.msec.2017.04.102.
[145] Raphel J, Holodniy M, Goodman SB, Heilshorn SC. Multifunctional coatings to simultaneously promote osseointegration and prevent infection of orthopaedic implants. *Biomaterials*. 2016;84:301–314. doi: 10.1016/j.biomaterials.2016.01.016.
[146] Zhou W, Peng X, Zhou X, Li M, Ren B, Cheng L. Influence of bio-aging on corrosion behavior of different implant materials. *Clin Implant Dent Relat Res*. 2019;21(6):1225–1234. doi: 10.1111/cid.12865.
[147] Haibin L, Lei W, Xueyang Z, et al. Effects of hydrocarbons contamination on initial responses of osteoblast-like cells on acid-etched titanium surface. *Rare Metal Materials and Engineering*. 2013;42(8):1558–1562.
[148] Ogawa T. Ultraviolet photofunctionalization of titanium implants. *Int J Oral Maxillofac Implants*. 2014;29(1):e95–e102. doi: 10.11607/jomi.te47.
[149] Chang LC. Clinical Applications of Photofunctionalization on Dental implant surfaces: A narrative review. *J Clin Med*. 2022;11(19):5823. doi: 10.3390/jcm11195823.
[150] Shen JW, Chen Y, Yang GL, Wang XX, He FM, Wang HM. Effects of storage medium and UV photofunctionalization on time-related changes of titanium

surface characteristics and biocompatibility. *J Biomed Mater Res B Appl Biomater*. 2016;104(5):932–940. doi: 10.1002/jbm.b.33437.
[151] Yamada Y, Yamada M, Ueda T, Sakurai K. Reduction of biofilm formation on titanium surface with ultraviolet-C pre-irradiation. *J Biomater Appl*. 2014;29(2):161–171. doi: 10.1177/0885328213518085.
[152] Henningsen A, Smeets R, Hartjen P, Heinrich O, Heuberger R, Heiland M, Precht C, Cacaci C. Photofunctionalization and non-thermal plasma activation of titanium surfaces. *Clin Oral Investig*. 2018;22(2):1045–1054. doi: 10.1007/s00784-017-2186-z.
[153] Pyo SW, Park YB, Moon HS, Lee JH, Ogawa T. Photofunctionalization enhances bone-implant contact, dynamics of interfacial osteogenesis, marginal bone seal, and removal torque value of implants: a dog jawbone study. *Implant Dent*. 2013;22(6):666–675. doi: 10.1097/ID.0000000000000003.
[154] Kim MY, Choi H, Lee JH, Kim JH, Jung HS, Kim JH, Park YB, Moon HS. UV Photofunctionalization effect on bone graft in critical one-wall defect around implant: A pilot study in beagle dogs. *Biomed Res Int*. 2016;2016:4385279. doi: 10.1155/2016/4385279.
[155] Funato A, Yamada M, Ogawa T. Success rate, healing time, and implant stability of photofunctionalized dental implants. *Int J Oral Maxillofac Implants*. 2013;28(5):1261–1271. doi: 10.11607/jomi.3263.
[156] Fridman A, Kennedy L A. *Plasma Physics and Engineering*. CRC Press. 2016.
[157] Duske K, Koban I, Kindel E, Schröder K, Nebe B, Holtfreter B, Jablonowski L, Weltmann KD, Kocher T. Atmospheric plasma enhances wettability and cell spreading on dental implant metals. *J Clin Periodontol*. 2012;39(4):400–407. doi: 10.1111/j.1600-051X.2012.01853.x.
[158] Danna NR, Beutel BG, Tovar N, Witek L, Marin C, Bonfante EA, Granato R, Suzuki M, Coelho PG. Assessment of atmospheric pressure plasma treatment for implant osseointegration. *Biomed Res Int*. 2015;2015:761718. doi: 10.1155/2015/761718.
[159] Wang L, Wang W, Zhao H, Liu Y, Liu J, Bai N. Bioactive effects of low-temperature argon-oxygen plasma on a titanium implant surface. *ACS Omega*. 2020;5(8):3996–4003. doi: 10.1021/acsomega.9b03504.
[160] Lee JH, Jeong WS, Seo SJ, Kim HW, Kim KN, Choi EH, Kim KM. Non-thermal atmospheric pressure plasma functionalized dental implant for enhancement of bacterial resistance and osseointegration. *Dent Mater*. 2017;33(3):257–270. doi: 10.1016/j.dental.2016.11.011.
[161] Dong Y, Long L, Zhang P, Yu D, Wen Y, Zheng Z, Wu J, Chen W. A chair-side plasma treatment system for rapidly enhancing the surface hydrophilicity of titanium dental implants in clinical operations. *J Oral Sci*. 2021;63(4):334–340. doi: 10.2334/josnusd.21-0090.
[162] Julák J, Scholtz V, Vaňková E. Medically important biofilms and non-thermal plasma. *World J Microbiol Biotechnol*. 2018;34(12):178. doi: 10.1007/s11274-018-2560-2.
[163] Huang J, Zhang X, Yan W, Chen Z, Shuai X, Wang A, Wang Y. Nanotubular topography enhances the bioactivity of titanium implants. *Nanomedicine*. 2017;13(6):1913–1923. doi: 10.1016/j.nano.2017.03.017.
[164] Diu T, Faruqui N, Sjöström T, Lamarre B, Jenkinson HF, Su B, Ryadnov MG. Cicada-inspired cell-instructive nanopatterned arrays. *Sci Rep*. 2014;4:7122. doi: 10.1038/srep07122.
[165] Cao Y, Su B, Chinnaraj S, Jana S, Bowen L, Charlton S, Duan P, Jakubovics NS, Chen J. Nanostructured titanium surfaces exhibit recalcitrance towards

Staphylococcus epidermidis biofilm formation. *Sci Rep*. 2018;8(1):1071. doi: 10.1038/s41598-018-19484-x.

[166] Hou C, An J, Zhao D, Ma X, Zhang W, Zhao W, Wu M, Zhang Z, Yuan F. Surface modification techniques to produce micro/nano-scale topographies on ti-Based implant surfaces for improved osseointegration. *Front Bioeng Biotechnol*. 2022; 10:835008. doi: 10.3389/fbioe.2022.835008.

[167] Huang YZ, He SK, Guo ZJ, Pi JK, Deng L, Dong L, Zhang Y, Su B, Da LC, Zhang L, Xiang Z, Ding W, Gong M, Xie HQ. Nanostructured titanium surfaces fabricated by hydrothermal method: Influence of alkali conditions on the osteogenic performance of implants. *Mater Sci Eng C Mater Biol Appl*. 2019; 94:1–10. doi: 10.1016/j.msec.2018.08.069.

[168] Su Y, Komasa S, Sekino T, Nishizaki H, Okazaki J. Nanostructured Ti6Al4V alloy fabricated using modified alkali-heat treatment: Characterization and cell adhesion. *Mater Sci Eng C Mater Biol Appl*. 2016;59:617–623. doi: 10.1016/j.msec.2015.10.077.

[169] Yu X, Xu R, Zhang Z, Jiang Q, Liu Y, Yu X, Deng F. Different cell and tissue behavior of micro-/nano-tubes and micro-/nano-nets topographies on selective laser melting titanium to enhance osseointegration. *Int J Nanomedicine*. 2021;16:3329–3342. doi: 10.2147/IJN.S303770.

[170] Ueno T, Tsukimura N, Yamada M, Ogawa T. Enhanced bone-integration capability of alkali- and heat-treated nanopolymorphic titanium in micro-to-nanoscale hierarchy. *Biomaterials*. 2011;32(30):7297–7308. doi: 10.1016/j.biomaterials.2011.06.033.

[171] Zhuang XM, Zhou B, Ouyang JL, Sun HP, Wu YL, Liu Q, Deng FL. Enhanced MC3T3-E1 preosteoblast response and bone formation on the addition of nano-needle and nano-porous features to microtopographical titanium surfaces. *Biomed Mater*. 2014;9(4):045001. doi: 10.1088/1748-6041/9/4/045001.

[172] Wang Q, Zhou P, Liu S, Attarilar S, Ma RL, Zhong Y, Wang L. Multi-scale surface treatments of titanium implants for rapid osseointegration: A review. *Nanomaterials*. 2020;10(6). doi: 10.3390/nano10061244

[173] Sisti KE, de Andrés MC, Johnston D, Almeida-Filho E, Guastaldi AC, Oreffo RO. Skeletal stem cell and bone implant interactions are enhanced by LASER titanium modification. *Biochem Biophys Res Commun*. 2016;473(3):719–725. doi: 10.1016/j.bbrc.2015.10.013.

[174] Li C, Yang Y, Yang L, Shi Z. Biomimetic Anti-Adhesive Surface Microstructures on Electrosurgical Blade Fabricated by Long-Pulse Laser Inspired by Pangolin Scales. *Micromachines*. 2019;10(12). doi: 10.3390/mi10120816.

[175] Beltrán AM, Giner M, Rodríguez Á, Trueba P, Rodríguez-Albelo LM, Vázquez-Gámez MA, Godinho V, Alcudia A, Amado JM, López-Santos C, Torres Y. Influence of femtosecond laser modification on biomechanical and biofunctional behavior of porous titanium substrates. *Materials (Basel)*. 2022;15(9):2969. doi: 10.3390/ma15092969.

[176] Oliveira V, Cunha A, Vilar R. Multi-scaled femtosecond laser structuring of stationary titanium surfaces. *Journal of Optoelectronics & Advanced Materials*. 2010;12(3):654–658. doi: 10.1166/jnn.2010.2128

[177] Cunha A, Zouani OF, Plawinski L, Botelho do Rego AM, Almeida A, Vilar R, Durrieu MC. Human mesenchymal stem cell behavior on femtosecond laser-textured Ti-6Al-4V surfaces. *Nanomedicine (Lond)*. 2015;10(5):725–739. doi: 10.2217/nnm.15.19.

[178] Dumas V, Guignandon A, Vico L, Mauclair C, Zapata X, Linossier MT, Bouleftour W, Granier J, Peyroche S, Dumas JC, Zahouani H, Rattner A.

Femtosecond laser nano/micro patterning of titanium influences mesenchymal stem cell adhesion and commitment. *Biomed Mater.* 2015;10(5):055002. doi: 10.1088/1748-6041/10/5/055002.
[179] Scarano A, Lorusso F, Inchingolo F, Postiglione F, Petrini M. The effects of erbium-doped yttrium aluminum garnet laser (Er: YAG) irradiation on sand-blasted and acid-etched (SLA) titanium, an In vitro study. *Materials (Basel).* 2020;13(18):4174. doi: 10.3390/ma13184174.
[180] Li T, Gulati K, Wang N, Zhang Z, Ivanovski S. Bridging the gap: Optimized fabrication of robust titania nanostructures on complex implant geometries towards clinical translation. *J Colloid Interface Sci.* 2018;529:452–463. doi: 10.1016/j.jcis.2018.06.004.
[181] Jayasree A, Ivanovski S, Gulati K. ON or OFF: Triggered therapies from anodized nano-engineered titanium implants. *J Control Release.* 2021;333:521–535. doi: 10.1016/j.jconrel.2021.03.020.
[182] Wang Q, Huang JY, Li HQ, Chen Z, Zhao AZ, Wang Y, Zhang KQ, Sun HT, Al-Deyab SS, Lai YK. TiO2 nanotube platforms for smart drug delivery: a review. *Int J Nanomedicine.* 2016;11:4819–4834. doi: 10.2147/IJN.S108847.
[183] Yang WE, Huang HH. TiO2 Nanonetwork on Rough Ti Enhanced osteogenesis in vitro and in vivo. *J Dent Res.* 2021;100(10):1186–1193. doi: 10.1177/00220345211001017.
[184] Huang J, Zhang X, Yan W, Chen Z, Shuai X, Wang A, Wang Y. Nanotubular topography enhances the bioactivity of titanium implants. *Nanomedicine.* 2017;13(6):1913–1923. doi: 10.1016/j.nano.2017.03.017.
[185] Lugovskoy A, Lugovskoy S. Production of hydroxyapatite layers on the plasma electrolytically oxidized surface of titanium alloys. *Mater Sci Eng C Mater Biol Appl.* 2014;43:527–532. doi: 10.1016/j.msec.2014.07.030.
[186] Bose S, Tarafder S. Calcium phosphate ceramic systems in growth factor and drug delivery for bone tissue engineering: a review. *Acta Biomater.* 2012;8(4):1401–1421. doi: 10.1016/j.actbio.2011.11.017.
[187] Gil J, Manero JM, Ruperez E, Velasco-Ortega E, Jiménez-Guerra A, Ortiz-García I, Monsalve-Guil L. Mineralization of titanium surfaces: Biomimetic implants. *Materials (Basel).* 2021;14(11):2879. doi: 10.3390/ma14112879.
[188] Qadir M, Li Y, Wen C. Ion-substituted calcium phosphate coatings by physical vapor deposition magnetron sputtering for biomedical applications: A review. *Acta Biomater.* 2019;89:14–32. doi: 10.1016/j.actbio.2019.03.006.
[189] Chi Y, An S, Xu Y, Liu M, Zhang J. In vitro biocompatibility of a sandblasted, acid-etched HA composite coating on ultrafine-grained titanium. *RSC Adv.* 2021;11(11):6124–6130. doi: 10.1039/d0ra10146j.
[190] Zhao SF, Dong WJ, Jiang QH, He FM, Wang XX, Yang GL. Effects of zinc-substituted nano-hydroxyapatite coatings on bone integration with implant surfaces. *J Zhejiang Univ Sci B.* 2013;14(6):518–525. doi: 10.1631/jzus.B1200327.
[191] Arcos D, Vallet-Regí M. Substituted hydroxyapatite coatings of bone implants. *J Mater Chem B.* 2020;8(9):1781–1800. doi: 10.1039/c9tb02710f.
[192] Geng Z, Cui Z, Li Z, Zhu S, Liang Y, Liu Y, Li X, He X, Yu X, Wang R, Yang X. Strontium incorporation to optimize the antibacterial and biological characteristics of silver-substituted hydroxyapatite coating. *Mater Sci Eng C Mater Biol Appl.* 2016;58:467–477. doi: 10.1016/j.msec.2015.08.061.
[193] Jagadeeshanayaka N, Awasthi S, Jambagi SC, Srivastava C. Bioactive surface modifications through thermally sprayed hydroxyapatite composite coatings: a review of selective reinforcements. *Biomater Sci.* 2022;10(10):2484–2523. doi: 10.1039/d2bm00039c.

[194] Raphel J, Holodniy M, Goodman SB, Heilshorn SC. Multifunctional coatings to simultaneously promote osseointegration and prevent infection of orthopaedic implants. *Biomaterials*. 2016;84:301–314. doi: 10.1016/j.biomaterials.2016.01.016.
[195] Shekaran A, García AJ. Extracellular matrix-mimetic adhesive biomaterials for bone repair. *J Biomed Mater Res A*. 2011;96(1):261–272. doi: 10.1002/jbm.a.32979.
[196] Hynes RO. Integrins: bidirectional, allosteric signaling machines. *Cell*. 2002;110(6):673–687. doi: 10.1016/s0092-8674(02)00971-6.
[197] Losic D. Advancing of titanium medical implants by surface engineering: Recent progress and challenges. *Expert Opin Drug Deliv*. 2021;18(10):1355–1378. doi: 10.1080/17425247.2021.1928071.
[198] Kimura Y, Miyazaki N, Hayashi N, Otsuru S, Tamai K, Kaneda Y, Tabata Y. Controlled release of bone morphogenetic protein-2 enhances recruitment of osteogenic progenitor cells for de novo generation of bone tissue. *Tissue Eng Part A*. 2010;16(4):1263–1270. doi: 10.1089/ten.TEA.2009.0322.
[199] Bellis SL. Advantages of RGD peptides for directing cell association with biomaterials. *Biomaterials*. 2011;32(18):4205–4210. doi: 10.1016/j.biomaterials.2011.02.029.
[200] Hammerschmidt S, Rohde M, Preissner KT. Extracellular Matrix Interactions with Gram-Positive Pathogens. *Microbiol Spectr*. 2019;7(2). doi: 10.1128/microbiolspec.GPP3-0041-2018.
[201] Bistolfi A, Massazza G, Verné E, Massè A, Deledda D, Ferraris S, Miola M, Galetto F, Crova M. Antibiotic-loaded cement in orthopedic surgery: a review. *ISRN Orthop*. 2011;2011:290851. doi: 10.5402/2011/290851.
[202] Hickok NJ, Shapiro IM. Immobilized antibiotics to prevent orthopaedic implant infections. *Adv Drug Deliv Rev*. 2012;64(12):1165–1176. doi: 10.1016/j.addr.2012.03.015.
[203] Goodman SB, Yao Z, Keeney M, Yang F. The future of biologic coatings for orthopaedic implants. *Biomaterials*. 2013;34(13):3174–3183. doi: 10.1016/j.biomaterials.2013.01.074.
[204] Damiati L, Eales MG, Nobbs AH, Su B, Tsimbouri PM, Salmeron-Sanchez M, Dalby MJ. Impact of surface topography and coating on osteogenesis and bacterial attachment on titanium implants. *J Tissue Eng*. 2018;9:2041731418790694. doi: 10.1177/2041731418790694.
[205] Neoh KG, Hu X, Zheng D, Kang ET. Balancing osteoblast functions and bacterial adhesion on functionalized titanium surfaces. *Biomaterials*. 2012;33(10):2813–2822. doi: 10.1016/j.biomaterials.2012.01.018.
[206] Harris LG, Tosatti S, Wieland M, Textor M, Richards RG. Staphylococcus aureus adhesion to titanium oxide surfaces coated with non-functionalized and peptide-functionalized poly(L-lysine)-grafted-poly (ethylene glycol) copolymers. *Biomaterials*. 2004;25(18):4135–4148. doi: 10.1016/j.biomaterials.2003.11.033.
[207] Hickok NJ, Shapiro IM. Immobilized antibiotics to prevent orthopaedic implant infections. *Adv Drug Deliv Rev*. 2012;64(12):1165–1176. doi: 10.1016/j.addr.2012.03.015.
[208] Chouirfa H, Bouloussa H, Migonney V, Falentin-Daudré C. Review of titanium surface modification techniques and coatings for antibacterial applications. *Acta Biomater*. 2019;83:37–54. doi: 10.1016/j.actbio.2018.10.036.
[209] Lundin PM, Fiser BL, Blackledge MS, Pickett HL, Copeland AL. Functionalized self-assembled monolayers: Versatile strategies to combat bacterial biofilm formation. *Pharmaceutics*. 2022;14(8):1613. doi: 10.3390/pharmaceutics14081613.

[210] Yang Y, Tao B, Gong Y, Chen R, Yang W, Lin C, Chen M, Qin L, Jia Y, Cai K. Functionalization of Ti substrate with pH-responsive naringin-ZnO nanoparticles for the reconstruction of large bony after osteosarcoma resection. *J Biomed Mater Res A*. 2020;108(11):2190–2205. doi: 10.1002/jbm.a.36977.

[211] Niihara K, Nakahira A, Sekino T. New Nanocomposite structural ceramics. *MRS Online Proceedings Library*. 1992;286:405–412. doi: 10.1557/PROC-286-405.

[212] Niihara K. New design concept of structural ceramics: Ceramic nanocomposites. *J Ceram Soc Japan*. 1991;99(10):974–982. doi: 10.2497/jjspm.44.887.

[213] Niihara K, Izaki K, Kawakami T. Hot-pressed $Si_3N_4$-32% SiC nanocomposite from amorphous Si-C-N powder with improved strength above 1200 °C. *Journal of Materials Science Letters*. 1991;10(2):112–114. doi: 10.1007/BF00721925.

[214] Pillai SKC, Baron B, Pomeroy MJ, Hampshire S. Effect of oxide dopants on densification, microstructure and mechanical properties of alumina-silicon carbide nanocomposite ceramics prepared by pressureless sintering. *Eur Ceram Soc*. 2004;24(12):3317–3326. doi: 10.1016/j.jeurceramsoc.2003.10.024.

[215] Tun KS, Gupta M. Effect of heating rate during hydrid microwave sintering on the tensile propertises of magnesium and $Mg/Y_2O_3$ nanocomposite. *Alloys Compd*. 2008;466(1-2):140–145. doi: 10.1016/j.jallcom.2007.11.047.

[216] Li XL, Ma HA, Zheng YJ, Liu Y, Zuo GH, Liu WQ, Li JG, Jia X. AlN ceramics prepared by high-pressure sintering with $La_2O_3$ as a sintering aid. *Alloys Compd*. 2008;463(1-2): 412–416. doi: 10.1016/j.jallcom.2007.09.050.

[217] Suzuki K, Kayanuma M, Tachikawa M, Ogawa H, Nishihara H, Kyotani T, Nagashima U. Nuclear quantum effect on hydrogen adsorption site of zeolite-templated carbon model using path integral molecular dynamics. *Journal of Alloys and Compounds*. 2011;509:868–871. doi: 10.1016/j.jallcom.2010.10.066.

[218] Kistner J, Chen X, Weng Y, Strunk HP, Schubert MB, Werner JH. Photoluminescence from silicon nitride—no quantum effect. *Journal of Applied Physics*. 2011;110:2352. doi: 10.1063/1.3607975.

[219] Matteazzip, BD, Miani F, Le Caër G. Mechanosynthesis of nanophase materials. *Nanostrudtured Materials*. 1993;2:217–229. doi: 10.1016/0965-9773(93)90149-6.

[220] Karagedov GR, Shatskaya SS, Lyakhov NZ. Role of metal additives in mechanochemical phase transformation of zirconia. *Journal of Materials Science*. 2007;42:7964–7971. doi: 10.1007/s10853-006-1245-7.

[221] Zhigachey AO, Umrikhin AV, Golovin YI. The effect of calcia content on phase composition and mechanical properties of Ca—TZP prepared by high—energy milling of baddeleyite. *Ceramics International*. 2015;41:13804–13809. doi: 10.1016/j.ceramint.2015.08.063.

[222] Fan Y, Song E, Mustafa T, Liu R, Qiu P, Zhou W, Zhou Z, Kawasaki A, Shirasu K, Hashida T, Liu J, Wang L, Jiang W, Luo W. Liquid-phase assisted engineering of highly strong SiC composite reinforced by multiwalled carbon nanotubes. *Adv Sci (Weinh)*. 2020;7(21):2002225. doi: 10.1002/advs.202002225.

[223] Kujawa J, Kujawski W, Koter S, Rozicka A, Cerneaux S, Persin M, Larbot A. Efficiency of grafting of $Al_2O_3$, $TiO_2$ and $ZrO_2$ powders by perfluoroalkylsilanes. *Colloids and Surfaces A: Physicochemical and Engineering Aspects 420, Complete*. 2013;420(5):64–73. doi: 10.1016/j.colsurfa.2012.12.021

[224] Bangi UKH, Park CS, Baek S, Park HH. Sol–gel synthesis of high surface area nanostructured zirconia powder by surface chemical modification. *Powder Technology*. 2013;239:314–318. doi: 10.1016/j.powtec.2013.02.014.

[225] Liu W, Xie Z, Cui J. Surface modification of ceramic powders by titanate coupling agent for injection molding using partially water soluble binder system. *Journal of the European Ceramic Society*. 2012;32(5):1001–1006. doi: 10.1016/j.jeurceramsoc. 2011.11.017.

[226] Liu W, Xie Z P, Yang XF. Surface modification mechanism of stearic acid to zirconia powders induced by Ball milling for water-based injection molding. *Journal of the American Ceramic Society*. 2011;94(5):1327–1330. doi: 10.1016/j.ceramint.2016.07.164.

[227] Wen JX, Zhu TB, Xie ZP, Cao WB, Liu W. A strategy to obtain a high-density and high-strength zirconia ceramic via ceramic injection molding by the modification of oleic acid. *International Journal of Minerals Metallurgy and Materials*. 2017; 24(6):718–725. doi: 10.1007/s12613-017-1455-9.

[228] Lermontov S, Malkova A, Yurkova L, Straumal E, Ivanov V. Hexafluoroisopropyl alcohol as a new solvent for aerogels preparation. *Journal of Supercritical Fluids*. 2014;89(5):28–32. doi: 10.1016/j.supflu.2014.02.011.

[229] Lu X, Xia Y, Liu M, Qian Y, Zhou X, Gu N, Zhang F. Improved performance of diatomite-based dental nanocomposite ceramics using layer-by-layer assembly. *Int J Nanomedicine*. 2012;7:2153–2164. doi: 10.2147/IJN.S29851.

[230] Kusunose T, Sekino T, Choa YH, Kondo H. Fabrication and microstructure of silicon nitride/boron nitride nanocomposites. *Journal of the American Ceramic Society*. 2002;85(11):2678–2688. doi: info:doi/10.1166/jnn.2002.124

[231] Vivekanandhan S, Venkateswarlu M, Rawls HR, Satyanarayana N. Facile fabrication and characterisation of $MoO_3$ coated nanocrystalline $ZrO_2$ by polymeric resin route. *Advances in Applied Ceramics*. 2015;112(8):460–465. doi: 10.1179/1743676113Y.0000000116.

[232] Fu XJ, Wang NX, Zhang SZ. Preparation of necklace-like $TiO_2$ nanoparticles templated with L-phenylalanine derivative based on supramolecular hydrogel. *Journal of Inorganic Materials*. 2008;23(2):393–397. doi: 10.3321/j.issn:1000-324X.2008.02.039.

[233] Strauss M, Maroneze CM, Silva JMS, Sigoli FA, Gushikem Y, Mazali IO. Annealing temperature effects on sol-gel nanostructured mesoporous TiO2/SiO2 and its photocatalytic activity. *Materials Chemistry & Physics*. 2010;126(1-2):188–194. doi: 10.1016/j.matchemphys.2010.11.041.

[234] Akovali G, Dilsiz N. Studies on the modification of interphase/interfaces by use of plasma in certain polymer composite systems. *Polymer Engineering and Science*. 1996;36(8):1081–1086. doi: 10.1002/pen.10498.

[235] Gonzalo-Juan I, Ferrari B, Colomer MT, Rodriguez MA, Sanchez-Herencia AJ, Koh PY, Teja AS. Synthesis and dispersion of yttria-stabilized zirconia (YSZ) nanoparticles in supercritical water. *Materials Chemistry and Physics*. 2012;134:451–458. doi: 10.1016/j.matchemphys.2012.03.016.

[236] Gil A, Gandia LM, Vicente MA. Recent advances in the synthesis and catalytic applications of pillared clays. *Catal. Rev—Sci. Eng*. 2000;42(1&2):145–212. doi: 10.1016/S1387-1811(99)00166-3.

[237] Zhu XB, Duan XC. Effects of pH value on co-precipitati on of nano-size ITO powder. *Rare Metals and Cemented Carbides*. 2006;34(2):8–11. doi: 10.1016/j.jmmm.2008.06.009.

[238] Kusunose T, Sekino T, Choa YH. Machinability of silicon nitride/boron nitride nanocomposites. *Journal of the American Ceramic Society*. 2002;85(11):2689–2695. doi: 10.1016/S0026-0657.

[239] Ohmori M, Deki S. Rise in transition temperature of GAMMA $Fe_2O_3$ particles to ALPHA $Fe_2O_3$ by $SiO_2$ coating. *Chemistry Letters*. 1994;23(8):1369–1370. doi: 10.1246/cl.1994.1369.

[240] Chen C, Trindade FZ, de Jager N, Kleverlaan CJ, Feilzer AJ. The fracture resistance of a CAD/CAM Resin Nano Ceramic(RNC)and a CAD ceramic at different thicknesses. *Dent Mater*. 2014;30(9):954–962. doi: 10.1016/j.dental.2014.05.018.

[241] Nguyen JF, Migonney V, Ruse ND, Sadoun M. Resin composite blocks via high-pressurehigh-temperature polymerization. *Dent Mater*. 2012;28(5):529–534. doi: 10.1016/j.dental.2011.12.003.

[242] Kelly JR, Benetti P. Ceramic materials in dentistry: Historical evolution and current practice. *Aust Dent J*. 2011;56 Suppl 1:84–96. doi:10.1111/j.1834-7819.2010.01299.x

[243] Bajraktarova-Valjakova E, Korunoska-Stevkovska V, Kapusevska B, Gigovski N, Bajraktarova-Misevska C, Grozdanov A. Contemporary dental ceramic materials, a review: Chemical composition, physical and mechanical properties, indications for use. *Open Access Macedonian Journal of Medical Sciences*. 2018;6(9): 1742–1755. doi: 10.3889/oamjms.2018.378.

[244] Coldea A, Swain MV, Thiel N. Mechanical properties of polymer-infiltrated-ceramic-network materials. *Dent Mater*. 2013;29(4):419–426. doi: 10.1016/j.dental.2013.01.002.

[245] Güngör MB, Nemli SK, Bal BT, Ünver S, Doğan A. Effect of surface treatments on shear bond strength of resin composite bonded to CAD/CAM resin-ceramic hybrid materials. *J Adv Prosthodont*. 2016;8(4):259–266. doi: 10.4047/jap.2016.8.4.259.

[246] Zhang Y, Kelly J R. Dental ceramics for restoration and metal veneering. *Dental Clinics of North America*. 2017;61(4): 797–819. doi: 10.1016/j.cden.2017.06.005.

[247] Hampe R, Theelke B, Lümkemann N, Eichberger M, Stawarczyk B. Fracture toughness analysis of ceramic and resin composite CAD/CAM material. *Operative Dentistry*. 2019;44(4): E190–E201. doi: 10.2341/18-161-L.

[248] Alshabib A, Silikas N, Watts DC. Hardness and fracture toughness of resin-composite materials with and without fibers. *Dent Mater*. 2019 Aug;35(8):1194–1203. doi: 10.1016/j.dental.2019.05.017.

[249] Nobuaki A, Keiichi Y, Takashi S. Effects of air abrasion with alumina or glass beads on surface characteristics of CAD/CAM composite materials and the bond strength of resin cements. *J Appl Oral Sci*. 2015;23(6): 629–636. doi: 10.1590/1678-775720150261.

[250] Stawarczyk B, Krawczuk A, Ilie N. Tensile bond strength of resin composite repair in vitro using different surface preparation conditionings to an aged CAD/CAM resin nanoceramic. *Clin Oral Investig*. 2015;19(2): 299–308. doi: 10.1007/s00784-014-1269-3.

[251] Silva PNFD, Martinelli-Lobo CM, Bottino MA, Melo RM, Valandro LF. Bond strength between a polymer-infiltrated ceramic network and a composite for repair: effect of several ceramic surface treatments. *Braz Oral Res*. 2018;32:e28. doi:10.1590/1807-3107bor-2018.vol32.0028.

[252] Ozcan M, Bernasconi M. Adhesion to zirconia used for dental restorations:a systematic review and meta-analysis. *J Adhes Dent*. 2015;17(1):7–26. doi: 10.3290/j.jad.a33525.

[253] Papia E, Larsson C, du Toit M, Vult von Steyern P. Bonding between oxide ceramics and adhesive cement systems:a systematic review. *J Biomed Mater Res B Appl Biomater*. 2014;102(2):395–413. doi: 10.1002/jbm.b.33013.

[254] Kirmali O, Barutcigil Ç, Ozarslan MM, Barutcigil K, Harorlı OT. Repair bond strength of composite resin to sandblasted and laser irradiated Y-TZP ceramic surfaces. *Scanning*. 2015;37(3):186–192. doi: 10.1002/sca.21197.

[255] Strasser T, Preis V, Behr M, Rosentritt M. Roughness, surface energy, and superficial damages of CAD/CAM materials after surface treatment. *Clin Oral Investig*. 2018;22(8):2787–2797. doi:10.1007/s00784-018-2365-6.

[256] Cengiz-yanardag E, Kurtulmus Yilmaz S, Karakaya I, Ongun S. Effect of different surface treatment methods on micro-shear bond strength Of CAD-CAM restorative materials to resin cement. *Journal of Adhesion Science and Technology*. 2018;33(3): 1–14. doi: 10.1080/01694243.2018.1514992.

[257] Reymus M, Roos M, Eichberger M, Edelhoff D, Hickel R, Stawarczyk B. Bonding to new CAD/CAM resin composites: influence of air abrasion and conditioning agents as pretreatment strategy. *Clin Oral Investig*. 2019;23(2):529–538. doi:10.1007/s00784-018-2461-7.

[258] Yoshihara K, Nagaoka N, Maruo Y, Nishigawa G, Irie M, Yoshida Y, Van Meerbeek B. Sandblasting may damage the surface of composite CAD-CAM blocks. *Dent Mater*. 2017;33(3):e124–e135. doi:10.1016/j.dental.2016.12.003.

[259] El-Damanhoury HMA, Elsahn N, Sheela S, Gaintantzopoulou MD. Adhesive luting to hybrid ceramic and resin composite CAD/CAM Blocks:Er:YAG Laser versus chemical etching and micro-abrasion pretreatment. *J Prosthodont Res*. 2021;65(2):225–234. doi:10.2186/jpr.JPOR_2020_50.

[260] Elsaka SE. Bond strength of novel CAD/CAM restorative materials to self-adhesive resin cement: The effect of surface treatments. *J Adhes Dent*. 2014;16(6): 531–540. doi: 10.3290/j.jad.a33198.

[261] Tekçe N, Tuncer S, Demirci M, Kara D, Baydemir C. Microtensile bond strength of CAD/CAM resin blocks to dual-cure adhesive cement: The effect of different sandblasting procedures. *J Prosthodont*. 2019;28(2):e485–e490. doi:10.1111/jopr.12737.

[262] Duzyol M, Sagsoz O, Polat Sagsoz N, Akgul N, Yildiz M. The effect of surface treatments on the bond strength between CAD/CAM blocks and composite resin. *J Prosthodont*. 2016;25(6):466–471. doi:10.1111/jopr.12322.

[263] Atsu SS, Kilicarslan MA, Kucukesmen HC, Aka PS. Effect of zirconium-oxide ceramic surface treatments on the bond strength to adhesive resin. *J Prosthet Dent*. 2006;95(6):430–436. doi:10.1016/j.prosdent.2006.03.016.

[264] Peumans M, Valjakova EB, De Munck J, Mishevska CB, Van Meerbeek B. Bonding effectiveness of luting composites to different CAD/CAM materials. *J Adhes Dent*. 2016;18(4):289–302. doi:10.3290/j.jad.a36155.

[265] Sismanoglu S, Yildirim-Bilmez Z, Erten-Taysi A, Ercal P. Influence of different surface treatments and universal adhesives on the repair of CAD-CAM composite resins: An in vitro study. *J Prosthet Dent*. 2020;124(2): 238.e1–238238.e9. doi: 10.1016/j.prosdent.2020.02.029.

[266] Park JH, Choi YS. Microtensile bond strength and micromorphologic analysis of surface-treated resin nanoceramics. *J Adv Prosthodont*. 2016;8(4): 275–284. doi: 10.4047/jap.2016.8.4.275.

[267] Elsaka SE. Repair bond strength of resin composite to a novel CAD/CAM hybrid ceramic using different repair systems. *Dent Mater J*. 2015;34(2): 161–167. doi: 10.4012/dmj.2014-159.

[268] Ho GW, Matinlinna JP. Insights on ceramics as dental materials. Part II: chemical surface treatments. *Silicon*. 2011;3:11–23. doi: 10.1007/s12633-011-9079-6.

[269] Akgungor G, Sen D, Aydin M. Influence of different surface treatments on the short term bond strength and durability between a zirconia post and a composite

resin core material. *Journal of Prosthetic Dentistry*. 2008;99(5): 388–399. doi: 10.1016/S0022-3913(08)60088-8.
[270] Della-Bona A. Characterizing ceramics and the interfacial adhesion to resin: II-the relationship of surface treatment, bond strength, interfacial toughness and fractography. *J Appl Oral Sci*. 2005;13(2): 101–109. doi: 10.1590/s1678-77572005000200002.
[271] Della B A, Anusavice KJ. Microstructure, composition, and etching topography of dental ceramics. *Int J Prosthodont* 2002;15(2): 159–167. PMID: 11951806
[272] Ferracane J L. Resin-based composite performance: are there some things we can't predict? *Dent Mater*. 2013;29(1):51–58. doi: 10.1016/j.dental.2012.06.013.
[273] Lise DP, Van Ende A, De Munck J, Vieira L, Baratieri LN, Van Meerbeek B. Microtensile bond strength of composite cement to novel CAD/CAM materials as a function of surface treatment and aging. *Oper Dent*. 2017;42(1):73–81. doi: 10.2341/15-263-L.
[274] Bajraktarova-Valjakova E, Grozdanov A, Guguvcevski L, Korunoska-Stevkovska V, Kapusevska B, Gigovski N, Mijoska A, Bajraktarova-Misevska C. Acid etching as surface treatment method for luting of glass-ceramic restorations, Part 1: Acids, application protocol and etching effectiveness. *Open Access Macedonian Journal of Medical Sciences*. 2018;6(3): 568–572. doi: 10.3889/oamjms.2018.147.
[275] Sagsoz O, Polat Sagsoz N, Yurtcan MT, Ozcelik N. Hydroxyapatite coating effect on the bond strength between CAD/CAM materials and a resin cement. *Odontology*. 2019;107(4): 491–499. doi: 10.1007/s10266-019-00420-y.
[276] Kim SH, Cho SC, Lee MH, Kim HJ, Oh NS. Effect of 9% hydrofluoric acid gel hot-etching surface treatment on shear bond strength of resin cements to zirconia ceramics. *Medicina (Kaunas)*. 2022;58(10):1469. doi:10.3390/medicina58101469.
[277] Özcan M, Bernasconi M. Adhesion to zirconia used for dental restorations: A systematic review and meta-analysis. *J Adhes Dent*. 2015;17(1): 7–26. doi: 10.3290/j.jad.a33525.
[278] Khan AA, Al Kheraif AA, Jamaluddin S, Elsharawy M, Divakar DD. Recent trends in surface treatment methods for bonding composite cement to zirconia: A reveiw. *J Adhes Dent*. 2017;19(1): 7–19. doi: 10.3290/j.jad.a37720.
[279] Duzyol M, Sagsoz O, Polat Sagsoz N, Akgul N, Yildiz M. The effect of surface treatments on the bond strength between CAD/CAM blocks and composite resin. *J Prosthodont*. 2016;25(6): 466–471. doi: 10.1111/jopr.12322.
[280] Cekic-Nagas I, Ergun G, Egilmez F, Vallittu PK, Lassila LV. Micro-shear bond strength of different resin cements to ceramic /glass-polymer CADCAM block materials. *J Prosthodont Res*. 2016;60(4): 265–273. doi: 10.1016/j.jpor.2016.02.003.
[281] Loomans BA, Mine A, Roeters FJ, Opdam NJ, De Munck J, Huysmans MC, Van Meerbeek B. Hydrofluoric acid on dentin should be avoided. *Dent Mater*. 2010;26(7):643–649. doi:10.1016/j.dental.2010.03.007.
[282] Söderholm KJ, Shang SW. Molecular orientation of silane at the surface of colloidal silica. *J Dent Res*. 1993;72(6):1050–1054. doi:10.1177/00220345930720061001.
[283] Matinlinna JP, Lung CYK, Tsoi JKH. Silane adhesion mechanism in dental applications and surface treatments: A review. *Dent Mater*. 2018;34(1):13–28. doi: 10.1016/j.dental.2017.09.002.

[284] Dimitriadi M, Zafiropoulou M, Zinelis S, Silikas N, Eliades G. Silane reactivity and resin bond strength to lithium disilicate ceramic surfaces. *Dent Mater*. 2019;35(8):1082–1094. doi:10.1016/j.dental.2019.05.002.

[285] Nihei T. Dental applications for silane coupling agents. *Journal of Oral Science*. 2016;58(2):151–155. doi: 10.2334/josnusd.16-0035.

[286] Chuenjit P, Suzuki M, Shinkai K. Effect of various surface treatments on the bond strength of resin luting agent and the surface roughness and surface energy of CAD/CAM materials. *Dent Mater J*. 2021;40(1): 16–25. doi:10.4012/dmj.2019-359.

[287] Maawadh AM, Almohareb T, Al-Hamdan RS, Al Deeb M, Naseem M, Alhenaki AM, Vohra F, Abduljabbar T. Repair strength and surface topography of lithium disilicate and hybrid resin ceramics with LLLT and photodynamic therapy in comparison to hydrofluoric acid. *J Appl Biomater Funct Mater*. 2020;18: 2280800020966938. doi:10.1177/2280800020966938.

[288] Fathy H, Hamama HH, El-Wassefy N, Mahmoud SH. Effect of different surface treatments on resin-matrix CAD/CAM ceramics bonding to dentin: in vitro study. *BMC Oral Health*. 2022;22(1):635. Published 2022 Dec 23. doi:10.1186/s12903-022-02674-5.

[289] Buyuk SK, Kucukekenci AS. Effects of different etching methods and bonding procedures on shear bond strength of orthodontic metal brackets applied to different CAD/CAM ceramic materials. *Angle Orthod*. 2018;88(2):221–226. doi:10.2319/070917-455.1.

[290] Barutcigil K, Barutcigil Ç, Kul E, Özarslan MM, Buyukkaplan US. Effect of different surface treatments on bond strength of resin cement to a CAD/CAM restorative material. *Journal of Prosthodontics*. 2019;28(1):71–78. doi:10.1111/jopr.12574.

[291] Deppe H, Warmuth S, Heinrich A, Körner T. Laser-assisted three-dimensional surface modifications of titanium implants: preliminary data. *Lasers Med Sci*. 2005;19(4):229–233. doi:10.1007/s10103-005-0327-0.

[292] Fornaini C. Er:YAG and adhesion in conservative dentistry: clinical overview. *Laser Ther*. 2013;22(1):31–35. doi:10.5978/islsm.13-or-04.

[293] Bayraktar Y, Arslan M, Demirtag Z. Repair bond strength and surface topography of resin-ceramic and ceramic restorative blocks treated by laser and conventional surface treatments. *Microsc Res Tech*. 2021;84(6):1145–1154. doi:10.1002/jemt.23672.

[294] Emsermann I, Eggmann F, Krastl G, Weiger R, Amato J. Influence of pretreatment methods on the adhesion of composite and polymer infiltrated ceramic CAD-CAM blocks. *J Adhes Dent*. 2019;21(5):433–443. doi:10.3290/j.jad.a43179

[295] Graves D B. The emerging role of reactive oxygen and nitrogen species in redox biology and some implications for plasma applications to medicine and biology. *Journal of Physics D Applied Physics*. 2012;45(26): 263001. doi:10.1088/0022-3727/45/26/263001.

[296] Kong M G, Liu D. Researches on the interaction between gas plasmas and aqueous solutions: Significance, challenges and new progresses. *High Voltage Engineering*. 2014;40(10): 2956–2965. doi:1003-6520(2014)40:10<2956:QTDLZT>2.0.TX;2-Y.

[297] Li HP, Zhang XF, Zhu XM, Zheng M, Liu SF, Qi X, Wang KP, Chen J, Xi XQ, Tan JG, Ostrikov K. Translational plasma stomatology: Applications of cold atmospheric plasmas in dentistry and their extension. *High Voltage*. 2017;2(3): 188–199. doi:10.1049/hve.2017.0066.

[298] Weltmann KD, Von Woedtke T. Plasma medicine-current state of research and medical application. *Plasma Physics & Controlled Fusion*. 2017;59(1): 014031. doi: 10.1088/0741-3335/59/1/014031.

[299] Vechiato Filho AJ, dos Santos DM, Goiato MC, de Medeiros RA, Moreno A, Bonatto Lda R, Rangel EC. Surface characterization of lithium disilicate ceramic after nonthermal plasma treatment. *J Prosthet Dent*. 2014;112(5):1156–1163. doi: 10.1016/j.prosdent.2014.02.021.

[300] Cho BH, Han GJ, Oh KH, Chung SN, Chun BH. The effect of plasma polymer coating using atmospheric pressure glow discharge on the shear Bond strength of composite resin to ceramic. *Journal of Materials Science*. 2011;46(8): 2755–2763. doi: 10.1007/s10853-010-5149-1.

[301] Henningsen A, Smeets R, Heuberger R, Jung OT, Hanken H, Heiland M, Cacaci C, Precht C. Changes in surface characteristics of titanium and zirconia after surface treatment with ultraviolet light or non-thermal plasma. *Eur J Oral Sci*. 2018;126:126–134. doi: 10.1111/eos.12400.

[302] Barquete CG, Simão RA, Almeida Fonseca SS, Elias AB, Antunes Guimarães JG, Herrera EZ, Mello A, Moreira da Silva E. Effect of cementation delay on bonding of self-adhesive resin cement to yttria-stabilized tetragonal zirconia polycrystal ceramic treated with nonthermal argon plasma. *J Prosthet Dent*. 2021;125(4):693.e1–693.e7. doi: 10.1016/j.prosdent.2020.11.032.

[303] Zhu X, Shi J, Ye X, Ma X, Zheng M, Yang Y, Tan J. Influence of cold atmospheric plasma on surface characteristics and bond strength of a resin nanoceramic. *Materials (Basel)*. 2022 Dec 21;16(1):44. doi: 10.3390/ma16010044.

[304] Choudhary N, Kamat S, Mangala T, Thomas M. Effect of pre-heating composite resin on gap formation at three different temperatures. *J Conserv Dent*. 2011;14(2):191–195. doi: 10.4103/0972-0707.82618.

[305] Brunthaler A, Konig F, Lucas T, Sperr W, Schedle A. Longevity of direct resin composite restorations in posterior teeth. *Clin Oral Investig*. 2003;7:63–70. doi: 10.1007/s00784-003-0206-7.

[306] Daronch M, Rueggeberg FA, De Goes MF. Monomer conversion of pre-heated composite. *J Dent Res*. 2005;84(7):663–667. doi: 10.1177/154405910508400716.

[307] Daronch M, Rueggeberg FA, Moss L, de Goes MF. Clinically relevant issues related to preheating composites. *J Esthet Restor Dent*. 2006;18(6):340–351. doi: 10.1111/j.1708-8240.2006.00046.x.

[308] Sousa SE, da Costa ES, Borges BC. Staining resistance of preheated flowable composites to drinking pigmented beverages. *Rev Port Estomatol Med Dent Cir Maxilofac*. 2015;56:221–225. doi: 10.1016/j.rpemd.2015.11.009.

[309] Taufin N, Priyadarshini BI, Shankarappa P, Ballullaya SV, Devalla S, Gavini S. Eighteen-month clinical performance of preheated nanoceramic resin-based composites in Class I occlusal cavities: A randomized clinical trial. *J Conserv Dent*. 2022;25(1):47–53. doi: 10.4103/jcd.jcd_492_21.

[310] Dionysopoulos D, Papadopoulos C, Koliniotou-Koumpia E. Effect of temperature, curing time, and filler composition on surface microhardness of composite resins. *J Conserv Dent*. 2015;18(2):114–118. doi: 10.4103/0972-0707.153071.

[311] Barreto MEV, Medeiros RP, Shearer A, Fook MVL, Montazerian M, Mauro JC. Gelatin and bioactive glass composites for tissue engineering: A Review. *J Funct Biomater*. 2022;14(1):23. doi: 10.3390/jfb14010023.

[312] Kargozar S, Mozafari M, Ghodrat S, Fiume E, Baino F. Copper-containing bioactive glasses and glass-ceramics: From tissue regeneration to cancer therapeutic strategies. *Mater Sci Eng C Mater Biol Appl*. 2021;121:111741. doi: 10.1016/j.msec.2020.111741.

[313] Kargozar S, Baino F, Hamzehlou S, Hill RG, Mozafari M. Bioactive glasses: Sprouting angiogenesis in tissue engineering. *Trends Biotechnol.* 2018;36(4):430–444. doi: 10.1016/j.tibtech.2017.12.003.

[314] Lalzawmliana V, Anand A, Roy M, Kundu B, Nandi SK. Mesoporous bioactive glasses for bone healing and biomolecules delivery. *Mater Sci Eng C Mater Biol Appl.* 2020;106:110180. doi: 10.1016/j.msec.2019.110180.

[315] Vallet-Regí M, Colilla M, Izquierdo-Barba I, Vitale-Brovarone C, Fiorilli S. Achievements in mesoporous bioactive glasses for biomedical applications. *Pharmaceutics.* 2022;14(12):2636. doi: 10.3390/pharmaceutics14122636.

[316] Kargozar S, Baino F, Hamzehlou S, Hill RG, Mozafari M. Bioactive glasses entering the mainstream. *Drug Discovery Today.* 2018;23(10):1700–1704. doi: 10.1016/j.drudis.2018.05.027.

[317] van Rijt S, de Groot K, Leeuwenburgh SCG. Calcium phosphate and silicate-based nanoparticles: History and emerging trends. *Tissue Eng Part A.* 2022;28(11-12):461–477. doi: 10.1089/ten.TEA.2021.0218.

[318] Jaimes ATC, Kirste G, de Pablos-Martín A, Selle S, de Souza E Silva JM, Massera J, Karpukhina N, Hill RG, Brauer DS. Nano-imaging confirms improved apatite precipitation for high phosphate/silicate ratio bioactive glasses. *Sci Rep.* 2021;11(1):19464. doi: 10.1038/s41598-021-98863-3.

[319] Ferraris S, Yamaguchi S, Barbani N, Cristallini C, Gautier di Confiengo G, Barberi J, Cazzola M, Miola M, Vernè E, Spriano S. The mechanical and chemical stability of the interfaces in bioactive materials: The substrate-bioactive surface layer and hydroxyapatite-bioactive surface layer interfaces. *Mater Sci Eng C Mater Biol Appl.* 2020;116:111238. doi: 10.1016/j.msec.2020.111238.

[320] Sirkiä SV, Nakamura M, Qudsia S, Siekkinen M, Smått JH, Peltonen J, Heino TJ, Hupa L, Vallittu PK. Structural and elemental characterization of glass and ceramic particles for bone surgery. *Dent Mater.* 2021;37(9):1350–1357. doi: 10.1016/j.dental.2021.06.004.

[321] Houaoui A, Szczodra A, Lallukka M, El-Guermah L, Agniel R, Pauthe E, Massera J, Boissiere M. New generation of hybrid materials based on gelatin and bioactive glass particles for bone tissue regeneration. *Biomolecules.* 2021;11(3):444. doi: 10.3390/biom11030444.

[322] Jafari N, Habashi MS, Hashemi A, Shirazi R, Tanideh N, Tamadon A. Application of bioactive glasses in various dental fields. *Biomater Res.* 2022;26(1):31. doi: 10.1186/s40824-022-00274-6.

[323] Zandi Karimi A, Rezabeigi E, Drew RAL. Aluminum-free glass ionomer cements containing 45S5 Bioglass® and its bioglass-ceramic. *J Mater Sci Mater Med.* 2021;32(7):76. doi: 10.1007/s10856-021-06553-3.

[324] Dittler ML, Unalan I, Grünewald A, Beltrán AM, Grillo CA, Destch R, Gonzalez MC, Boccaccini AR. Bioactive glass (45S5)-based 3D scaffolds coated with magnesium and zinc-loaded hydroxyapatite nanoparticles for tissue engineering applications. *Colloids Surf B Biointerfaces.* 2019;182:110346. doi: 10.1016/j.colsurfb.2019.110346.

[325] Lopes JH, Souza LP, Domingues JA, Ferreira FV, de Alencar Hausen M, Camilli JA, Martin RA, de Rezende Duek EA, Mazali IO, Bertran CA. In vitro and in vivo osteogenic potential of niobium-doped 45S5 bioactive glass: A comparative study. *J Biomed Mater Res B Appl Biomater.* 2020;108(4):1372–1387. doi: 10.1002/jbm.b.34486.

[326] Min Y, Luo J, Liu C. Viscosity and related structure transformation of fluorine bearing silicate melt under ultrasonic field. *Ultrason Sonochem.* 2019;55:289–296. doi: 10.1016/j.ultsonch.2019.01.015.

[327] Possolli NM, da Silva DF, Vieira J, Maurmann N, Pranke P, Demétrio KB, Angioletto E, Montedo ORK. Arcaro S. Dissolution, bioactivity behavior, and cytotoxicity of $19.58Li_2 O{\cdot}11.10ZrO_2 {\cdot}69.32SiO_2$ glass-ceramic. *J Biomed Mater Res B Appl Biomater*. 2022;110(1):67–78. doi: 10.1002/jbm.b.34889.
[328] Zhao R, Shi L, Gu L, Qin X, Song Z, Fan X, Zhao P, Li C, Zheng H, Li Z, Wang Q. Evaluation of bioactive glass scaffolds incorporating SrO or ZnO for bone repair: In vitro bioactivity and antibacterial activity. *J Appl Biomater Funct Mater*. 2021;19:22808000211040910. doi: 10.1177/22808000211040910.
[329] Simila HO, Boccaccini AR. Sol-gel bioactive glass containing biomaterials for restorative dentistry: A review. *Dent Mater*. 2022;38(5):725–747. doi: 10.1016/j.dental.2022.02.011.
[330] Mukundan LM, Nirmal S R, Kumar N, Dhara S, Chattopadhyay S. Engineered nanostructures within sol-gel bioactive glass for enhanced bioactivity and modulated drug delivery. *J Mater Chem B*. 2022;10(48):10112–10127. doi: 10.1039/d2tb01692c.
[331] Tiskaya M, Shahid S, Gillam D, Hill R. The use of bioactive glass (BAG) in dental composites: A critical review. *Dent Mater*. 2021;37(2):296–310. doi: 10.1016/j.dental.2020.11.015.
[332] Zheng K, Sui B, Ilyas K, Boccaccini AR. Porous bioactive glass micro- and nanospheres with controlled morphology: Developments, properties and emerging biomedical applications. *Mater Horiz*. 2021;8(2):300–335. doi: 10.1039/d0mh01498b.
[333] El-Okaily MS, El-Rafei AM, Basha M, Abdel Ghani NT, El-Sayed MMH, Bhaumik A, Mostafa AA. Efficient drug delivery vehicles of environmentally benign nanofibers comprising bioactive glass/chitosan/polyvinyl alcohol composites. *Int J Biol Macromol*. 2021;182:1582–1589. doi: 10.1016/j.ijbiomac.2021.05.079.
[334] Oudadesse H, Najem S, Mosbahi S, Rocton N, Refifi J, El Feki H, Lefeuvre B. Development of hybrid scaffold: Bioactive glass nanoparticles/chitosan for tissue engineering applications. *J Biomed Mater Res A*. 2021;109(5):590–599. doi: 10.1002/jbm.a.37043.
[335] Zheng W, Bai Z, Huang S, Jiang K, Liu L, Wang X. The effect of angiogenesis-based scaffold of mesoporous bioactive glass nanofiber on osteogenesis. *Int J Mol Sci*. 2022;23(20):12670. doi: 10.3390/ijms232012670.
[336] Zhu H, Zheng K, Boccaccini AR. Multi-functional silica-based mesoporous materials for simultaneous delivery of biologically active ions and therapeutic biomolecules. *Acta Biomater*. 2021;129:1–17. doi: 10.1016/j.actbio.2021.05.007.
[337] Hosseini M, Hassani Besheli N, Deng D, Lievens C, Zuo Y, Leeuwenburgh SCG, Yang F. Facile post modification synthesis of copper-doped mesoporous bioactive glass with high antibacterial performance to fight bone infection. *Biomater Adv*. 2023;144:213198. doi: 10.1016/j.bioadv.2022.213198.
[338] Pajares-Chamorro N, Wagley Y, Maduka CV, Youngstrom DW, Yeger A, Badylak SF, Hammer ND, Hankenson K, Chatzistavrou X. Silver-doped bioactive glass particles for in vivo bone tissue regeneration and enhanced methicillin-resistant Staphylococcus aureus (MRSA) inhibition. *Mater Sci Eng C Mater Biol Appl*. 2021;120:111693. doi: 10.1016/j.msec.2020.111693.
[339] Kurtuldu F, Kaňková H, Beltrán AM, Liverani L, Galusek D, Boccaccini AR. Anti-inflammatory and antibacterial activities of cerium-containing mesoporous bioactive glass nanoparticles for drug-free biomedical applications. *Mater Today Bio*. 2021;12:100150. doi: 10.1016/j.mtbio.2021.100150.
[340] Canales DA, Reyes F, Saavedra M, Peponi L, Leonés A, Palza H, Boccaccini AR, Grünewald A, Zapata PA. Electrospun fibers of poly (lactic acid) containing

bioactive glass and magnesium oxide nanoparticles for bone tissue regeneration. *Int J Biol Macromol.* 2022;210:324–336. doi: 10.1016/j.ijbiomac.2022.05.047.
[341] Karadjian M, Essers C, Tsitlakidis S, Reible B, Moghaddam A, Boccaccini AR, Westhauser F. Biological properties of calcium phosphate bioactive glass composite bone substitutes: Current experimental evidence. *Int J Mol Sci.* 2019;20(2):305. doi: 10.3390/ijms20020305.
[342] Zeimaran E, Pourshahrestani S, Fathi A, Razak NABA, Kadri NA, Sheikhi A, Baino F. Advances in bioactive glass-containing injectable hydrogel biomaterials for tissue regeneration. *Acta Biomater.* 2021;136:1–36. doi: 10.1016/j.actbio.2021.09.034.
[343] Thamma U, Kowal TJ, Falk MM, Jain H. Nanostructure of bioactive glass affects bone cell attachment via protein restructuring upon adsorption. *Sci Rep.* 2021;11(1):5763. doi: 10.1038/s41598-021-85050-7.
[344] Costa RC, Souza JGS, Cordeiro JM, Bertolini M, de Avila ED, Landers R, Rangel EC, Fortulan CA, Retamal-Valdes B, da Cruz NC, Feres M, Barão VAR. Synthesis of bioactive glass-based coating by plasma electrolytic oxidation: Untangling a new deposition pathway toward titanium implant surfaces. *J Colloid Interface Sci.* 2020;579:680–698. doi: 10.1016/j.jcis.2020.06.102.
[345] Borden M, Westerlund LE, Lovric V, Walsh W. Controlling the bone regeneration properties of bioactive glass: Effect of particle shape and size. *J Biomed Mater Res B Appl Biomater.* 2022;110(4):910–922. doi: 10.1002/jbm.b.34971.
[346] Batool SA, Ahmad K, Irfan M, Ur Rehman MA. Zn-Mn-doped mesoporous bioactive glass nanoparticle-loaded zein coatings for bioactive and antibacterial orthopedic implants. *J Funct Biomater.* 2022;13(3):97. doi: 10.3390/jfb13030097.
[347] Shikimaka O, Bivol M, Sava BA, Dumitru M, Tardei C, Sbarcea BG, Grabco D, Pyrtsac C, Topal D, Prisacaru A, Cobzac V, Nacu V. Hydroxyapatite-bioglass nanocomposites: Structural, mechanical, and biological aspects. *Beilstein J Nanotechnol.* 2022;13:1490–1504. doi: 10.3762/bjnano.13.123.
[348] Salinas AJ, Esbrit P. Mesoporous bioglasses enriched with bioactive agents for bone repair, with a special highlight of María Vallet-Regí's contribution. *Pharmaceutics.* 2022;14(1):202. doi: 10.3390/pharmaceutics14010202.
[349] Das MP, Pandey G, Neppolian B, Das J. Design of poly-l-glutamic acid embedded mesoporous bioactive glass nanospheres for pH-stimulated chemotherapeutic drug delivery and antibacterial susceptibility. *Colloids Surf B Biointerfaces.* 2021;202:111700. doi: 10.1016/j.colsurfb.2021.111700.
[350] Ilyas K, Singer L, Akhtar MA, Bourauel CP, Boccaccini AR. Boswellia sacra extract-loaded mesoporous bioactive glass nano particles: Synthesis and biological effects. *Pharmaceutics.* 2022;14(1):126. doi: 10.3390/pharmaceutics14010126.
[351] Hum J, Boccaccini AR. Collagen as coating material for 45S5 bioactive glass-based scaffolds for bone tissue engineering. *Int J Mol Sci.* 2018;19(6):1807. doi: 10.3390/ijms19061807.
[352] Rifane TO, Cordeiro KEM, Silvestre FA, Souza MT, Zanotto ED, Araújo-Neto VG, Giannini M, Sauro S, de Paula DM, Feitosa VP. Impact of silanization of different bioactive glasses in simplified adhesives on degree of conversion, dentin bonding and collagen remineralization. *Dent Mater.* 2023;39(2):217–226. doi: 10.1016/j.dental.2023.01.005.
[353] Wu Q, Hu L, Yan R, Shi J, Gu H, Deng Y, Jiang R, Wen J, Jiang X. Strontium-incorporated bioceramic scaffolds for enhanced osteoporosis bone regeneration. *Bone Res.* 2022;10(1):55. doi: 10.1038/s41413-022-00224-x.

[354] Ding X, Shi J, Wei J, Li Y, Wu X, Zhang Y, Jiang X, Zhang X, Lai H. A biopolymer hydrogel electrostatically reinforced by amino-functionalized bioactive glass for accelerated bone regeneration. *Sci Adv.* 2021;7(50):eabj7857. doi: 10.1126/sciadv.abj7857.

[355] Chen J, Chen X, Yang Z, Tan X, Wang J, Chen Y. Preparation and characterization of folic acid functionalized bioactive glass for targeted delivery and sustained release of methotrexate. *J Biomed Mater Res A.* 2019;107(2):319–329. doi: 10.1002/jbm.a.36471.

[356] Moonesi Rad R, Alshemary AZ, Evis Z, Keskin D, Tezcaner A. Cellulose acetate-gelatin-coated boron-bioactive glass biocomposite scaffolds for bone tissue engineering. *Biomed Mater.* 2020;15(6):065009. doi: 10.1088/1748-605X/ab8d47.

[357] Ghorbani F, Ghalandari B, Sahranavard M, Zamanian A, Collins MN. Tuning the biomimetic behavior of hybrid scaffolds for bone tissue engineering through surface modifications and drug immobilization. *Mater Sci Eng C Mater Biol Appl.* 2021;130:112434. doi: 10.1016/j.msec.2021.112434.

[358] Jiménez-Holguín J, Sánchez-Salcedo S, Cicuéndez M, Vallet-Regí M, Salinas AJ. Cu-doped hollow bioactive glass nanoparticles for bone infection treatment. *Pharmaceutics.* 2022;14(4):845. doi: 10.3390/pharmaceutics14040845.

[359] Farag MM, Abd-Allah WM, Ahmed HYA. Study of the dual effect of gamma irradiation and strontium substitution on bioactivity, cytotoxicity, and antimicrobial properties of 45S5 bioglass. *J Biomed Mater Res A.* 2017;105(6):1646–1655. doi: 10.1002/jbm.a.36035.

[360] Abd-Allah WM, Fathy RM. Gamma irradiation effectuality on the antibacterial and bioactivity behavior of multicomponent borate glasses against methicillin-resistant Staphylococcus aureus (MRSA). *J Biol Inorg Chem.* 2022;27(1):155–173. doi: 10.1007/s00775-021-01918-z.

[361] Mesquita-Guimarães J, Detsch R, Souza AC, Henriques B, Silva FS, Boccaccini AR, Carvalho O. Cell adhesion evaluation of laser-sintered HAp and 45S5 bioactive glass coatings on micro-textured zirconia surfaces using MC3T3-E1 osteoblast-like cells. *Mater Sci Eng C Mater Biol Appl.* 2020;109:110492. doi: 10.1016/j.msec.2019.110492.

[362] Qian G, Zhang L, Liu X, Wu S, Peng S, Shuai C. Silver-doped bioglass modified scaffolds: A sustained antibacterial efficacy. *Mater Sci Eng C Mater Biol Appl.* 2021;129:112425. doi: 10.1016/j.msec.2021.112425.

[363] Sharifi E, Bigham A, Yousefiasl S, Trovato M, Ghomi M, Esmaeili Y, Samadi P, Zarrabi A, Ashrafizadeh M, Sharifi S, Sartorius R, Dabbagh Moghaddam F, Maleki A, Song H, Agarwal T, Maiti TK, Nikfarjam N, Burvill C, Mattoli V, Raucci MG, Zheng K, Boccaccini AR, Ambrosio L, Makvandi P. Mesoporous bioactive glasses in cancer diagnosis and therapy: Stimuli-responsive, toxicity, immunogenicity, and clinical translation. *Adv Sci (Weinh).* 2022;9(2):e2102678. doi: 10.1002/advs.202102678.

[364] Kim TH, Kang MS, Mandakhbayar N, El-Fiqi A, Kim HW. Anti-inflammatory actions of folate-functionalized bioactive ion-releasing nanoparticles imply drug-free nanotherapy of inflamed tissues. *Biomaterials.* 2019;207:23–38. doi: 10.1016/j.biomaterials.2019.03.034.

[365] Garello F, Svenskaya Y, Parakhonskiy B, Filippi M. Micro/nanosystems for magnetic targeted delivery of bioagents. *Pharmaceutics.* 2022;14(6):1132. doi: 10.3390/pharmaceutics14061132.

[366] Gupta S, Majumdar S, Krishnamurthy S. Bioactive glass: A multifunctional delivery system. *J Control Release.* 2021;335:481–497. doi: 10.1016/j.jconrel.2021.05.043.

# 16 Nanotechnology

## *Nanocomposites Processing*

*Sadasivam Kannan, Vadivel Siva, Anbazhagan Murugan, Abdul Samad Shameem, Arumugam Thangamani, Subramanian Thangarasu, and Arumugam Raja*

### 16.1 INTRODUCTION

In order to offer a longer shelf life with higher food sensory and safety quality and traceability, nanotechnology developed as an inventive solution that is increasingly used in the meat production chain. It can be characterized as a technological field with the objective of developing novel and distinctive features in nanomaterials with a size of less than 100 nm [1]. Nano-sized materials act and reach their target more effectively at very low concentrations due to their high surface area to volume ratio [2]. The use of nanotechnology in meat processing and packaging was carried out to achieve a number of goals, including (i) safeguarding products against microbial spoilage, (ii) enhancing sensory properties, (iii) upgrading the functional and dietary aspects of the meat products, and (iv) monitoring the quality during storage. Due to a lack of information and laws, nanotechnology is still a contentious topic for the general population, who had more questions than answers. As a result, the economic value of implementing nanotechnology in the meat processing industry, customer acceptance, and consideration of various legislation relevant to the implementation of this technology are all important factors [3]. Nature has mastered the usage of nanocomposites, and scientists are once again taking cues from nature. The size scale of a nanocomposite's component phases and the degree of mixing between the two phases have a significant impact on the material's characteristics. Significant changes in composite properties may be obtained depending on the type of components utilized (layered silicate or nanofiber, cation exchange capacity, and polymer matrix), as well as the technique of synthesis [4].

Depending on the diameters of the distributed nanoscale fillers, there are three main forms of polymer nanocomposites. In the first kind, polymeric matrices contain two-dimensional (2D) nanoscale fillers such as layered silicate, graphene, or MXene in the form of sheets that range in thickness from one

 DOI: 10.1201/9781003364948-16

nanometer to a few nanometers and in length from hundreds to thousands of nanometers. Layered polymer nanocomposites are a subcategory of the analogous polymer nanocomposites. These nanoscale fillers include nanofibers or nanotubes, for example, carbon nanofibers and nanotubes or halloysite nanotubes as reinforcing nanofillers to create materials with exceptional properties. In the second type, two dimensions are in the nanometer scale and the third is larger, forming an elongated one-dimensional structure. The third type is made up of nanocomposites, which are three-dimensional nanoscale fillers. Iso-dimensional low aspect ratio nanoparticles such spherical silica, semiconductor nanoclusters, and quantum dots serve as these nanoscale fillers [5].

## 16.2 OVERVIEW OF MICRO MANUFACTURING PROCESS

Technology for micro-manufacturing is being developed to not only satisfy these manufacturing needs but also to create new values. For instance, the creation of standard mechanical and mechatronic parts/elements for micro-machines or miniature equipment, the production of high-quality 3D electrical interconnections for 3D micro-assemblies, the creation of micro-/nano-structural surface/parts for batteries and micro-fuel cells, etc., would all add value to the associated products. The manufacturing processes and approaches used in micro-manufacturing may differ from those used in the production of macro-products. The process of creating individual components or parts, then removing, deforming, or adding materials, can be used to create macro-products.

Either a single industrial site or several sites might be used to conduct these operations. Within a single machine/manufacturing platform, patterning, deposition, and stacking processes may be used to manufacture micro-products. To fully solve difficulties linked to manufacturing in the micro-world, micro-manufacturing mostly uses non-conventional manufacturing processes as well as scaling down or changing the classic methods, as needed. In addition, manufacturing chains may differ from traditional manufacturing because to material characteristics, structural strength and stiffness, form and size scale factors, challenges with clamping and releasing, residual stress, and surface integrity. It is important to comprehend the fundamentals of the roles that reduced length scales may play in various processing mechanisms, such as the functions of surfaces at various length scales and in various manufacturing processes with regard to surface fabrication and micro-/nano-manipulation, surface metrology, etc. These are crucial in choosing industrial processes and in streamlining production networks. In order to achieve high efficiency, lower production errors, and do away with needless handling, transporting, and packing of micro-components between various operations, a process chain for micro-manufacturing should ideally be short. While some of the inherited drawbacks, such as lengthening the process chains, may be mitigated or removed, hybrid manufacturing processes may benefit from the advantages of specific micro-manufacturing methods/processes [6].

Different types of energy are used in micromachining operations at the same time and are influenced at the same machining zone. One of the goals of future research is to comprehend the mechanics of the hybrid setup and the material removal mechanism. Modelling and simulation of hybrid micromachining processes is another area of future research. Such processes can be simulated using a combination of techniques, including multiscale modelling, molecular dynamic simulation, and finite element analysis. The hybrid processes can be better understood and then optimized with the help of modelling and simulation. There is a need for specific multi-axis ultraprecision machines with high stiffness characteristics. In order to do jobs in a variety of directions and angles, machines with multi-axis ultraprecision capabilities are also necessary. Additionally, for the tool to move and travel smoothly, there needs to be improved numerical control, which will guarantee excellent accuracy in the manufactured microstructures [7].

## 16.3 NANOSTRUCTURED MATERIALS AND NANOCOMPOSITES

### 16.3.1 Carbon Nanotubes and Their Nanocomposites

Nanotubes of carbon CNTs are tubes made of coiled graphene sheets (one atomic layer of graphite). Single-walled carbon nanotubes (SWCNTs), double-walled carbon nanotubes (DWCNTs), and multi-walled carbon nanotubes are created via concentric arrangements of the tubes (MWCNTs). As biological and chemical sensors, probe tips for scanning probe microscopy, nano-electromechanical systems (NEMS), and reinforcement in nanocomposites, CNTs are attractive candidates for usage in a variety of applications [8]. CNTs are divided into different categories based on how many walls they have and are made up of rolled graphene sheets organized in a concentric manner. The nanotubes have a length that ranges from a few hundred nanometers to a few micrometers. They tangle because of their length SWCNTs are often defect-free, but MWCNTs have flaws.

High electrical and thermal conductivities and high strength (130 GPa) are all characteristics of a single graphene sheet. CNTs can be utilized as reinforcement in any form of matrix, however, polymer matrix composites use them more frequently. Due to qualities including low density, reasonable strength, flexibility, and ease of processing, typical polymer matrix composites have found use in a variety of industries. However, the development of CNTs reinforced nanocomposites has been prompted by the hunt for materials that can enhance the performance of advanced components. A lot of research has been done on CNT/epoxy nanocomposites because of the industrial and technological uses for them. In order to create these nanocomposites, melt mixing or solution mixing techniques are used. Zhou et al. [9] showed that adding 0.3 weight percent of CNTs to the epoxy matrix can increase strength and fracture toughness.

### 16.3.2 Zeolites and Composites

A group of crystalline microporous aluminium silicates known as zeolites can be found on the surface of the planet in a variety of habitats, such as sediments, soils, seafloor deposits, hydrothermal alteration products, altered volcanic deposits, and so on. Zeolites are used as absorbents, water filters, ion exchange beds, catalysts, optically active materials, polymerization science, separation technology, microelectronics, photoelectrochemical applications in solar cells, thin-film sensors, and drug and biomolecule encapsulation for targeted or controlled release applications [10]. $TiO_2$/zeolite sheets were created by Fukahori et al. using a papermaking procedure, and the finished product was used as a photocatalyst in the breakdown of bisphenol A. (BPA). In comparison to $TiO_2$ sheets, the $TiO_2$/zeolite sheets demonstrated a greater effectiveness for the removal of BPA. The increased efficiency is attributed to reversible adsorption, BPA molecules moving freely on the composite due to the cage-like structure, and zeolite pore connectivity [11].

### 16.3.3 Metal–Organic Frameworks (MOFs) and Their Composites

Metal-organic frameworks (MOFs) are a class of highly porous, crystalline hybrid materials created by combining metallic ions with organic ligands. MOFs could be employed as sensors and have intriguing features for gas storage, gas separation, and medication delivery. The investigation of polyimide (Matrimid) and polysulfone polymer-based synthetic polymer-MOF composite membranes. By using a solution blending technique, composite membranes made of MOF-5 and ZIF-8 (Zeolitic Imidazolate Framework) and Matrimid were created. Pre-treating the MOF crystals with the silylating chemical N-methylN-(trimethylsilyl)trifluoroacetamide allowed for the formation of composite membranes of higher quality. ZIF-8-polysulfone membranes were also made using the solution blending method in a related investigation.

The ZIF-8 crystals were found to enhance the transport of the gas through the membrane when the $CO_2$ diffusion properties of the composite membranes were examined. Because of the remarkable mechanical and hydrophobicity features of CNTs, one of the modification procedures involves adding CNTs to MOFs in order to produce improved composite performance. Doping MOFs or COFs with electropositive metals is an alternative strategy. According to recent research, a substrate's surface carboxylate functional groups may serve as nucleation sites to create MOFs through heterogeneous nucleation and crystal development. Due to their very high surface areas and controllable pore sizes, MOFs have been considered as viable Hydrogen storage media. Furthermore, ZIF-8 nanocrystals and its composites are used as excellent electrode materials for energy storage applications [12,13]. MOFs also have the benefit of being easy to prepare because they are created using "one-pot" solvothermal techniques in mild circumstances. These materials have attracted a lot of interest due to their huge surface areas, low framework densities, and high pore volumes when compared to other porous matrices. They have the potential to be used in a wide range of applications, from gas storage to heterogeneous catalysis [14].

## 16.4 CLASSIFICATIONS OF NANOCOMPOSITES

There are essentially two ways to categorize nanocomposites. They are nanocomposites, both organic and inorganic, Researchers make numerous attempts to

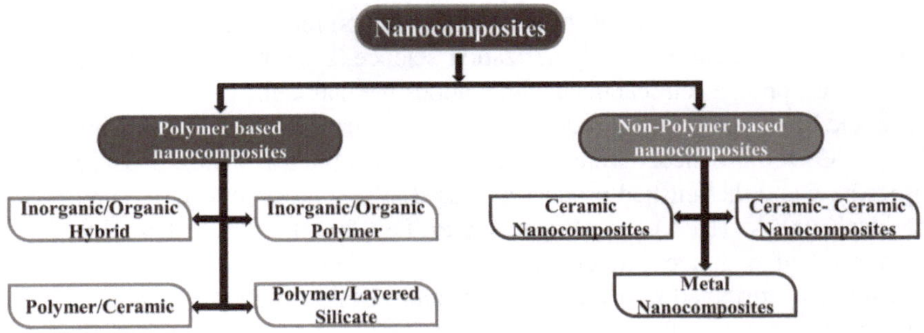


**FIGURE 16.1** Classifications of nanocomposites.

manipulate nanostructures using synthetic methods. The shape and interfacial characteristics of the nanocomposites as well as their separate parent compositions affect the nanocomposites' attributes. In the classification of nanocomposites, inorganic components can include one-dimensional and zero-dimensional materials like $(Mo_3Se_3\text{-})n$, chains, and clusters as well as three-dimensional framework systems like zeolites, two-dimensional layered materials like clays, metal oxides, metal phosphates, and chalcogenides. When compared to their macrocomposite counterparts, experimental work has typically demonstrated that almost all types and classes of nanocomposite materials result in new and better properties. Nanocomposites, therefore, have the promise of novel uses in a variety of industries, including mechanically reinforced lightweight components, non-linear optics, battery cathodes and ionics, nanowires, sensors, and other systems. The classifications of nanocomposites are given in Figure 16.1.

### 16.4.1 Metallic Nanocomposites

Recently, material scientists and device researchers have developed a strong interest in nanocomposites made by combining elastic polymers and metallic nanoparticles, particularly for human-friendly electronics [15]. Due to the tremendous stretchability that elastomers naturally possess as well as the exceptional electrical properties of metallic nanoparticles, such nanocomposites have the potential to achieve high intrinsic stretchability as well as high conductivity. The proper application of such stretchable metallic nanocomposites based on comprehension of the intricate technological strategies for materials and fabrications is thus crucial for eco-friendly electronic devices that require material softness comparable to human tissue, low power consumption, and high signal-to-noise ratio. The majority of elastomers are essentially insulators; hence they are unable to transport electrons. Conductive filler networks are integrated into the elastomeric matrix to create a nanocomposite, which gives these elastomers conductivity. Because of their high inherent conductivity, metallic nanomaterials such as Au nanoparticles (NPs),

Ag nanoparticles, Ag flakes, Ag nanowires (NWs), and Cu NWs have been frequently employed as conductive fillers in the composite [16]. Transition metal-based metal oxides are promising materials for supercapacitor electrodes. Siva et. al., reports the facile hydrothermal synthesis of CuO/ZnO nanocomposite as electrode material for energy storage applications [17]. Lee et al. [18] looked into how the conductivity was impacted by the length of the Ag NWs. By using a consecutive multistep growth approach, the authors have created very long Ag NWs (VAg NWs) with a length greater than 500 m, in contrast to normal Ag NWs. The findings indicate that the higher conductivity is correlated with longer NWs. The electrons can go farther within the NWs without crossing the junction as the NWs grow longer. Thus, the percolated network's overall contact resistance is reduced, resulting in good conductivity even under flexed situations. He also looked at how the conductivity was affected by the diameter of the Ag NWs. Ag NWs of various diameters were created specifically by varying the polyvinylpyrrolidone (PVP) content while keeping their length constant. The Ag NWs with a smaller diameter created a denser percolation network when combined with the elastomeric media under a fixed NW concentration, resulting in reduced resistance.

### 16.4.2 Ceramic Nanocomposites

Research to develop multi-functional engineering materials by creating structures at the nanometric scale has been driven by recent improvements in the manufacturing of nanocrystalline ceramic powders with unique features. One of the most quickly developing fields in composites research, nanocomposite ceramics, was inspired by the current passion for nanotechnology. In ceramics, reducing the material grain size can have a major impact on both mechanical and physical properties. Niihara first presented the idea of structural ceramic nanocomposites in 1991, and it may be viewed as an application of the nanocomposite strategy for the microstructural customization of structural ceramic composites. The main inspiration for this work came from findings in the $Si_3N_4/SiC_6$ and $Al_2O_3/SiC$ systems [19]. The fracture mode is another distinction between monolithic ceramics and similar nanocomposites. Regarding the $Al_2O_3/SiC$ system once more, it was noted that the fracture mode changed from a mixed inter/transgranular fracture mode in monolithic $Al_2O_3$ to a pure transgranular mode in the composites, highlighting a strengthening of the grain boundary in the latter instance. According to Ferroni and Pezzotti [20], monolithic $Al_2O_3$ had a 10% to 15% transgranular fracture mechanism. Depending on the amount of SiC present, this value increased to between 55% and 85% in $Al_2O_3/SiC$ nanocomposites. In instance, the largest fraction of transgranular fracture mode was found at the lowest SiC content (5 vol.%). The amount of SiC particles in the $Al_2O_3$ matrix that were inter- or transgranular, however, directly correlated with this behavior. When SiC grains were primarily found within the $Al_2O_3$ matrix, the primary transgranular fracture route was seen [20].

However, a good design and fabrication procedure that results in customized sintered micro/nanostructures is essential to achieving excellent mechanical characteristics in nanocomposite ceramics. A proper evaluation of the quantity, distribution, and shape of the reinforcement phase should be made when building a composite structure in addition to taking the chemical composition of the matrix and second phase into account. Achieving a uniform distribution of the second phase in the matrix material is crucial, but when nanocrystalline particles are utilized, this is still difficult because of their high specific surface and consequently high inclination to aggregate. This means that every production process involved in the creation of nanocomposite ceramics, from the synthesis of the composite powders through the creation of green bodies and their sintering, should be carried out with great care. $Al_2O_3$/SiC and $Si_3N_4$/SiC are the structural ceramic nanocomposites that have been studied the most. However, a variety of additional phases, including TiN, TiC, $TiO_2$, $ZrO_2$, $Cr_3C_2$, and YAG, can be employed as nanoreinforcements in ceramic matrixes made of $Al_2O_3$, $Si_3N_4$, MgO, mullite, or SiAlON. With restrictions on the distribution of the ultrafine particles in the composite feedstock, the majority of nanocomposite ceramic powders are made by the conventional mixing and milling of the constituent oxides. Many cutting-edge techniques, including vapor-phase reaction synthesis, combustion synthesis, wet chemical procedures, spray decomposition, and solution combustion methods, have been used to directly create composite nanopowders. There are now many literary examples of compositions created using these techniques, but most of them have not yet been used in an industrial production. An innovative technology suggests that ceramic particles' surfaces be altered by the inorganic second-phase precursors. This approach allows for better control of the final microstructure and can be seen as a balance between chemical and powder mixing techniques. This technique is far more straightforward and might be applied to industrial scale manufacturing because the process just calls for the use of inorganic precursors and aqueous media [21].

### 16.4.3 Polymer-Based Nanocomposites

Due to their attractive mechanical strength, controlled pore space, and surface chemistry, polymer-based materials are seen as suitable candidates. In contrast to polymer-based electrode materials, which offer superior durability, capability, and synergetic effects caused by the interaction of nanoparticle and matrix, the resulting polymer-based nanocomposites (PNCs) hold the characteristic phenomenon of nanoparticles. The most often utilized nanoparticles are made of single enzyme particles, biopolymers, metallic oxides, and zero-valent metals. Various types of porous resins, including cellulose, chitosan, and alginate, can be loaded with these nanoparticles. The mechanical and thermal characteristics of the polymeric supports are often what guide the choice. Potential applications for this approach might include engineering plastics, polymer-based materials, rubbers, adhesives, and coatings. PNCs are exceptional in that they exhibit astounding hybrid behaviors that can be seen in conjunction with two different components. This chapter will explore several PNC types, energy storage methods, and environmental applications. The energy storage capabilities of PNCs in batteries and supercapacitors will also be discussed.

Researchers are now thinking about developing clean energy systems for energy conversion and storage devices due to the increasing demand for energy, environmental pollution, and resource depletion. Electrochemical energy generation is another alternative energy source. In recent years, electrochemical energy storage systems have mostly consisted of batteries and electrochemical supercapacitors in order to address these Polymer-based Nanocomposites for Energy and Environmental Applications challenges. To get superior behaviors, it is required to design PNC materials based on these dissimilar materials. Due to the combination of the excellent mechanical and conducting properties of carbonaceous materials and the high pseudocapacitance of the conducting polymers, these nanocomposite materials exhibit improved stability (CP). The general information on the various CP nanocomposite materials for energy-related applications is provided in this part along with recent advancements in the field. Research on electrochemical supercapacitors technology led to the development of pseudocapacitors and asymmetrical supercapacitors as a result of the ongoing advancement of electronic technologies. The energy storage behaviors of traditional conventional capacitors and other electrochemical supercapacitors are covered in this segment.

Redox intercalative polymerization (RIP) of Polythiophenes (PT) derivatives in $V_2O_5$ produces a well-ordered hybrid nanocomposite with beneficial qualities like enhanced electric and Li+ -ion diffusion characteristics, according to Vadivel et al. report [22]. Additionally, nanometer-sized fillers with conductive route structures and large surface areas should be needed in order to take advantage of the electric behavior of PTs NCs. Because of their better surface area and improved overall behaviors, which are caused by the synergistic interactions between components, graphene oxide (GO) is acknowledged as a good choice [23]. Polymer-based nanocomposites are widely employed in the packaging, energy, transportation, electromagnetic shielding, military systems, sensors, catalysis, and information industries and are significant materials for both industrial and research applications. Finding new electrode materials to increase the specific capacitance of supercapacitors has been the focus of recent research. Metal oxides and conductive polymers have both undergone considerable testing as supercapacitor electrode materials [24,25].

## 16.5 SYNTHESIS METHODS

According to the properties that need to be improved, fillers are added to a matrix to create nanocomposites. The various types of synthetic methods are displayed in Figure 16.2.

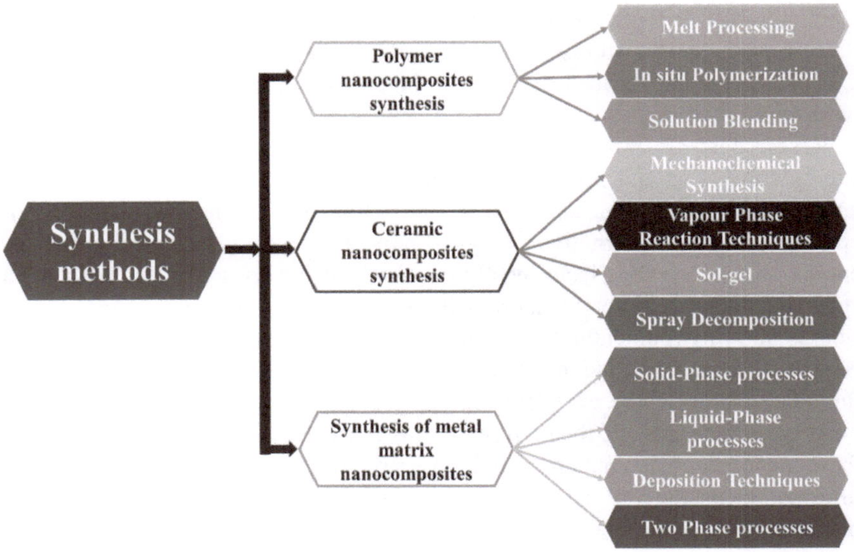


**FIGURE 16.2** Synthesis methods of nanocomposites.

### 16.5.1 Polymer Nanocomposites Synthesis

Due to their versatility, low weight, and ductility, polymers are frequently employed in a variety of industries. However, compared to ceramics and metals, polymers have some disadvantages, including low strength, weak mechanical characteristics, and limited stability. These restrictions are solved by polymer nanocomposites with nanoparticle- or nanofiber-based fillers, which are becoming effective materials for use in solar cells, supercapacitors, sports equipment, and aerospace components.

### 16.5.2 Melt Processing

Melt processing, also known as melt compounding or melt mixing, is a relatively quick and easy method that eliminates the need for hazardous solvents. However, because of the high viscosity of the thermoplastic polymers, it is challenging to manage the filler distribution in the matrix. In order to melt the polymer and evenly distribute the fillers, this technique needs high temperatures as well as mechanical or shear forces. Low-quality nanocomposites were discovered to be produced by higher screw speeds, longer residence durations, and different mixing conditions. This approach has also been used to create numerous publications on graphene/polymer, CNT/polymer, and clay polymer nanocomposites [26].

### 16.5.3 In Situ Polymerization

This process produces a consistent distribution of fillers throughout the polymer matrix while typically avoiding the use of solvents. The process starts with the creation of a combination of monomer and filler particles, which is subsequently polymerized using conventional polymerization methods such mechanical means (ultrasonication, stirring, etc.) or photoinitiation. According to reports, this method can create nanocomposites with covalent or noncovalent bonds between the components. This technique can be used to process polymers that cannot be handled by solution mixing and melt blending. This method can also be used to treat polymers that are insoluble or thermally unstable. Using this technique, Bingzhen Li et al. [27] created binary and ternary $Fe_3O_4$/PPy/PANI nanocomposite materials. Pyrrole and aniline were polymerized in situ under the right circumstances and in the presence of coprecipitated $Fe_3O_4$ nanoparticles. A magnetic decantation process was used to create the nanocomposite. It was discovered that carbonyl groups‘ affinity for PPy-PANI caused a powerful contact between them on the surface of nanoparticles. The ternary nanocomposites shown an improved capacity for absorbing microwave energy.

### 16.5.4 Solution Blending

A straightforward and popular technique for creating nanocomposite materials is solution blending. By mixing the solutions of several polymers, including polystyrene, polyimides, polymethyl methacrylate, and polyacrylamides, graphene or graphene oxide nanoplatelets have been created [28]. For instance, the innovative polymer nanocomposite synthesis using polyacrylamide/graphene-based nanocomposite for usage in water purification was highlighted in recent research on their applications. The monomers were chain polymerized in an aqueous environment to create the acrylamide-acrylic acid copolymer. For efficient dispersion and to encourage covalent bonds with the matrix, the graphene was acid functionalized. In these nanocomposites, exfoliation of the functionalized graphene layers is a crucial characteristic.

### 16.5.5 Ceramic Nanocomposites Synthesis

Metal makes up the second component of a ceramic matrix nanocomposite, whose primary component is ceramic (a chemical compound made of oxides, nitrides, silicides, and related substances). In general, a variety of processes are used to create ceramic matrix nanocomposites. Methods known as the “Powder Process” are frequently employed to create ceramic nanocomposites.

### 16.5.6 Mechanochemical Synthesis

This technique can be used to create nanocomposite powders made of oxide, nonoxide, and mixed oxide/nonoxide materials. The two main limitations of this synthesis process are the inability to generate separate nanoparticles in the tiniest size range and the result being contaminated by the milling media. By mechanically combining the component phases, nearly any ceramic composite powder can be created. The type of raw materials, purity, particle size, size distribution, and degree of agglomeration are the primary determinants of the attributes of the final nanocomposite products. To prevent the creation of a secondary phase during sintering, the powders must be kept pure. The creation of a homogenous powder combination can be accomplished using wet ball milling or attrition milling processes. $Al_2O_3$/SiC composites are frequently made using this traditional method of powder mixing, which involves ball milling [29]. The downside of the milling process is that it could result in some contamination from the milling media.

### 16.5.7 Vapor Phase Reaction Technique

Using this method, a solid is evaporated to create a supersaturated vapor, which is then condensed to produce a nanomaterial with a variety of particle sizes, shapes, and compositions [30]. Joule's heating, electron beam evaporation, arc discharge, and laser ablation can all produce supersaturated vapor. Chemical reaction, mass transfer, nucleation, coagulation, and condensation are the processes at play. In this process, there is no requirement for calcination.

### 16.5.8 Sol-Gel Method

One of the straightforward wet chemical methods utilized frequently for the creation of ceramics and nanocomposites (mainly oxides) is called sol-gel. The sol-gel process is the chemical conversion of a sol into a gel (three-dimensional polymer) state, followed by a post-treatment-induced transition into solid oxide material. This approach is founded on the inorganic polymerization reaction, which includes hydrolysis, polycondensation, gelation, ageing, drying, and calcination or sintering (densification) [31].

### 16.5.9 Spray Decomposition

A versatile method for the synthesis of ceramic materials with a variety of compositions and morphologies is spray decomposition. The processes in the synthesis are: (1) creation of the precursor aerosol; (2) solvent evaporation and solute precipitation; and (3) solute pyrolytic breakdown. The ability to produce highly homogeneous, high-purity nanoparticles is one of this method's benefits. The necessity for copious amounts of solvents and the challenge of scaling up production are drawbacks.

## 16.6 SYNTHESIS OF METAL MATRIX NANOCOMPOSITES

Metal matrix nanocomposites are nanomaterials with a strong metallic matrix and soft reinforcement or hard ceramics. In comparison to monolithic metals, metal matrix nanocomposites have better wear resistance, higher specific strength and modulus, and lower coefficients of thermal expansion [32].

### 16.6.1 Solid-Phase Processes

The particulate-reinforced MMCs are made from mixed elements powders in solid-phase procedures. The principal method that fits into this category is powder metallurgy. The main phases in this approach include powder preparation, powder blending, compacting, sintering, sizing, and machining.

### 16.6.2 Liquid-Phase Processes

These procedures use a variety of exclusive methods to insert ceramic particles into a molten metallic matrix. The final processes are casting the finished composite mixture after mixing. The following elements are included in the selection criteria for a ceramic reinforcement:

- matrix alloy,
- melting temperature,
- tensile strength,
- thermal stability,
- elastic modulus,
- size and form of the reinforcing particle, and
- cost.

### 16.6.3 Deposition Techniques

The crucial deposition methods are immersion plating, electroplating, spray deposition, CVD, and PVD. Individual fibers are covered with the matrix material required to create the composite using deposition processes. After that, a structural shape or consolidated composite plate is formed through diffusion bonding.

### 16.6.4 Two-Phase Processes

A region of the phase diagram where the matrix contains both solid and liquid phases is where ceramic and matrix are mixed in two-phase processes. The two main processes that fall under the heading of two-phase processes are spray deposition and compocasting/rheocasting.

### 16.6.5 Size Effect in Micro-Manufacturing Process

In the highly popular field of microtechnology, size effects are crucial. Reviews on micro shaping emphasize how crucial size effects are in this area, where long-term basic research is necessary to realize industrial applications. Size effects are another issue to be overcome in the construction of microsystems. Furthermore, the recent review on micromachining makes clear that no significant efforts have been made to address size effects, and no new interdisciplinary review on size effects has been reported since then. The purpose of this chapter is to conduct a thorough analysis of previous research on size effects [33]. When reducing or increasing all important length dimensions of the workpiece, tool, and/or process parameter, deviations from extrapolated values of process characteristics occur during the fabrication of metallic components. The ratio of critical features that cannot be maintained constant in accordance with process requirements results in the size impact. The definition also emphasizes the fact that size effects are initially somewhat perplexing in nature since they take place even though the length relationships between all geometric aspects are maintained constant.

Size impact in machining is mainly caused by the effects of small ratios of the undeformed chip thickness to the tool edge radius for which the assumptions of tool edge sharpness as well as the homogeneity and isotropic nature of the workpiece material cannot be valid. A review of the studies done on mechanical micro machining by earlier writers [34,35] revealed areas that could help with process improvement. The need for a paradigm change from macro machining to micro domain cutting was underlined by a number of important considerations. Among these elements are: When machining is done at undeformed chip thicknesses less than the tool edge radius, specific cutting energy increases in a non-linear manner. In addition, the microstructure of the material of the work-piece has also been acknowledged for the size effect phenomenon seen in the particular cutting energy. The lower boundary of a workable process window for a specific workpiece and tool combination is determined by the minimum chip thickness effect. Surface roughness produced at low feeds, which are typical in the micro-cutting process, differs noticeably from theoretical and measured surface roughness. Effects of the ploughing procedure negatively affecting burr size and surface polish of components.

## 16.7 ADDITIVE MANUFACTURING PROCESS

Our collective imagination has been captured by additive manufacturing (AM) technology, also known as "3D printing," which has given rise to fantastical images of printed organs and aircraft. Although the technology is already having a significant impact on our immediate surrounds and has great promise for the future of manufacturing, those visions are still a long way from being a reality. AM is expected to radically alter the way things are manufactured, whether the effects are immediate or delayed. A procedure known as 3D printing involves building up an object layer by layer in an additive manner. A digital 3D model created with computer-aided design (CAD) software or by 3D scanning is divided into layers that serve as the source code for a 3D printer's toolpath. The machine uses a certain procedure to duplicate the model in the real world, building it up from the base slice to the top depending on the technology used.

Binder jetting, directed energy deposition (DED), material extrusion, material jetting, powder bed fusion, sheet lamination, and vat photopolymerization are the seven different types of processes that an AM system may use to 3D print, according to the ISO/ASTM52900-15 standard of process categorization [36].

### 16.7.1 Binder Jetting

A method for applying a liquid bonding agent to a powder bed, it can be utilized with many materials, including gypsum, sand, glass, metal, etc.

### 16.7.2 Directed Energy Deposition

In which a multiaxis robotic arm-mounted energy source, such as an electron or laser beam, is fed metal as a powder or wire feedstock. Layer by layer, the substance is melted onto a substrate, used with metals like cobalt-chrome and titanium.

### 16.7.3 Material Extrusion

An extruder deposits a substance onto a substrate. Typically, a heating device melts a thermoplastic filament before extruding it via a hot end. However, viscous materials like concrete, clay, organic tissue, and even food can be processed using the same method.

### 16.7.4 Material Jetting

Specialty printheads spray a liquid substance onto a substrate, such as piezoelectric printheads like those used in 2D inkjet printers. Most frequently, this substance is a photosensitive plastic resin (sometimes called a photopolymer) that is then UV-hardened.

### 16.7.5 Powder Bed Fusion

Here, a powder bed is heated by an energy source-like a laser or electron beam-until the individual powder particles melt into one another. This technology is typically linked to materials like nylon and metals like titanium.

### 16.7.6 Sheet Lamination

In this method, the required shape is etched into each shape before the material sheets are fused together. The last item is then taken out of the stack of bound papers. Currently, this uncommon 3D printing technique is utilized most frequently not only with paper but also with metal and plastic.

### 16.7.7 Vat Photopolymerization

Layer-by-layer hardening of the material occurs when a vat of photopolymer resin is subjected to an energy source, such as a laser beam or digital light projector. Thermoset polymers are typically linked to this process.

## 16.8 APPLICATIONS

A matrix to which nanoparticles have been added to enhance a particular property of the material is called a nanocomposite. Researchers are already considering employing nanocomposites in a variety of industries because of their unique features. Batteries with higher power output are being created using nanocomposite. Faster charging or discharging of power is possible thanks to closer contact between the silicon-carbon nanocomposite anodes and the lithium electrolyte. It expedites the recovery time for shattered bones. Researchers have demonstrated that the implantation of a nanotube-polymer nanocomposite as a form of scaffold that directs the growth of replacement bone speeds up the growth of that bone. Study after study is being done by the experts to learn more about how this nanocomposite promotes bone formation. High strength-to-weight structural components are being produced by this method. For instance, nanotube-polymer composite windmill blades can be made using an epoxy that contains carbon nanotubes. Longer windmill blades are practical as a result of the sturdy yet lightweight blade that is produced. Each windmill produces more electricity thanks to these longer blades.

Researchers discovered that epoxy composites employing a same weight of carbon nanotubes may produce weaker components than epoxy composites using graphene. The polymers in the epoxy appear to adhere to graphene more tightly, allowing for a more effective coupling of the material into the composite's structural elements. This characteristic might lead to the creation of parts with higher strength-to-weight ratios for applications like windmill blades or aviation parts. Nanocomposites are used to create light-weight sensors. The degree to which a polymer-nanotube nanocomposite conducts electricity is determined by the nanotubes' spacing. Because of this characteristic, patches of polymer-nanotube nanocomposite can monitor tension on windmill blades. The nanocomposite will bend when powerful wind gusts bend the blades. An alarm is set off when the electrical conductivity of the nanocomposite sensor changes due to bending. This alarm would enable the windmill to be turned off before significant harm takes place.

A conductive paper could be created by combining nanotubes and cellulosic components in a nanocomposite. A flexible battery is created when this conductive paper is bathed in an electrolyte, making it simpler to detect and eliminate cancers. In order to create a nanocomposite particle that is magnetic and fluorescent, researchers are aiming to combine fluorescent and magnetic nanoparticles. During an MRI test performed before surgery, the tumor is more visible due to the nanocomposite particle's magnetic property. The surgeon may be able to see the tumor more clearly while performing surgery thanks to the nanocomposite particle's fluorescent quality. The technological applications of nanocomposites are presented in Figure 16.3.

### 16.8.1 Nanocomposites for Optical Applications

Numerous optical applications exist for nanocomposite structures built on embedded functional components in workable matrixes. Particulate phases can be

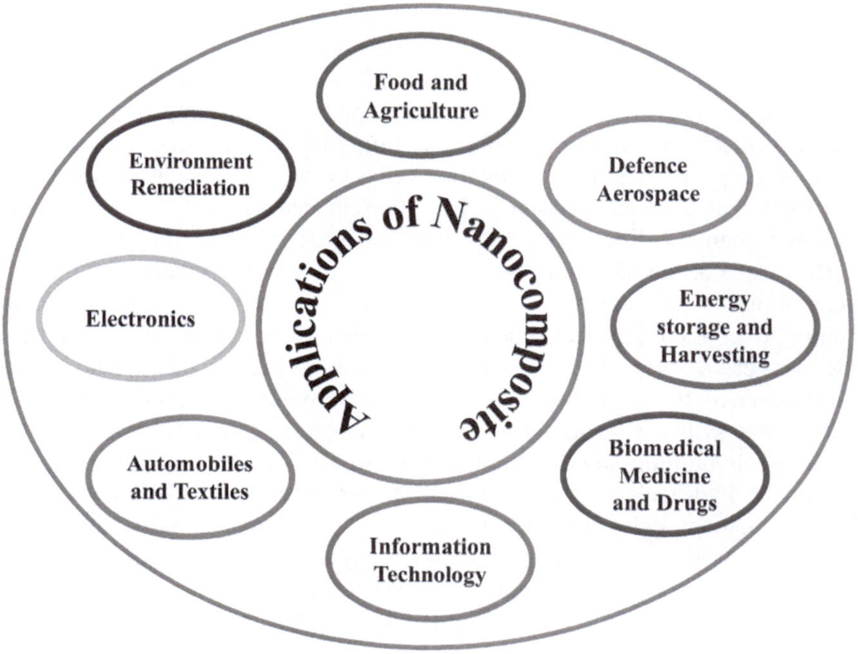


**FIGURE 16.3** Technological applications of nanocomposites.

created in-situ or ex-situ and can have a variety of characteristics, such as enhanced hardness or laser amplification. Polymers, glasses, and ceramics can all be used as matrix materials. These kinds of composites have been utilized to create materials with high refractive index, nonlinear optical properties, laser properties, and magnetic capabilities. Semiconductor particles can be created in polymers, ceramics, and glasses and have a wide range of optical properties. Their third-order nonlinear optical features, which might be helpful for all optical switches, are of particular interest. These devices will be simpler to integrate into intended applications if processing is done in film or fiber forms with a composite structure. In terms of stability and processability, composites combining optical polymers and tiny molecules outperform bulk materials. In the near-IR and beyond, solid-state laser materials have a lot of potential as amplification films. Last but not least, unique transparent magnetic materials have also been created employing nanocomposite principles and have applications in magnetooptics, printing, and data storage [37].

### 16.8.2 Nanocomposite Hydrogels for Biomedical Applications

Advanced biomaterials known as nanocomposite hydrogels have a variety of biological and pharmacological uses. These hydrogel networks have applications in biomedical devices such as sensors, actuators, medication delivery, stem cell engineering, and regenerative medicine. Nanocomposite hydrogels have better physical, chemical, electrical, and biological properties than traditional polymeric hydrogels. Examples of nanocomposite hydrogels include injectable matrices, controlled drug delivery depots, mechanically rigid and highly elastomeric networks, cell and tissue adhesion matrices, and scaffolds that conduct electricity. The strengthened interactions between the polymer chains and nanoparticles are primarily responsible for the enhanced performance of the nanocomposite hydrogel network. The creation of the next generation of nanocomposite hydrogels will necessitate not only precise control over their physical, chemical, and electrical characteristics but also the network's integration of the appropriate biological cues. Due to the limitations of two-component systems in incorporating numerous functionalities, the synthesis, and fabrication of such hydrogel networks will concentrate on building multi-component networks. Future research on nanocomposite hydrogels will also concentrate on understanding the interactions between nanoparticles and polymeric chains at various length scales. As a result, the properties of the nanocomposite hydrogels will be tailored for the necessary applications. To further replicate the cellular microenvironment of native tissues within the nanocomposite hydrogels, new fabrication technologies will also be developed [38].

## 16.9 CHALLENGES AND FUTURE PERSPECTIVE

In the last ten years, investigations have demonstrated that polymer nanocomposites have unique physicochemical properties that cannot be quantified when isolated components are used in membranes and adsorbents for water treatment and gas separation. Multiphase solid materials, such as gel, porous media, colloids, and copolymers, make up the majority of nanocomposites. The choice of hosts for nanocomposites is extremely important since it affects how well they perform [39]. Therefore, it is essential to conduct research on polymer nanocomposites based on the creation of customized materials that have the capacity to regulate nanocomposite performance in gas separation and water filtration. Additionally, it is crucial to research polymer nanocomposite based on reusability and sustainability for high performance in gas separation and water treatment for commercial viability. This will make it possible for it to be used commercially due to how simple, flexible, and adaptable it is to utilize.

Furthermore, over the past 10 years, concerns have been raised about the potential extreme environmental repercussions of breakthroughs in polymer nanocomposites. The difficulty of removing graphene from garbage is one example of this problem with regard to the adverse environmental impact. As a result of nanocomposite graphene's noxious characteristics and potential for fire outbreaks due to its thermal conductivity and fire-retardant qualities, this has happened. Additionally, there are concerns about the harmful effects of nanomaterials as well as their unknown negative effects [40]. For the confirmation of their practical and commercial usage, appropriate methods of measuring toxicity in polymer nanocomposite are crucial. Additionally, criteria for comparing and evaluating the efficacy of various polymer nanocomposite materials in the treatment of water exist [41]. However, it is challenging to compare the abilities of various nanoparticles and identify excellent nanocomposites with merit for long-term use. As a result, scientists ought to think about developing a performance evaluation instrument for water purification.

## 16.10 SUMMARY AND CONCLUSIONS

With a scientific basis combining chemistry, physics, biology, materials science, and engineering, the topic of nanocomposites is still relatively young. The information from several disciplines will ultimately produce new science, especially new materials, technologies, and engineering. The ability of nanocomposites to change numerous industries, including the aerospace and transportation sectors, as well as the fields of medicine, health care, and environmental protection, is a crucial feature. The field of nanocomposites is becoming more exciting every day due to its potential to improve our quality of life. Materials like nanocomposites can be used to meet future demands for advancement.

Different processing techniques are now available, however, their downsides and restrictions force engineers to look for innovative new engineering solutions. They outperform their monolithic and microcomposite counterparts in terms of performance, and they are a good choice for overcoming the shortcomings of many materials and gadgets that are already on the market. Although nanocomposites have applications today, their full potential has not yet been realized. Potential applications and markets for these materials are created by characteristics like very high mechanical endurance even at minimal loading of reinforcements, gas barrier, and flame-related properties. Thus, all three of the forms of nanocomposites discussed in this work have the potential to usher in a new era of materials.

## REFERENCES

[1] M. Taherimehr, H. Yousefnia Pasha, R. Tabatabaeekoloor, and E. Pesaranhajiabbas, Trends and challenges of biopolymer-based nanocomposites in food packaging, *Compr. Rev. Food Sci. Food Saf.* 20(6) (2021), pp. 5321–5344.

[2] I. Khan, K. Saeed, and L. Khan, Nanoparticles: Properties, applications and toxicities, *Arab. J. Chem.* 12(7) (2019), pp. 908–931.

[3] J. C. Buzby, Nanotechnology for food applications: more questions than answers, *J. Consum. Aff.* 44(3) (2010), pp. 528–545.

[4] C. I. Park, O. O. Park, J. G. Lim, and H. J. Kim, The fabrication of syndiotactic polystyrene/organophilic clay nanocomposites and their properties, *Polymer*, 42(17) (2001), pp. 7465–7475.

[5] P. Huang, H. Q. Shi, S. Y. Fu, H. M. Xiao, N. Hu, and Y. Q. Li, Greatly decreased red shift and largely enhanced refractive index of mono-dispersed ZnO-QD/silicone nanocomposites, *J. Mater. Chem. C.* 4(37) (2016), pp. 8663–8669.
[6] B. Lauwers, F. Klocke, A. Klink, A. E. Tekkaya, R. Neugebauer, and D. Mcintosh, Hybrid processes in manufacturing, *CIRP annals*. 63(2) (2014), pp. 561–583.
[7] S. Borra, R. Thanki, N. Dey (2019). [SpringerBriefs in Applied Sciences and Technology] Satellite Image Analysis: Clustering and Classification || Overview of Hybrid Micro-manufacturing Processes. 10.1007/978-981-13-6424-2(Chapter 1), 1–12. doi:10.1007/978-3-030-13039-8_1.
[8] H.-S. Philip Wong and D. Akinwande (2011) Carbon Nanotube and Graphene Device Physics, Cambridge University Press, USA. H.S.P Wong and D. Akinwande, 2011 *Carbon nanotube and graphene device physics,* Cambridge University Press.
[9] Y. Zhou, F. Pervin, L. Lewis and S. Jeelani, Fabrication and characterization of carbon/epoxy composites mixed with multi-walled carbon nanotubes, *Mater. Sci. Eng., A.* 475(1-2) (2008), pp. 157–165.
[10] H. Heinemann, Technological applications of zeolites in catalysis, *Catal. Rev. Sci. Eng.* 23(1-2) (1981), pp. 315–328.
[11] S. Fukahori, H. Ichiura, T. Kitaoka and H. Tanaka, Photocatalytic decomposition of bisphenol A in water using composite TiO2-zeolite sheets prepared by a papermaking technique, *Environ. Sci. Technol.* 37(5) (2003), pp. 1048–1051.
[12] V. Siva, A. Murugan, A. Shameem, S. Thangarasu and S. A. Bahadur, A Simple Synthesis Method of Zeolitic Imidazolate Framework-8 (ZIF-8) Nanocrystals as Superior Electrode Material for Energy Storage Systems, *Journal of Inorganic and Organometallic Polymers and Materials*, (2022), pp. 1–8. 10.1007/s10904-022-02475-x.
[13] V. Siva, A. Murugan, A. Shameem, S. Thangarasu, S. Kannan, S. Asath Bahadur, In situ encapsulation of V 2 O 5@ ZIF-8 nanocomposites as electrode materials for high-performance supercapacitors with long term cycling stability, *Journal of Materials Chemistry C*, (2023), 11, 3070–3085. 10.1039/D2TC03996F
[14] J. R. Long and O. M. Yaghi, The pervasive chemistry of metal–organic frameworks, *Chem. Soc. Rev.* 38(5) (2009), pp. 1213–1214.
[15] J. Hu, Y. A. Tang, A. H. Elmenoufy, H. Xu, Z. Cheng and X. Yang, Nanocomposite-based photodynamic therapy strategies for deep tumor treatment, *Small*, 11(44) (2015), pp. 5860–5887.
[16] H. Joo, D. Jung, S. H. Sunwoo, J. H. Koo and D. H. Kim, Material design and fabrication strategies for stretchable metallic nanocomposites, *Small*, 16(11) (2020) p.1906270.
[17] V. Siva, A. Murugan, A. Shameem and S. Asath Bahadur, One-step hydrothermal synthesis of transition metal oxide electrode material for energy storage applications, *Journal of Materials Science: Materials in Electronics*, 31 (2020), 20472–20484.
[18] J. Lee, P. Lee, H. Lee, D. Lee, S. S. Lee and S. H. Ko, Very long Ag nanowire synthesis and its application in a highly transparent, conductive and flexible metal electrode touch panel, *Nanoscale*, 4(20) (2012), pp. 6408–6414.
[19] K. Niihara and A. Nakahira, Strengthening of oxide ceramics by SIC and S&N, dispersion, In Proc. 3rd. Znt. Symp. on Ceramic Materials and Components for Engines, ed. V. J. Tennery. Westerville Ohio, (1988), pp. 919–926.

[20] T. Ohji, Y. K. Jeong, Y. H. Choa and K. Niihara, Strengthening and toughening mechanisms of ceramic nanocomposites, *J. Am. Ceram. Soc.* 81(6) (1998), pp. 1453–1460.
[21] P. Palmero, Structural ceramic nanocomposites: a review of properties and powders' synthesis methods, *Nanomaterials.* 5(2) (2015), pp. 656–696.
[22] C. W. Kwon, A. V. Murugan, G. Campet, J. Portier, B. B. Kale, K. Vijaymohanan and J. H. Choy, Poly (3, 4-ethylenedioxythiophene) $V_2O_5$ hybrids for lithium batteries, *Electrochem Commun.* 4(5) (2002), pp. 384–387.
[23] C. Bora, R. Pegu, B. J. Saikia and S. K. Dolui, Synthesis of polythiophene/graphene oxide composites by interfacial polymerization and evaluation of their electrical and electrochemical properties, *PolymInt.* 63(12) (2014), pp. 2061–2067.
[24] V. Siva, A. Murugan, A. Shameem, S. Thangarasu and S. A. Bahadur, A facile microwave-assisted combustion synthesis of $NiCoFe_2O_4$ anchored polymer nanocomposites as an efficient electrode material for asymmetric supercapacitor application, *Journal of Energy Storage.* 48 (2022), p. 103965.
[25] W. Ramya, V. Siva, A. Murugan, A. Shameem, S. Kannan, K. Venkatachalam, A Novel Biodegradable Polymer-Based Hybrid Nanocomposites for Flexible Energy Storage Systems, *Journal of Polymers and the Environment* (2022) 10.1007/s10924-022-02695-9.
[26] G. Mittal, V. Dhand, K. Y. Rhee, S. J. Park and W. R. Lee, A review on carbon nanotubes and graphene as fillers in reinforced polymer nanocomposites, *Ind. Eng. Chem.* 21 (2015), pp. 11–25.
[27] B. Li, X. Weng, G. Wu, Y. Zhang, X. Lv and G. Gu, Synthesis of Fe3O4/polypyrrole/polyaniline nanocomposites by in-situ method and their electromagnetic absorbing properties, *J. Saudi Chem. Soc.* 21(4) (2017), pp. 466–472.
[28] J. R. Potts, D. R. Dreyer, C. W. Bielawski, R.S. Ruoff, Graphene-based polymer nanocomposites, *Polymer* 52 (2011), pp. 5–25.
[29] M. Sternitzke, Review: structural ceramic nanocomposites, *J. Eur. Ceram. Soc.* 17(9) (1997), pp. 1061–1082.
[30] L. S. Abovyan, H. H. Nersisyan, S.L. Kharatyan, R. Orrù, R. Saiu, G. Caon, and D. Zedda, Synthesis of alumina–silicon carbide composites by chemically activated self-propagating reactions, *Ceram. Int.* 27(2) (2001), pp. 163–169.
[31] C. J. Brinker and G. W. Scherer, *Sol-Gel Science: The Physics and the Chemistry of Sol Gel Processing*, Academic Press, Inc, London, 1990.
[32] G. S. Shiva Shankar and S. Basavarajappa, Synthesis and characterization of aluminium 2219 reinforced with boron carbide and molybdenum disulfide metal matrix hybrid composites, *Adv. Mater. Res.* 1101 (2015), pp. 28–31.
[33] T. Masuzawa, *State-of-the-art on micromachining*, *Annals of CIRP* 49(2) (2000), pp. 473–488.
[34] J. Chae, S. S. Park and T. Freiheit, Investigation of micro-cutting operations, *International Journal of Machine Tools and Manufacture*, 46 (3-4) (2006), pp. 313– 332.
[35] D. Dornfeld, S. Min and Y. Takeuchi, Recent advances in mechanical micromachining, *CIRP Annals – Manufacturing Technology*, 55(2) (2006), pp. 745– 768.
[36] ISO/ASTM52900-15. *Standard Terminology for Additive Manufacturing – General Principles – Terminology. West Conshohocken*, PA: ASTM International; 2015.
[37] L. L. Beecroft and C. K. Ober, Nanocomposite materials for optical applications, *Chem. Mater.* 9(6) (1997), pp. 1302–1317.

[38] A. K. Gaharwar, N. A. Peppas and A. Khademhosseini, Nanocomposite hydrogels for biomedical applications, *Biotechnology and Bioengineering*, 111(3) (2014), pp. 441–453.
[39] A. M. Abu-Dief and S. O. Alzahrani, Sustainability nanocomposite for water treatment, *Int. J. Org. Inorg. Chem.* 2 (2020), pp. 1–12.
[40] D. E. Babatunde, I. H. Denwigwe, O. M. Babatunde, S. L. Gbadamosi, I. P. Babalola and O. Agboola, Environmental and societal impact of nanotechnology, *IEEE Access*, 8 (2019), pp. 4640–4667.
[41] H. D. Beyene and T. G. Ambaye, Application of sustainable nanocomposites for water purification process, *Sustainable polymer composites and nanocomposites*, (2019), pp. 387–412.

# 17 Opportunities and Challenges in Miniaturized Product Manufacturing Processes

*Pawan Kumar Rakesh and Rajiv Kumar Garg*

## 17.1 INTRODUCTION

Manufacturing processes are rapidly changing depending upon the product requirement. The concept of the miniaturized product has come into force by the expression "smaller is better" [1–3]. The miniaturized product can be produced by the development of micro/nanoscale manufacturing machine. A miniaturised machine tool system is conceived to achieve higher quality and quantity with a reduced amount of space (compact machine) and reduced consumption of both energy and capital via economizing the fabrication processes. At the initial stage, the investment cost of the nanofabrication machine is very high, but it decreases the operating cost, overhead cost, and material requirements; and creates a more user-friendly production environment than the conventional machine. Furthermore, since the equipment is very small, it may reduce the cutting tool speed and feed speed, resulting in improving the manufacturing cycle time. As the number of processes decreased, resulting in the potential for energy saving, and the resulting systems can be called energy-efficient machining systems. The effect of miniaturization in fabrication systems was investigated by Mechanical Engineering Laboratory (MEL), Japan. It was estimated that a reduction of 1/10th in the size of a production machine decreases the overall energy consumption to approx. 1/100 as associated to that of the conventional machine [4–6]. The advantage of a miniaturized machine tool is its competence for manufacturing parts with dimensions of less than 1–100 nm; compare that with the thinness of a natural human hair at 2.5 nm [7,8]. At a microscopic scale, the least disparity in the engineering process instigated by beneficial material or cutting tool, geometrical variation, and vibration will directly impact the production of the product.

DOI: 10.1201/9781003364948-17

Manufacturing processes can be divided in two ways, (a) the top-down approach—the art of carving nanomaterial out of a raw block, for example, a raw block of stone is chiseled with the help of chisel to make a shaped remarkable sculpture. The block of stone is reduced to half of its actual volume and leaves the other half of the stone as waste material during this process. This method is time-consuming and complicated, (b) the bottom-up approaches—the process of adding particles at the molecular level, for example, direct digital manufacturing, additive fabrication process, rapid prototyping, and 3D printing. In this process, 3D objects are fabricated by adding layer-upon-layer of material. The raw material may be used in liquid, powder, and sheet material [9–12].

The challenges in miniaturized manufacturing processes are precision handling, product quality, and mechanical properties of the finished product. The micro-indentation tester, atomic force microscopy, and scratch testing are methods to measure the nanoindentation of the product. The application of the miniature product is found in aerospace, defense, electronics, bio-medical, microheat exchangers, microchemical reactors, microfuel cells, power generation, and propulsion [13–16].

## 17.2 MICROMANUFACTURING

The manufacturing of a product at the microscale level is known as micro-manufacturing. Traditional lithography-based technology, such as Ion lithography, x-ray lithography, optical lithography, and electron lithography, forms the basis of silicon-based microelectronics device fabrication, but also precision engineering technologies, such as precision machines, metrology, machine structures, and sources of error, have been developed to maintain microscale dimensional accuracy.

### 17.2.1 Micromanufacturing Methods

The research trend is more focused on producing microproducts by down-scaling both traditional and non-traditional techniques. Besides, the hybrid manufacturing processes, which may combine two or more than two processes, are also emerging trends. The micromanufacturing processes can be classified mainly based on the type of energy used, such as thermal, radiant, chemical, electrical, and motion energy. The working principles of each process depend upon the type of energy used to manufacture the parts. Generally, the manufacturing processes can also be categorized according to the technique of fabricating products, such as forming, additive, subtractive, assembly, and hybrid, as shown in Figure 17.1.

### 17.2.2 Subtractive Machining Processes

In micromachining processes, the quantifiable materials are removed from the workpiece in the form of microchips to achieve the desired dimensions and shape. Micro-mechanical cutting (drilling, turning, milling, grinding, and polishing, etc.); micro-electric discharge machining; micro-electrochemical

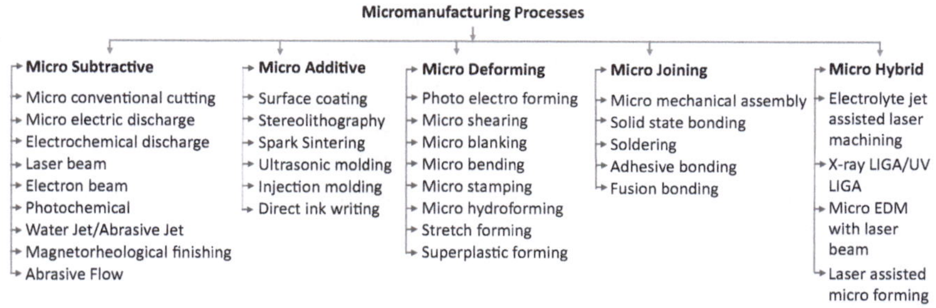


**FIGURE 17.1** Classification of micromanufacturing processes.

machining; electron beam machining; laser beam machining; photochemical machining; abrasive jet micromachining, magnetorheological finishing, and abrasive flow finishing, are the examples of subtractive machining processes. The digital control microdrilling process is one in which the microdrilling tool can have translatory and rotary motion. The microdrilling tool has a diameter of 50 μm, length of 500 μm (length to diameter ratio of 4 to 14), and a spindle speed of 2000 to 4000 rpm recommended for the drilling process. The carbide micro drills can be used to drill materials, such as iron, stainless steel, aluminium and non-ferrous alloys, and superalloys and hard steels.

### 17.2.3 Additive Manufacturing Processes

A transformative approach to manufacture objects by layer by layer creation can convey more design flexibility and productivity in the current scenario. With the application of computer-aided-design software or 3D scanner, the three-dimensional geometry can be created, after that the created geometry will be discretized into different shapes (triangle, square, tetragonal, hexagonal, etc.) before printing of original objects. The material is deposited by layer upon layer in an accurate symmetrical shape. The surface coating (chemical vapour deposition and physical vapour deposition); direct ink writing (ink-jet and laser-guided); micro-injection molding; micro-casting; sintering; and photo-electro-forming are also similar examples of additive manufacturing technologies. The fused deposition method is one of the additive manufacturing technologies. It is used for modeling and prototyping three-dimensional objects directly from a CAD design. Any thermoplastic materials such as nylon, acrylonitrile butadiene styrene, polycarbonates, polylactic acid, polystyrene, and urethane can be used as filament wire for FDM printers.

### 17.2.4 Deforming Processes

There are various applications of formed sheet-metal parts such as microsprings for microswitches, electrical connectors, and lead-frames, microcups for electron guns, micromeshes for masks and optical devices, and micropackaging, microgears for micro-mechanical objects, microlaminates for micromotor and fluidic devices, exteriors/coverings for microdevice assemblage/wrapping, microknives for surgical treatment, etc. The basic approach for the forming of macro- and micro-products include blanking, shearing, stamping, bending, spinning, deep drawing, stretch forming, hydro-forming, superplastic forming, age forming, explosive forming, incremental forming, hot-embossing, and imprinting, etc. It is generally difficult to forecast and control the deformation behavior of micro parts. The following factors, namely, workpiece size, grain size, flow stress, fracture behavior, elastic recovery, and friction, affect the performance of microforming products such as deformation load, dimensional accuracy, and surface finish.

### 17.2.5 Joining Processes

The joining of miniaturized products, such as micro-electro-mechanical systems (MEMS) is a very challenging task. With the increasing demands of innovative products in the marketplace, the microjoining of products/components becomes more difficult. The different types of microjoining methods, such as micro-mechanical assembly, soldering/brazing, adhesive bonding, solid-state bonding, and fusion welding are applied to join the micro components. Laser stands for light amplification by stimulated emission of radiation. Laser beam is utilized to form a weld between two similar or dissimilar metals or non-metals to join together in a process called laser welding process. The different types of lasers such as solid-state lasers, gas lasers, and fiber lasers are most commonly used in laser welding. Being a concerted heat source of laser beam in tiny materials, it can be performed at high speed of welding, and thicker materials can produce narrow and deep welds. Generally, single-pass welding is recommended for the laser joining process. Multi-pass layers create a more uniform microstructure in each layer, but it may exhibit imperfections in the form of porosity, lack of fusion, solidification cracking, and humping on the layer. Therefore, a minimal amount of material has been wasted during the laser welding process.

### 17.2.6 Hybrid Process

This is the combination of two or more micromachining processes taking the double advantages of the single process to produce fast and more accurate components/parts. The process has the combined effect of two processes in terms of energy and processing steps. The hybrid process has been classified into two categories (a) Mixed processes—in which both the processes are involved directly in the removal of material, and (b) Assisted-type processes—in which one of the processes is assisting the other process in material removal. Examples of hybrid processes are electrolyte jet-assisted laser machining, LIGA (lithography galvanoformung electroplating abformung molding) and LIGA mixed with laser beam machining; laser beam with micro-EDM assembly; laser beam with shape deposition machining; laser beam with micro-forming; laser beam with micro injection molding assembly; etc. In the case of electrolyte jet-assisted laser machining, a laser beam combined with an electrolyte jet are applied coaxially to a workpiece. The mechanism of material removal from the workpiece is electrochemical termination which is assisted by the laser beam conducted to the workpiece. The laser thermal energy improves the electrochemical reactions process by enhancing the kinetic energy to the localised heating zone. Hence, the material removal rate, surface finish, and dimensional stability of the workpiece is enhanced compared with the sole process.

## 17.3 COST ANALYSIS OF MICROMANUFACTURING PROCESSES

The manufacturing cost of microproducts greatly varies depending on the engineering processes used for the making of products. The manufacturing cost is classified into three categories, i.e., direct materials, direct labor, and overhead. These costs are varied depending upon the processes used to make the microparts. The manufacturing cost per part vs. a number of parts for different manufacturing processes, such as forming, subtractive, additive, and joining, is very high compared to conventional machining.

## 17.4 MICROMANUFACTURING MACHINES

The aim is to develop a micromanufacturing machine for the production of micro parts/components by the down-scaling of outdated machines. For instance, in the year 2000, the National Institute of Advanced Industrial Science and Technology, Japan, designed and developed a micro-lathe machine with a highly precise alphanumeric controller system. Further, another micro-lathe machine was designed by Nagano Techno Foundation consortium in 2003 with enhanced micro manufacturing features. The sizes of micro-lathe machine was 297 mm * 420 mm with a unique desktop-sized control panel. FANUC developed a micro-lathe cum compatible machine software with a high-precision multi-functional machine called ROBOnano in the year 2004, which can work in five-axis (three linear axes, and two rotational axes). Later, a five-axis milling, grinding, shaping machine was designed and developed. The University of Strathclyde, United Kingdom, developed a micro-forming machine in 2006. The micro-forming machine was driven by a servo motor controller. Nowadays, micromachines have the features of automatic machining with automatic inspection of parts and integration with assembly system/transfer lines.

## 17.5 NANOMANUFACTURING

The fabrication of parts/components, assembly of parts (structures), devices, and processing capabilities at the nanoscale is called nanomanufacturing; for example, nano-SIM, integrated circuits, and biomedical devices. It encompasses scaled-up, consistent, robust, and profitable nanomaterials manufacturing and integrating with top-down methods. The nanomaterials are processed at 10 to

1000 nanometers or less (for example, a human hair is about 2.5 nm in diameter and 1,00,000 nm in length) based on their small size (high surface to volume ratio). At this stage, the material properties (strength, durability, flexibility, self-lubrication, and also resistance to different environmental conditions with exciting temperatures and pressure) are getting further enhanced by avoiding/minimizing the defects in the final products.

The main purpose of nanomanufacturing is to improve the performance of materials by as much as ten times, as compared with the process by conventional manufacturing. It has an application in lightweight automobile vehicles and aerospace. Nanomaterials are especially in demand in the automotive and aerospace fields, with top applications being lightweight vehicles and spacecraft. According to National Nanotechnology Initiative, the reduction of overall weight of aircraft is directly proportional to its fuel consumption. A preliminary investigation accomplished by NASA shows that the replacement of conventional materials with innovative nanomaterials would reduce the gross weight of an aircraft by 63%. Nanomaterials are being used in electronics devices and sensors, solar panels, rechargeable battery, and also in electromagnetic shielding, and many more. Carbon nanotubes are being reinforced in polymer matrix for the development of light-weight composite materials for aerospace applications.

## 17.6 CHALLENGES IN MICROMANUFACTURING

The design of a miniaturized product is different from a product manufactured by a conventional machine. Products should be designed based on their functional requirements, simplicity, aesthetics, reliability, and high-volume production. Nearly distinctive problems need to be considered at the product design phase of micromanufacturing as follows:

### 17.6.1 Manufacturing Parameters

The manufacturing parameters depends upon the types of processes used to manufacture the microproducts. The conventional machine can be downsized at certain dimensions. Beyond certain dimensions, micromanufacturing machine plays a significant role such as the rigidity of the tools, vibration, the structure of the machines, tool-offset, temperature, and chip removal, in the development of microproducts.

### 17.6.2 High-volume Production and Automation

Process automation and control are also important challenges in microproduct manufacturing. The material loading and unloading on the machine, tool positioning and alignment, and machining processes on the raw materials are difficult to control manually. However, the manual process is also time-consuming and suitable for low volume production processes. In addition, manual adjustment of machine tools gives greater parallax error than computerized closed-loop and programmable regulators with error-compensation structures.

### 17.6.3 Limitation on Machinable Materials

Generally, ductile and soft materials are used to make the microproducts. The acceptable machining performance of ductile materials is effortlessly achieved due to their mechanical properties and they tend to deform straightforwardly under applied force by the cutting tool. On the other hand, the application of soft materials is very limited, and not meeting the increasing demands of steadiness and the long term service conditions. The machining of soft materials requires extra care/special attachment during operations. Thus, machining on the harder materials is the only option.

### 17.6.4 Tooling Dimension

The machine tool dimension is also one of the limitations of micromanufacturing. Milling and drilling tools of 10–50 μm have been successfully designed and developed for printed circuit board. Aspect ratios of 5 to 10 have been found suitable for micromachining. A deeper hole will result in tooling rupture.

### 17.6.5 Micro-part Handling

The main problem encountered is the precise positioning and control of microparts during processing. The exterior loads applied in physical contacts, such as the sticking or adhesion effect, electrostatic, and Van Der Waals force, have turned out to be key problems, and plenty of lessons have been learned to eliminate these forces. However, sensors of excessively large size are difficult to be positioned correctly on a workstation, therefore, controlling the level of accuracy is not achievable for microparts.

## 17.7 SUMMARY

The selection of materials and manufacturing processes for product development is a very challenging task for engineers/scientists. The development of materials is happening very fast, but manufacturing processes have not kept up with the existing materials. A lot of effort has been made to carry out machining at micro level like microforming, microjoining, micro-turning center, but this process is very time consuming and costly. A fully automatic machine is developed to fulfil the manufacturing demands of industrial need in terms of processes, tooling, materials handling, and transfer lines/assembly.

## REFERENCES

[1] Y. Zhou, A. Hu, From microjoining to nanojoining, *The Open Surface Science Journal*, 3 (2011) 32–41.

[2] V. K. Jain, Sidpara, R. Balasubramaniam, G. S. Lodha, V. P. Dhamgaye, R. Shukla, Micromanufacturing: A review- part I, Proc I Mech E Part B, *J Engineering Manufacture* (2014) 1–22.

[3] V. K. Jain, U. S. Dixit, C. P. Paul, A. Kumar, Micromanufacturing: A review-part II, Proc I Mech E Part B, *J Engineering Manufacture*, 228 (9) (2014) 995–1014.

[4] M. Geiger, M. Kleiner, R. Eckstein, et al., Microforming. *CIRP Ann: Manuf Techn*, 50 (2001) 445–462.

[5] A. R. Razali, Y. Qin, A review on micro-manufacturing, micro-forming and their key issues. *Procedia Eng*, 53 (2013) 665–672.

[6] Y. Zhou, *Microjoining and nanojoining*. Cambridge: Woodhead Publishing Ltd; Boca Raton, FL: CRC Press, (2008).

[7] M. Koc, T. Ozel, *Micro-manufacturing: design and manufacturing of micro-products*. Hoboken, NJ: John Wiley & Sons, (2011).

[8] Michael F. Ashby, Materials selection in mechanical design, Butterworth-Heinemann, ISBN 0750643579, (2000).

[9] I. Mys, M. Schmidt, Laser micro welding of copper and aluminium. *Proc SPIE*, 6107 (2006) 25–26.

[10] V. K. Jain, Micromachining: an introduction. In: V. K. Jain (ed.) *Introduction to micromachining*. New Delhi, India: Narosa Publishing House, (2010), pp.1.1–1.27.

[11] D. Dornfeld, S. Min, Y. Takeuchi, Recent advances in mechanical micro machining. *CIRP Ann: Manuf Techn*, 55(2) (2006) 745–768.

[12] T. J. Ko Abrasive jet micromachining. In: V. K. Jain (ed.) *Introduction to micromachining*. New Delhi, India: Narosa Publishing House, (2010), pp.4.1–4.28.

[13] R. S. Jadoun, Ultrasonic micromachining. In: V. K. Jain (ed.) *Introduction to micromachining*. New Delhi, India: Narosa Publishing House, (2010), pp.7.1–7.36.

[14] P. Cardoso, J. P. Davim, A brief review on micro-machining of materials. *Rev. Adv. Mater. Sci.*, 30 (2012) 98–102.

[15] X. Luo, K. Cheng, D. Web, F. Wardle, Design of ultraprecision machine tools with applications to manufacture of miniature and micro components, *Journal of Materials Processing Technology*, 167 (2005) 515–528.

[16] K. M. Li, S. L. Wang, Effect of tool wear in ultrasonic vibration-assisted micro-milling, *Proc IMechE Part B: J Engineering Manufacture*, 228(6) (2014) 847–855.

# 18 COMSOL Multiphysics

## *A Simulation Tool to Analyze the Processing of Biodegradable Biomaterials Used in Arthroplasty*

*Shivani Gupta and Apurbba Kumar Sharma*

## 18.1 INTRODUCTION

In the current scenario, the development of new artificial biomaterials is demanded for the curing of various severe diseases. These demands lead to many advancements in the processing of biomaterials. Due to technological development, highly advanced software is available, which assists in materials development by providing simulation as a virtual analyzing tool before their actual manufacturing. In biomaterial manufacturing, simulation software is truly required for doing various analyses to save material, energy, and time.

In the modern competitive era, many software have been developed for analyzing the properties of materials and their response under their working environment. Various software and simulation techniques including Ansys, MAT Lab, Molecular Dynamics, Abaqus, Finite Element Method, COMSOL Multiphysics, etc. are widely used in modeling and simulation purposes. COMSOL Multiphysics software is one of the advanced simulation tools for analyzing various biomaterials using various modules. Various modules assist to analyze the interaction of the material with the system at different interfaces. For example, a heat transfer module is used for analyzing heat interaction with various materials at different processing conditions. Similarly, fluid flow with heat transfer, electromagnetics, structural mechanics and acoustics, chemical, multipurpose, and interfacing modules are available to study by various analyses in material development.

Fundamentally, this chapter focuses only on artificial biodegradable materials used in arthroplasty. Magnesium, zinc, and their selected alloys, polyglycolic acid (PGA), poly lactic acid (PLA) and their copolymers, hydroxyapatite, bioglass, and their composites are used as biodegradable artificial materials. Magnesium and its

DOI: 10.1201/9781003364948-18

alloys are extensively used in biomedical applications owing to their particular characteristics such as high strength-to-weight ratio, low density, biodegradability, biocompatibility, and non-toxic metal-based biomaterials [1]. The amount of Mg present in the human body is 760 mg at birth, and 30–40 % of magnesium is preserved in muscles and soft tissues; subsequently, 1% is in body fluid and remains in the human skeleton [2]. Mg has osteoblast characteristics that help in the human tissue ingrowth process by providing many osteoconductive reactions between natural tissue and artificial implant. It has a low density approximately similar to human bone density, and low Young's modulus can eliminate the stress shielding problem. However, pure magnesium has poor corrosion resistance, and the degradation rate is much high; further, various metals are being added, such as aluminum, zinc, calcium, zirconium, etc. [3,4].

However, the processing of magnesium, and its alloys through advanced manufacturing techniques has become popular among researchers and technocrats. The reasons behind this are the challenges that ensued during the processing under various conditions (i.e., atmospheric and inert). Magnesium alloys have an affinity with oxygen present in the atmosphere and result in the formation of oxides [5]. Generally, inert gases and a mixture of $N_2$ and $H_2$ are used to avoid the formation of oxides of these alloys while casting and sintering [6]. Many studies have been done on the processing of magnesium, magnesium alloys, and its composites using conventional sintering and reported their outcomes in terms of mechanical properties, and microstructural and biological characteristics [7–10]. Other than this, many studies have been done on spark plasma sintering of magnesium alloys and their composites and demonstrated better mechanical and metallurgical properties than conventional sintering [11–13]. These emerging processes are needed for the advancement or development of new economic processes in the field of biomaterials. In recent years, microwave energy has emerged as an alternative source of heat for biomaterials processing. Synthesis and sintering of various biomaterials have been demonstrated using microwave energy at different frequency ranges (i.e. 915 MHz, 2.45 GHz, 5.8 GHz, and 24.124 GHz) and various power levels (700 W, 900 W, 1.45 kW, 3 kW, etc.) [14–17]. Moreover, microwave sintering has been used to process pure magnesium, its alloys, Mg/SiC/$Al_2O_3$, Mg/HA, and Mg/Zr/Ti composites, some of that are widely used as biomaterial [18–20]. However, hardly any report is available on the simulation of microwave sintering of pure magnesium and its alloys.

In the present work, microwave energy at 2.45 GHz was applied to sinter magnesium alloy AZ31 (green compact) inside a multi-mode microwave applicator. The microwave heating technique was used for heating the green compact. The surface temperature of the compact was monitored using a built-in infrared pyrometer during the irradiation. It was observed that the compact catches fire as it reacts with the oxygen present in the atmosphere, as a result of the burning of the compact due to excessive heating and oxide formation. The inert gas environment was used during the microwave exposure of the compact. This inert ambiance appreciably reduces the possibility of catching fire during heating. This processing technique was evaluated concerning the time reduction and energy competence of magnesium and its alloy heating. Further, the physics of microwave heating is explained through simulation work using COMSOL Multi-physics software. This software analyzed the temperature variation in the green compact at different times.

## 18.2 NECESSITY OF SIMULATION AND MODELING TOOL

Simulation and modeling tools play an important role in materials manufacturing industries due to their advantages of being able to simulate any manufacturing process as well as material interaction. Various simulation software provide a virtual platform to understand a material's behavior under various sets of stress conditions, temperature, heat transfer modes, chemical environment, etc. COMSOL multiphysics has several modules that are proven needed for industrial applications. The simulation tool can generate similar virtual conditions and apply various loads in different working conditions for analyzing the behavior of material without applying the material in the real working environment. COMSOL multiphysics can perform various studies at a time. Due to the presence of various modes in this software, the user can select more than two modules at a time and study the influence of different parameters, working mechanisms, and physics together.

## 18.3 METHODOLOGY

A 3-D model of the manufacturing process is required to design using a multiphysics COMSOL software tool. 3-D model design should contain the virtual process set up with some assumptions. Later, environmental conditions like inert or vacuum need to be considered. For example, Mg has a high affinity with oxygen present in the atmosphere, which is why in the simulation study inert atmosphere should be considered. On the other hand, the processing of polymers and HA/bioglass does not require any special environment. The model for microwave heating of biomaterial is built by integrating multi-domains like heat transfer and electromagnetic wave frequency modules. Simulation for microwave heating of the AZ31 is performed using frequency and time-dependent domain studies. The finite element method (FEM) is used to mesh the model and solve the simulation problem. Maxwell's equations are used as governing equations to analyze electromagnetic wave propagation inside the microwave cavity. These equations are used to evaluate the electric field (E) inside the microwave cavity and temperature distribution within the green compact. The following equations are used as the governing equation:

$$\nabla \times \mu_r^{-1}(\nabla \times E) - k_0^2\left(\varepsilon_r - \frac{j\sigma}{\omega\varepsilon_0}\right)E = 0 \tag{18.1}$$

Where, $\omega$, $\varepsilon_r$, $\varepsilon_o$, $\mu_r$, and $\sigma$ are angular frequency, relative permittivity, the permittivity of vacuum, relative permeability, and electrical conductivity of the target material, respectively. The $k_o$ is the wavenumber of free space, which is defined as:

$$k_o = \frac{\omega}{c_o} \tag{18.2}$$

Where $c_o$ is the speed of light in free space. The complex permittivity of any material is expressed as:

$$\varepsilon^* = \varepsilon' - j\varepsilon'' \tag{18.3}$$

Where $\varepsilon'$and $\varepsilon''$ is the real part and imaginary part of the permittivity, and specifically, in the case of metals, complex permittivity is expressed as:

$$\varepsilon_m^* = 1 - \frac{4\pi\sigma_m}{\gamma\left[\left(\frac{\omega^2}{\gamma^2}\right) + 1\right]} + \mathrm{j}\frac{4\pi\sigma_m}{\omega\left[\left(\frac{\omega^2}{\gamma^2}\right) + 1\right]} \tag{18.4}$$

Where $\gamma$ and $\sigma_m$ are the oscillation frequency and electrical conductivity of the metal, respectively. Electrical conductivity changes with the temperature condition of the metal expressed in Equation 18.5.

$$\sigma_m(T) = \frac{\sigma_o}{1 + \beta(T - T_o)} \tag{18.5}$$

Where $\sigma_0$ and $\beta$ are metal electrical conductivity at $T_0$ temperature and metal temperature coefficient, respectively. The 3D Cartesian plane Fourier energy equation can be expressed as Equation (18.6), i.e. used to estimate microwave energy dissipation inside the green pellet.

$$\rho c_P \frac{dT}{dt} + \rho c_P \mu \nabla T = Q + \nabla.\,(k\nabla T) \tag{18.6}$$

Where $\rho$, $c_{p,}$ k, and Q are density, the specific heat of the material, thermal conductivity, and energy dissipated, respectively. Microwave energy for the metallic material can be expressed as:

$$Q(r, T) = \frac{1}{2}[\omega\varepsilon_m''(r, T)\,|E(r, T)|^2 + \omega\mu_m''(r, T)\,|H(r, T)|^2 + \sigma_m|E(r, T)|^2 + \sigma_l|E(r, T)|^2] \tag{18.7}$$

Where E (r, T) and H (r, T) are the electric and magnetic field vectors of the electromagnetic spectrum at point r and temperature T. Whereas $\varepsilon''$ and $\mu''$ are dielectric and magnetic loss factors of metallic material and $\sigma_i$ the conductivity of ions present in metallic material.

In the modeling and simulation of any process, boundary conditions are required for calculating expected results. Subsequently, boundary conditions are implemented using two modules, i.e. heat transfer and electromagnetic frequency, to obtain estimated experimental results. For microwave cavity and waveguide, impedance boundary conditions are applied, i.e. expressed the minimum penetration:

$$\sqrt{\frac{\mu_o\mu_r}{\varepsilon_o\varepsilon_r - j\frac{\sigma}{\omega}}}n \times H + E - (n.\ E_s)m - E_s \tag{18.8}$$

Thermal analysis of the model is achieved using the module of heat transfer in solids and defined the zero flux at the thermally insulated boundary by equation 18.9, where n is the unit vector perpendicular to the metal surface.

$$-n.\ (-k\nabla T) = 0 \tag{18.9}$$

Further, boundary conditions for heat flux and surface diffusion are applied on the outer layer of the green pellet and susceptor, respectively to consider the heat transfer through convection and radiation from the surfaces. These heat transfers are expressed as;

$$-n.\ (-k\nabla T) = h(T_o - T) \tag{18.10}$$

$$-n.\ (-k\nabla T) = \varepsilon_{rad}\sigma_m(T_o^4 - T^4) \tag{18.11}$$

Where h and $\varepsilon_{rad}$ are the convective heat transfer coefficient and emissivity of radiant surfaces, respectively. These governing equations and boundary conditions are used to model and simulate the microwave heating of magnesium alloy. In this study, green pellets and surrounding crucible material are considered heat sources for modeling. The unavailability of simulation data related to the heating temperature of magnesium alloy; electric field (E) is considered as regulating parameter. Normal elemental size with a triangular shape is used in the meshing of the model. The meshing of the model is shown in Figure 18.1.

Thermal and electromagnetic properties of magnesium alloy, alumina, and refractory brick were placed in the model for starting time (t = 0) of the simulation. The simulation of heating of green pellet was carried out using frequency domain and time-dependent studies using microwave applicator of 2.45 GHz frequency at 1.4 kW. A time-dependent solver is run to monitor the temperature distribution of the green pellet at each time step to achieve the convergence criteria (Error = 0.02).

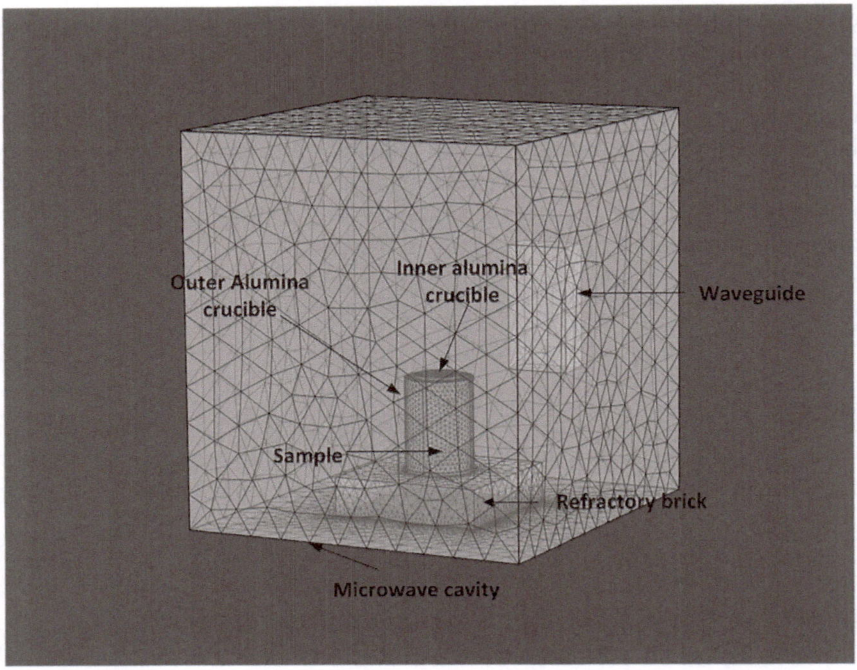


**FIGURE 18.1** Generated mesh in a 3D model of the microwave heating process.

## 18.4 DISCUSSION

In this study, magnesium alloy; AZ31 where A and Z represent aluminum and zinc and their weight percentage, i.e. 3 and 1, respectively, was used. AZ31 powder of particle size 40 ± 5 μm was used as the targeted material; the chemical composition of the material is given in Table 18.1. The green compact of this alloy powder was used as the sample for modeling. The dimensions of the compact are 13 mm in diameter and 10 mm in height. Other than this, Figure 18.2 shows the SEM micrograph of AZ31 powder.

**TABLE 18.1**
**Elemental Composition of Magnesium Alloy AZ31**

| | Chemical Composition (Wt %) | | | | |
|---|---|---|---|---|---|
| Material | Al | Zn | Si | Mn | Mg |
| AZ31 | 2.5–3.5 | 0.6–1.4 | 0.1 | 0.15–0.25 | Balance |

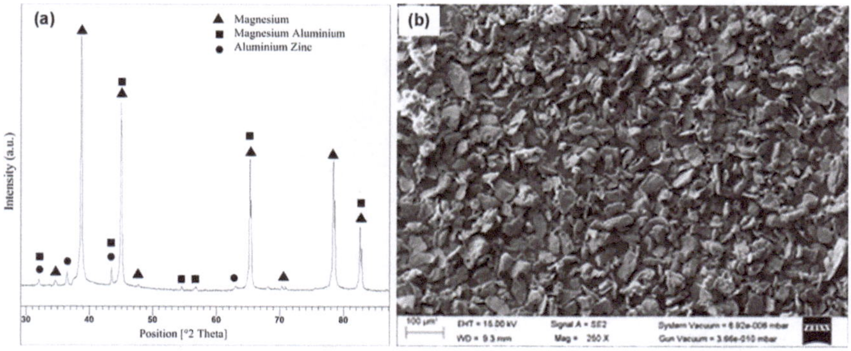


**FIGURE 18.2** (a) SEM micrograph and (b) XRD spectrum of AZ31 alloy powder.

A microwave applicator (Mode: multi-mode, Make: Enerzi Microwave Systems Pvt. Ltd. India) was used for experimentation. The process schematic comprised a base, alumina crucible, insulation, and green pellet. An alumino-silicate brick was used as the base and alumina crucibles are good absorbers of microwave above 700°C but below that temperature behave like transparent. The green pellet was heated in direct heating mode without using any susceptor. The simulation results also reveal that the heating of the compact is slow as compared to microwave absorber materials as the targeted material is a metal alloy (reflective in microwave). Therefore, it is needed to add some microwave absorber material to enhance the heating of the compact resulting in a reduction of the heating time and power consumption. Atmospheric condition is not suitable for the heating of magnesium alloy due to the high affinity of magnesium with oxygen. Therefore, an inert atmosphere (i.e. argon gas) was used for the safe processing of AZ31 alloy.

The distribution of the electromagnetic field and temperature in the green pellet was obtained through simulation. Moreover, the simulated time-temperature profile of microwave heating of AZ31 was drawn through the modeling software COMSOL multiphysics. Argon gas ambient condition during heating was used, and these simulated results of the simulation study are presented in the following sections.

### 18.4.1 Electric Field and Temperature Distribution in the AZ31 Green Compact

The electromagnetic field distribution inside the microwave applicator is shown as a column bar in Figure 18.3(a-d). It is evident from the results that with the variation in the exposure time, there are no changes in the electromagnetic field value. This may occur due to keeping the whole setup in the same place to the varying exposure time. Moreover, the same processing material is also the reason behind the electromagnetic field constant value

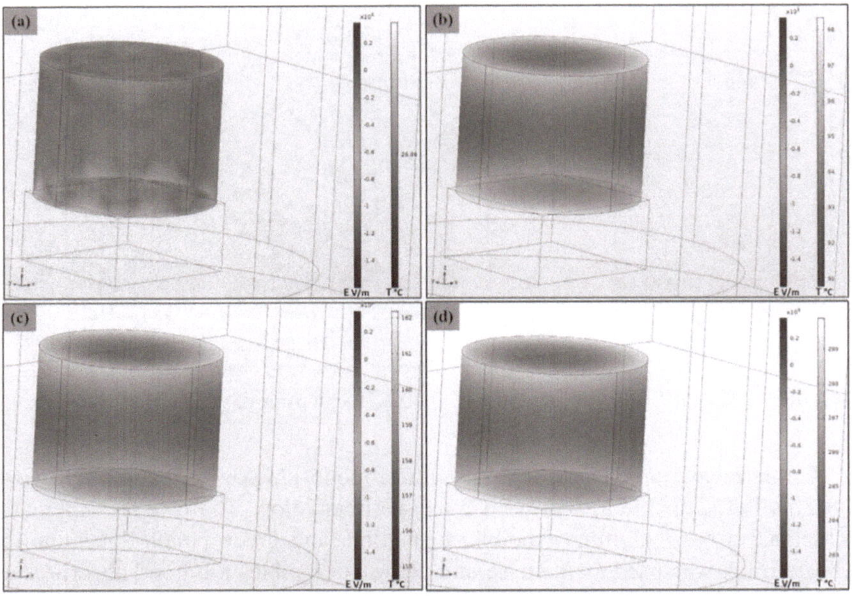

**FIGURE 18.3** Distribution of electric field and temperature inside the AZ31 metal green pellet at different exposure times; (a) t = 0 s, (b) t = 400 s, (c) t = 800 s, and (d) t = 1200 s.

inside the applicator. Although field intensities are almost similar in all cases, field distribution may be varied.

### 18.4.2 Time-Temperature Profile in Microwave Heating of AZ31 Green Compact

Time and temperature profile is an important factor in the heating of material as it deliberates the temperature of the material concerning the exposure time. The simulation result of the time-temperature profile is obtained from simulation as well as experimental study, as shown in Figure 18.4. From this profile, it can be said that it is a moderate heating process due to the high conductivity of the targeted material. It reduces microwave absorption and leads to modifying the material by adding some other material (microwave absorbing material like hydroxyapatite) for achieving uniform rapid heating during microwave heating process that is reported by Gupta et al. [21]. These all results were obtained from simulation work as the heating temperature of AZ31 is below 600°C. In the experimental study, the temperature is recorded after 250°C with the help of an IR pyrometer (temperature ranges from 250°C to 1800°C) assembled in microwave furnace. The time-temperature profile is data up to this temperature, the sample exploded due to non-uniform heating.

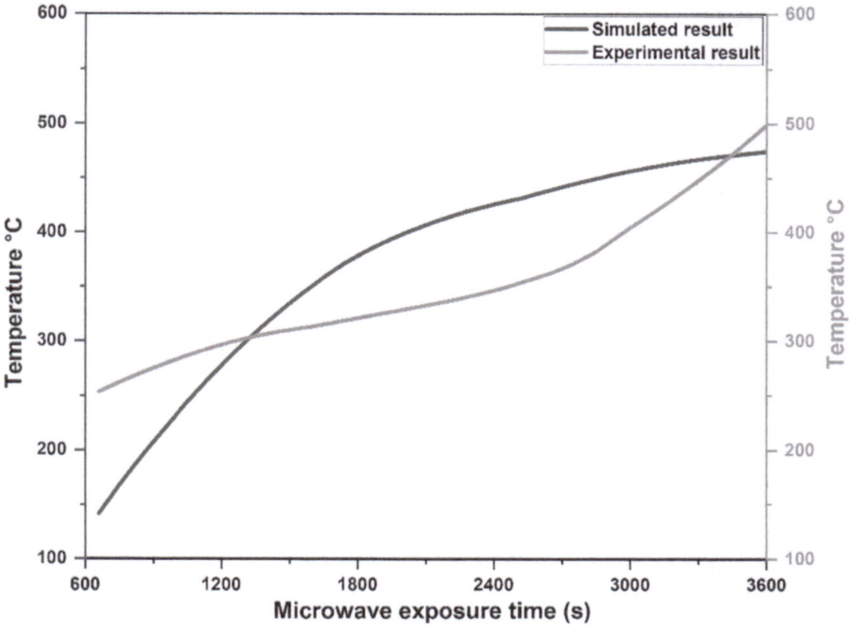


**FIGURE 18.4** Time-temperature profile of microwave heating of AZ31 green pellet.

## 18.5 CONCLUSIONS

Modeling and simulation are very important for the development of new artificial biomaterials and manufacturing processes. The simulation tool provides a basic idea of the behavior of the material in its actual application without implementation in the real environment. This chapter focused on the modeling and simulation of the development of biodegradable biomaterial used in arthroplasty, and the importance of requirement of simulation and modeling software in the material manufacturing field. In addition, simulation and experimental analysis of microwave heating of AZ31 alloy were studied. This alloy is widely used as a biodegradable material in arthroplasty due to its similar characteristics to human bone. The chapter outlaid ideas about tooling and setup designing of microwave heating of various magnesium alloys and their composites. The electromagnetic field has no variation if the setup or processing material is kept the same. Another idea about the selection of other heating processes like selective heating and hybrid heating is the use of susceptor. The effect of susceptor heating in hybrid mode microwave heating process on targeted material can be studied. In hybrid heating, low melting point material is complicated due to high heat generation in very short exposure time resulting in overheating and thermal gradients. Therefore, simulation and modeling tools can be very helpful in designing setups for the processing of material with various melting points as well as composites and functionally graded materials.

## REFERENCES


[1] S. Agarwal, J. Curtin, B. Duffy and S. Jaiswal, Biodegradable magnesium alloys for orthopaedic applications: A review on corrosion, biocompatibility and surface modifications. *Materials Science and Engineering: C* 68 (2016): 948–963.

[2] M. E. Shils and M. Shike, *Modern nutrition in health and disease*. 10th Eds. (Lippincott Williams & Wilkins, 2006).

[3] G. E. J. Poinern, S. Brundavanam and D. Fawcett, Biomedical magnesium alloys: a review of material properties, surface modifications and potential as a biodegradable orthopaedic implant. *American Journal of Biomedical Engineering* 2, no. 6 (2012): 218–240.

[4] M. Gupta and S. N. M. Ling, Magnesium, Magnesium Alloys, and Magnesium Composites, 1st ed. (New York: Wiley, 2010): 113–195.

[5] Y. Wan, C. Teng, L. Wei, L. Chunzhi, X. Jian, Z. Yong, J. Dehui, X. Guangyao and L. Honglin, Mechanical and biological properties of bioglass/magnesium composites prepared via microwave sintering route. *Materials & Design* 99 (2016): 521–527.

[6] R. R. Mishra and A. K. Sharma, On melting characteristics of bulk Al-7039 alloy during in-situ microwave casting. *Applied Thermal Engineering* 111 (2017): 660–675.

[7] X. Gu, Z. Weirui, Z. Yufeng, D. Limin, X. Yulin and C. Donglang, Microstructure, mechanical property, bio-corrosion and cytotoxicity evaluations of Mg/HA composites. *Materials Science and Engineering: C* 30, no. 6 (2010): 827–832.

[8] M. Gupta and E. W. W. Leong, *Microwaves and metals*. (New York: John Wiley & Sons, 2008).

[9] R. R. Mishra and A. K. Sharma, Microwave material interaction phenomena: heating mechanisms, challenges and opportunities in material processing. *Composites Part A: Applied Science and Manufacturing* 81 (2016): 78–97.

[10] R. R. Mishra and A. K. Sharma, A review of research trends in microwave processing of metal-based materials and opportunities in microwave metal casting. *Critical Reviews in Solid State and Materials Sciences* 41 (2016): 217–255.

[11] A. K. Sharma and R. R. Mishra, Role of particle size in microwave processing of metallic material systems. *Materials Science and Technology* 34 (2018): 123–137.

[12] S. Gupta and A. K. Sharma, Sintering of biomaterials for arthroplasty: A comparative study of microwave and conventional sintering techniques. *In Applied Mechanics and Materials* 895 (2019): 83–89.

[13] S. Gupta and A. K. Sharma. Microstructure and microhardness of Mg/SiC metal matrix composites developed by microwave sintering. *Journal of The Institution of Engineers (India): Series C* (2020): 1–6.

[14] R. R. Mishra and A. K. Sharma, Multi-physics simulation of in situ microwave casting of 7039 Al alloy inside different applicators and cast microstructure. *Proceedings of the Institution of Mechanical Engineers, Part E: Journal of Process Mechanical Engineering* 233(2019): 617–629.

[15] B. Basu, D. S. Katti and A. Kumar, *Advanced Biomaterials: Fundamentals, Processing, and Applications*, 1st ed. (New York: Wiley, 2010), 1–18.

[16] H. Zhou, J. F. L. Timothy and S. B. Bhaduri, Microwave assisted synthesis of amorphous magnesium phosphate nanospheres. *Journal of Materials Science: Materials in Medicine* 23 (2012): 2831–2837.

[17] D. Zhao, W. Frank, L. Faqiang, W. Jiali, L. Junlei and Q. Ling, Current status on clinical applications of magnesium-based orthopaedic implants: A review from clinical translational perspective. *Biomaterials* 112 (2017): 287–302.
[18] M. Gupta and W. L. E. Wong, Magnesium-based nanocomposites: Lightweight materials of the future. *Materials Characterization* 105 (2015): 30–46.
[19] K. Zhou, D. Chunfa, Z. Xianglin, S. Lei, C. Zhichao, X. Yanlin and C. Haoi, Preparation and characterization of nanosilver-doped porous hydroxyapatite scaffolds. *Ceramics International* 41 (2015): 1671–1676.
[20] A. K. Sharma and S. Gupta, Microwave processing of biomaterials for orthopedic implants: Challenges and possibilities. *JOM* 72, no. 3 (2020): 1211–1228.
[21] S. Gupta, A. K. Sharma, D. Agrawal, M. T. Lanagan, E. Sikora, and I. Singh. Characterization of AZ31/HA biodegradable metal matrix composites manufactured by rapid microwave sintering. *Materials* 16, no. 5 (2023): 1905.

# 19 Elastically Embedded Abrasives for Micro Finishing

## *Theory and State-of-the-art*

*V. S. Sooraj and V. Radhakrishnan*

## 19.1 BACKGROUND

Topology and roughness of surface play an important role in tribological performance, fluid flow, heat transfer, optical characteristics, corrosion and fatigue resistance, bio-compatibility, etc., of a manufactured component. Micron/sub-micron sized abrasive grits, relatively harder than the work material, have been identified as a good choice for fine finishing of engineering surfaces. State-of-the-art abrasive finishing methodologies and practices mainly include the application of abrasives either in rigidly bonded or loose (free) form [1]. Generally, they follow two-body or three-body abrasion principles to remove material from the surface.

Ultra precision grinding and honing are typical examples of using abrasives in such bonded form to impart fine finish on engineering surfaces. The abrasives are in the form of a rigidly bonded wheel for grinding and in the stick form in honing. Grinding wheels bonded with clay, resin, glass, and metal are widely used in many applications [2,3]. One of the major technical issues associated with bonded abrasive wheel is its accessibility limitations imparted by the work piece geometry, making them ideal for surfaces of revolution, flat surface, or geometrically well-defined surfaces like the gear tooth surface. There is also a technical challenge when addressing meso/micro scale components with intricate profiles and stringent finish requirements. Development of dedicated abrasive tool for each type of surface becomes the bottleneck. Surface/sub-surface level damage and probability of form error are other concerns during the interaction of bonded abrasives. Control of grinding force to eliminate the form error becomes an essential requirement in these situations to yield ultra-fine finishing.

Use of loose abrasives is also demonstrated to be effective for fine finishing of surfaces, classical examples of which include three-body abrasion in polishing operations. However, handling of fine abrasive powder is always a topic of concern. Based on the intricacy and geometrical requirements of end-use

 DOI: 10.1201/9781003364948-19

components, different modes were attempted for employing loose abrasives in action. Typical example include semi-solid mixture of abrasive and polymer media used in Abrasive Flow Finishing [4,5]. Such a media is prepared as a non-Newtonian mixture of viscoelastic polymer, abrasives, and processing oil with flowability and deformability. Typical types of media reported for abrasive flow finishing include (a) Silicon polymer based media, (b) Rubber based media, (c) Natural Polymer based media, and (d) Hydrogel based media. In the first type, abrasives mixed with silicon based polymer, metallic soap gel, and hydrocarbon oil has been attempted as the semi-solid carrier medium. Abrasion performance and finish achieved on the target surface were found highly sensitive to viscosity of such flowable polymer medium, where an increase in viscosity showed better performance [5]. The type 2 composition of silicon rubber-based media became a choice due to its higher viscosity in comparison with pure silicone. A mixture of silicon rubber, silicon oil, and abrasives has been reported as a typical selection in this category. Further, natural polymers (typically ester-based polymers) and appropriate processing oils (such as naphthenic based oil) as in type 3 and hydrogels (guar gem, etc.) mixed with rheology enhancers (such as silica and borax) as in type 4, were also attempted to carry abrasives to the finishing zone [5]. All these media compositions were found to be effective for various types of metallic surfaces including titanium alloys, additively manufactured (SLM) components, ceramics, metal matrix composites, etc. [5]. With the advancement of hybrid abrasive flow finishing techniques, attempts were also reported on the development of specialized abrasive flow media. An electrolytic abrasive media with the inclusion of poly propylene/poly ethylene glycol, which is capable of contributing to anodic dissolution of work material in addition to its flow based abrasion is a typical example [5]. Magnetorheological polishing media with carbonyl iron particles, which behave as a Bingham plastic in the presence of magnetic field or as a Newtonian fluid without magnetic field is another good example of custom-designed flow finishing media.

Though there are great contributions in the media development and hybrid approaches, the major question is its multi-utility. Major requirements from the end-user perspective can be summarized as follows;

- Development of a multiple application oriented, cost-effective, and environmental friendly medium, which can be customized for different mode of applications.
- Development of a media capable of fine finishing internal, external, flat, cylindrical, curved, grooved, and any other complex profiled surfaces without altering the geometric form of target surface and without damaging the sub-surface.
- Development of a media suitable for both abrasion as well as erosion based approaches for fine finishing.

For example, the semi-solid type flowing plastic may not be suitable for *erosion* based abrasive processing such as fluidized abrasive machining. Fluidization of loose abrasives and material removal through low-velocity impingement of work surface with these fluidized particles have been demonstrated as promising surface finishing methodology [6,7]. Though very fine abrasives are generally capable of yielding smooth finish, direct employment of such loose abrasives, especially with the size range less than 35 μm (which is classified as Geldart group C particles [8]), is nearly impossible to fluidize due to the restrictions of cohesiveness and agglomeration. A suitable mode to deploy fine abrasives becomes a necessity in this situation. However, the plastic medium described above may not be suitable for such applications, and thus there are demands for new solutions.

Another critical requirement identified is the necessity of controlling cutting (finishing) forces to refine the surface profile without altering the basic surface form and by minimizing the surface/ sub-surface defects [9]. Direct application of loose abrasives under a processing pad/tool may not be a good option to meet such a requirement in all type of work geometries, as the loading/penetration of abrasive edges into the work surface may depend fully on the pressure exerted by the pad/tool. If the abrasives can be carried by a medium that can conveniently deform in conformity to the work surface and control the cutting edge penetration, the concerns of cutting force as well as surface/sub-surface damage can be addressed effectively, especially in brittle materials.

The present chapter reviews the hypothesis of bonding loose abrasives using an elastomeric carrier medium, *referred as "elastically embedded abrasives or elastic abrasives" in literature*. Elastically embedded abrasives, in the form of meso scale spheres have been employed as the medium for multi-purpose fine finishing applications. The theory of such an elastic connection, its merits and state-of-the art applications have been covered in the chapter with appropriate examples and case studies from prior arts.

## 19.2 ELASTICALLY EMBEDDED ABRASIVE SPHERES: THEORY AND ILLUSTRATION

The idea of elastically bonded abrasive is to use fine abrasives embedded over an elastomeric polymer bead to form spherical shaped particles, referred as **Elastic Abrasives** in previous literature [10–21]. In its simplest form, elastically bonded abrasives comprise two major ingredients; (i) an elastomeric polymer base in the form of spherical bead and (ii) abrasive grits in the form of embedded layer. Though the present chapter considered meso scale (millimeter sized) spheres, they are not limited to the said shape and size ranges.

In general, the state-of-the-art abrasive finishing operation adopts either two-body abrasion, three-body abrasion, or erosion as its working principle. A conventional abrasive grit in such situations may act like a hard indenter getting forced/penetrated into the work surface, subsequently removing the material during its cutting motion. The maximum pressure ($P_c$) at the contact zone and its dependence on interaction parameters can be identified as follows [22–24].

$$P_c \propto \left[ \frac{F_n^{\frac{1}{3}} \; E_{eq}^{\frac{2}{3}}}{R_g^{\frac{2}{3}}} \right] \tag{19.1}$$

Where the abrasive grit of radius $R_g$ is pushed into the surface due to normal force $F_n$ in the contact zone. The term $E_{eq}$ denotes equivalent elastic modulus, which can be expressed as

$$\frac{1}{E_{eq}} = \frac{1 - \upsilon_w^2}{E_w} + \frac{1 - \upsilon_g^2}{E_g} \tag{19.2}$$

Here, "$E$" and "$\upsilon$" are the elastic modulus and Poisson's ratio with suffix "$w$" and "$g$" representing the work piece and abrasive grit, respectively.

The depth-of-cut (depth of penetration) established by a typical abrasive grit into the work surface depends on the contact pressure mentioned above.

Due to the presence of elastomeric polymer in the proposed sphere, their modulus of elasticity is significantly lower than conventional abrasive grit. As a result, the application of elastically embedded abrasive spheres causes significant reduction in equivalent elastic modulus and thereby reduces the contact pressure and penetration depth of embedded abrasive grains.

This action of elastomeric medium is similar to the action of a spring, which makes the elastic spheres deform in accordance with the surface form of target work piece, significantly minimizing the penetration depth for embedded abrasive grains. This is a unique feature of the proposed elastic abrasive, which will impart a fine finish on the target surfaces without altering their geometric form. An ideal model for such an elastically embedded abrasive sphere is shown in Figure 19.1, where the deformation of sphere can be simulated using a spring. Upon unloading, the deformed spring will get released and the elastic sphere will fully regain its original shape.

The ideal spring model may not be fully valid for all the possible options of polymer media. To account this factor, the elastic spheres can be considered as a Hookean Spring-Newtonian dashpot system as shown in Figure 19.2, where the spring characterizes the elastomeric action and damper symbolizes viscous behavior. In this hypothesis, the behavior of elastic abrasive is represented as a Kelvin-Voigt material [26]. As the elastic abrasive gets loaded to work surface, it will deform as shown in segment A-B of the characteristics curve where the strain on spring and damper symbolizes the energy absorption. Unlike the rigid action of conventional abrasive grit of same size, such an action will facilitate the deformation of elastic abrasive in conformity to work surface. The deformation and energy absorption by polymer media results in significant reduction of equivalent elastic modulus and depth of penetration by the grain. Thus the material removal will be gentle in comparison with conventional grit, which enables ultra-fine surface finish without altering original surface form.

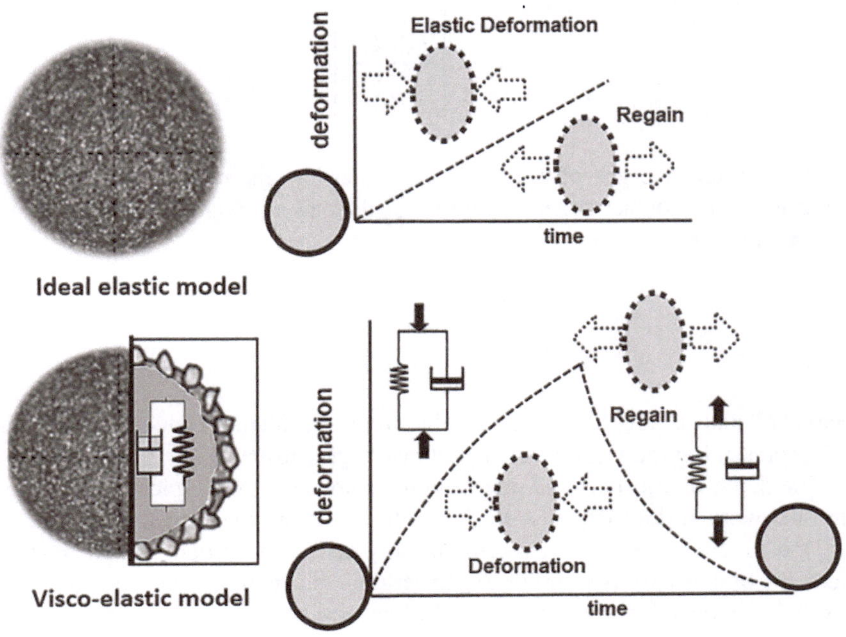


**FIGURE 19.1** Theory of elastically embedded abrasives.

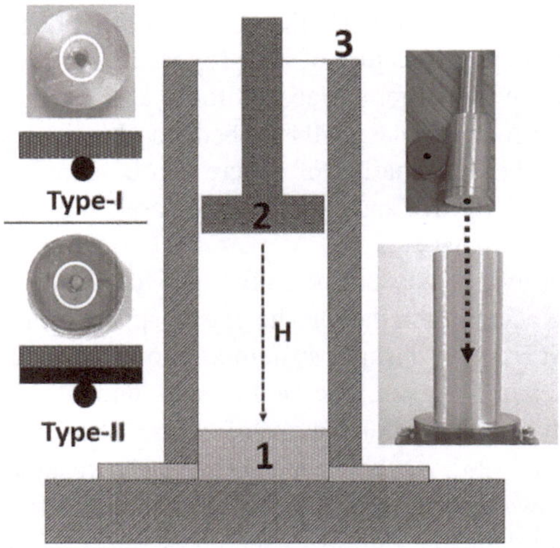


**FIGURE 19.2** Schematic representation of grit dropping experiment.

To demonstrate the reduction of equivalent elastic modulus, contract pressure and depth of penetration by the application of elastomeric medium, a simple grit-dropping experiment is presented here. An abrasive grit ($Al_2O_3$ grit) attached to a holder, either in Type-I or Type-II configuration, is dropped to a metallic target from a fixed height through a guide tube as shown in Figure 19.2 [18].

Type-I presents the direct attachment of abrasive grit on holder, whereas Type-II is with the backup of a rubber pad to simulate the elastomeric action. Stainless steel SS304 has been considered as the target, and the holder was dropped from two different heights, H = 0.1 and H = 0.2 m. The depth of crater created by the abrasive contact was measured using non-contact optical 3D profiler (Taylor Hobson Make Talysurf CCI Lite), and typical results are presented in Table 19.1.

The depth of crater measured during this experiment was clearly indicating the effectiveness of elastomeric medium. There was a significant reduction in erosion depth, while the abrasive grit was backed by a rubber pad. The hypothesis of spring-damper analogy and energy absorption has been illustrated clearly in this experiment.

**TABLE 19.1**
**Effect of Elastomeric Medium During Grit Dropping Experiment**

| Condition | Depth of Crater (μm) | Surface Image |
|---|---|---|
| Without elastomeric backup Dropping height = 0.2 m | 33.3 ± 2 μm | 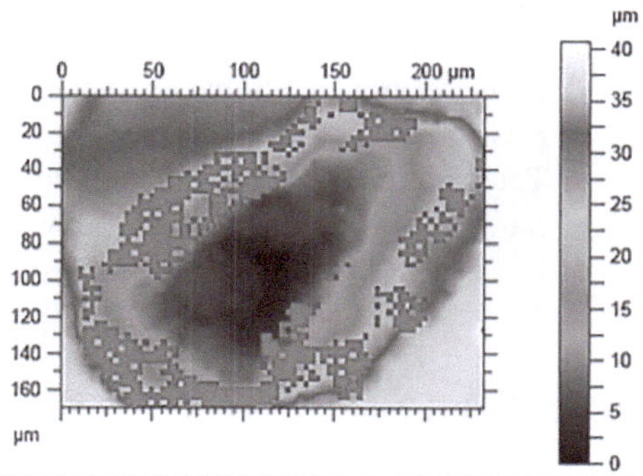 |
| With elastomeric (rubber) backup Dropping height = 0.2 m | 18.0 ± 1.5 μm | 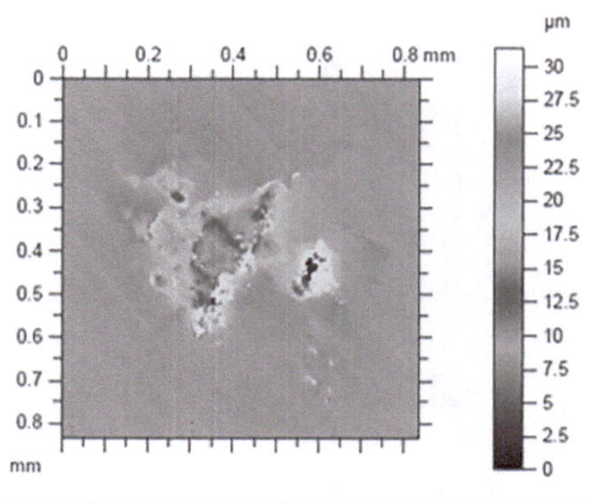 |

## 19.3 ELASTICALLY EMBEDDED ABRASIVE SPHERES: COMPOSITION AND PREPARATION

As mentioned earlier, elastically embedded abrasive spheres (Elastic Abrasives) are prepared by embedding abrasive grains on the surface of an appropriate elastomeric polymer bead. Typical method adopted to prepare elastically embedded abrasive spheres include the following 4 steps, which are shown schematically in Figure 19.3 [10–21,25].

Step 1. Selection/preparation of appropriate elastomeric polymer beads
Step 2. Surface softening of polymer beads using an appropriate chemical solvent
Step 3. Embedding of abrasives over the surface softened polymer beads
Step 4. Appropriate heating/ drying cycle to facilitate the embedding process

### 19.3.1 Selection of Elastomeric Polymer [10–21,25]

Desired features that guide the selection of polymer for elastic abrasive include;

- High resilience and elastomeric characteristics
- Low shore hardness and flexural modulus
- Ability to surface soften using an appropriate organic solvent
- Reasonably good impact resistance, abrasion resistance, and durability
- Resistance to corrosion
- Temperature stability to overcome fusing/melting, during the application
- Safety, biocompatibility, and environmental acceptance

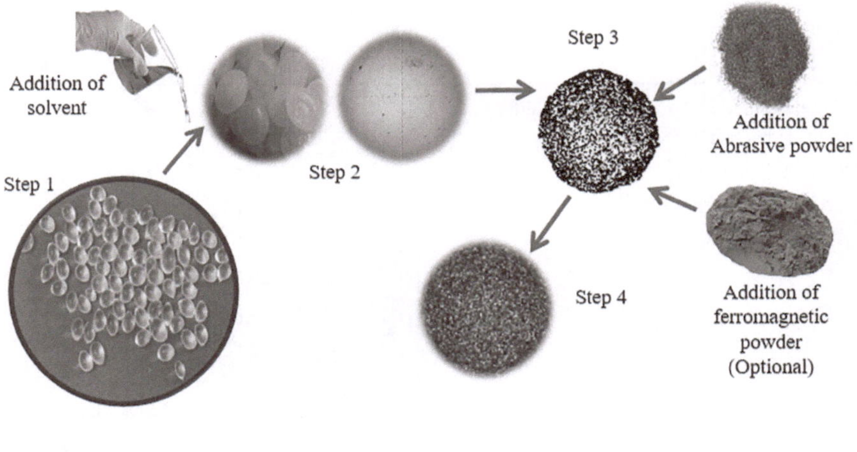


**FIGURE 19.3** Preparation of elastically embedded abrasives.

Typical choice of polymer for the experiments presented in this chapter include a medical grade-polyether based thermoplastic polyurethane with relatively low shore (shore A) hardness.

### 19.3.2 Selection of Organic Solvent [10–21,25]

The organic solvent is expected to produce surface softening of polymer beads, to facilitate effective embedding of abrasive grains. An environmentally safe, non-health hazardous, relatively high volatile organic solvent; capable of softening the selected polymer bead at normal operating environment; is generally recommended for the preparation of elastically embedded abrasive spheres.

### 19.3.3 Selection of Abrasives [10–21,25]

Any type of abrasive, from the group of natural and synthetic abrasives that are commonly employed for finishing applications, can be selected for the preparation of elastically embedded abrasive spheres, based on the combination of polymer and organic solvent. The size of abrasive grits can be chosen as per the size of polymer beads, depending on the level of finish requirements; with the hypothesis of selecting fine sized grains for better surface finish. Typical size range attempted for the experiments presented in this chapter include 10 to 250 μm.

## 19.4 CLASSIFICATION OF ELASTICALLY EMBEDDED ABRASIVE SPHERES

The main classification of elastic abrasive is based on its magnetic response, which include *non-magnetic (traditional)* and *magnetic type* elastic abrasive spheres. The magnetic version is prepared using the same methodology described above, with some ferromagnetic particles (typically, fine iron powder) as additional ingredient to make the particles magnetically responsive. The application of such a classification is discussed later in the processing methodology section 19.5.

In addition, the elastic abrasive spheres can be of *meso* or *micro* scale depending on their size range. Meso-scale elastically embedded abrasive spheres of diameter range 3 to 4 mm (with density nearly 1.8 g/cc) can be prepared as per the methodology described in section 19.3. Attempts were also made to develop micro scale elastic abrasive (in the size range 700–900 μm) using specially devised precipitation method, which is covered in the detailed specifications of Indian Patent granted for elastic abrasive spheres [25].

Further, it can also be noticed that the elastically embedded abrasive spheres can be prepared using special ingredients in the form of additives to control the elastomeric characteristics. Specially processed rice-husk is a natural additive attempted to alter the elastomeric characteristics in a typical situation. Though this ingredient, overall elastomeric action was altered and the particles were used for some special texturing/ deburring applications. Hence the elastically embedded abrasive spheres can also be classified based on the ingredients used for their preparation.

## 19.5 SCOPE AND APPLICATIONS OF ELASTICALLY EMBEDDED ABRASIVES: STATE-OF-THE-ART

Major modes of using elastically embedded abrasive spheres described in this chapter have been presented in Table 19.2 [10–21].

### 19.5.1 Mode-1: Abrasion via Direct Loading and Cutting Motion {Reference: [10,14,15]}

This mode of processing is generally recommended for internal surfaces, where the elastically embedded abrasive spheres can be filled/loaded inside the work-space followed by an axial squeezing of this filling as shown in Figure 19.4. This axial squeezing will make the elastically embedded abrasive spheres deform in conformity to internal surface form, which imparts radial contract pressure ($P_{radial}$) that penetrate embedded abrasive grains on the deformed elastomeric

**TABLE 19.2**
**Modes of Application for Elastic Abrasive Spheres**

| | |
|---|---|
| Mode 1 | Controlled abrasion via direct loading and cutting motion of elastically embedded abrasive spheres<br>Principle: Combined two body and three body abrasion |
| Mode 2 | Erosion mode via fluidization of elastically embedded abrasive spheres<br>Principle: Erosive wear caused by low velocity impingement in a fluidized bed |
| Mode 3 | Abrasion using flexible brush of elastically embedded abrasive spheres<br>Principle: Three body abrasion using magnetic type elastically embedded abrasive spheres |

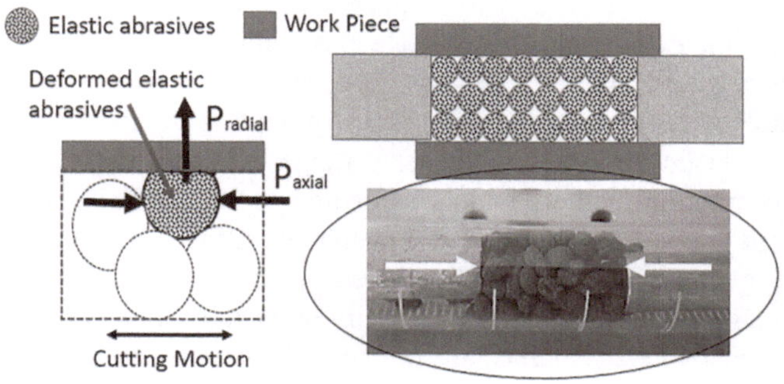


**FIGURE 19.4** Mode-1 application of elastic abrasive spheres: Abrasion based processing.

beads into the work surface. Such a package (filling) of penetrated grains while undergoing linear cutting motion will remove the material in finer quantities. Mechanics of material removal associated with this mode is combined effect of two-body and three-body abrasion. The package (filling) of elastically embedded abrasive spheres will act as an abrasive body-1 interacting with work surface (body-2) in two-body mode, along with some possible free movements of elastic spheres in three-body mode within the package (depending on axial pressure).

In this mode, the penetration depth imparted by an embedded abrasive grain will be a function of *radial contact pressure ($P_{radial}$), type and size of embedded grains*, as well as *hardness/other mechanical characteristics of work material.* As described in section 19.2 equation 19.1, the radial contact pressure depends on equivalent elastic modulus established by elastically embedded abrasive spheres, which is a function of elastomeric characteristics of selected polymer bead.

Volume of material getting removed by single embedded abrasive grain in such a mode can be expressed as;

$$V_{w/g} = \alpha\left(\frac{F_r}{H_w}\right)^2 \tag{19.3}$$

Where $F_r$ is the radial force on embedded abrasive grit and $H_w$ is the hardness of work material [9,14].

After evaluating the material removal volume per grain, the count of active grains in the full package (filling) of elastically embedded abrasive spheres can be assessed, and total volume of material removal for a given processing time can be calculated. A detailed mathematical model for material removal and related analysis for this mode-1 elastic abrasive processing have been presented by the authors in previous publications [14,15].

As a summary, major process variables that may influence the depth of grain penetration and material removal rate in the proposed methodology include *(i) Axial squeezing pressure, (ii) Type and elastomeric characteristics of polymer beads, (iii) Selection and size of abrasive grits, (iv) Cutting velocity, (v) Work material characteristics and Initial surface topography.*

#### 19.5.1.1 Possible Machine Configuration for Mode-1 Processing {Reference: [10,14,15]}

Pneumatic finishing setups consisting of a work holding fixture (W) with piston-cylinder arrangement (double acting type) on both sides of it (C1 and C2), as shown in Figure 19.5, have been developed to illustrate mode-1 processing methodology. The elastically embedded abrasive spheres filled inside work piece is squeezed using pistons from either side at appropriate axial pressure, and then reciprocating the whole package (filling) of elastically embedded abrasive spheres to and fro for the desired processing time.

Figure 19.5(a) shows the pneumatic machine developed for different types of internal surfaces such as geometrically symmetric internal bores, geometrically

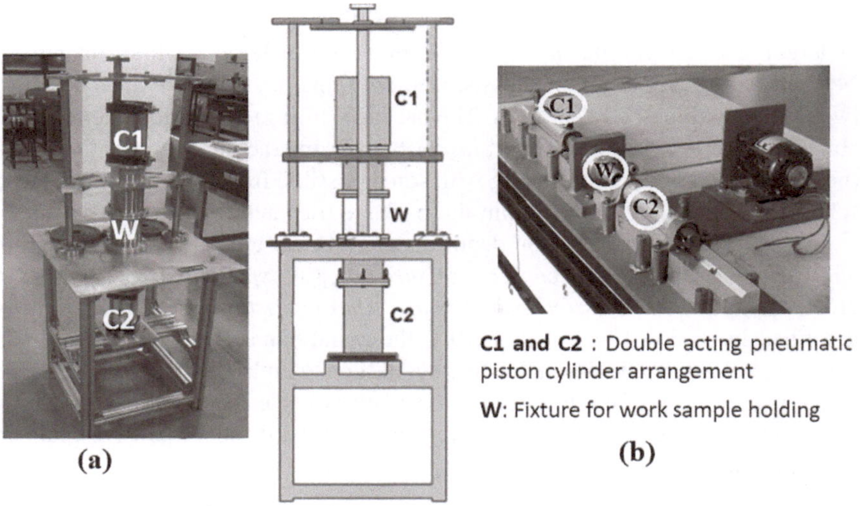


**FIGURE 19.5** Finishing machines for mode-1 processing using elastic abrasive spheres.

asymmetric internal bores, internal circumferential grooves etc. Figure 19.5(b) shows a lab scale version of similar finishing system with an additional feature of rotating the work piece as and when required. Typical case studies of experimentation using mode-1 processing methodology is presented below.

#### 19.5.1.2 Mode-1: Experimental Case Study 01: Fine Finishing of Geometrically Symmetric Internal Surface of Steel Tubes/Bushes {Reference: [10,14,15]}

Finishing experiments were performed using the setup shown in Figure 19.5, for the aforementioned sleeve and the results obtained are presented below. Here the experiments were performed using the principle mentioned in section 19.5.1, with and without rotation of work piece, to illustrate the capability of proposed methodology and machines.

Typical experiment considered the samples of hardness 58 HRC with initial roughness Ra = 0.18 μm, Rp = 0.60 μm, and Rt = 1.96 μm. Elastically embedded abrasive spheres with SiC particles of 10 μm were considered with an axial squeezing pressure of 0.3 MPa and linear cutting velocity of 8.0 m/min. The roughness after 45 minutes of processing without rotation was of the order Ra = 0.038 μm, Rp = 0.11 μm and Rt = 0.73 μm. As the work piece was rotated at 600 rpm, the roughness values were typically Ra = 0.02 μm, Rp = 0.06 μm and Rt = 0.25 μm, showing further improvement in finish [15]. Thus the average roughness (Ra), Peak height (Rp), and Peak to valley roughness (Rt) were showing significant reduction from its initial value as presented above. Typical photograph of sample surface after mode-1 finishing is shown in Figure 19.6.

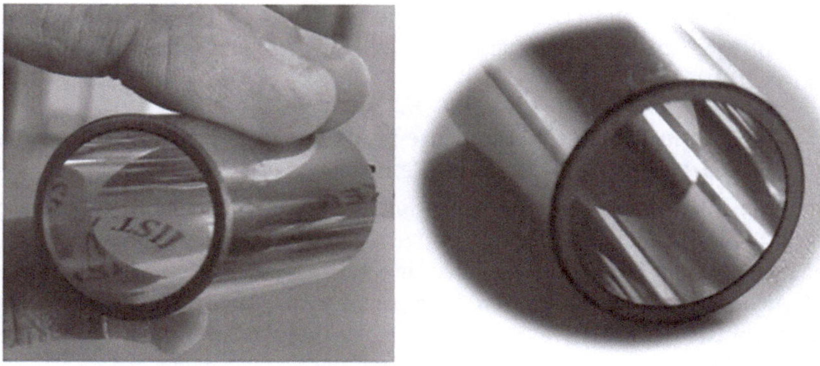

**FIGURE 19.6** Typical internal surface after mode-1 processing using elastically embedded abrasives.

A detailed discussion on process variables, range of selection, and influence on surface finish associated with this mode of processing in the machine configuration shown above is discussed by the authors in previous literature [10,14,15].

#### 19.5.1.3 Mode-1: Experimental Case Study 02: Fine Finishing of Geometrically Asymmetric Internal Surface {Reference: [10,19,20]}

The processing mode-1 can also be utilized for geometrically non-symmetric surfaces shown in Figure 19.7. In these samples, the fixture to accommodate work samples and the extension rods for reciprocating pistons (to squeeze the filing of elastically embedded abrasive spheres) are custom made as per the cross-section of selected sample part. For such type of cross-sections, *only reciprocating linear motion of axially squeezed filling of elastically embedded abrasive spheres need to be used, without switching on the rotation of work piece*. Typical experimental results for sample part-2 (with initial Ra = 0.25 μm) shown an average roughness of 0.03 μm after 50 minutes of processing, with

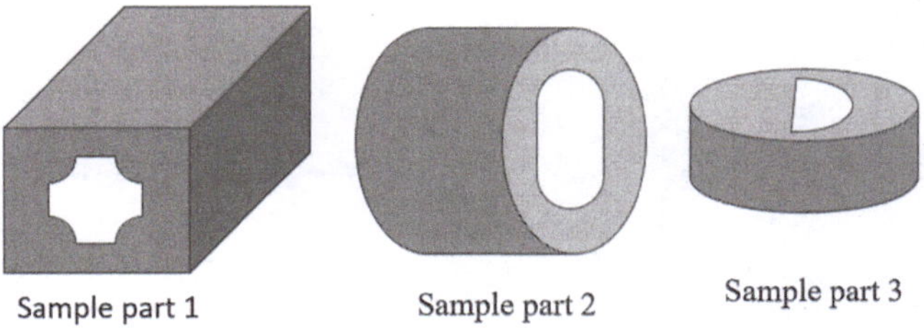


**FIGURE 19.7** Geometrically asymmetric test specimens for mode-1 elastic abrasive processing.

significant reduction in Rt (2.5 μm to 0.28 μm) and Rz (1.2 μm to 0.16 μm). More details of these experiments can be referred from related publication by the authors [10,19,20].

#### 19.5.1.4 Mode-1: Experimental Case Study 03: Fine Finishing of Internal Circumferential Groove {Reference [10,21]}

One of the unique capabilities of processing mode-1 is its adaptability for micro-finishing of internal circumferential groove as shown in Figure 19.8, which is not easy in traditional methodologies. For this kind of a surface, ***only rotation of work piece without linear motion*** is imparted to the filling of elastically embedded abrasive spheres (reverse to case-2). A typical result on such a test sample is given in Figure 19.9. More details of these experiments can be referred from related publication by the authors [10,21].

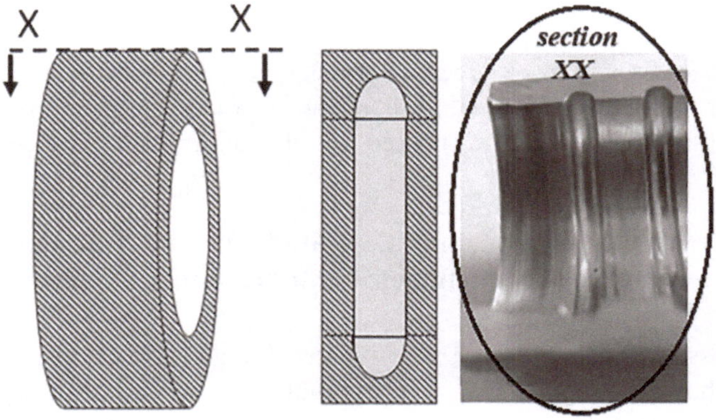


**FIGURE 19.8** Typical internal circumferential groove profile for Mode-1 elastic abrasive processing.

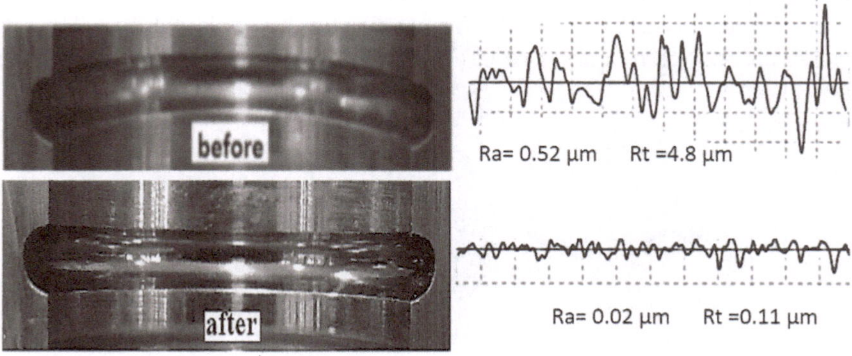


**FIGURE 19.9** Typical results of mode-1 processing applied to internal grooves shown in Figure 19.8.

### 19.5.2 Mode-2: Erosion via Fluidization of Elastically Embedded Abrasive Spheres {Ref: [10,16–18]}

In this mode of processing, elastically embedded abrasive spheres are used in a fluidized state and the target work surface is exposed to such a fluidized chamber to facilitate low velocity impingement of elastically embedded abrasive spheres on work surface (Figure 19.10). Low velocity impingement of conventional abrasive grits has already been reported as a promising methodology for finishing engineering surfaces [6]. However, it is highly limited to higher size ranges of abrasive grits, as it is difficult or almost impossible to fluidize abrasive grains of size less than 35µm (referred as Geldart group C particles) due to their high cohesiveness and agglomeration. Use of elastically embedded abrasive spheres helps to overcome this barrier, as it is capable of deploying very fine abrasive grits in the form of embedded grains on a carrier bead with an overall size greater than Geldart group C category. Such an option facilitates easy fluidization of meso-scale elastically embedded abrasive spheres, at the same time employ very fine micro-scale abrasives for impingement and erosive cutting action. The contact characteristics of elastic abrasive in an erosive environment has already been demonstrated in section 19.2.

The embedded abrasive spheres will reduce the equivalent elastic modulus during impact due to their high energy absorbing capability. The energy stored by elastomeric medium will be released during the rebound, effectively increasing the restitution coefficient (rebound velocity to impingement velocity). With this increase in coefficient of restitution, energy transferred to work piece for erosion will get decreased and corresponding material removal will also be

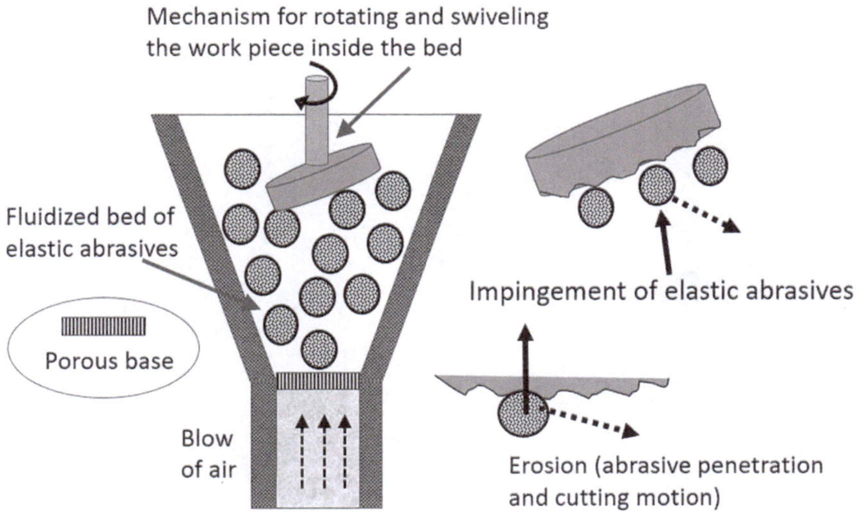


**FIGURE 19.10** Schematic of mode-2 elastic abrasive processing: Erosion via Fluidization.

delicate. Thus the elastically embedded abrasive spheres may produce very mild erosion compared to a standard abrasive grit of similar size and properties, which is the key merit of this mode-2 processing approach. Coefficient of restitution of the elastically embedded abrasive spheres will be a function of their size and characteristics, velocity of impingement, mechanical properties of target work material, etc. A detailed mathematical model on this impact and material removal aspects have been covered in reference [17].

Parameters that may influence the surface finish in this mode of processing include *(i) the ratio of fluidized bed height to static bed height, (ii) rotational speed of work specimen, (iii) size and density of elastically embedded abrasive spheres, (iv) work material characteristics, (v) size of embedded abrasive grain, (vi) angle and velocity of impingement etc.*

#### 19.5.2.1 Possible Machine Configuration for Mode-2 Processing {Reference [10,16–18]}

A fluidization chamber maintained by air supplied from an appropriate blower unit as shown in Figure 19.11 is the major part of proposed machine configuration for Mode-2 processing. The work piece can be rotated/swiveled inside the fluidized

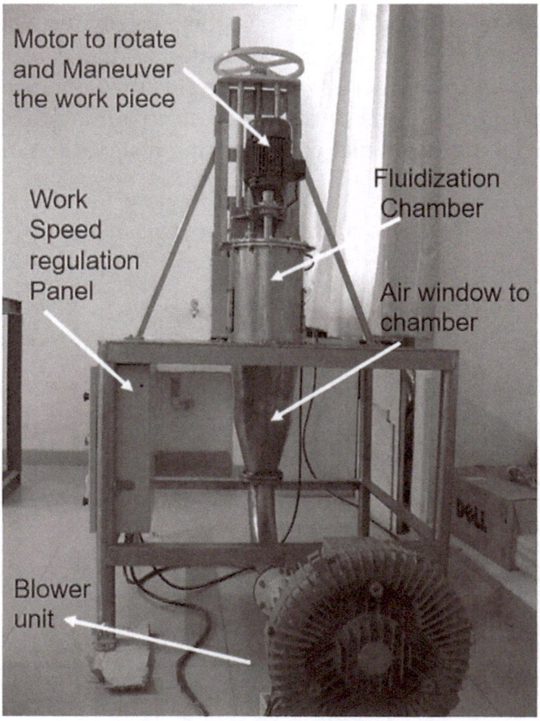


**FIGURE 19.11** Finishing machine for Mode-2 processing of elastically embedded abrasives.

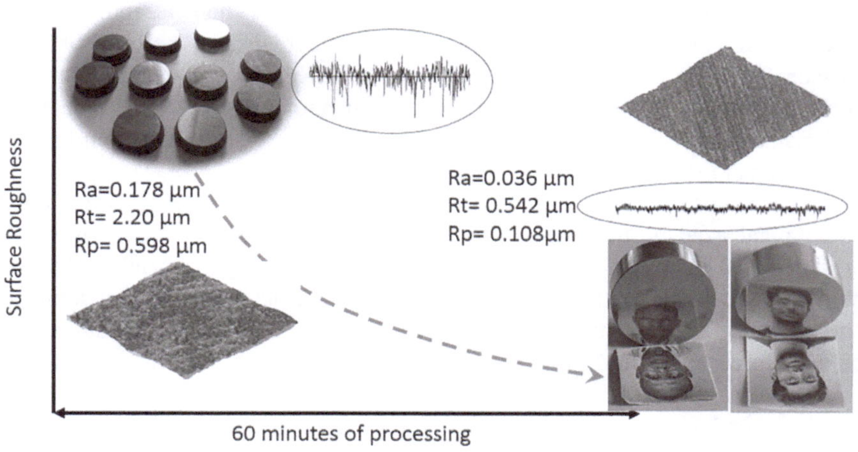


**FIGURE 19.12** Results of mode-2 elastic abrasive processing via fluidized erosion.

chamber using an appropriate clamping/holding mechanism driven by an electric motor, facilitating the selection of variable speed ranges for work rotation.

A detailed discussion on process variables, range of selection, and influence on surface finish associated with this mode of processing in the machine configuration shown above can be referred from previous publications of the authors [10,16–18].

#### 19.5.2.2 Mode-2: Experimental Case Study 01 {Reference [10,16–18]}

Attempt was made to finish flat hardened steel specimens using abrasive embedded elastomeric spheres of size 3.0 mm in a fluidized chamber. Blow of air was controlled in such a way that the ratio of fluidized bed to static bed was maintained as 2.0, with work piece rotated at 400 rpm inside this fluidized bed. After 60 minutes of processing, the improvements in roughness parameters observed are shown in Figure 19.12.

### 19.5.3 Mode-3: Abrasion Using Magnetic Type Elastically Embedded Abrasive Spheres {Reference [10–13]}

In this mode of processing, magnetic type elastically embedded abrasive spheres are held together like a flexible brush which is loaded and moved over the surface to be finished. Magnetic type elastically embedded abrasive spheres include ferromagnetic particles in addition to polymer-abrasive combinations. Chain of such magnetic balls formed like a flexible brush in the present case is referred as *elasto-magnetic abrasive brush* [10–13]. After establishing the contact between flexible brush and work surface, it can be rotated and moved over the surface to impart the cutting motion. When the magnetic abrasive brush is in contact with work surface, the cumulative action of mechanical and magnetic forces makes the elastomeric ball to deform in accordance to the form of work surface. This action will make the embedded abrasives to get penetrated into the surface, as equivalent to depth of cut. As the brush rotates, combined mechanical and magnetic force in tangential direction will create the shearing action leading to micro/nano removal of materials.

As mentioned already in equation 19.1: section 19.2, normal force imparted by a typical sized elastic abrasive will be a function of equivalent elastic modulus at the contact interface, which depends on the type of polymer bead, embedded abrasive and work material characteristics. As elastically embedded abrasive spheres pose unique capability of reducing equivalent elastic modulus, the contact force in normal direction and thus the depth-of penetration are relatively low.

The material removal in this mode is a combination of two-body and three-body abrasion. Since the elastic abrasive balls are held like a flexible brush between magnet and work piece, it is free to move/rotate as a third body during the finishing action. Such a movement/ rotation will facilitate embedded abrasives grains on different locations of elastic sphere to come in the finishing action. Holding strength of elastically embedded abrasive spheres in the brush depends on the flux density of magnet and thus the two-body abrasion depends on the characteristics of magnetic field. The abrasion in this mode of processing also depends on the rotational speed of elasto-magnetic brush.

Key variables that may affect the finishing characteristics in mode-3 processing include *(i) flux density of magnet selected for magnetic abrasive brush, (ii) rotational speed of magnetic abrasive brush, (iii) magnetic characteristics of elastically embedded abrasive spheres, (iv) characteristics of work material, (v) size of embedded abrasive grains etc.*

#### 19.5.3.1 Possible Machine Configuration for Mode-3 Processing {Reference [10–13]}

The magnetic mode of elastic abrasive processing was attempted using a vertical milling machine, with a simple modification of tool holding unit. A magnet which can offer appropriate flux density can be mounted on the machine spindle using a suitably designed mandrel allowing its rotation at selected speeds. Such a tool unit can attract magnetic type elastically embedded abrasive spheres to form the elasto-magnetic abrasive brush.

A detailed discussion on process variables, range of selection, and influence on surface finish associated with this mode of processing in the machine configuration shown above can be referred from the authors' publications [10–13].

#### 19.5.3.2 Mode-3: Experimental Case Study 01 {Reference [10,13]}

Attempts were made to finish a flat surface using an elasto-magnetic abrasive brush as shown in Figure 19.13. A magnet of flux density 0.3 Tesla was mounted at the tip of tool mandrel, and the same was rotated at 400 rpm. The flexible

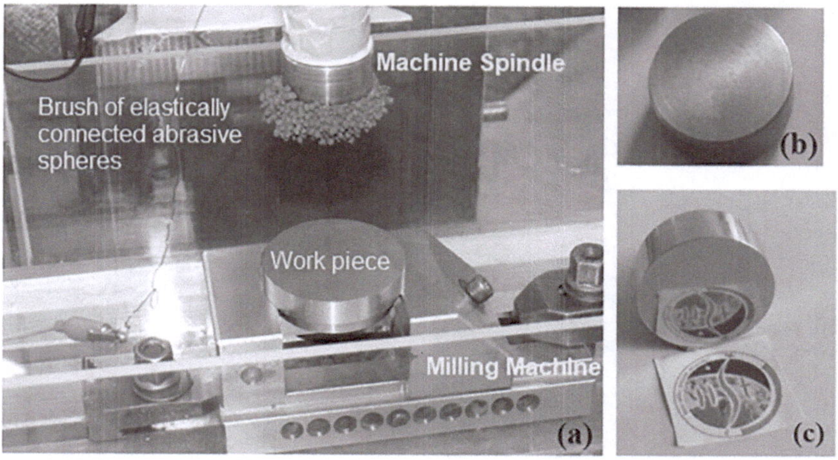


**FIGURE 19.13** Magnetic field assisted elastic abrasive processing (mode-3) via abrasion.

brush was formed (Figure 19.13(a)) using elastically embedded abrasive spheres prepared with iron powder of 10 μm size, SiC powder of size 10 μm and elastomeric polymer. Initial roughness of the work sample was Ra = 0.21 μm, Rt = 2.03 μm, Rp = 0.38 μm (Figure 19.13(b)). After a processing time of 60 min, at a feed rate of 0.1 mm/min for the flexible brush against the work surface, the roughness was reduced down to Ra = 0.05 μm, Rt = 0.42 μm, Rp = 0.10 μm (Figure 19.13(c).

### 19.5.3.3 Mode-3: Experimental Case Study 02 {Reference [10–12]}

A deployable tool mechanism was developed as shown in Figure 19.14, which can be attached to the spindle of a vertical milling machine. When a geometrically symmetric internal round surface to be finished, such a tool can be located inside the work piece and deployed like an umbrella mechanism. The magnet on the tool will hold elastically embedded abrasive spheres and the same will touch the work surface during the deployed position of this tool. As this tool rotates and reciprocate up and down, similar to a honing tool, the material removal can be achieved.

As the penetration of embedded grain into the work surface during radial loading via deployment is controlled by elastomeric characteristics, the depth of penetration will be low. The reciprocating motion of the tool will act as cutting motion, and the material will be removed in fine quantity without altering the surface form due to elastomeric action. Typical experiment using such a tool on steel bores shown more than 45% reduction in average roughness, with nearly 60% reduction in peak to valley roughness.

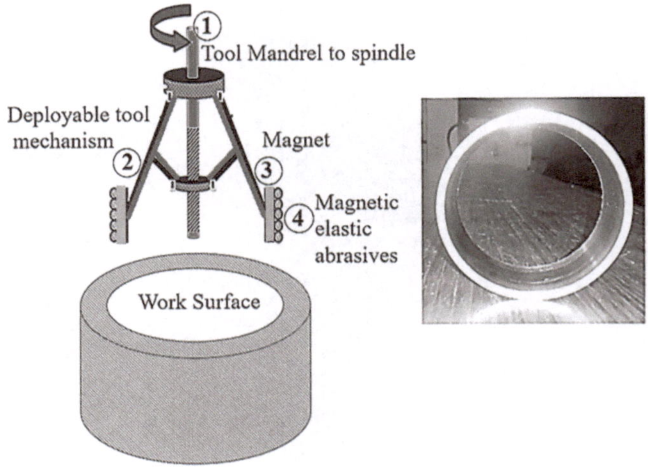


**FIGURE 19.14** Magnetic assisted processing (mode-3) via deployable magnetic elastic abrasives.

## 19.6 SUMMARY

Loose abrasive particles embedded elastically in the form of meso-scale spheres have been considered for fine finishing of surfaces, highlighting its multi-utility perspective both in erosion as well as abrasion mode. The contact theory of such an embedded sphere has been covered using spring-damper analogy, with practical demonstration of reduced equivalent elastic modulus at contact interface. In mode-1, these elastic abrasives have been employed for controlled abrasion via direct loading and cutting motion, which adopts the principle of combined two and three-body abrasion for surface finishing. In the second mode, finishing via fluidization of embedded spheres have been illustrated which adopts low velocity impingement erosion as the working principle. In its third mode, the embedded spheres with an additional magnetic ingredient was formed like a brush to work for magnetic abrasive finishing via three-body abrasion. Multi-mode application capabilities of elastically embedded abrasive spheres were practically demonstrated using case studies from prior arts, such as fine finishing of geometrically symmetric internal surfaces, internal circumferential grooves, asymmetric internal surfaces, flat surfaces, etc.

## REFERENCES

[1] Komanduri, R., Lucca, D.A., Tani, Y. (1997). Technological advances in fine abrasive processes, *CIRP Annals*, 46(2): 545–597.

[2] Malkin, S., Guo, C. (2008). *Grinding Technology: Theory and Applications of Machining with Abrasives*, Industrial Press Inc.: New York

[3] Marinescu, L.D., Hitchiner, M., Uhlmann, E., Rowe, B.W., Inasaki, I. (2007). *Handbook of Machining with Grinding Wheels*, CRC Press (Taylor & Francis): USA.

[4] Jain, V.K. (2017). *Nano finishing Science and Technology*, CRC Press (Taylor & Francis): Boca Raton.
[5] Dixit, N., Sharma, V., Kumar, P. (2021). Research trends in abrasive flow machining: A systematic review, *Journal of Manufacturing Processes*, 64: 1434–1461.
[6] Jaganathan, R., Radhakrishnan, V. (1997). A preliminary study on fluidized bed abrasive polishing, *Transactions of NAMRI/SME*, 25: 189–194.
[7] Barletta, M. (2009). Progress in abrasive fluidized bed machining. *Journals of Material Processing Technology*, 209: 6087–6102.
[8] Geldart, D. (2005). The Characterization of bulk powders. In D. McGlinchey (Ed.), *Characterization of Bulk Solids* (pp. 132–149), CRC Press. Blackwell publishing Ltd: UK
[9] Jain, V.K. (2013). *Micromanufacturing Processes*, CRC Press (Taylor & Francis): New York.
[10] Sooraj V.S., Radhakrishan V. (2017). Elasto-Abrasive Finishing. In Jain V K (Ed.), *Nano finishing Science and Technology* (pp. 111–133), CRC Press (Taylor & Francis): Boca Raton.
[11] Sooraj V.S. (2019). Contact behaviour of elastic abrasive spheres during self-centring type magneto-mechanical deployment for internal bore finishing, *International Jouranal of Precision Technology*, 8(2/3/4): 158–173.
[12] Sooraj V.S. (2017). Concept and mechanics of fine finishing circular internal surfaces using deployable magneto-elastic abrasive tool, *Journal of Manufacturing Science and Engineering (Int. J./ ASME)*, 139: 081001-081001.
[13] Sooraj V.S., Radhakrishan V. (2015). Investigations on the application of elasto magnetic abrasive balls for fine finishing, *Journal of Manufacturing Science and Engineering (Int. J./ ASME)*, 137 (2): 201018.
[14] Sooraj V.S., Radhakrishnan V. (2014). Fine finishing of internal surfaces using elastic abrasives, *International Journal of Machine Tools and Manufacture* (Elsevier), 78: 30–40.
[15] Sooraj V.S. (2017). On the Process and Mechanics of rotary elasto-abrasive finishing, *Machining Science and Technology: Int. J.* (Taylor & Francis), 21(3): 474–492.
[16] Sooraj V.S., Radhakrishnan V. (2014). Prospective methodologies to use impact wear for micro/nano finishing of surfaces, *International Journal of Manufacturing Technology and Management* (Inderscience), 28 (1/2/3): 94–113.
[17] Sooraj V.S., Radhakrishnan V. (2014). A study on fine finishing of hard workpiece surfaces using fluidized elastic abrasives, *International Journal of Advanced Manufacturing Technology* (Springer), 73(9): 1495–1509.
[18] Sooraj V.S., Radhakrishnan V. (2013). Elastic impact of abrasives for controlled erosion in fine finishing of surfaces, *Manufacturing Science and Engineering (Int.J./ASME)*, 135(5): 051019.
[19] Sooraj V.S., Radhakrishnan V. (2014). Ultra-high finishing of oval bores using elastic abrasive balls, 5th International and 26th All India Manufacturing Technology Design and Research (AIMTDR) Conference, India, December 2014.
[20] Sooraj V.S., Radhakrishnan V. (2015). Sizing and Finishing of non-circular internal bores using elasto-abrasives, *International Journal of Precision Technology* (Inderscience), 5(3/4): 261–276.
[21] Sooraj V.S., Radhakrishnan V. (2013). Feasibility study on fine finishing of internal grooves using elastic abrasives, *Materials and Manufacturing Processes* (Taylor & Francis), 28: 1110–1116.

[22] Johnson, K.L. (1985). Contact Mechanics, Cambridge University Press: UK.
[23] Kogut, L., Etsion. I. (2002). Elastic-plastic contact analysis of a sphere and a rigid flat, *Journal of Applied Mechanics* (ASME), 69: 657–662.
[24] Lin, Y.-L., Wang, D.-M., Lu, W.-M., Lin, Y.-S., Tung, K.-L. (2008). Compression and deformation of soft spherical particles, *Chemical Engineering Science*, 63: 195–203.
[25] Radhakrishnan V., Sooraj V.S., Nirmala R.J. Multipurpose Resilient Elastomagnetic-Abrasive Spheres for fine finishing of Surface, Indian Patent No: 346350.
[26] Marques, S., Severino, P.C., Creus, G.J. (2012). *Computational Viscoelasticity*, pp. 11–21, Springer: New York.

# Index